New Perspectives on Mineral Nucleation and Growth

Alexander E.S. Van Driessche
Matthias Kellermeier • Liane G. Benning
Denis Gebauer
Editors

New Perspectives on Mineral Nucleation and Growth

From Solution Precursors to Solid Materials

Springer

Editors
Alexander E.S. Van Driessche
University Grenoble Alpes
CNRS, ISTerre
Grenoble, France

Liane G. Benning
German Research Center
for Geosciences, GFZ
Interface Geochemistry Section
Potsdam, Germany

Matthias Kellermeier
Material Physics
BASF SE
Ludwigshafen
Germany

Denis Gebauer
Department of Chemistry
Physical Chemistry
University of Konstanz
Konstanz, Germany

ISBN 978-3-319-45667-6 ISBN 978-3-319-45669-0 (eBook)
DOI 10.1007/978-3-319-45669-0

Library of Congress Control Number: 2016956851

Printed on acid-free paper

This Springer imprint is published by Springer Nature
The registered company is Springer International Publishing AG
The registered company address is: Gewerbestrasse 11, 6330 Cham, Switzerland

Contents

Introduction

Crystals surround us in our daily life. They form integral parts of our body (e.g. bones), they make up all the ground we stand on (our, as well as any other rocky planet consists of crystals), and they play important roles in a large array of industrial processes and technologies (from table salt over concrete to biomedical nanoparticles). Accordingly, the formation mechanisms and properties of crystals have been the subject of extensive studies since the early days of science, not least due to the aesthetic beauty and fascinating structural perfection of single crystals like those shown on the cover page of this book. While there are long-standing theories on both the birth of crystals (nucleation) and their later evolution (growth, recrystallization and/or transformation), many central questions remain unanswered, and meanwhile, there is a vast amount of evidence suggesting that the traditional view on crystallization is way too simplified. This notion has been corroborated by numerous excellent studies over the past two decades, witnessing a true renaissance in the field and providing completely new perspectives on the underlying physical processes. Today, the scientific interest in the nucleation and growth of mineral phases appears to be greater than ever, due to the relevance of crystallization phenomena across various disciplines such as chemistry, physics, biology, geology, medicine or material science.

In this context, the aim of the present book is to illustrate, based on a number of exemplary spotlights, the current state of the art in crystallization research. By bringing together contributions from leading experts in their particular areas, we aim to guide the reader through the complex world of crystallizing systems, highlighting only recently discovered aspects that moved into the focus of ongoing investigations. The individual chapters outline the results of studies related to:

1. Different types of materials (from classical minerals like calcium carbonate over iron oxides all the way to organic crystals and smoke particles)
2. Different stages of the crystallization process (homogeneous solution, nucleation, phase transformation, aggregation, growth)
3. Different possible phases (clusters, liquid and/or amorphous precursors, crystalline intermediates)

4. A range of advanced methods (both experimental and theoretical) to study the occurring processes at unprecedented levels of detail
5. The relevance of crystallization for different fields of research including biomineralization, geochemistry and industrial applications

Chapter 1, authored by De Yoreo and colleagues, sets the stage for this discussion by introducing a variety of experimentally observed pathways for mineral nucleation from solution. The authors compare the concepts of classical nucleation theory (CNT) with alternative mechanisms such as spinodal decomposition or the aggregation of preformed ion clusters and indicate the conditions under which one or the other pathway becomes operative on a thermodynamic level. They further show that the formation of stable crystals often involves multiple steps, which can be influenced dramatically by polymeric additives or the presence of surfaces. Despite these, and many other, observations evidencing that the classical model of nucleation may not adequately describe the real situation, the theoretical community still mainly uses CNT as dominant conceptual framework. In Chap. 2, Lutsko addresses this issue and presents a novel theoretical description, which is based on the understanding that nucleation is a non-equilibrium process for which equilibrium approximations (as employed by CNT) are of limited applicability. This alternative theory reproduces the results of CNT in appropriate limits, but at the same time, it also allows integrating the new paradigms resulting from experimental studies and simulation work.

The following four contributions deal with the crystallization of calcium carbonate – arguably the most extensively studied mineral in the recent past – with each of the chapters focusing on different stages of the precipitation process. Wolf and Gower (Chap. 3) start out by introducing liquid-like mineral phases and discuss the so-called polymer-induced liquid-precursor (PILP) process from a nonclassical point of view, where solution crystallization is replaced by pseudomorphic solidification. This pathway provides efficient means for controlling mineral morphologies in a bio-inspired manner, ranging from thin films to hierarchical superstructures. The authors address mechanistic aspects of PILP formation and transformation, aiming at an improved physicochemical understanding and emphasizing the relevance of classical concepts of colloid chemistry in any such liquid-precursor pathway. In Chap. 4, Fernandez-Martinez and co-authors continue with non-crystalline solid precursors and address the important case of amorphous calcium carbonate (ACC). Based on a combination of advanced experimental and modelling techniques, they provide insight into the static structure of ACC and highlight the crucial role of water in this context. The authors further describe how impurities such as magnesium ions can influence the structure and thus the stability of ACC, thereby introducing us to the fascinating field of "polyamorphism". Finally, it is discussed whether and how structural variations in ACC may affect the characteristics of crystalline $CaCO_3$ polymorphs formed from these amorphous precursors. This latter aspect is taken up by Rodriguez-Blanco et al. (Chap. 5), who provide an overview of the formation of ACC and vaterite (a metastable crystalline $CaCO_3$ polymorph) as key intermediates on the way to calcite (the stable polymorph). The authors describe the influence of important parameters like pH, temperature, supersaturation and the presence of (in)organic additives on the structures, compositions and morphologies of both phases. In addition, possible mechanisms for the transformation of ACC into vaterite and/or calcite are explained. In Chap. 6, at the end of our $CaCO_3$ journey, another interesting example for the structural variety and complexity of calcium carbonate phases – even in the crystalline state – is presented. Based on ab initio modelling, Demichelis et al. show that vaterite can exist in multiple polytypic forms that are energetically very similar and thus can interconvert readily under ambient conditions. This finding rationalizes seemingly contradictory experimental observations and clears up long-standing confusion about the actual structure(s) of vaterite. Apart from that and on a more general level, the authors introduce the reader to the approach and use of atomistic simulations in predicting the structures and properties of crystalline phases.

Chapter 7 leads us away from mineral nucleation and metastable precursor/intermediate phases towards the later stages of crystal growth. As in the case of nucleation, "classical" and "nonclassical" concepts have been proposed for growth processes. Andreassen and Lewis outline the differences between both paradigms and demonstrate how these are often confusedly used or applied. By presenting a number of illustrative examples from various inorganic systems, the authors show that the two mechanisms are differently influenced by process variables such as supersaturation, temperature or additives, making one or the other operative (much like in the case of nucleation). It is concluded that detailed knowledge of the growth mechanism is key to predict crystallization rates in industrial applications as well as biomineralization processes. In Chap. 8, Rao and Cölfen follow up on this discussion and focus their attention on nonclassical crystallization mechanisms, especially particle-based growth pathways leading to superstructures such as mesocrystals in the presence of organic additives. The authors emphasize the importance of a combined physicochemical and biochemical perspective on such processes and identify key aspects for regulating nucleation and growth so as to achieve ordered particle assemblies. Furthermore, they highlight evidence for the relevance of mesocrystal formation in biomineralization.

Most of the fundamentals of mineral nucleation and growth discussed in the previous chapters merge together in the exciting field of biomineralization, which is at the heart of Chap. 9, authored by Falini and Fermani. This contribution provides a summary of the most common crystallization strategies observed in living organisms, along with a discussion of a few representative cases for the most frequently studied biominerals. In addition, the authors present a comprehensive overview of the effects of ocean acidification on the synthesis and fate of biominerals, a topic of great relevance considering the ever-increasing emission of CO_2 into the atmosphere. After this general introduction to biomineralization, the next two chapters deal with one of its major players, calcium phosphate, discussing the crystallization of apatites from different perspectives. In Chap. 10, Birkedal outlines the various stages observed on the way from dissolved ions to the final mineral, including ion association phenomena prior to nucleation, initially precipitated amorphous phases and intermediate crystalline states. It is shown that experimental conditions, in particular, the pH level of the medium and confinement effects, have a crucial influence on the crystallization mechanism and the properties of the resulting material. This leads the author to propose that physiological conditions may not be representative for biomineralization, as the local environment in compartments like vesicles could be completely different. In Chap. 11, Delgado-Lopez and Guagliardi draw the reader's attention to the most familiar example of biogenic calcium phosphate mineralization – the formation of bone. This highly sophisticated process relies on platelike apatite nanocrystals as building units, along with a complex matrix of bio(macro)molecules. The authors discuss the particular role of citrate as a small-molecule additive that is believed to assist the formation of these nanobricks, both in biogenic and bio-inspired apatite. A second major theme in the two contributions on calcium phosphate is the great value of advanced synchrotron-based techniques – such as in situ X-ray diffraction (Chap. 10) or X-ray total scattering (Chap. 11) – in the characterization of crystallization mechanisms and the properties of nanocrystalline materials in terms of structure and defects, stoichiometry, size and morphology. The use of such powerful in situ methods has also significantly improved our understanding of the solution-mediated formation of calcium sulphate, another interesting – but much less frequently studied – mineral system. In Chap. 12, Van Driessche et al. first summarize the vast and controversial body of literature available on the solubility and corresponding stability fields of the different $CaSO_4$ phases as a function of key variables like temperature or salinity. In the second part of their contribution, the authors turn to consider more recent work on the mechanisms of $CaSO_4$ crystallization, revealing "nonclassical" pathways with a surprisingly high level of complexity and suggesting a new and seemingly converging picture for the precipitation process.

At this point, we abandon the realm of purely ionic minerals and direct our attention to inorganic compounds whose crystallization involves, at least partially, covalent interactions. This renders speciation schemes and the analysis of mineral nucleation and growth inherently more complicated. One such example is the rich family of iron oxide and (oxy)hydroxide phases, which are the main subject of Chaps. 13 and 14, authored by Penn et al. and Reichel and Faivre, respectively. The authors discuss that well-known iron oxides such as hematite often grow via oriented attachment of preformed (nano)particles and that multiple – at times competing or complementing – mechanisms can operate simultaneously in these systems (Chap. 13). Penn and colleagues raise the important issue that transformations between different iron (oxy)(hydr)oxide phases happen to occur after the structure of the final product is set, yet to a degree that the characteristics of the intermediate states are not anymore recognizable. Reichel and Faivre, on the other hand, provide the link of such processes to biomineralization, where magnetotactic bacteria or molluscs produce magnetite in a highly controlled fashion. The authors contrast these biogenic iron oxide particles with those obtained with the help of organic or inorganic additives in biomimetic syntheses. An equally or perhaps even more complex situation is found for two other predominantly covalent minerals, silica and alumina. Despite their great relevance, the formation of these materials is still poorly understood. In Chap. 15, Tobler et al. face this challenge and give a detailed overview of what is known about the solution chemistry, nucleation and growth pathways as well as the structure and composition of solid phases occurring in the aqueous silica and alumina systems. In particular, they focus on processes at the nanoscale and highlight implications of the behaviour of the two minerals for both geochemical environments and industrial applications.

Even though this book is dedicated to recent advances made in the nucleation and growth of minerals, we considered it worthwhile to also include contributions that reach out to other fields, not least since we strongly believe that crystallization follows general concepts that apply for a larger variety of systems. The two examples chosen here relate to crystals of small organic molecules (Chap. 16), which bear immense importance for the pharmaceutical area, and to the more exotic but equally fascinating case of cosmic dust particles (Chap. 17), vital components of astrophysical processes and relevant to environmental science (every year, 40,000 tons of cosmic dust reach the Earth's surface). The former topic is covered by the contribution of Yang and Ter Horst, who review the nucleation of small molecules within the framework of CNT. The authors emphasize the difficulty of a molecular interpretation of nucleation rate measurements and show that the basic reason is the rather ill-defined nature of the nucleating species. However, in the absence of more advanced theories, the concepts of CNT still prove invaluable for the prediction of polymorphism and particle size distributions of organic crystals. The second example is presented by Kimura and Tsukamoto, who report on nucleation experiments of smoke particles that produce analogues of cosmic dust. The authors first give a general introduction on the formation of cosmic dust particles, which in essence are the building units of planetary systems. Subsequently, they describe how this process can be mimicked in the lab by the nucleation of smoke particles, how relevant physical properties of these particles can be extracted and how singularities observed with such particles can help to understand the formation of cosmic dust.

Finally, the last contribution of this book is meant to highlight the potential of emerging new analytical techniques to observe nucleation and growth processes in situ. In Chap. 18, Nielsen and De Yoreo describe recent advances made in the area of liquid-phase transmission electron microscopy (LP-TEM) and its use for investigating the early stages and mechanisms of mineral formation. The authors first address technical aspects of the method and then summarize the results of recent studies on the precipitation pathways of calcium carbonate, including the influence of additives on the occurring processes. As a second example, the formation of metal nanoparticles from precursor solutions and their development into faceted nanocrystals in the presence of an organic ligand are discussed. The chapter is concluded by a perspective on future developments that will further improve the utility of LP-TEM.

We end this book by presenting an outlook about the fascinating field of crystallization and ask key questions that in our opinion need to be addressed in the years to come, so that ultimately a more complete understanding of mineral nucleation, growth and transformation processes can be achieved.

Chapter 1
Nucleation Pathways in Electrolyte Solutions

James J. De Yoreo, Nico A.J.M. Sommerdijk, and Patricia M. Dove

1.1 Introduction

The formation of a solid crystalline phase from a solution has been the subject of scientific investigation for centuries (Gibbs and Smith 1874). This first-order phase transition is generally referred to as nucleation, but strictly speaking nucleation is just one of two classical mechanisms of phase separation. The second mechanism, which is known as spinodal decomposition (Bray 2002; Scheifele et al. 2013), is common in alloys, polymer melts, and protein solutions. This process differs from nucleation in both its underlying thermodynamic description and the resulting formation dynamics. In both cases, the pathway of phase separation can be complex, because the free-energy landscape traversed by a material system typically possesses many local minima representing different structural states, each of which is separated from the others by barriers of varying height (Navrotsky 2004; De Yoreo et al. 2015). Thus, systems can follow multistage pathways in which the phase of the solid evolves as it grows. When local free-energy minima exist as a function

J.J. De Yoreo (✉)
Physical Sciences Division, Pacific Northwest National Laboratory, Richland, WA 99352, USA

Departments of Materials Science and Engineering and of Chemistry, University of Washington, Seattle, WA 98195, USA
e-mail: james.deyoreo@pnnl.gov

N.A.J.M. Sommerdijk
Laboratory of Materials and Interface Chemistry and Center of Multiscale Electron Microscopy, Department of Chemical Engineering and Chemistry, Eindhoven University of Technology, PO Box 513, 5600 MB Eindhoven, The Netherlands

Institute for Complex Molecular Systems, Eindhoven University of Technology, PO Box 513, 5600 MB Eindhoven, The Netherlands

P.M. Dove
Department of Geosciences, Virginia Tech, Blacksburg, VA 24061, USA

A.E.S. Van Driessche et al. (eds.), *New Perspectives on Mineral Nucleation and Growth*,
DOI 10.1007/978-3-319-45669-0_1

of particle size, formation pathways can involve the aggregation of particles with or without changes in the solid phase (Baumgartner et al. 2013; Habraken et al. 2013; Hu et al. 2012; De Yoreo et al. 2015). (Note that throughout this article, the term particle refers to any species more complex than simple ions or monomers and makes no assumption about their stability or metastability (De Yoreo et al. 2015).) Moreover, both the minima and the barriers can be altered through the introduction of surfaces (Hu et al. 2012; Hamm et al. 2014; Giuffre et al. 2013) or changes in the solvation media by, for example, addition of additives (Gebauer et al. 2009; Wang et al. 2006; Gower and Odom 2000) or polymer globules (Smeets et al. 2015). This chapter provides a brief introduction to these various mechanisms, characteristics, and controls of crystal formation from solution and illustrates each using the results of recent experimental and computational studies.

1.2 Classical Nucleation Theory

The modern theory of nucleation, commonly referred to as classical nucleation theory (CNT), originates with the work of Gibbs in the late 1800s (Gibbs and Smith 1874), but reflects later refinements due to Stransky and others in the mid-twentieth century (Gutzow 1997). According to CNT, nucleation comes about through the addition of monomeric solution species—whether atoms, molecular complexes, or colloids—to a growing particle (Fig. 1.1a), whose origin lies in the inherent microscopic thermodynamic fluctuations of the solution (De Yoreo and Vekilov 2003; Kashchiev 2003). As long as the chemical potential of the solid phase is less than that of the solution—i.e., the solution is supersaturated—the particle can grow to macroscopic size. However, the CNT analysis shows that only a fraction of the particles formed through fluctuations will become stable nuclei of the solid phase. This is because there is surface tension between the solid and the parent solution. Consequently, formation of the surface comes with an energetic cost, which is given by the interfacial free energy times the surface area of the particle. In contrast, the decrease in free energy due to the applied chemical potential $\Delta\mu$ is proportional to the number of monomers that move from solution to solid state and thus scales with the volume of the particle. The area-dependent term exceeds the volume-dependent term when the particle is small, but the reverse is true at large particle size. Thus these opposing terms create a free-energy barrier (Fig. 1.1b) with a maximum at the "critical particle size." On average, below this critical size particles will dissolve, while above it they will grow to macroscopic dimensions through the continued addition of monomeric species (Kashchiev 2003; Gibbs and Smith 1874).

The existence of fluctuations and the stochastic nature of nucleation are illustrated in Fig. 1.1c–n, which shows liquid-phase (LP)-TEM images that capture the formation of gold nanoparticles in gold chloride solution containing citric acid as a capping agent (Nielsen et al. 2014b; Nielsen and de Yoreo 2017, Chap. 18). While numerous successful particle formation events are observed, there are many instances in which a gold particle begins to form, fluctuates in size, and eventually

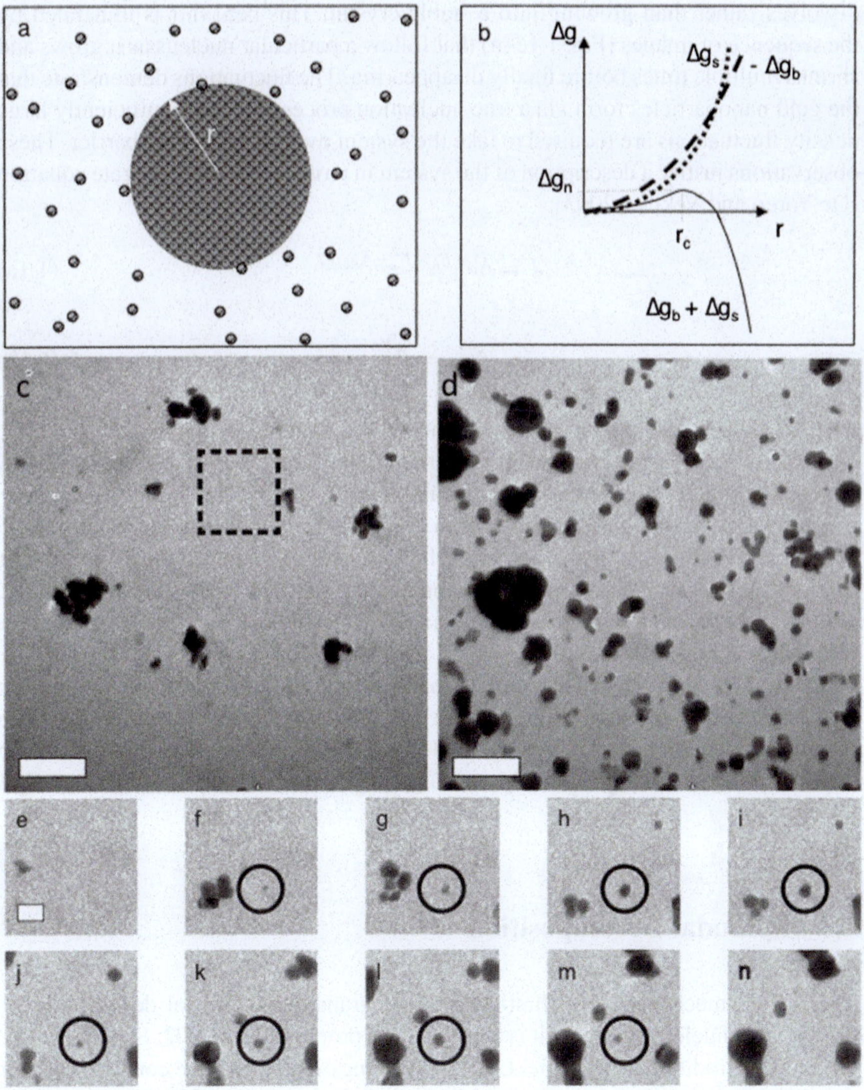

Fig. 1.1 (**a**) Formation of a spherical nucleus of radius r from a solution leads to the free-energy changes shown in (**b**). The crossover of the bulk (Δg_b) and surface (Δg_s) terms combined with their opposing signs leads to a free-energy barrier. (**c–n**) Nucleation of Au nanoparticles. (**c, d**) Sequence of LP-TEM images showing the nucleation of Au nanoparticles from a gold chloride solution containing citrate as a capping agent. As seen in the example highlighted by the *circle* in (**e–n**), which were taken from the region delineated by the rectangular box in (**c**), many of the nascent nuclei fail to reach the point of spontaneous growth and instead fluctuate in size until they disappear. This behavior demonstrates particle formation results from unstable fluctuations, the hallmark of classical nucleation. Times in s: (**c**) 0.0, (**d**) 103.2, (**e**) 2.1, (**f**) 8.0, (**g**) 16.4, (**h**) 34.9, (**i**) 36.6, (**j**) 44.8, (**k**) 60.5, (**l**) 73.1, (**m**) 93.3, and (**n**) 103.2. *Scale bars*: (**c, d**) 500 nm, (**e**) 200 nm. Scale in (**f–n**) is same as (**e**) (Panels **a** and **b**: (De Yoreo and Vekilov 2003) used by permission of the Mineralogical Society of America. Panels **c–n**: (Nielsen et al. 2014b) used by permission of the Microscopy Society of America)

dissolves rather than growing into a stable crystal. This behavior is illustrated by the sequence of images (Fig. 1.1e–n) that follow a particular nucleus as it grows and shrinks multiple times before finally disappearing. The fluctuations demonstrate that the gold nanoparticles form via a true nucleation process in which sufficiently large density fluctuations are required to take the system over a free-energy barrier. These observations justify a description of the system in terms of the classical rate equation (De Yoreo and Vekilov 2003):

$$J = Ae^{-E_A/kT}e^{-\Delta g_c/kT} \tag{1.1a}$$

$$\Delta g_c = B\frac{\gamma^3}{\sigma^2} \tag{1.1b}$$

where J is the nucleation rate per volume, A is a material-dependent constant, E_A is the activation energy associated with the atomic processes such as desolvation, attachment, and/or structural rearrangement within the nucleus prior to or during phase separation that must occur to form the nucleus, k is Boltzmann's constant, T is the temperature, Δg_c is the height of the free-energy barrier, B is a constant that depends on the shape and density of the nucleating solid, as well as the temperature, γ is the interfacial free energy, and σ is the supersaturation ($\sigma = -\Delta\mu/kT$) (De Yoreo and Vekilov 2003).

In the classical view of crystallization, post nucleation growth largely occurs in the absence of particle-coalescence events and results in a population of stable macroscopic particles (Andreassen and Lewis 2017, Chap. 7), though if true equilibrium could be reached, only one large particle would be present at the end of the process.

1.3 Spinodal Decomposition

The second mechanism of a first-order phase transition, spinodal decomposition, differs from nucleation in that it is a barrier-free process (Bray 2002; Scheifele et al. 2013). The traditional mean-field picture assumes that when the concentration of monomers exceeds a certain threshold, the bulk free-energy density of the system exhibits negative curvature (Reif 1965). In this situation there is no penalty for creating a new interface, and the parent phase becomes inherently unstable thereby allowing all microscopic fluctuations to grow spontaneously. In a more modern view of this mechanism, the condition for spinodal decomposition is simply that the supersaturation is increased to the point where the free-energy barrier to phase change is comparable with kT (Scheifele et al. 2013). Once the spinodal is crossed, particles are generated in such large numbers that they can grow by direct collision and coalescence with other particles. When the products of spinodal decomposition consist of two liquids, a secondary event is required to produce the first solid.

Evidence for spinodal decomposition in electrolyte solutions is limited. Experimental work documents the formation of liquid $MgSO_4$ at high temperature

(Wang et al. 2013), as well as the persistence of a polymer-stabilized liquid calcium carbonate phase under ambient conditions (Gower and Odom 2000). A recent study that used NMR to quantify the diffusion of ions in solution following the introduction of calcium concluded that a bicarbonate-rich liquid phase may also exist in pure $CaCO_3$ solutions at pH = 8.5 (Bewernitz et al. 2012; Wolf and Gower 2017, Chap. 3). This conclusion was supported by two lines of evidence. First, light-scattering measurements revealed the presence of nanoparticles (or droplets) 10s of nm in size, despite the fact that the NMR showed no evidence for a solid $CaCO_3$ phase. Second, calorimetry measurements (Fig. 1.2a) demonstrated the occurrence of a first-order phase transition (Fig. 1.2a inset) at a concentration equal to that at which the nanoparticles appeared.

Molecular dynamics (MD) simulations of cluster formation in moderately concentrated $CaCO_3$ solutions (15 mM Ca^{2+} and 15 mM $CO_3{}^{2-}$) based on the commonly utilized force fields (Raiteri and Gale 2010) predicted the appearance of polymeric species of small size (<10 ion pairs) that continued to grow with little or no free-energy barrier into compact clusters (Fig. 1.2b) exhibiting the ion diffusivities of liquids (Fig. 1.2c) (Wallace et al. 2013). Kinetic models of particle formation and growth in a solution possessing a binodal based on an Ising lattice gas model (Fig. 1.2d, e) show that, as the binodal is crossed, the spinodal is approached, and the barrier to nucleation decreases toward kT (red to blue to green symbols in Fig. 1.2d corresponding to the red blue and green curves in Fig. 1.2e), a wide range of particle sizes is generated (Fig. 1.2f). These particles grow both through monomer addition and particle–particle coalescence resulting in a bimodal distribution of small particles (blue through green in Fig. 1.2f) and larger aggregates (yellow and red in Fig. 1.2f). The small particles form rapidly and persist throughout the spinodal process. Following an initial phase change, their size distribution is stable in a statistical sense, though the lifetime of an individual species is finite. These are essentially the same particles that would be present in a solution without a spinodal that behaves according to CNT. In contrast, large particles are unique to a system undergoing spinodal decomposition, because all microscopic fluctuations are stable against dissolution and can grow by coalescence. Although they have no special thermodynamic status, they exhibit a mean radius that should evolve rapidly as $R(t) \sim (t)^{1/3}$ through diffusion-limited coalescence. As in the case of CNT, in the true equilibrium state, the system should contain just one large particle when phase separation is complete.

To investigate the formation of solids from the liquid particles generated by the MD simulations, Wallace et al. (2013) simulated random aggregation followed by dehydration through random removal of water molecules until the Ca/CO_3 ratio was equal to unity, which is the value typically measured for hydrated amorphous calcium carbonate (ACC). The pair distribution function (pdf) for the resulting solid was calculated and found to give a good match to the measured pdf for ACC, but provided a poor match to all other $CaCO_3$ polymorphs. Based on these findings, authors concluded that liquid–liquid separation followed by dehydration was a plausible scenario for producing the first solid precipitate from $CaCO_3$ solutions. They also proposed that the particle distributions that appear as the dynamic

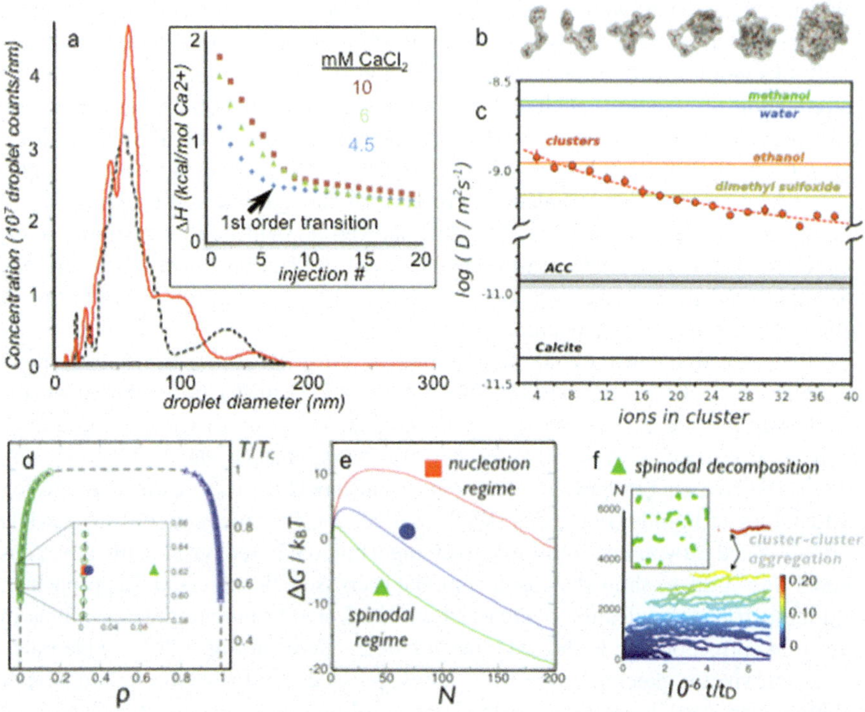

Fig. 1.2 (**a**) Nanoparticle tracking analysis of a calcium carbonate solution during sequential injections of 6 mM $CaCl_2$(aq) solution into a carbonate/bicarbonate buffer. Nanoparticles develop at about the tenth injection, which is concurrent with the emergence of a new phase as determined by calorimetry. At the phase transition (tenth injection, *black curve*), droplets with a distributed diameter averaging ~60 nm emerge. The addition of more Ca^{2+} to the solution (13th injection, *red curve*) yields more detectable particles emerging at 60–70 nm diameter, and they seem to grow larger as well. *Inset* to (**a**) the enthalpy of reaction during the titration of $CaCl_2$(aq) into 20 mM carbonate buffer, pH 8.5. The data shows that there is an endothermic phase transition occurring, indicating that the phase transition is entropically driven. (**b**) Snapshots taken from replica-exchange molecular dynamics simulations show the evolution of polymeric cluster configurations toward denser structures at larger sizes. (**c**) The diffusivity of calcium ions within the cluster species at various stages of growth compared to two solid phases of calcium carbonate, calcite, and ACC (from simulation) and the self-diffusivities (experimental) of several common solvents. *Error bars* represent the mean +/− standard error for $N=6$ simulation trajectories at each cluster size. (**d**) Temperature-density phase diagram for the 2-D Ising lattice gas with a spinodal region. (**e**) Free-energy barrier to phase change diminishes with supersaturation where the *red, blue,* and *green curves* correspond to the free energy vs. cluster size (N monomers) at the similarly colored points in the phase diagram in panel **d**. *Red* lies just before crossing the binodal, *blue* is in the binodal region, and *green* is in the spinodal region where the barrier is $\leq kT$. (**f**) In the small-barrier (spinodal) regime, an evolving cluster population is generated (see *inset* snapshot). The distribution of cluster sizes (color scheme indicates amount of system mass contained in the clusters) versus time where t_D is the characteristic time for a monomer to diffuse a length equal to its diameter (Panel **a**: (Bewernitz et al. 2012) used by permission of the Royal Society of Chemistry. Panels **b**–**f**: (Wallace et al. 2013) used by permission of the American Association for the Advancement of Science)

products of the incipient liquid–liquid phase separation in the Ising models provide a potential explanation for the multi-ion clusters reported in experimental studies of supersaturated $CaCO_3$ solutions, as discussed below.

1.4 Aggregation of Stable or Metastable Particles

An increasing number of experimental studies (Gebauer et al. 2008, 2014; Habraken et al. 2013; Pouget et al. 2009) have reported nucleation via aggregation of solution species comprised of ion complexes or multi-ion clusters—generally referred to as "pre-nucleation clusters"—that exist in equilibrium with the free monomers (Fig. 1.3). Three lines of evidence for this pathway in the $CaCO_3$ system were cited:

First, the development of a sedimentation peak in analytical ultracentrifugation (AUC) data that corresponded to species much larger than that of the ions (Fig. 1.3a) (Gebauer et al. 2008).

Second, observation of Ca^{2+} complexation during titration of $CO_3{}^{2+}$ with a dependence on carbonate addition that is consistent with a particular model of multi-ion binding (Gebauer et al. 2008).

Third, molecular dynamics simulations that predict the development of small clusters (<10 ion pairs) with dynamic polymeric structures (Demichelis et al. 2011).

The multi-ion-binding analysis led to the argument that these clusters are stable relative to the free ions. That is, their formation lowers the free energy of the system, and their aggregation may eliminate the need to overcome the large free-energy barrier associated with CNT (Fig. 1.3b).

Subsequent studies have argued these features in the data are equally understandable from purely classical considerations coupled with a transition from a regime of nucleation to spinodal decomposition (Wallace et al. 2013). For example, the multi-ion peak in the AUC data reflects a time-averaged value for an evolving particle population created by high supersaturations (Gebauer 2012) and that the polymeric clusters seen in earlier MD studies represent transient states (Wallace et al. 2013). Although cryo-TEM data was initially interpreted as evidence for $CaCO_3$ pre-nucleation clusters averaging about 0.7 nm in size (Pouget et al. 2009), subsequent analysis led to the conclusion that these features were below the direct interpretable size range of the measurement, and that the observations also could be reconciled with a distribution of cluster sizes as predicted by classical nucleation models.

Evidence based on cryo-TEM for nucleation by aggregation of pre-nucleation clusters was also reported for the calcium phosphate system (Dey et al. 2010; Nudelman et al. 2010). The pre-nucleation clusters were shown to aggregate at organic interfaces forming amorphous calcium phosphate, which subsequently transformed into oriented apatite crystals. A subsequent more detailed study that combined cryo-TEM with chemical analysis and ab initio calculations showed that the multistage pathway leading to hydroxyapatite began with aggregation of

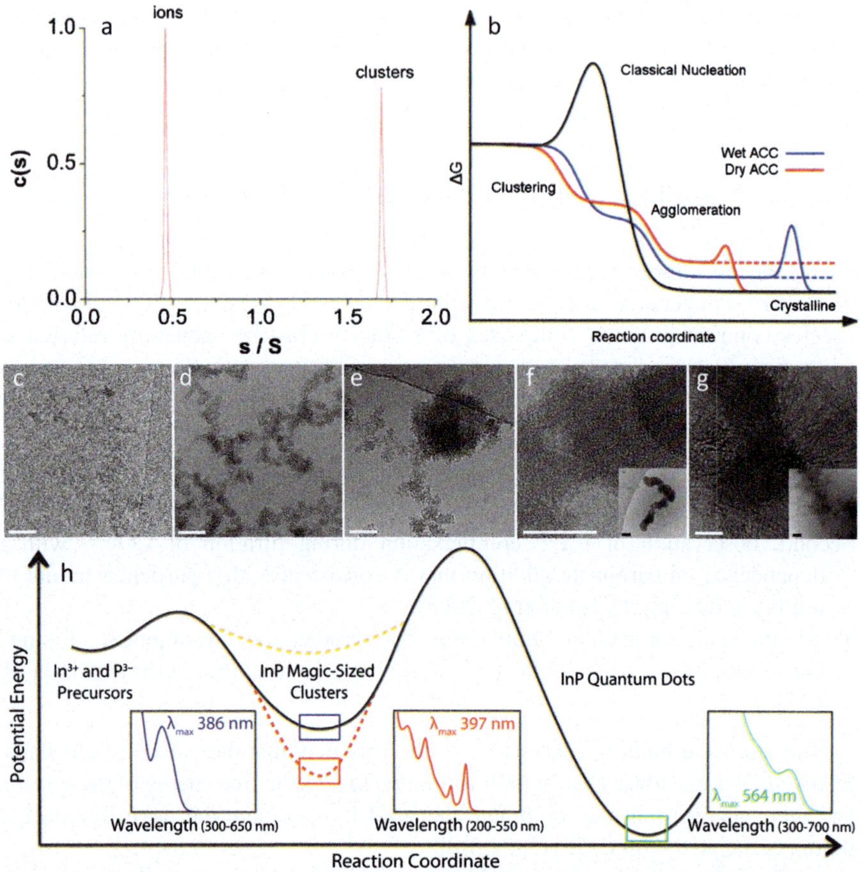

Fig. 1.3 (**a**) Diffusion-corrected distribution of sedimentation coefficients for a $CaCO_3$ solution after 90 min reaction time at pH 9.00. Besides the free ions, multi-ion clusters can be detected as indicated. (**b**) Schematic representation of the free-energy profile proposed by Gebauer et al. (2008) for the nonclassical formation of calcium carbonate from ion pairs in solution through aggregation of multi-ion clusters. The two lines for ACC represent the boundaries of a distribution of possible curves that depend on water content. The final transformation of ACC to a crystalline polymorph can occur at variable reaction coordinate depending on the environmental conditions. (**c–g**) Morphological transformations of calcium phosphate during time as observed by cryo-TEM. (**c**) Polymeric strands form from nanometer-sized units (<2 min), (**d**) branched polymeric assemblies (2–20 min), (**e**) nodules of ACP (10–20 min), (**f**) aggregated spheres of ACP (15–60 min), and (**g**) ribbons of OCP (80–110 min). (**h**) Illustration of the two-step nucleation mechanism for the growth of InP QDs from In^{3+} and P^{3-} precursors depicts the magic-sized clusters (MSC) as an isolable 1–2 nm diameter intermediate species that acts as a bottleneck for InP monomers. Amines destabilize the MSC (increase potential energy, *yellow dotted line*), whereas phosphonate ligands increase the stability of the MSC (*red dotted line*) relative to a carboxylate ligand set. The luminescence peaks for the MSCs and the InP QDs are specific to their diameters and enable the assignment of size to the MSCs (Panel **a**: (Gebauer et al. 2008) used by permission of the American Association for the Advancement of Science. Panel **b**: (Gebauer and Cölfen 2011) used by permission of Elsevier. Panel **c**: Habraken et al. (2013). Panel **d**: (Gary et al. 2015) used by permission of the American Chemical Society)

charged calcium triphosphate complexes to form a polymeric network (Fig. 1.3c, d) (Habraken et al. 2013). Titration analysis led to the conclusion that the collapse to a compact, crystalline structure was accompanied by Ca-binding in three successive steps that led to the formation of amorphous calcium phosphate (ACP) (Fig. 1.3e, f), octacalcium phosphate (Fig. 1.3g) and, finally, hydroxyapatite (not shown). In situ AFM data on the relative rates of ACP and hydroxyapatite formation, as well as the critical sizes, were found to be inconsistent with the predictions of CNT, but could be reconciled if a particle aggregation model were invoked, though a rigorous quantitative theory for this mechanism was not provided. Moreover, as the complexes that purportedly aggregate contain only a single calcium ion, the often-used term "cluster" was found to be of limited applicability in this system, and the term multi-ion complex was adopted instead.

Particle-based models of nucleation are not restricted to aqueous solution systems, but rather have also been proposed for semiconductor materials grown through high-temperature colloidal synthesis using organic solvents (Gary et al. 2015). Taking advantage of the dependence of semiconductor luminescence wavelength on nanoparticle size, Gary et al. (2015) showed that the formation of indium phosphide quantum dots (InP QDs) from solutions of molecular precursors was preceded by the appearance of "magic-sized clusters" (MSCs) with a diameter of 1–2 nm (Fig. 1.3h). The MSCs disappeared as the InP QDs formed, with no evidence for particles having sizes intermediate between these two. While the study concluded that the MSCs represent a metastable state in the formation pathway, whether the MSCs aggregated to form the InP QDs or simply dissolved as the latter nucleated remains uncertain, as does the stability of the magic clusters relative to the monomers themselves.

At least one of the above studies on particle-based pathways concluded that the particles are metastable with respect to the monomers (Habraken et al. 2013). That is, while they lie in a local free-energy minimum as a function of particle size, they are nonetheless uphill in free energy from the monomers. Other studies maintain that the particles are stable with respect to the monomers (Demichelis et al. 2011; Gebauer et al. 2008, 2014). That is, their formation is favored over their disaggregation, even in an undersaturated state. The latter picture defines a thermodynamic landscape, that is, a clear departure from those described both by CNT and the theory of spinodal decomposition.

Regardless of stability of precritical particles or the extent to which their aggregation plays a role in nucleation from electrolyte solutions, their common presence would not be surprising (De Yoreo et al. 2015). Even in the framework of CNT, one expects any solution, whether undersaturated or supersaturated, to contain a population of particles that decreases exponentially in number with increasing particle radius (Kashchiev 2003). Moreover, energy landscapes are unlikely to be flat. That is, the free energy of a cluster of ions exhibits local minima and maxima depending on the size, configuration, and composition, including the configuration of the solvent. The resulting peaks and valleys in that landscape will naturally lead to deviations from the exponential size distribution of CNT and create anomalously large populations of clusters with sizes that correspond to the minima (Hu et al. 2012; De Yoreo et al. 2015).

Hu et al. (2012) attempted to estimate the impact of these cluster populations on the nucleation rate equation and free-energy barrier given in Eq. 1.1. For calcite nucleation by aggregation of precritical particles that occupy a minimum in free energy vs. size, they concluded the expression for the nucleation rate remains the same, but the dependence of the barrier on supersaturation and bulk interfacial free energy can be estimated by:

$$\Delta g_c = B\frac{\gamma^3}{(\sigma \pm C)^2} \tag{1.2}$$

where C is a constant that depends on the shape factor, the particle radius, and the excess free energy of the particles. The plus or minus sign depends on whether the minimum in Δg is local (plus) or global (minus). Baumgartner et al. (2013) reported a similar analysis and applied it directly to cryo-TEM observations of iron oxide formation from disordered precursor particles, though those particles were not considered to be precritical.

One interesting implication of Eq. 1.2 is that the impact of precritical particles—whether stable or metastable relative to the free ions—is to effectively alter the supersaturation from σ to $\sigma \pm C$. This has the consequence of increasing the effective supersaturation with respect to metastable clusters and decreasing supersaturation for stable clusters. Thus, considering just the contribution of the free-energy barrier in Eq. 1.2, the existence of metastable clusters *increases* the probability of nucleation, while the existence of stable clusters *decreases* that probability. This conclusion is inconsistent with the picture presented in Fig. 1.3b and highlights the current lack of a quantitative model for particle-based nucleation.

1.5 Multistep Pathways

Spinodal decomposition of a solution due to an increase in solute concentration does not create a solid. Rather it generates two distinct liquid states defined by the intersection of the spinodal curve and the line of constant temperature: the monomer-rich solution and the dilute solution (Fig. 1.2d). Hence the formation of a crystal still requires nucleation of a solid–liquid phase boundary within (homogeneous) or on the surface of (heterogeneous) the dense liquid droplets. Moreover, regardless of whether formation of the solid occurs via the monomer-based mechanism described by CNT or via aggregation of precritical particles, the first phase to form may be amorphous or consist of a crystal phase that is distinct from the stable bulk crystal (Baumgartner et al. 2013; Beniash et al. 2009; Gebauer et al. 2008; Gower and Odom 2000; Habraken et al. 2013; Nudelman et al. 2010; Pouget et al. 2009; Smeets et al. 2015; Van Driessche et al. 2012; Wolf et al. 2008; Hunger and Benning 2007; Rodriguez-Blanco et al. 2017, Chap. 5; Birkedal 2017, Chap. 10; Penn et al. 2017, Chap. 13). These multistep pathways may occur due to kinetic limitations on the formation of the stable phase, as is likely to be the case, for example, in the calcium phosphate system described above (Habraken et al. 2013). However, they may also occur due to the commonly observed inversion in phase stability with decreasing particle size, which arises from the increasing importance of surface free energy (Navrotsky 2004; Gribb and Banfield 1997). This inversion may mean the bulk phase has lower stability at the nucleation stage.

Macroscopic evidence for nucleation in solution by two-step pathways has been obtained through optical microscopy (Galkin et al. 2002; Gower and Odom 2000), light scattering (Bewernitz et al. 2012), and small angle X-ray scattering (Bots et al. 2012; Hunger and Benning 2007) for a number of organic, protein, and colloidal systems. Direct molecular-scale observation of this process was achieved using in situ AFM for the crystallization of S-layer proteins (Chung et al. 2010). In the case of the proteins, the evidence points to liquid–liquid separation as the first step in the two-step process (Vekilov 2005), and various analyses argue that proteins, as well as colloids, should commonly follow this pathway. For example, ten Wolde and Frenkel (tenWolde and Frenkel 1997) argued that in systems for which the characteristic length scale of inter-monomer interaction is small compared to the size of the monomer, the phase diagram should exhibit a spinodal (and binodal) line buried beneath the solution–solid phase boundary. Thus, when the barrier to nucleation of the crystalline solid is sufficiently high, as the solution is driven to increasing levels of supersaturation, the spinodal line will be crossed before nucleation can occur. This results in phase separation of the solution into two liquids. Nucleation of the crystalline phase is then more likely within the monomer-rich droplets.

Numerous examples of two-step pathways in electrolyte solutions have also been reported. Of particular interest are the calcium carbonate and calcium phosphate systems, in which the amorphous phase is often the first to form and is followed by the appearance of one or more of the crystalline phases (Dey et al. 2010; Habraken et al. 2013; Nudelman et al. 2010; Pouget et al. 2009, 2010; Pichon et al. 2008; Birkedal 2017, Chap. 10; Rodriguez-Blanco et al. 2017, Chap. 5; Wolf and Gower 2017, Chap. 3; Fernandez-Martinez et al. 2017, Chap. 4; Demichelis et al. 2017, Chap. 6; Falini and Fermani 2017, Chap. 9; Delgado-Lopez and Guagliardi 2017, Chap. 11). Similar results are reported for iron oxides (Baumgartner et al. 2013; Towe and Lowenstam 1967; Reichel and Faivre 2017, Chap. 14; Penn et al. 2017, Chap. 13), zeolites (Burkett and Davis 1994; Kumar et al. 2008), and other inorganic materials (Van Driessche et al. 2012, Chap. 12; Tobler et al. 2017, Chap. 15). However, proof that the crystalline phase forms *directly* from the amorphous precursor rather than through a separate nucleation event is rare. The LP-TEM data in Fig. 1.4a–d show an example from the $CaCO_3$ system in which an amorphous precursor particle transforms into vaterite (Fig. 1.4a–d) (Nielsen et al. 2014a; Nielsen and De Yoreo 2017, Chap. 18). Following nucleation and growth of an amorphous calcium carbonate (ACC) particle, numerous vaterite nuclei appear at or just below the surface of the precursor particle. The amorphous particle is then consumed as the vaterite crystals continue to grow. Cryo-TEM data provide similar evidence for direct transformations in both the $CaCO_3$ and calcium phosphate systems (Habraken et al. 2013; Pouget et al. 2010). Whether or not any of the

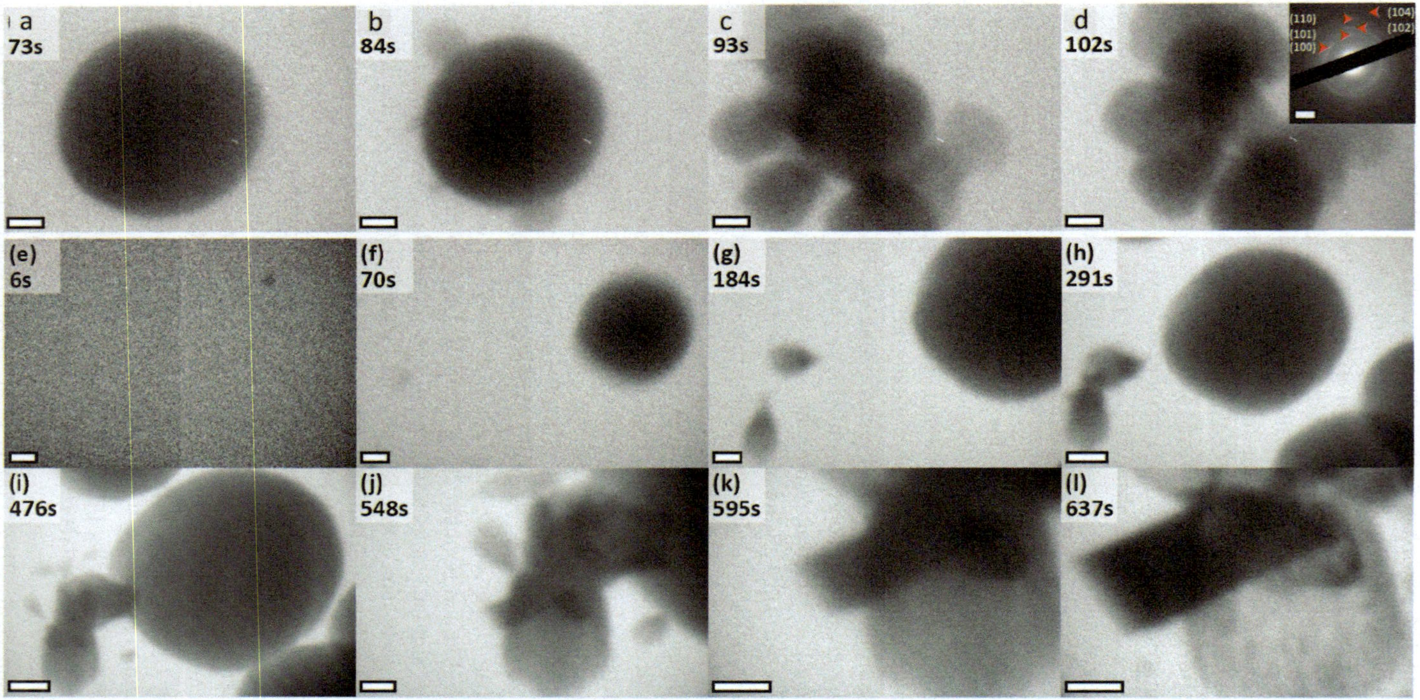

Fig. 1.4 (**a–d**) Series of LP-TEM images show an ACC particle that nucleates early in the experiment (**a**) with secondarily nucleated vaterite plates forming on or in the amorphous particle (**b**). These plates grow at the expense of the ACC (**c, d**). Diffraction from the resulting plates identifies them as vaterite (**d***inset*). (**e–l**) Series of LP-TEM images showing direct nucleation of vaterite (single particle in **e**) and aragonite (two new particles in **f**) and subsequent growth (**g–i**). Formation of calcite occurs on aragonite (**i**), followed by calcite growth, and concomitant dissolution of the aragonitic bundle (**j–l**). *Scale bars* are 500 nm (**a–g, j–l**), 1 micron (**h–i**), and 2 nm^{-1} (*inset* to **d**) (From Nielsen et al. (2014a) used by permission of the American Association for the Advancement of Science)

documented examples involve liquid precursor states is uncertain at this time. However, the mounting evidence that phase diagrams of electrolyte systems exhibit spinodal lines suggests this two-step pathway is likely (as discussed above).

Many multistep nucleation events occur along *indirect* pathways. By this process, metastable phases form and grow until more stable phases appear, after which time the metastable phases dissolve back into solution as the solute activity product drops below their respective solubility limits. Such dissolution–reprecipitation events have been well documented for $CaCO_3$ (Han and Aizenberg 2008; Hu et al. 2012; Bots et al. 2012; Lee et al. 2007; Donners et al. 2000, 2002). The reason for this behavior is clear when one considers the consequence of creating a highly supersaturated solution. Once the solute activity product exceeds the solubility limits of all the possible phases, every one of them will nucleate at a rate given by Eq. 1.1. Although the value of σ will be largest with respect to the most stable phase, the values of interfacial free energy and E_A are likely to be smaller for the metastable phases, particularly those that are amorphous and hydrated. Consequently, one or more metastable phases may form and do so at a higher rate than the most stable phase, as postulated in the Ostwald step rule. The LP-TEM data in Fig. 1.4e–l illustrate this phenomenon for $CaCO_3$ (Nielsen et al. 2014a). Vaterite is the first phase to appear, but aragonite soon appears through independent nucleation events. Eventually, a calcite crystal nucleates, leading to the eventual dissolution of the other two phases.

1.6 Heterogeneous Nucleation

Substrates can alter the probability of nucleation because the interfacial energy between a crystal nucleus and a solid substrate typically differs from that of the same crystal in contact with solution (Chernov and Givargizov 1984; Mutaftschiev 1993). The expressions for the nucleation rate and the free-energy barrier given by Eq. 1.1 remain intact; however, the value of γ for the crystal–fluid interface must be replaced by an effective interfacial energy γ_{het}, which is a composite of the interfacial energies for the three interfaces created or destroyed during nucleation—the crystal–substrate, crystal–liquid, and liquid–substrate interfaces. This relationship is given by (De Yoreo and Vekilov 2003):

$$\gamma_{\text{het}} = \gamma_{cl} + h\,(\gamma_{cs} - \gamma_{ls}) \tag{1.3}$$

where h is a factor that depends on the aspect ratio of the nucleus (see SI of Hu et al. (2012) for details). When $\gamma_{cs} < \gamma_{ls}$, the value of γ_{het} will be reduced from that for the homogeneous nucleus as shown for calcite nucleation in Fig. 1.5a (Hu et al. 2012). The calculations for calcite also show that even when the effective γ_{ls} equals γ_{cs}—that is, the interfacial energies for the crystal–substrate and fluid–substrate interfaces are equal—the barrier is reduced (Fig. 1.5a, $\gamma_{\text{het}}/\gamma_{cl} = 1.0$). This is because a surface that would have been generated during homogeneous nucleation is now a crystal–substrate interface that carries no energy penalty. A further reduction in γ_{het} by only

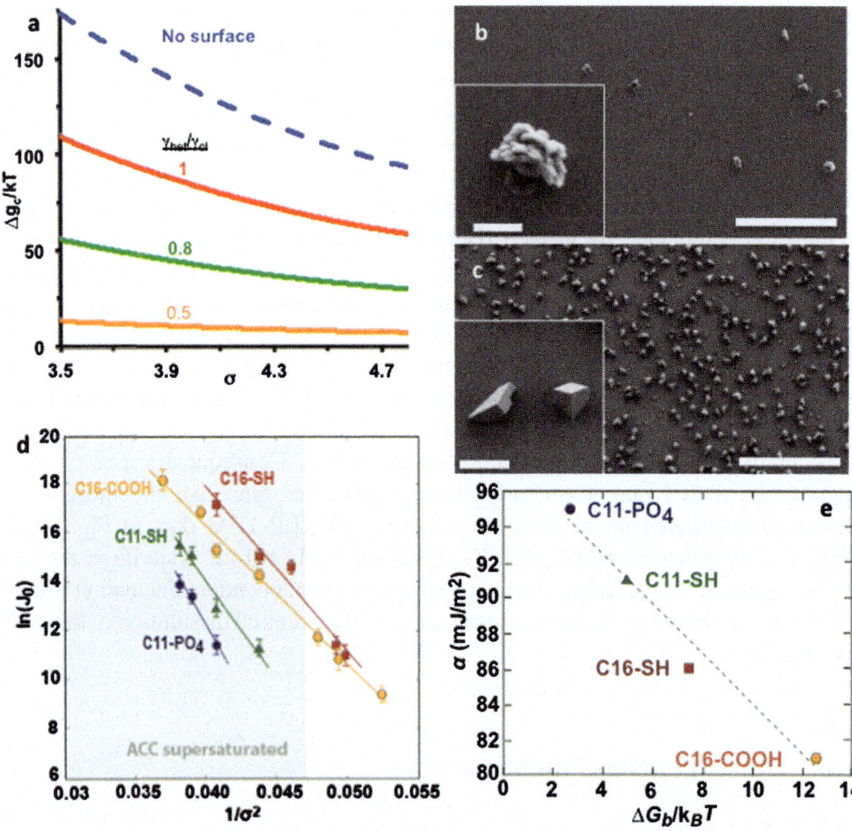

Fig. 1.5 (**a**) Dependence of free-energy barrier Δg_c on σ for homogeneous nucleation of a calcite rhomb with $\gamma = 109$ mJ/m^2 (*blue line*) and heterogeneous nucleation of a calcite rhomb on an (012) face with the indicated values of $\gamma_{\text{het}}/\gamma_{cl}$. (**b, c**) SEM images showing number density and orientation of calcite crystals on (**b**) bare gold and (**c**) MHA surfaces. *Scale bars* are: main images – 100 μm, *insets* – 10 μm. (**d**) Dependence of nucleation rate on supersaturation relative to calcite show agreement with CNT prediction. (**e**) Interfacial energies (α) derived from nucleation rate measurements in (**d**) vs. free energies of binding (ΔG_b) for alkanethiol SAM molecules on calcite (104) surfaces as measured by dynamic force spectroscopy. Lower α correlates with strong calcite–SAM interactions (large ΔG_b) (Panels **a**, **b**: Hu et al. (2012) used by permission of the Royal Society of Chemistry. Panels **c**, **d**: Hamm et al. (2014))

20–50 % due to $\gamma_{cs} < \gamma_{ls}$ leads to a further decrease in the barrier by a factor of 3–13 (Fig. 1.5a, $\gamma_{\text{het}}/\gamma_{cl} = 0.8$ and 0.5). Given that nucleation rate depends exponentially on this barrier, these large reductions mean that surfaces have the potential to completely alter the dynamics and pathways of formation.

The impact of surfaces on nucleation can be readily seen by comparing the outcome of $CaCO_3$ nucleation on a bare gold surface to that on alkyl thiol self-assembled monolayers (SAMs) that are formed on an otherwise identical surface (Aizenberg et al. 1999; Hu et al. 2012; Lee et al. 2007). When CO_2 and NH_3 from an ammonium carbonate source are allowed to diffuse into a calcium chloride solution, the nucleation density on bare gold is low, and the products are a mixture of phases (Fig. 1.5b). In contrast, a 16-mercaptohexadecanoic (MHA) acid SAM, which is terminated with carboxyl groups, produces calcite almost exclusively with a much higher areal density. Moreover, nearly all crystals nucleate on the calcite (012) plane (Fig. 1.5c). In situ observations show that, for this well-ordered carboxyl-terminated SAM, nucleation first occurs on the film, and the products consist of calcite at the earliest time of observation by either optical or AFM techniques (Hu et al. 2012). In contrast, if the SAM is instead OH-terminated, nucleation of ACC occurs first in the solution above the SAM, and calcite crystals appear only sporadically at later times.

Measurements of nucleation rates on surfaces as a function of supersaturation were used to obtain a quantitative measure of the impact of alkyl thiol SAMs on calcite interfacial energies (Giuffre et al. 2013; Hamm et al. 2014; Hu et al. 2012) by analyzing the data with Eqs. 1.1 and 1.3 written in the form:

$$\ln J = \ln A' - \mathrm{B}\frac{\gamma_{\mathrm{het}}^3}{\sigma^2} \tag{1.4}$$

where $A' = Ae^{-E_A/kT}$. Hu et al. (2012) measured J for a range of supersaturations on both MHA and 11-mercaptoundecanoic acid (MUA), which are both carboxyl terminated but differ in chain length. Hamm et al. (2014) extended this study to a range of headgroup chemistries including COOH, SH, and PO_4 groups (Fig. 1.3d). Both studies observed dependencies of J on σ that were in accordance with Eq. 1.4, despite the fact that the supersaturation range straddled the solubility limit of ACC given in the literature (Brecevic and Nielsen 1989). Although the range of supersaturation over which statistically meaningful data could be obtained was limited, the data from all substrates produced γ_{het} values that were lower than estimates for the homogeneous nucleation of calcite. For the most extreme case, the MHA surface, the data yielded an effective interfacial energy that was only 66 % of that estimated for calcite nucleation in bulk solution (72 vs. 109 mJ/m^2). Thus, the barrier for nucleation on MHA is five times less than for nucleation in solution. All other factors being equal, this difference in barriers would correspond to relative rates of heterogeneous nucleation on an MHA film to homogeneous nucleation in solution of 1.0: 4.5×10^{-38}. Note that the outcome of these studies does not depend on whether the fundamental units that comprise the nucleus are individual ions or clusters—stable or metastable. This is because, as Eq. 1.2 indicates, the form of the rate equations is the same in either case and, in fact, for the range of supersaturations used in these studies, the value of C in Eq. 1.2 never exceeds 10 % of σ. Thus these experiments did not have sufficient sensitivity to provide insights to the importance of cluster aggregation during $CaCO_3$ nucleation.

The physical basis for the reduced interfacial energy was investigated by Hamm et al. using dynamic force spectroscopy (DFS) (Hamm et al. 2014). Considering the expression for γ_{het} in terms of the individual interfacial energies given by Eq. 1.3, they assumed the crystal–liquid interfacial energy should be similar for all films. They also assumed similar SAM–fluid interfacial energies given the low pK_A of the films and the comparatively high pH (10.8 ± 0.1) of the experiments. Thus, the only term that should differ significantly from one film to another is the interfacial energy between the SAM and the crystal (γ_{cs}) with smaller values leading to lower barriers. Moreover, γ_{cs} should be largely controlled by the binding free-energy ΔG_b between the crystal and the SAM. In fact, because DFS determines the difference in free energy between the bound and unbound state, the ΔG_b can be written in terms of the same interfacial energies that appear in Eq. 1.3 where

$$\Delta G_b = a\left(\gamma_{cl} + \gamma_{ls}\right) - a\gamma_{cs} \tag{1.5}$$

where a is the tip–substrate contact area. Rewriting Eq. 1.5 in terms of γ_{het} predicts:

$$\gamma_{\text{het}} = -\frac{h}{a}\Delta G_b + (1+h)\,\gamma_{cl}. \tag{1.6}$$

Because γ_{cl} depends only upon the contact area between the crystal and the liquid, this term is independent of template chemistry. Consequently, γ_{het} and ΔG_b should be linearly related, and, from Eq. 1.1b, an explicit relationship between the barrier to nucleation and the strength of the crystal–substrate binding energy should exist.

To test these relationships, AFM tips were gold coated and functionalized with the same SAM monomers used in the measurements of nucleation rates on SAMs. These tips were then used to perform DFS measurements on flat (104) faces of calcite, which served as a proxy for the (012) faces that typically define the plane of nucleation in the nucleation rate experiments on SAMs. Figure 1.5e reveals there is indeed a direct, linear relationship between these γ_{het} and ΔG_b, and the slope was shown to be in reasonable agreement with the expected value. Thus, this study found that a large crystal–SAM binding free energy leads to a small crystal–SAM interfacial energy, resulting in a small barrier to nucleation. These insights provide a physical basis for the common belief that good binders are good nucleators.

1.7 Macromolecular Matrices

In many biomineralizing systems, the formation of mineral components often occurs within a hydrated macromolecular matrix (Nudelman et al. 2007; Young et al. 1999) that is rich in biopolymers with numerous charged side groups. Based on in vitro evidence, these molecules are assumed to play a role in stabilizing amorphous precursor phases from which many crystalline biominerals are formed (Nudelman et al. 2007; Young et al. 1999). The control over crystal orientation often seen in biominerals suggests a role for surfaces that reduce interfacial energies, as in the biomimetic examples presented above. However, immobilized macromolecules, in particular polysaccharides, have also been proposed to act as concentrators of mineral ions by cation binding prior to crystallization (Marsh 1994; Addadi et al.

1987). Within the framework of CNT, one would expect ion binding to enhance nucleation, because concentrating the cations should increase the kinetic pre-factor and/or the local supersaturation in Eq. 1.1.

The idea that an ion-binding mechanism promotes nucleation was recently substantiated by Smeets et al. (2015), who used LP-TEM to examine the effects of the macromolecule polystyrene sulfonate (PSS) on $CaCO_3$ nucleation. With an abundance of sulfonate groups, the PSS served as a surrogate for highly charged biomolecular matrices. To conduct this study, control experiments were performed in which CO_2 and NH_3 were diffused into $CaCl_2$ solutions within silicon nitride cells in the absence of PSS. These experiments resulted in the nucleation and growth of vaterite, which formed randomly throughout the experimental volume. In contrast, when PSS was introduced to the $CaCl_2$ solution prior to introducing CO_2, the Ca^{2+} complexed with the PSS to form Ca–PSS globules. These were approximately 10–20 nm in diameter and were evident in TEM, AFM, and dynamic light-scattering measurements (Fig. 1.6a, b). FTIR and zeta potential data demonstrated that the Ca^{2+} interacted with the negatively charged sulfonate groups (Fig. 1.6c, d), and analysis of titration calorimetry data showed that 56 % of the Ca^{2+} in the solution was complexed within the globules.

Further experiments found that when CO_2 was diffused into the TEM fluid cells containing the globules, the first phase to appear was ACC rather than vaterite (Smeets et al. 2015). Moreover, the ACC nanoparticles nucleated exclusively within the globules (Fig. 1.6e). At much later times, after growth of the ACC particles had ceased, vaterite formed sporadically in regions away from the globules. Analysis of the ACC and vaterite growth rates showed that ACC nucleation in the presence of PSS occurred at a higher supersaturation (σ = 1.0–1.8 *relative to ACC*) than the nucleation of vaterite in the absence of PSS (σ = 0.5–0.6 *relative to vaterite*). Because ACC has a lower surface energy (De Yoreo and Vekilov 2003; Hamm et al. 2014), the free-energy barrier to ACC nucleation should be lower than that opposing vaterite nucleation *at equal supersaturations*. These results show, however, that the addition of PSS significantly *inhibits* ACC nucleation. These results suggest that, while PSS can control the location of nucleation and stabilize the amorphous phase, it does not act by lowering the interfacial energy, which would instead lead to a reduction in the supersaturation needed for nucleation.

The results of Smeets et al. (2015) show that the negatively charged polymer acts to control nucleation through a mechanism that is distinct from the SAM surfaces described above. Instead of providing a surface with low interfacial free energy to promote nucleation, the PSS binds to Ca^{2+} and locally concentrates it. The localized formation of the metastable ACC then results from both the locally high supersaturation and interaction with the stabilizing polymer matrix, while the reduced calcium concentration in solution prevents immediate vaterite crystallization outside the globules. Thus, the results demonstrate the significant role that ion binding can play in directing nucleation by effects that are predominant over controls by interfacial free energy.

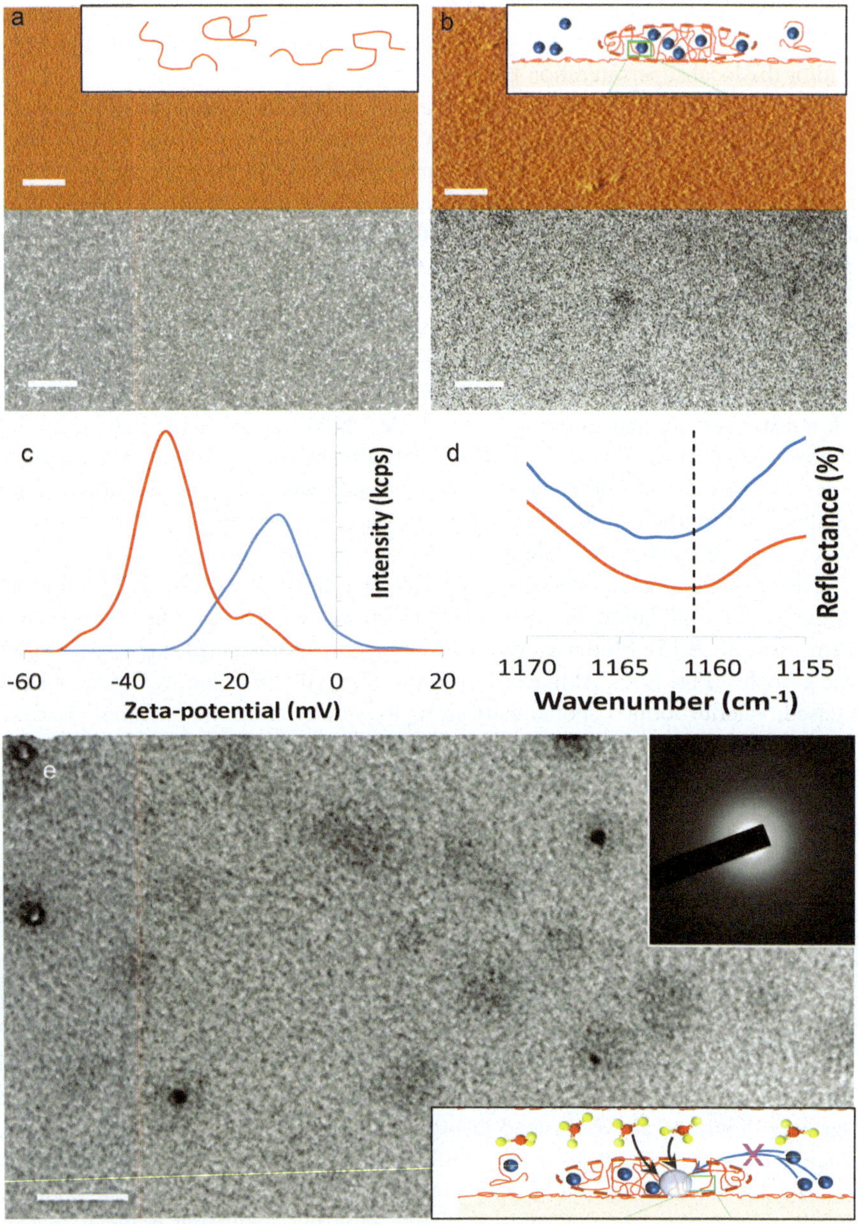

Fig. 1.6 (**a**) and (**b**) in situ AFM (*upper*) and LP-TEM (*lower*) images of (**a**) PSS solution and (**b**) Ca–PSS solution showing globules are only observed in presence of Ca. (**c**) Zeta potential for the Ca–PSS solution (*blue*) and the solution containing only PSS (*red*) (**d**) FTIR spectrum showing the shift in the peak associated with the symmetric stretch of the sulfonate group upon addition of CaCl to PSS solution. The shift shows that the Ca^{2+} ions bind to the sulfonate groups. *Red*: Ca-free PSS solution, *blue*: Ca–PSS solution. (**e**) LP-TEM image from a time-lapse series exhibiting many

1.8 Outlook

Advances in analytical techniques are providing new insights into nucleation that reveal the many pathways to producing crystalline phases from solution. The studies discussed above show that, in a number of cases, the behavior of nucleating systems is at least in qualitative agreement with CNT and, in the case of heterogeneous nucleation on SAM surfaces, exhibits quantitative behavior that is in reasonable agreement with the predictions of CNT. Nonetheless, in a number of systems, the involvement of species of higher order than monomers is evident, and, as Eq. 1.2 makes clear, the analysis of nucleation rate data can in no way rule out precritical particle aggregation even in the cases where CNT appears to be consistent with the data. This is because, as explicitly shown by Eq. 1.2, CNT does not differentiate between ion-by-ion and particle-based nucleation processes. Whether the particle populations that lead to nucleation are always metastable or can be stable relative to the monomers is yet to be determined.

Although all systems supersaturated relative to any phase, even at infinitesimal levels, must eventually nucleate that phase, the high barriers encountered in most systems at low supersaturation—particularly for sparingly soluble systems—force researchers to perform nucleation experiments at high supersaturations where nucleation occurs on laboratory timescales. However the consequence of working in this regime is that multiple pathways of nucleation become operative. There is strong evidence that one of these pathways involves passage into a spinodal region even in simple electrolyte solutions at supersaturations achievable in the laboratory. Whether the multistep nucleation pathways passing through precursor phases that are commonly reported for systems like calcium carbonate or phosphate involve these liquid states remains uncertain, as does the relative importance of direct vs. indirect transformation during multistep nucleation events. However, the process is well documented in protein systems. Moreover, data on heterogeneous nucleation rates show unequivocally that, in cases where high supersaturation opens multiple pathways, a single pathway and outcome can be reestablished through the introduction of surfaces that reduce interfacial free energies. This level of control

Fig. 1.6 (continued) nuclei in Ca–PSS globules and an electron diffraction pattern (*inset*) showing they are amorphous. *Insets* to **a**, **b**, and **e** illustrate stage of mineral formation process probed in each panel: (**a**) In pure water, PSS (*red* free-form lines) exists in a dissolved state with a loose chain conformation. (**b**) Introduction of Ca^{2+} (*blue dots*) results in Ca^{2+}-binding with the SO_3^- groups of the PSS leading to chain collapse to form Ca–PSS globules with a locally high Ca^{2+} concentration deposited on the surface with low free Ca^{2+} concentration in the surrounding solution (*left*) where dissolved PSS is also bound to Ca^{2+} (*right*). (**e**) CO_3^{2-} (*red and yellow dots*) from the ammonium carbonate source diffuses into the globules (*black arrows*) where it binds with Ca^{2+}, replacing the weaker SO_3^-/Ca^{2+} interaction and creating a supersaturated state from which ACC (*light blue sphere*) nucleates and grows until Ca^{2+} in globules is depleted to the solubility limit of ACC. The depleted concentration of Ca outside of the globules limits additional inward diffusion of Ca (*red crosses over blue arrows*). *Scale bars* (**a, b**) 100 nm and (**e**) 50 nm (From Smeets et al. (2015) used by permission of Macmillan Publisher Limited)

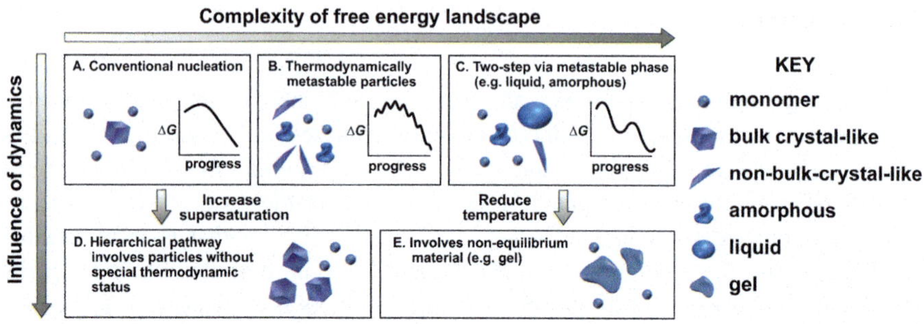

Fig. 1.7 The possible pathways by which monomers form a stable bulk crystal, and the physical mechanisms that give rise to them, can have thermodynamic (**a**, **b**, **c**) and kinetic (**d**, **e**) origins. (**a**) Classical monomer-by-monomer addition. (**b**) Aggregation of metastable particles, such as liquid, amorphous, or poorly crystalline particles, or of oriented (and nearly oriented) attachment of metastable nanocrystals. (**c**) Crystallization via the formation of a metastable bulk phase, such as a liquid or solid polymorph. (**d**) Kinetically dominated aggregation of clusters or oligomers. (**e**) Aggregation of unstable particles whose internal structures are not those of equilibrium phases (From De Yoreo et al. (2015) used by permission of the American Association for the Advancement of Science)

also occurs with polymeric matrices that locally concentrate the solute ions and interact with the mineralizing constituents to define the phase that forms.

The rich diversity of nucleation processes that are evident in the examples presented here demonstrates that the classical picture of nucleation needs to be expanded. Beyond the relatively simple process of monomer addition to a growing particle to overcome a smooth barrier (Fig. 1.7a), additional pathways exist that require considering both thermodynamic and kinetic factors. The former includes energy landscapes that allow for metastable (or stable) precritical particle populations created through microscopic fluctuations and which render barriers bumpy rather than smooth (Fig. 1.7b), as well as landscapes that allow for metastable bulk phases (Fig. 1.7c). Overlain on top of these thermodynamic complexities are the dynamic factors of supersaturation and temperature, which lead, respectively, to hierarchical pathways involving particles that have no special thermodynamic status (Fig. 1.7d) and nonequilibrium bulk materials that are nonetheless kinetically stabilized indefinitely (Fig. 1.7e). By developing a mechanistic understanding of these competing—and sometimes—simultaneous phenomena, a more accurate predictive and quantitative picture of crystal nucleation will emerge.

Acknowledgments This work was performed, in part, at Pacific Northwest National Laboratory (PNNL) through support from the U.S. Department of Energy, Office of Basic Energy Sciences, Division of Materials Science and Engineering. PNNL is operated by Battelle for the U.S. Department of Energy under Contract DE-AC05-76RL01830 (JDY). The material is also based upon work supported by the National Science Foundation NSF OCE-1061763 (PMD) and the U.S. Department of Energy Office of Science, Office of Basic Energy Sciences under Award Number BES-FG02-00ER15112 (PMD), as well as on work performed at Eindhoven University of Technology supported by the Dutch Science Foundation – Chemical Sciences (NWO–CW), and by NanoNextNL, a micro and nanotechnology consortium of the government of The Netherlands and 130 partners (NAJMS).

References

Addadi L, Moradian J, Shay E, Maroudas NG, Weiner S (1987) A chemical-model for the cooperation of sulfates and carboxylates in calcite crystal nucleation—relevance to biomineralization. Proc Natl Acad Sci U S A 84:2732–2736

Aizenberg J, Black AJ, Whitesides GM (1999) Control of crystal nucleation by patterned self-assembled monolayers. Nature 398:495–498

Andreassen JPL, Lewis A (2017) Classical and non-classical theories of crystal growth. In: Van Driessche AES, Kellermeier M, Benning LG, Gebauer D (eds) New perspectives on mineral nucleation and growth: from solution precursors to solid materials. Springer, Cham, pp 137–154

Baumgartner J, Dey A, Bomans PHH, Le Coadou C, Fratzl P, Sommerdijk N, Faivre D (2013) Nucleation and growth of magnetite from solution. Nat Mater 12:310–314

Beniash E, Metzler RA, Lam RSK, Gilbert PUPA (2009) Transient amorphous calcium phosphate in forming enamel. J Struct Biol 166:133–143

Bewernitz MA, Gebauer D, Long J, Cölfen H, Gower LB (2012) A metastable liquid precursor phase of calcium carbonate and its interactions with polyaspartate. Faraday Discuss 159: 291–312

Birkedal H (2017) Phase transformations in calcium phosphate crystallization. In: Van Driessche AES, Kellermeier M, Benning LG, Gebauer D (eds) New perspectives on mineral nucleation and growth: from solution precursors to solid materials. Springer, Cham, pp 199–210

Bots P, Benning LG, Rodriguez-Blanco JD, Roncal-Herrero T, Shaw S (2012) Mechanistic insights into the crystallization of amorphous calcium carbonate (ACC). Cryst Growth Des 12: 3806–3814

Bray AJ (2002) Theory of phase-ordering kinetics. Adv Phys 51:481–587

Brecevic L, Nielsen AE (1989) Solubility of amorphous calcium-carbonate. J Cryst Growth 98:504–510

Burkett SL, Davis ME (1994) Mechanism of structure direction in the synthesis of Si-Zsm-5—an investigation by intermolecular H-1-Si-29 Cp Mas Nmr. J Phys Chem 98:4647–4653

Chernov AA, Givargizov EI (1984) Modern crystallography III: crystal growth. Springer, Berlin

Chung S, Shin SH, Bertozzi CR, De Yoreo JJ (2010) Self-catalyzed growth of S layers via an amorphous-to-crystalline transition limited by folding kinetics. Proc Natl Acad Sci U S A 107:16536–16541

De Yoreo JJ, Vekilov PG (2003) Principles of crystal nucleation and growth. In: Dove PM, Deyoreo JJ, Weiner S (eds) Biomineralization. Mineralogical Soc America, Washington

De Yoreo JJ, Gilbert PUPA, Sommerdijk NAJM, Penn RL, Whitelam S, Joester D, Zhang HZ, Rimer JD, Navrotsky A, Banfield JF, Wallace AF, Michel FM, Meldrum FC, Cölfen H, Dove PM (2015) Crystallization by particle attachment in synthetic, biogenic, and geologic environments. Science 349:498

Delgado-Lopez JM, Guagliardi A (2017) Control over nanocrystalline apatite formation: what can the X-ray total scattering approach tell us. In: Van Driessche AES, Kellermeier M, Benning LG, Gebauer D (eds) New perspectives on mineral nucleation and growth: from solution precursors to solid materials. Springer, Cham, pp 211–226

Demichelis R, Raiteri P, Gale JD, Quigley D, Gebauer D (2011) Stable prenucleation mineral clusters are liquid-like ionic polymers. Nat Commun 2:8

Demichelis R, Raiteri P, Gale JD (2017) Ab Initio modelling of the structure and properties of crystalline calcium carbonate. In: Van Driessche AES, Kellermeier M, Benning LG, Gebauer D (eds) New perspectives on mineral nucleation and growth: from solution precursors to solid materials. Springer, Cham, pp 113–136

Dey A, Bomans PHH, Muller FA, Will J, Frederik PM, De With G, Sommerdijk N (2010) The role of prenucleation clusters in surface-induced calcium phosphate crystallization. Nat Mater 9:1010–1014

Donners JJJM, Heywood BR, Meijer EW, Nolte RJM, Roman C, Schenning APHJ, Sommerdijk NAJM (2000) Amorphous calcium carbonate stabilised by poly(propylene imine) dendrimers. Chem Commun 2000:1937–1938

Donners JJJM, Heywood BR, Meijer EW, Nolte RJM, Sommerdijk NAJM (2002) Control over calcium carbonate phase formation by dendrimer/surfactant templates. Chem Eur J 8: 2561–2567

Falini G, Fermani A (2017) Nucleation and growth from a biomineralization perspective. In: Van Driessche AES, Kellermeier M, Benning LG, Gebauer D (eds) New perspectives on mineral nucleation and growth: from solution precursors to solid materials. Springer, Cham, pp 185–198

Fernandez-Martinez A, Lopez-Martinez H, Wang D (2017) Structural characteristics and the occurrence of polyamorphism in amorphous calcium carbonate. In: Van Driessche AES, Kellermeier M, Benning LG, Gebauer D (eds) New perspectives on mineral nucleation and growth: from solution precursors to solid materials. Springer, Cham, pp 77–92

Galkin O, Chen K, Nagel RL, Hirsch RE, Vekilov PG (2002) Liquid-liquid separation in solutions of normal and sickle cell hemoglobin. Proc Natl Acad Sci U S A 99:8479–8483

Gary DC, Terban MW, Billinge SJL, Cossairt BM (2015) Two-step nucleation and growth of InP quantum dots via magic-sized cluster intermediates. Chem Mater 27:1432–1441

Gebauer D (2012) General discussions. Faraday Discuss 159:139–180

Gebauer D, Cölfen H (2011) Prenucleation clusters and non-classical nucleation. Nano Today 6:564–584

Gebauer D, Völkel A, Cölfen H (2008) Stable prenucleation calcium carbonate clusters. Science 322:1819–1822

Gebauer D, Cölfen H, Verch A, Antonietti M (2009) The multiple roles of additives in $CaCO_3$ crystallization: a quantitative case study. Adv Mater 21:435–439

Gebauer D, Kellermeier M, Gale JD, Bergström L, Cölfen H (2014) Pre-nucleation clusters as solute precursors in crystallisation. Chem Soc Rev 43:2348–2371

Gibbs JW, Smith AW (1874) On the equilibrium of heterogeneous substances, Transactions of the Connecticut Academy of Arts and Sciences, Volume III, Chapter V. and Chapter IX. Tuttle, Morehouse & Taylor, New Haven

Giuffre AJ, Hamm LM, Han N, De Yoreo JJ, Dove PM (2013) Polysaccharide chemistry regulates kinetics of calcite nucleation through competition of interfacial energies. Proc Natl Acad Sci U S A 110:9261–9266

Gower LB, Odom DJ (2000) Deposition of calcium carbonate films by a polymer-induced liquid-precursor (PILP) process. J Cryst Growth 210:719–734

Gribb AA, Banfield JF (1997) Particle size effects on transformation kinetics and phase stability in nanocrystalline TiO2. Am Mineral 82:717–728

Gutzow IS (1997) Prof. Dr. Ivan N. Stranski (1897–1979): one of the founders of present-day theory of crystal forms, crystal nucleation and growth. Cryst Res Technol 32:753–758

Habraken W, Tao JH, Brylka LJ, Friedrich H, Bertinetti L, Schenk AS, Verch A, Dmitrovic V, Bomans PHH, Frederik PM, Laven J, Van Der Schoot P, Aichmayer B, De With G, Deyoreo JJ, Sommerdijk N (2013) Ion-association complexes unite classical and non-classical theories for the biomimetic nucleation of calcium phosphate. Nat Commun 4:12

Hamm LM, Giuffre AJ, Han N, Tao J, Wang D, De Yoreo JJ, Dove PM (2014) Reconciling disparate views of template-directed nucleation through measurement of calcite nucleation kinetics and binding energies. Proc Natl Acad Sci U S A 111:1304–1309

Han TYJ, Aizenberg J (2008) Calcium carbonate storage in amorphous form and its template-induced crystallization. Chem Mater 20:1064–1068

Hu Q, Nielsen MH, Freeman CL, Hamm LM, Tao J, Lee JRI, Han TYJ, Becker U, Harding JH, Dove PM, De Yoreo JJ (2012) The thermodynamics of calcite nucleation at organic interfaces: classical vs. non-classical pathways. Faraday Discuss 159:509–523

Hunger S, Benning LG (2007) Greigite: a true intermediate on the polysulfide pathway to pyrite. Geochem Trans 8:1

Kashchiev D (2003) Thermodynamically consistent description of the work to form a nucleus of any size. J Chem Phys 118:1837–1851

Kumar S, Wang ZP, Penn RL, Tsapatsis M (2008) A structural resolution cryo-TEM study of the early stages of MFI growth. J Am Chem Soc 130:17284–17286

Lee JRI, Han TYJ, Willey TM, Wang D, Meulenberg RW, Nilsson J, Dove PM, Terminello LJ, Van Buuren T, De Yoreo JJ (2007) Structural development of mercaptophenol self-assembled monolayers and the overlying mineral phase during templated $CaCO_3$ crystallization from a transient amorphous film. J Am Chem Soc 129:10370–10381

Marsh ME (1994) Polyanion-mediated mineralization—assembly and reorganization of acidic polysaccharides in the golgi system of a coccolithophorid alga during mineral deposition. Protoplasma 177:108–122

Mutaftschiev B (1993) Nucleation. In: Hurle D (ed) Handbook on crystal growth. North-Holland, Amsterdam

Navrotsky A (2004) Energetic clues to pathways to biomineralization: precursors, clusters, and nanoparticles. Proc Natl Acad Sci U S A 101:12096–12101

Nielsen MH, De Yoreo JJ (2017) Liquid phase TEM investigations of crystal nucleation, growth and transformation. In: Van Driessche AES, Kellermeier M, Benning LG, Gebauer D (eds) New perspectives on mineral nucleation and growth: from solution precursors to solid materials. Springer, Cham, pp 353–374

Nielsen MH, Aloni S, De Yoreo JJ (2014a) In situ TEM imaging of $CaCO_3$ nucleation reveals coexistence of direct and indirect pathways. Science 345:1158–1162

Nielsen MH, Li DS, Zhang HZ, Aloni S, Han TYJ, Frandsen C, Seto J, Banfield JF, Cölfen H, De Yoreo JJ (2014b) Investigating processes of nanocrystal formation and transformation via liquid cell TEM. Microsc Microanal 20:425–436

Nudelman F, Chen HH, Goldberg HA, Weiner S, Addadi L (2007) Spiers memorial lecture: lessons from biomineralization: comparing the growth strategies of mollusc shell prismatic and nacreous layers in Atrina rigida. Faraday Discuss 136:9–25

Nudelman F, Pieterse K, George A, Bomans PHH, Friedrich H, Brylka LJ, Hilbers PAJ, De With G, Sommerdijk N (2010) The role of collagen in bone apatite formation in the presence of hydroxyapatite nucleation inhibitors. Nat Mater 9:1004–1009

Penn RL et al (2017) A perspective on the particle-based crystal growth of ferric oxides, oxyhydroxides, and hydrous oxides. In: Van Driessche AES, Kellermeier M, Benning LG, Gebauer D (eds) New perspectives on mineral nucleation and growth: from solution precursors to solid materials. Springer, Cham, pp 257–274

Pichon BP, Bomans PHH, Frederik PM, Sommerdijk NAJM (2008) A quasi-time-resolved CryoTEM study of the nucleation of CaCO3 under langmuir monolayers. J Am Chem Soc 130:4034–4040

Pouget EM, Bomans PHH, Goos J, Frederik PM, De With G, Sommerdijk N (2009) The initial stages of template-controlled $CaCO_3$ formation revealed by Cryo-TEM. Science 323: 1455–1458

Pouget EM, Bomans PHH, Dey A, Frederik PM, De With G, Sommerdijk N (2010) The development of morphology and structure in hexagonal vaterite. J Am Chem Soc 132: 11560–11565

Raiteri P, Gale JD (2010) Water is the key to nonclassical nucleation of amorphous calcium carbonate. J Am Chem Soc 132:17623–17634

Reichel V, Faivre DDA, et al (2017) Magnetite nucleation and growth. In: Van Driessche AES, Kellermeier M, Benning LG, Gebauer D (eds) New perspectives on mineral nucleation and growth: from solution precursors to solid materials. Springer, Cham, pp 275–298

Reif F (1965) Fundamentals of statistical and thermal physics. McGraw-Hill Inc, New York

Rodriguez-Blanco JD, Sand KK, Bennin LG (2017) ACC and vaterite as intermediates in the solution based crystallization of $CaCO_3$. In: Van Driessche AES, Kellermeier M, Benning LG, Gebauer D (eds) New perspectives on mineral nucleation and growth: from solution precursors to solid materials. Springer, Cham, pp 93–112

Scheifele B, Saika-Voivod I, Bowles RK, Poole PH (2013) Heterogeneous nucleation in the low-barrier regime. Phys Rev E 87:042407

Smeets PJM, Cho KR, Kempen RGE, Sommerdijk NAJM, De Yoreo JJ (2015) Calcium carbonate nucleation driven by ion binding in a biomimetic matrix revealed by in situ electron microscopy. Nat Mater 14:394–399

Tenwolde PR, Frenkel D (1997) Enhancement of protein crystal nucleation by critical density fluctuations. Science 277:1975–1978

Tobler DJ, Stawski TM, Benning LG (2017) Silica and alumina nanophases: natural processes and industrial applications. In: Van Driessche AES, Kellermeier M, Benning LG, Gebauer D (eds) New perspectives on mineral nucleation and growth: from solution precursors to solid materials. Springer, Cham, pp 293–316

Towe KM, Lowenstam H (1967) Ultrastructure and development of iron mineralization in radular teeth of Cryptochiton stelleri (Mollusca). J Ultrastruct Res 17:1–13

Van Driessche AES, Benning LG, Rodriguez-Blanco JD, Ossorio M, Bots P, Garcia-Ruiz JM (2012) The role and implications of bassanite as a stable precursor phase to gypsum precipitation. Science 336:69–72

Van Driessche AES, Stawski TM, Benning LA, Kellermeier M (2017) Calcium sulfate precipitation throughout its phase diagram. In: Van Driessche AES, Kellermeier M, Benning LG, Gebauer D (eds) New perspectives on mineral nucleation and growth: from solution precursors to solid materials. Springer, Cham, pp 227–254

Vekilov PG (2005) Two-step mechanism for the nucleation of crystals from solution. J Cryst Growth 275:65–76

Wallace AF, Hedges LO, Fernandez-Martinez A, Raiteri P, Gale JD, Waychunas GA, Whitelam S, Banfield JF, De Yoreo JJ (2013) Microscopic evidence for liquid-liquid separation in supersaturated $CaCO_3$ solutions. Science 341:885–889

Wang LJ, Zhang W, Qiu SR, Zachowicz WJ, Guan XY, Tang RK, Hoyer JR, De Yoreo JJ, Nancollas GH (2006) Inhibition of calcium oxalate monohydrate crystallization by the combination of citrate and osteopontin. J Cryst Growth 291:160–165

Wang XL, Chou IM, Hu WX, Burruss RC (2013) In situ observations of liquid-liquid phase separation in aqueous $MgSO_4$ solutions: geological and geochemical implications. Geochim Cosmochim Acta 103:1–10

Wolf SE, Gower LB (2017) Challenges and perspectives of the polymer-induced liquid-precursor process: the pathway from liquid-condensed mineral precursors to mesocrystalline products. In: Van Driessche AES, Kellermeier M, Benning LG, Gebauer D (eds) New perspectives on mineral nucleation and growth: from solution precursors to solid materials. Springer, Cham, pp 43–76

Wolf SE, Leiterer J, Kappl M, Emmerling F, Tremel W (2008) Early homogenous amorphous precursor stages of calcium carbonate and subsequent crystal growth in levitated droplets. J Am Chem Soc 130:12342–12347

Young JR, Davis SA, Bown PR, Mann S (1999) Coccolith ultrastructure and biomineralisation. J Struct Biol 126:195–215

Chapter 2
Novel Paradigms in Nonclassical Nucleation Theory

James F. Lutsko

2.1 Introduction

The process of nucleation is of fundamental importance in many areas of physics, chemistry, and materials science. Nucleation involves the formation of macroscopic objects—droplets, crystals, etc.—based on microscopic mechanisms such as molecular attachment and detachment and thermal fluctuations. As such, it is an intrinsically challenging problem of coupled length and time scales. Furthermore, only the initial and final states are equilibrium states: indeed, the initial state by definition is actually only metastable. Hence, the process is *nonequilibrium* in nature further complicating its description. It is therefore remarkable that a well-established theoretical paradigm exists, namely, classical nucleation theory (CNT), which is based on a deft combination of physical intuition and a certain dexterity in skirting technical difficulties. In fact, as discussed below, the real key to the success of CNT is that it is applicable in a regime of very weak driving forces (e.g., low supersaturation) so that processes are slow and the key object of discussion, the critical cluster, is macroscopically large, thus minimizing the importance of the details of microscopic mechanisms. However, today, nucleation is studied experimentally, and even imaged, at the molecular level, for cases of high supersaturation in which the critical cluster consists of fewer than 100 molecules (Sleutel et al. 2014). At the same time, computer simulation is well suited to the study of nucleation at these length and time scales, and modern, so-called "rare-event" techniques extend this to the regimes intermediate between the molecular and the macroscopic (see, e.g., Van Erp 2012). All of this new focus has not only pushed

J.F. Lutsko (✉)
Center for Nonlinear Phenomena and Complex Systems, Code Postal 231, Université Libre de Bruxelles, Blvd. du Triomphe, 1050 Brussels, Belgium
e-mail: jlutsko@ulb.ac.be

A.E.S. Van Driessche et al. (eds.), *New Perspectives on Mineral Nucleation and Growth*,
DOI 10.1007/978-3-319-45669-0_2

us well beyond the bounds of applicability of CNT but has revealed new phenomena such as two-step nucleation that lies outside of the conceptual framework of CNT altogether.

2.2 Classical Nucleation Theory

In order to set the context for an alternative approach to nucleation, the standard classical nucleation theory is reviewed and critiqued. The aim here is not to give a comprehensive review—for that there are many excellent books, e.g., Kashchiev (2000). Instead, the goal is simply to set out the basic conceptual elements and to indicate how they are linked together. For this reason, the discussion will be limited to homogeneous nucleation in a single-component system in three dimensions. Specifically, we will have in mind liquid-vapor or solid-liquid diffusion-limited nucleation. The same basic ideas are readily extended to multicomponent systems, two-dimensional (e.g., surface) nucleation, etc.

2.2.1 *Thermodynamics of CNT*

Nucleation involves the transition from an initial, metastable phase A (the mother phase) to a new phase B, e.g., from vapor to liquid, liquid to solid, weak solution to crystal, etc. In CNT, we presume that we know two things: the free energy per unit volume for the initial and final phases, ω_A and ω_B, and the planar surface tension for coexisting A and B phases, γ_{AB}. In the classical nucleation theory approach, we need to be able to calculate the excess free energy of a cluster of phase B surrounded by a large (in principle, infinite) volume of mother phase A. CNT makes use of the capillary approximation based on the assumption that (a) the cluster-mother phase interface has zero width so all molecules are either inside or outside the cluster, (b) the material inside the cluster has the same thermodynamic properties as in the bulk, and (c) the interfacial free energy per unit volume is the same as for the planar interface. This gives a total free energy for a cluster of size N of

$$\Omega_N = V_N \omega_B + (V - V_N)\omega_A + S_N \gamma_{AB} \tag{2.1}$$

where V_N is the volume of the cluster, S_N is its surface area, and V is the total (very large) volume of the system. Here and below, we work in the grand-canonical ensemble wherein the independent variables are volume, chemical potential, and temperature: since we always have in mind the thermodynamic limit and since we ultimately are only interested in free energy differences, all results will be independent of the choice of ensemble. The excess free energy of the cluster (i.e., the difference in free energy with and without the cluster) is then

$$\Delta\Omega_N \equiv \Omega_N - \omega_A V = V_N (\omega_B - \omega_A) + S_N \gamma_{AB} \tag{2.2}$$

The number of molecules in the cluster N and its volume are related by the bulk density of the new phase, ρ_B, as $N = V_N \rho_B$. If the cluster is assumed to be spherical, this allows us to relate the radius (and hence, also, the surface area) to N. The excess free energy is of course zero for clusters of size zero and increases as the size increases until reaching a maximum at the *critical radius* R_c:

$$R_c = -\frac{2\gamma_{AB}}{\omega_B - \omega_A} \tag{2.3}$$

where the excess free energy is

$$\Delta\Omega_c = \frac{16\pi}{3}\frac{\gamma_{AB}^3}{(\omega_B - \omega_A)^2} \tag{2.4}$$

For radii larger than R_c, the free energy decreases as R increases. Thus, in the nucleation stage, $R = 0$ to $R = R_c$, growth is driven by thermal fluctuations against the gradient of the free energy, whereas in the deterministic growth stage, $R > R_c$, the free energy favors ever-increasing cluster size.

2.2.2 *Kinetics of CNT*

Since it is assumed in CNT that the properties of the material inside the cluster are the same as those of the new phase in the bulk, the only property of the cluster which can change as a function of time is its size: the number of molecules comprising the cluster is the only dynamic variable. In CNT it is assumed that the dominant mechanism for evolution of the cluster is the attachment and detachment of individual growth units (atoms, molecules, colloids, etc.) or *monomers*. Neglected is the addition or removal of larger structures from dimers to mesoscopic clusters. All of these processes are possible, but it is assumed that they occur so rarely that they can be safely ignored. Thus, a cluster consisting of N growth units can only undergo transitions to either size $N + 1$ or $N - 1$. Let the concentration (number per unit volume) of clusters of size N be c_n. The dynamics of the model is then expressed as

$$\frac{dc_N}{dt} = f_{N-1}c_1c_{N-1} - f_Nc_1c_N + g_{N+1}c_{N+1} - g_Nc_N,\ N > 1 \tag{2.5}$$

$$\frac{dc_1}{dt} = -2f_1c_1^2 - \sum_{N=2}^{\infty} f_Nc_1c_N + 2g_2c_2 + \sum_{N=3}^{\infty} g_Nc_N$$

where f_N is the probability per unit time that a cluster of size N will gain a monomer—the *monomer attachment frequency*—and g_N is the probability per unit time that it will give up a monomer, the *monomer detachment frequency*.

The second equation arises because monomers are special: they participate in every event. Note that this model applies to both an undersaturated system and to supersaturated systems. At undersaturation, we can assume that the system is in equilibrium so that the distribution of cluster sizes is stationary and one expects to have detailed balance whereby transition probabilities between sizes N and $N+1$ are equal: $f_{N-1}c_1c_{N-1} = g_Nc_N$. If one further assumes that the distribution of clusters follows a Boltzmann-like distribution, $c_N = c_1e^{-\beta\Delta\Omega_N}$, where $\beta = 1/(k_BT)$ is the inverse of the temperature, then one obtains the relation $g_N = c_1e^{-\beta(\Omega_{N-1}-\Omega_N)}f_{N-1}$.

The total number of clusters per unit volume is $c(t) \equiv \sum_{n=1}^{\infty} c_N(t)$: summing Eq. (2.5) over N gives an equation for its evolution for a closed system:

$$\frac{d}{dt}\sum_{N=1}^{\infty} c_N = c_1\sum_{N=1}^{\infty} f_N\left(e^{-\beta(\Omega_N-\Omega_{N+1})}c_{N+1} - c_N\right)$$

In equilibrium, we are assuming that $c_N \propto e^{-\beta\Omega_N}$ in which case the right-hand side vanishes, as one would expect. If the system is open, the number of clusters can vary. For example, if the concentration of monomers is artificially held fixed at some given value, $c_1(t) = c_1$, then we find that

$$\frac{dc(t)}{dt} = f_1c_1\left(c_1 - e^{-\beta(\Omega_1-\Omega_2)}c_2(t)\right) \tag{2.6}$$

so that $c(t)$ will, in general, vary with time. However, even for a supersaturated system, it is usually assumed that $c_N(t) \approx c_1e^{-\beta(\Omega_1-\Omega_N)}$ for small N so that again the number of clusters would be, for all intents and purposes, constant. Finally, it is easy to verify that the quantity $n(t) \equiv \sum_{n=1}^{\infty} Nc_N(t)$, which is simply the total number of particles per unit volume, i.e., the average density, is in all cases constant with time, thus verifying conservation of particle number.

2.2.3 *Zeldovich Equation*

For large clusters, it makes sense to suppose that the concentration of clusters of size N, c_N, can be viewed as a continuous function, $c(N)$, with similar generalizations of the free energy, the monomer attachment frequencies, and so forth. We can formally replace quantities such as $c(N \pm 1)$ by $c(N \pm \epsilon)$ and Taylor expand around $\epsilon = 0$ to get

$$\frac{\partial c(N;t)}{\partial t} = \epsilon^2\frac{\partial}{\partial N}\left(f(N;t)\frac{\partial\beta\Omega(N;t)}{\partial N} + f(N;t)\frac{\partial}{\partial N}\right)c(N;t) + O(\epsilon^3). \tag{2.7}$$

When the higher-order terms are neglected and ϵ is set to its true value of one, the result is known as the Zeldovich equation. If we integrate the Zeldovich equation over N from N_0 to infinity and note that $c(N > N_0; t) \equiv \int_{N_0}^{\infty} c(N'; t)\, dN'$ is the concentration of clusters of size N or larger, $c(N > N_0; t)$, then

$$\frac{\partial c(N > N_0; t)}{\partial t} = -\left(f(N_0; t) \frac{\partial \beta \Omega (N_0; t)}{\partial N_0} c(N_0; t) + f(N_0; t) \frac{\partial c(N_0; t)}{\partial N_0} \right) \tag{2.8}$$

is the number of clusters per unit volume and per unit time that become larger than N_0 which, for $N_0 = N_c$, is just the nucleation rate.

A somewhat different perspective is possible if we note that $c(N; t)$ is the number of clusters of size N per unit volume. Integrating this quantity over all N therefore gives the total number of clusters per unit volume. The probability to find a cluster of size N will then be $P(N; t) = c(N; t) / c(t)$ which satisfies

$$\frac{\partial P(N; t)}{\partial t} = \frac{\partial}{\partial N} \left(f(N; t) \frac{\partial \beta \Omega (N; t)}{\partial N} + f(N; t) \frac{\partial}{\partial N} \right) P(N; t) - c(t)^{-1} P(N; t) \frac{\partial c(t)}{\partial t}. \tag{2.9}$$

Now, if the average number of clusters is constant, e.g., in an undersaturated, equilibrium solution or in a steady-state nucleation experiment as described below, the last term will be zero. What remains has the form of a *Fokker-Planck* equation for the probability to find a cluster of a given size. This allows us to make contact with the extensive theory of stochastic processes (Gardiner 2004; Hänggi and Borkovec 1990).

2.2.4 Nucleation Rate

A time-independent equilibrium state is only possible in an undersaturated system. As the saturation is increased, e.g., by lowering the temperature, the system eventually reaches the supersaturated state in which nucleation will take place. The key practical question is the rate at which nucleation occurs. The classical method used to determine the nucleation rate is to imagine a thought experiment in which one fixes the concentration of monomers at some prescribed value, c_1, and one removes post-critical clusters once they reach a given size, say $N_{\max}$. In this way, a steady nonequilibrium state is established, and the rate at which clusters are removed is a measure of the nucleation rate.

In the steady state, the kinetic equation, Eq. (2.5), can be written as

$$f_{N-1} c_1 c_{N-1} - g_N c_N = f_N c_1 c_N - g_{N+1} c_{N+1}, 2 \leq N \leq N_{\max -1} \tag{2.10}$$

with boundary conditions c_1 fixed at its prescribed value and $c_{N_{\max}} = 0$. The rate at which critical nuclei are formed is $J_{N_c} = f_{N_c} c_1 c_{N_c} - g_{N_c+1} c_{N_c+1}$, and this turns out to be (Kashchiev 2000)

$$J_{N_c} = \left(\sum_{n=1}^{N_{\max}-1} \frac{1}{f_n c_1 e^{-\beta(\Omega_n - \Omega_1)}} \right)^{-1}. \tag{2.11}$$

Similarly, we could start with the Zeldovich equation with boundary conditions $c(1) = c_1$ and $c(N_{\max}) = 0$ which gives the stationary solution (Kashchiev 2000)

$$c(N) = c_1 e^{-\beta \Delta\Omega(N)} - \frac{\int_1^N \frac{1}{f(N')} e^{\beta \Delta\Omega(N')} dN'}{\int_1^{N_{\max}} \frac{1}{f(N')} e^{\beta \Delta\Omega(N')} dN'} c_1 e^{-\beta \Delta\Omega(N)}, \tag{2.12}$$

so that using Eq. (2.8), the nucleation rate is then

$$J_{N_c} = \left(\int_1^{N_{\max}} \frac{1}{c_1 f(N')} e^{\beta \Delta\Omega(N')} dN' \right)^{-1}. \tag{2.13}$$

The similarity between the two results, Eqs. (2.11) and (2.13), is evident. Assuming that the dominant contribution to the integral comes from the neighborhood of the maximum of the excess free energy, which is to say at the critical cluster, we can make a saddle-point approximation using $\Delta\Omega(N) \simeq \Delta\Omega(N_c) + \frac{1}{2}\Delta\Omega''(N_c)(N - N_c)^2$, to get the classic expression

$$J_{N_c} \simeq c_1 f(N_c) \sqrt{\frac{1}{2\pi} |\beta \Delta\Omega''(N_c)|} e^{-\beta \Delta\Omega(N_c)}. \tag{2.14}$$

2.2.5 The Monomer Attachment Rate

The final quantity needed to complete CNT is a specification of the monomer attachment frequency, $f(N)$. This is typically assumed to be the rate at which molecules strike the surface of the cluster times a heuristic "sticking probability" (a quantity between zero and one). The rate at which molecules strike the cluster depends on the physics governing the movement of matter in the system. For example, for colloids and macromolecules in solution, the movement of molecules is generally diffusive, and the rate is determined by solving the diffusion equation. For phase transitions involving simple fluids, kinetic theory can be used to determine the rate. In any case, the calculations inevitably involve the assumption of stationarity, i.e., that the growth of the cluster can be ignored in determining the rate (the "quasi-static" approximation). This leads to the classic expression $f(N) = \gamma_N 4\pi R(N) D c_1$ where γ_N is the sticking probability, $R(N)$ is the cluster radius, and D is the diffusion constant for the colloid particles. Many other examples can be found in Kashchiev (2000).

2.2.6 *Nucleation Theorems*

Finally, this brief survey of CNT is concluded by mentioning the so-called nucleation theorems which are widely used in the interpretation of data from both experiments and computer simulation. The first nucleation theorem is a relation between the free energy of the critical cluster, the chemical potential, and the size of the critical cluster (Kashchiev 2000):

$$\frac{d\Delta\Omega_c}{d\mu} = -\Delta N_c. \tag{2.15}$$

Here, we have formulated the theorem in the grand-canonical ensemble as is appropriate when the independent variable is chemical potential—hence, the use of the grand potential $\Omega = F - \mu N$. The quantities occurring in this relation are the excess free energy (relative to the original, mother phase) of the critical cluster and the excess number of molecules. This result follows trivially from equilibrium statistical mechanics using density functional theory techniques (see Appendix) and is therefore free of any model-based assumptions. For a more thermodynamic proof, see Bowles et al. (2001). It should also be noted that in the thermodynamic limit, differences in free energy are the same no matter which ensemble is used, so the result automatically holds, e.g., in the canonical ensemble where the grand potential would be replaced by the Helmholtz free energy. The second nucleation theorem relates the derivative of the nucleation rate, J, with respect to temperature to thermodynamic quantities characteristic of the two phases (Ford 1997):

$$\left(\frac{d\ln J}{dT}\right)_{\Delta\mu} = \frac{1}{k_B T^2}\left(L - k_B T - E_x\left(N^*\right)\right)$$

where L is the latent heat and $E_x\left(N^*\right)$ is the excess internal energy. However, its derivation is not as clean as that for the first nucleation theorem in that it is predicated on particular assumptions concerning the relation between the monomer attachment and detachment rates.

2.3 Criticisms and Refinements of CNT

2.3.1 *Criticisms of CNT*

From the preceding summary of CNT, several deficiencies are immediately apparent. Some of the most important are:

1. The capillary approximation is very crude: real clusters have diffuse interfaces with finite interfacial widths as is well attested from computer simulations (Wolde et al. 1998). Only for very large clusters can the finite interfacial width be ignored: for small clusters, virtually all molecules may be part of the interface.

2. Even if the zero-width interface is accepted, it is also not true that the properties of the material inside the cluster are the same as in the bulk.
3. The free energy plays a crucial role: particularly the assumption of detailed balance used to relate the monomer attachment and detachment frequencies. But free energy is only defined for equilibrium systems: what is the free energy of an unstable cluster?
4. If the cluster interface is diffuse, what exactly is meant by monomer attachment and detachment?
5. Some transitions are characterized by more than cluster size: the most important example is crystallization where both density and some characterization of molecular-scale order, "crystallinity," are critical. How can CNT account for more than just cluster size?
6. In order to build a cluster, mass must be moved from one part of the system to another. In post-critical cluster growth, this is described by transport equations (e.g., the diffusion equation) coupled to equations describing the growth of a cluster. Shouldn't transport be similarly accounted for in the nucleation stage? Is it really adequate to treat transport in the quasi-static approximation?

These and other criticisms have been widely known and have driven work on the refinement of CNT which can mostly be classified in two categories: refinements of the capillary approximation and refinements of the monomer attachment-detachment rates.

2.3.2 Refinements of the Capillary Model

In CNT, the nucleation rate scales as the exponential of the free energy barrier so an accurate determination of the barrier is obviously crucial. The capillary approximation is based on macroscopic concepts such as surface tension and bulk free energies and so, at best, can only be expected to be accurate for large clusters. There have therefore been many suggestions at improvements on this estimate.

The extended liquid drop model (ELDM) of Reiss, Reguera, and co-workers (Reguera et al. 2003; Reguera and Reiss 2004) is based on a more microscopic concept of a cluster than the macroscopic ideas used in the capillary model including the fact that clusters constantly fluctuate in size. It still makes use of the concept of the free energy of nonequilibrium clusters and the assumption that the probability to observe a cluster of a given size is given by a Boltzmann-like distribution. One of its most important consequences is that the free energy barrier (correctly) vanishes at the spinodal (e.g., for sufficiently high supersaturation), whereas the capillary model always gives a finite barrier. In the generalized Gibbs approach (Schmelzer et al. 2006), it is recognized that the surface tension is not a constant but depends on the properties of both the cluster and the ambient phase. Dillman and Meier (1991) and Laaksonen et al. (1994) generalized the capillary model by, among other things, including size-dependent corrections to the surface tension and treating all constants as adjustable parameters, some of which are fixed by demanding agreement, e.g., with the virial expansion for the pressure. Other parameters are fit to experimental data yielding the so-called semi-phenomenological model. Similarly, Prestipino, Laio, and Tosatti (2012) showed that better estimates could be obtained by allowing for the dependence of the surface tension on the cluster size (e.g., the Tolman length) and taking into account nonspherical clusters. Kalikmanov (2006) has also tried to refine the notion of surface tension by trying to determine cluster free energies directly from the canonical configuration integral in the so-called mean-field kinetic nucleation theory. Classical density functional theory is more complex than most of these proposals but has been shown to give the most realistic qualitative and quantitative description of cluster structure and free energy (see, e.g., Lutsko (2008), for a comparison with simulation).

2.3.3 Refinements of the Becker-Döring Equation

One way to avoid the issue of free energies altogether is to use other means to determine the monomer attachment and detachment frequencies required in the kinetic equation, Eq. (2.5). In dynamical nucleation theory (Schenter et al. 1999; Kathmann et al. 1999), this is done by using transition state theory and equilibrium statistical mechanics to develop expressions for the monomer attachment and detachment rates based on the first principles. This effectively requires evaluating many-body configuration integrals and is in practice a method of extracting the monomer rates from simulation. Another well-known attempt to improve on the determination of the monomer detachment frequency is the Ruckenstein-Narsimhan-Nowakowski (RNN) theory (Narsimhan and Ruckenstein 1989; Ruckenstein and Nowakowski 1990; Djikaev and Ruckenstein 2005). The idea in this approach is to determine the detachment frequency by treating the movement of a particle on the surface of the cluster as being Brownian motion under the mean field of the cluster. The escape rate of surface particles can then be determined and used to fix the detachment frequency.

2.4 Beyond CNT

As we have seen, the theory of nucleation began with the Becker-Döring model which is based on the cluster rate equations and the capillary approximation, and a great deal of work has been done to refine, extend, and clarify these concepts. So, the question arises: Is there any other way to understand nucleation? In particular, is it possible to remedy some of the shortcomings noted above for CNT? In fact, the closely related problem of critical phenomenon—that is, the behavior of systems near the critical point—has long been framed in the alternative language of field theory. The recently introduced mesoscopic nucleation theory (MeNT) makes use of similar concepts and ideas to develop an alternative to CNT (Lutsko 2011, 2012a,b,c; Lutsko and Durán-Olivencia 2013). The basic idea is a shift of viewpoint: rather than treating clusters as distinct, independent objects, they are viewed as local increases (e.g., for droplet nucleation) or decreases (e.g., for bubble nucleation) in the density of the mother phase. Instead of trying to describe the statistics of clusters, the shift in paradigm requires us to think in terms of the dynamics of density fluctuations. Happily, such a theory already exists and has played a central role in nonequilibrium statistical mechanics for at least half a century. It was first formulated by Landau and is called fluctuating hydrodynamics. It has long been used to study such phenomena as light scattering from fluids (Lutsko and Dufty 1985), fluctuation renormalization of transport coefficients (Lovesey 1986), the glass transition (Kawasaki 1998), and long-time tails (Fox 1976).

The term "hydrodynamics" might seem to imply that the theory is only applicable to fluid-fluid transitions. However, in nonequilibrium statistical mechanics, the term has a more general meaning and simply indicates that the theory is developed for long-wavelength fluctuations. The quantity that is fluctuating could be anything: not only fluid quantities such as density, momentum, and temperature but also, say, magnetization, some measure of local order (e.g., crystallinity), etc. What is meant by "long"? That depends on the problem, but, e.g., in fluids, it has the precise meaning of larger than the mean free path. In dense fluids and solids, the mean free path can be less than the typical particle size so that a "hydrodynamic" description remains valid down to molecular length scales. (Note that it is similar to, but more complex than, the familiar Navier-Stokes equations.) Thus, it is not unreasonable for the density to include molecular-length-scale structure allowing the theory to account for spatial ordering (e.g., in crystallization). Even though attention has so far been focused on the density field, one naturally also includes temperature and velocity fields so that heat transport and flow can also be included.

In the following, a brief sketch of the theory will be presented. Most mathematical developments will only be described, and the reader can refer to the original literature for detailed derivations and proofs. The main question we wish to address, beyond the basic description of the theory, is as follows: How can such an approach have anything in common with the Becker-Döring theory? What happens to concepts like the critical cluster and monomer attachment-detachment?

2.4.1 Fluctuating Hydrodynamics

The original form of fluctuating hydrodynamics as formulated by Landau looked like the Navier-Stokes equations of fluid dynamics but with some additional, stochastic contributions to the stress tensor and the heat flux vector (Lifshitz and Pitaevskii 1980). In order to describe colloids and large molecules in solution, we will begin with a slightly different version:

$$\frac{\partial \rho}{\partial t} + \frac{1}{m}\nabla \cdot \mathbf{j} = 0 \tag{2.16}$$
$$\frac{\partial \mathbf{j}}{\partial t} + \nabla \cdot \mathbf{jj}/\rho + \rho \nabla P + \nabla \cdot \mathbf{S} = -\gamma \mathbf{j} + \sqrt{2m\gamma k_B T \rho}\,\boldsymbol{\xi}$$

The macromolecules are characterized by their local number density $\rho(\mathbf{r};t)$ and local momentum flux $\mathbf{j}(\mathbf{r};t) = m\rho(\mathbf{r};t)\,\mathbf{v}(\mathbf{r};t)$ with $\mathbf{v}(\mathbf{r};t)$ being the average local velocity of the large molecules. Also appearing here are the mass of the molecules, m; the local partial pressure of the macromolecules, $P(\mathbf{r};t)$; and the dissipative stress tensor, $\mathbf{S}(\mathbf{r};t)$, which accounts for viscosity due to the interaction of the molecules. The dissipative stress tensor also has a stochastic part as introduced by Landau. Finally, since the large molecules are in solution, some account must be given of the interaction between these two components, and this is the role of the terms on the right-hand side. The first term on the right in the momentum equation is an additional friction due to transfer of momentum from the large molecules to the bath. The second term is a fluctuating force which accounts for the effect of the small solute molecules colliding with the large molecules giving them a Brownian motion. There are no additional terms in the first (density) equation since it simply described for the conservation of mass which is unchanged by the interaction with the solution. Finally, there is no equation for the temperature as we assume that the bath acts as a thermostat so that the large molecules are effectively in an isothermal state. To describe nucleation of a system of small molecules (with no bath), the two terms on the right in the momentum equation must be dropped and the usual equation for temperature fluctuations included.

Three additional elements are needed to fully specify the dynamics. First is an expression for the local pressure. Here, ultimately we will use a localized Gibbs-Duhem expression to relate pressure gradients to gradients of the local chemical potential which, following ideas from density functional theory (Lutsko 2010), can be expressed in terms of the functional derivative of the local free energy with respect to the density. Thus, we replace

$$\nabla P(\mathbf{r};t) \rightarrow \nabla \rho(\mathbf{r};t)\,\nabla \frac{\delta F[\rho]}{\delta \rho(\mathbf{r};t)} \tag{2.17}$$

which is the local equilibrium approximation and is the main assumption of the theory. Even cruder approximations are widely used in nonequilibrium statistical mechanics because it proves very difficult to do better. The virtue of using this expression is that it brings in the free energy functional in a very natural way and allows us to make contact with the well-developed machinery of DFT (Lutsko 2010). The second element needed is a specification of the dissipative stress tensor. It would be quite reasonable to use the usual Navier-Stokes form for this, supplemented with the usual stochastic contribution, but we will work in the "overdamped" regime in which it is assumed that the friction due to the bath is much more important than the internal viscosity of the solute, so this term can be safely neglected. Finally, we must specify the remaining fluctuating term which is usually taken to be Gaussian, white noise with correlation $\langle \boldsymbol{\xi}(\mathbf{r};t)\,\boldsymbol{\xi}(\mathbf{r}';t')\rangle = \delta(\mathbf{r}-\mathbf{r}')\,\delta(t-t')$. All of these elements can be used, together with the assumption of "overdamping," to give a single equation for the density field:

$$\frac{\partial \rho}{\partial t} = D\,\nabla\cdot\rho\nabla\frac{\delta \beta F[\rho]}{\delta \rho} - \nabla\cdot\sqrt{2D\rho}\boldsymbol{\xi} \tag{2.18}$$

Here, $D = \frac{k_B T}{\gamma m}$ which is the diffusion constant for a single large molecule in solution. Note that at low density, the free energy functional is known exactly (it is just the result for the ideal gas) and this reduces to

$$\frac{\partial \rho}{\partial t} = D\nabla^2\rho - \nabla\cdot\sqrt{2D\rho}\boldsymbol{\xi} \tag{2.19}$$

which is simply diffusion with noise. We can specialize to the formation of clusters by imposing spherical symmetry. It turns out that when we do so, a better variable than the local density is the cumulative mass defined simply by summing the density out to a given radius:

$$m(r) = 4\pi \int_0^r \rho(r')\, r'^2 dr', \tag{2.20}$$

in terms of which we find

$$\frac{\partial m(r)}{\partial t} = -D\frac{\partial m(r)}{\partial r}\frac{\delta F[\rho]}{\delta m(r)} - \varepsilon\sqrt{2D\frac{\partial m(r)}{\partial r}}\xi(r;t) \tag{2.21}$$

with the scalar noise having correlations

$$\left\langle \xi(r;t)\,\xi(r';t')\right\rangle = \delta(r-r')\,\delta(t-t')\,. \tag{2.22}$$

2.4.2 Interpretation

Aside from their potential utility for performing stochastic simulations, what can we say about these stochastic models, Eqs. (2.18) and (2.21)? Clearly, the free energy plays an important role in the deterministic dynamics. In fact, due to their rather simple structure, one can prove that in the so-called weak-noise limit (which is expected to become more accurate as the temperature decreases), the *most likely pathway* (MLP) for nucleation passes through the critical cluster and that the energy barrier that must be overcome is exactly that determined by the critical cluster (Lutsko 2011, 2012a). This important result demonstrates that the phenomenology of nucleation is in fact contained within this framework. Furthermore, the MLP for nucleation can be obtained in this limit simply by starting at the critical cluster and integrating the deterministic part of Eqs. (2.18) or (2.21) (i.e., by setting $\xi = 0$) (Lutsko 2011, 2012a). We thus potentially obtain a complete picture of cluster formation, from initial density fluctuation to the post-nucleation growth phase, within a single, unified model. And there is no difficulty in using the driving force, the free energy functional, and the most sophisticated microscopic models available.

What have we learned from performing these calculations? The exploration of this approach is still in its early stages, but already there have been some surprising insights related to the role played by mass conservation (Lutsko 2011, 2012a,c; Lutsko and Durán-Olivencia 2015). A dense cluster must be formed by bringing together an excess (relative to the average) of mass, and this mass must come from somewhere. In CNT, the picture of a cluster is that it starts with very small (effectively zero) radius and grows steadily with the interior mass fixed at that of the bulk liquid (i.e., the capillary approximation). Due to the role of mass conservation in MeNT, the MLP is quite different: clusters start out with very large radii and densities very close to the average, e.g., as long-wavelength, small-amplitude density fluctuations. This small excess of mass then gathers, due to random fluctuations, so that in effect the radius of the cluster becomes **smaller**, while the density inside it increases as illustrated schematically in Fig. (2.1). At a certain point, once the density is high enough which typically means close to the bulk value, the radius begins to increase, and the growth of the cluster begins to match that assumed in CNT. Note that this implies that clusters with large interior density and very small radius are rare and this in fact agrees with certain observations from simulation (Trudu et al. 2006).

2.4.3 Order Parameters and Connection to CNT

A pressing question has to be how can one have two such different approaches to the description of the same phenomena? If MeNT is really more general than CNT, one would like that the latter result from some approximation to the former, and this is indeed the case. The key is to consider parameterized density fields. Rather than deal with the quantity $\rho(r;t)$, in the spherically symmetric case, we guess some form for the shape and approximate the density as $\rho(r;t) \approx \phi(r;\mathbf{x}(t))$ where ϕ is some function of r, e.g., a sigmoidal function, that is parameterized with one or more quantities $x_1, \ldots$, such as a radius, an interfacial width, etc., and it is assumed that all of the change in the density as a function of time can be accounted for by changes in the parameters. From the stochastic model for the density, we develop a corresponding (approximate) model for the parameters which turns out to have the form (Lutsko 2012a)

$$\frac{dx_i}{dt} = -DK_{ij}(\mathbf{x})\frac{\partial \beta F}{\partial x_j} + 2DA_i(\mathbf{x}) - \sqrt{2D}M_{ij}(\mathbf{x})\,\xi_j(t)\,,\ \left\langle \xi_i(t)\,\xi_j(t')\right\rangle = \delta_{ij}\delta(t-t') \tag{2.23}$$

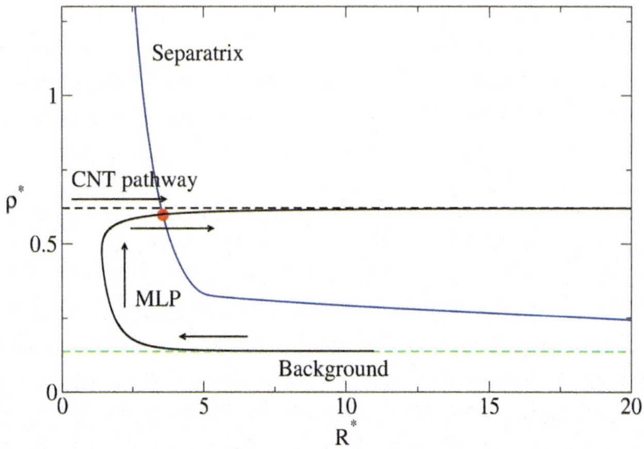

Fig. 2.1 Schematic illustration of nucleation pathways using density (*vertical axis*) and radius (*horizontal axis*) as the variables used to characterize a cluster. The initial, or "background," state is a low-density gas indicated by the *dashed line* at constant (low) density. In CNT, clusters begin with zero radius but high density and simply increase in size (radius). In MeNT, mass conservation leads to a more complex picture where the nucleation pathway (as characterized by the most likely path or MLP) begins with a large cluster with small excess density that first decreases in radius as mass is localized and then increases as density during the nucleation event and eventually joins the classical pathway. In both cases—CNT and MeNT—the pathways pass through the critical cluster (indicated by the *circle*) which, in MeNT, is just the most likely, but not the only, place to cross the separatrix which divides the plane into the stochastic regime (to its *left*) and the deterministic regime (to its *right*)

where we have used the Einstein summation convention wherein repeated indices are summed. Note that once we introduce the parameterization, the free energy becomes an ordinary function of the parameters, $F[\rho] \rightarrow F(\mathbf{x})$. The matrix of kinetic coefficients, $K_{ij}(\mathbf{x})$, is calculated solely from the function $\phi(r; \mathbf{x}(t))$ (Lutsko 2011, 2012a), and the amplitude of the noise term is related to it via $K = MM^T$. The "anomalous force," A_i, is technical in nature and not present in the weak-noise limit (Lutsko 2011, 2012a). Finally, from this stochastic equation, one immediately gets the corresponding equation for the probability to observe a given set of parameters, $P(\mathbf{x}; t)$, which is

$$\frac{\partial}{\partial t}P(\mathbf{x};t) = -\frac{\partial}{\partial x_i}\left\{DK_{ij}(\mathbf{x})\frac{\partial \beta F(\mathbf{x})}{\partial x_j} + 2DA_i(\mathbf{x}) - \frac{\partial}{\partial x_l}K_{il}(\mathbf{x})\right\}P(\mathbf{x};t) \quad (2.24)$$

Now it is possible to make a connection with CNT. We have only to assume the capillary density profile, $\rho(r) = \rho_B$, for $r < R$ and $\rho(r) = \rho_A$ for $r > R$ so that we have a parameterized density profile with one parameter—the radius. Similarly, we use the capillary approximation for F. One then finds that when the kinetic coefficient is calculated from this simple profile and the results used in the equation for $P(\mathbf{x}; t)$, which is now $P(R; t)$, the result in the weak-noise limit exactly matches the Zeldovich equation with the monomer attachment frequency for diffusion-limited nucleation. We thus recover CNT without ever having explicitly introduced concepts like monomer attachment and detachment or steady-state diffusion. The framework then allows one to go on to generalize CNT to multiple-order parameters in a natural manner (Lutsko and Durán-Olivencia 2015).

2.5 Conclusion and Perspectives

The goal of this chapter has been threefold. First, an overview of classical nucleation theory, including its deficiencies, was presented. It was emphasized that the key elements of CNT are the Becker-Döring rate equations for the evolution of the cluster population and the capillary approximation for the free energy. This was followed by a brief survey of some of the many efforts that have been made to improve on CNT. These largely aim to improve one or the other of the two CNT pillars. Finally, the elements of mesoscopic nucleation theory were discussed as an illustration of an alternate paradigm for nucleation which nevertheless is able to reproduce at least certain key aspects of CNT. Although MeNT has already provided new qualitative insights into nucleation, it is still a relatively new approach to the problem, and many applications remain to be explored such as the effect of flows, unsteady conditions, heterogeneous nucleation, crystallization, application to single-component systems with heat transport, etc. At the theoretical level, most of the work has so far involved the weak-noise regime, and the highly nonclassical strong-noise regime remains to be explored especially now that some preliminary simulation results are becoming available.

Appendix: Proof of the First Nucleation Theorem

To prove the first nucleation theorem, we first recall some elementary—but exact—results from finite-temperature density functional theory (Lutsko 2010). From the theory behind DFT, we learn that the statistical mechanics of a single-component classical system can be entirely formulated in terms of the local number density $\rho(\mathbf{r})$. There exists a functional of this density of the form $\Psi[\rho] = F[\rho] - \mu N$ such that the equilibrium density distribution, $\rho_{eq}(\mathbf{r})$, minimizes $\Psi[\rho]$, e.g.,

$$0 = \left.\frac{\delta \Psi[\rho]}{\delta \rho(\mathbf{r})}\right|_{\rho_{eq}(\mathbf{r})} \tag{2.25}$$

In the expression for Ψ, the first term F has no dependence on the chemical potential, while in the second term, it is important to remember that N is simply the integral of the density over all space, so that the only independent quantities are the local density, the chemical potential, and the temperature (which does not change and so is not indicated). Furthermore, the free energy is simply the functional evaluated at the equilibrium density $\Omega_{eq} = \Psi\left[\rho_{eq}\right]$. Although it is not an equilibrium state, this is assumed to hold true as well for the critical cluster which has density $\rho_c(\mathbf{r})$ and free energy $\Omega_c = \Psi[\rho_c]$. Hence, we have that

$$\frac{d\left(\Omega_c - \Omega_{eq}\right)}{d\mu} = \int \left(\left.\frac{\delta \Psi[\rho]}{\delta \rho(\mathbf{r})}\right|_{\rho_c(\mathbf{r})} \frac{\partial \rho_c(\mathbf{r})}{\partial \mu} - \left.\frac{\delta \Psi[\rho]}{\delta \rho(\mathbf{r})}\right|_{\rho_{eq}(\mathbf{r})} \frac{\partial \rho_{eq}(\mathbf{r})}{\partial \mu} \right) d\mathbf{r} - \left(N_c - N_{eq}\right) \tag{2.26}$$

Since the left-hand side demands the total derivative with respect to the chemical potential, there are two contributions: the first due to the change in the equilibrium density distribution when the chemical potential changes and the second due to the explicit factor of chemical potential in the functional Ψ. Now, since both $\rho_c(\mathbf{r})$ and $\rho_{eq}(\mathbf{r})$ satisfy Eq. (2.25), the first term is identically zero leaving

$$\frac{d\left(\Omega_c - \Omega_{eq}\right)}{d\mu} = -\left(N_c - N_{eq}\right) \tag{2.27}$$

which is the desired result.

Acknowledgements This work was supported in part by the European Space Agency under contract number ESA AO-2004-070.

References

Bowles RK, Reguera D, Djikaev Y, Reiss H (2001) A theorem for inhomogeneous systems: The generalization of the nucleation theorem. J Chem Phys 115:1853

Dillmann A, Meier GEA (1991) A refined droplet approach to the problem of homogeneous nucleation from the vapor phase. J Chem Phys 94:3872

Djikaev YS, Ruckenstein E (2005) Kinetic theory of nucleation based on a first passage time analysis: improvement by the density-functional theory. J Chem Phys 123:214503

Ford I (1997) Nucleation theorems, the statistical mechanics of molecular clusters, and a revision of classical nucleation theory. Phys Rev E 56:5615–5629

Fox RF (1976) Fluctuating hydrodynamics explanation of the Alder & Wainwright velocity autocorrelation computer experiments. J Chem Phys 64:5307

Gardiner CW (2004) Handbook of stochastic methods, 3rd edn. Springer, Berlin

Hänggi P, Borkovec M (1990) Reaction-rate theory: fifty years after Kramers. Rev Mod Phys 62:251–341

Kalikmanov VI (2006) Mean-field kinetic nucleation theory. J Chem Phys 124:124505

Kashchiev D (2000) Nucleation: basic theory with applications. Butterworth-Heinemann, Oxford

Kathmann SM, Schenter GK, Garrett BC (1999) Dynamical nucleation theory: calculation of condensation rate constants for small water clusters. J Chem Phys 111:4688

Kawasaki K (1998) Microscopic analyses of the dynamical density functional equation of dense fluids. J Stat Phys 93:527–546

Laaksonen A, Ford I, Kulmala M (1994) Revised parametrization of the Dillmann-Meier theory of homogeneous nucleation. Phys Rev E 49:5517–5524

Lifshitz E, Pitaevskii LP (1980) Statistical physics: theory of the condensed state. Course of theoretical physics, vol 9. Butterworth-Heinemann, Oxford

Lovesey SW (1986) Condensed matter physic: dynamic correlations, 2nd edn. Benjamin/Cummings, Reading
Lutsko J, Dufty JW (1985) Hydrodynamic fluctuations at large shear rate. Phys Rev A 32:3040
Lutsko JF (2008) Density functional theory of inhomogeneous liquids. III. Liquid-vapor nucleation. J Chem Phys 129:244501+
Lutsko JF (2010) Recent developments in classical density functional theory. Adv Chem Phys 144:1–92
Lutsko JF (2011) Communication: a dynamical theory of homogeneous nucleation for colloids and macromolecules. J Chem Phys 135:161101
Lutsko JF (2012a) A dynamical theory of nucleation for colloids and macromolecules. J Chem Phys 136:034509
Lutsko JF (2012b) Nucleation of colloids and macromolecules: does the nucleation pathway matter? J Chem Phys 136:134502
Lutsko JF (2012c) Nucleation of colloids and macromolecules in a finite volume. J Chem Phys 137:154903
Lutsko JF, Durán-Olivencia Ma (2013) Classical nucleation theory from a dynamical approach to nucleation. J Chem Phys 138:244908
Lutsko JF, Durán-Olivencia MA (2015) A two-parameter extension of classical nucleation theory. J Phys Cond Matt 27:235101
Narsimhan G, Ruckenstein E (1989) A new approach for the prediction of the rate of nucleation in liquids. J Colloid Interface Sci 128:549–565
Prestipino S, Laio A, Tosatti E (2012) Systematic improvement of classical nucleation theory. Phys Rev Lett 108:225701
Reguera D, Bowles RK, Djikaev Y, Reiss H (2003) Phase transitions in systems small enough to be clusters. J Chem Phys 118:340
Reguera D, Reiss H (2004) Fusion of the extended modified liquid drop model for nucleation and dynamical nucleation theory. Phys Rev Lett 93:165701
Ruckenstein E, Nowakowski B (1990) A kinetic theory of nucleation in liquids. J Colloid Interface Sci 137:583–592
Schenter G, Kathmann S, Garrett B (1999) Dynamical nucleation theory: a new molecular approach to vapor-liquid nucleation. Phys Rev Lett 82:3484–3487
Schmelzer JWP, Boltachev GS, Baidakov VG (2006) Classical and generalized Gibbs' approaches and the work of critical cluster formation in nucleation theory. J Chem Phys 124:194503
Sleutel M, Lutsko JF, Driessche AESV, Durán-Olivencia MA, Maes D (2014) Observing classical nucleation theory at work by monitoring phase transitions with molecular precision. Nat Commun 5:5598
Trudu F, Donadio D, and Parrinello M (2006) Freezing of a Lennard-Jones fluid: from nucleation to spinodal regime. Phys Rev Lett 97:105701
Van Erp TS (2012) Dynamical rare event simulation techniques for equilibrium and nonequilibrium systems. Adv Chem Phys 151:27–60
Wolde PT, Frenkel D, ten Wolde PR (1998) Computer simulation study of gas-liquid nucleation in a Lennard-Jones system. J Chem Phys 109:9901

Chapter 3
Challenges and Perspectives of the Polymer-Induced Liquid-Precursor Process: The Pathway from Liquid-Condensed Mineral Precursors to Mesocrystalline Products

Stephan E. Wolf and Laurie B. Gower

Crystallization and phase separation is a key topic in physical chemistry which greatly impacts a plurality of scientific disciplines, be they of applied or fundamental nature. Our increasing insight into realistic and non-idealized systems is clearly becoming manifest in the rising number of publications challenging the simplicity of classical theories. These developments recently condensed into the concept of so-called nonclassical crystallization which embraces all crystallization pathways taking place outside of the framework of classical nucleation theories (Cölfen et al. 2008; De Yoreo et al. 2017, Chap. 1; Lutsko 2017, Chap. 2; Andreassen and Lewis 2017, Chap. 7; Yang and Ter Horst 2017, Chap. 16). Research has revealed three important phenomena, all of which were observed in crystallizations in real systems and have profoundly altered our perception of crystallization in general: the pathway of oriented attachment, the existence of prenucleation clusters and the occurrence of a liquid-condensed mineral precursor.

Following the pathway of oriented attachment (Penn et al. 2017) to its final destination, we can end up with so-called mesocrystals (Cölfen et al. 2008; Rao and Cölfen 2017). In this chapter, we stick to the recently proposed definition of a mesocrystal given by Bergström et al. (2015) which states that it is a "nanostructured material with a defined long-range order on the atomic scale, which can be inferred

S.E. Wolf (✉)
Institute for Glass and Ceramics, Department of Materials Science and Engineering, Friedrich-Alexander University Erlangen-Nürnberg (FAU), Haberstrasse 9a, 91058 Erlangen, Germany

Interdisciplinary Center for Functional Particle Systems (FPS), Friedrich-Alexander University Erlangen-Nürnberg (FAU), Haberstrasse 9a, 91058 Erlangen, Germany
e-mail: stephan.e.wolf@fau.de

L.B. Gower (✉)
Department of Materials Science and Engineering, University of Florida, 210A Rhines Hall, Gainesville, FL 32611-6400, USA
e-mail: lgower@mse.ufl.edu

A.E.S. Van Driessche et al. (eds.), *New Perspectives on Mineral Nucleation and Growth*,
DOI 10.1007/978-3-319-45669-0_3

from the existence of an essentially sharp wide angle diffraction pattern (with sharp Bragg peaks) together with clear evidence that the material consists of individual nanoparticle building units".

The notion of prenucleation clusters challenged classical nucleation theory profoundly (Gebauer et al. 2008). These clusters, formed from constituents of the mineral to be precipitated, are considered to be thermodynamically relatively stable as they exist even in non-saturated conditions. These highly dynamic entities exhibit solute character and are not thought to be a new phase. It was recently proposed that the phase separation of these solutes is accompanied/triggered by a distinct decrease in their dynamics and leads to the formation of a condensed liquid phase comprised of these dispersed nanoconstituents (Wallace et al. 2013; Gebauer et al. 2014); this behaviour may then be used to discriminate between solutes (i.e. prenucleation entities) and separate phases (i.e. post-nucleation entities). But as of yet, their impact on crystallization and phase separation remains to be understood.

The scope of this chapter is the last but not least of the aforementioned three phenomena, the formation of liquid-condensed phases (LCP) in supersaturated solutions of small, sparingly soluble, inorganic and mineral-forming compounds such as calcium carbonate. Laurie Gower and co-workers were the first to propose the occurrence of liquid-like amorphous intermediates during the precipitation of calcium carbonate in a set of seminal contributions from 1997 to 2000 (Gower 1997; Gower and Odom 2000). They reported formation of mineral tablets and other non-equilibrium-shaped yet crystalline bodies of calcium carbonate and explained their genesis by the precipitation and coalescence of highly hydrated and hence liquid-like colloids of a transient amorphous mineral precursor phase. The formation of this exceptional phase was assumed to be induced by the presence of tiny amounts of a small polyanionic polymer (e.g. polyaspartate) and, hence, was dubbed the polymer-induced liquid-precursor (PILP) process.

Due to the liquid-like state of the transient amorphous phase, a multitude of mineral morphologies can be generated, illustrating the outstanding potential of the PILP process for the morphosynthesis of non-facetted, non-equilibrium morphologies. If the transient droplets are kept dispersed while ripening, solid spheres with monomodal size distribution, hollow and hedgehog-like structures of crystalline calcium carbonate can be obtained (Faatz et al. 2004; Loges et al. 2006). Sedimentation and subsequent coalescence leads to the formation of films and tablets (Gower and Tirrell 1998; Gower and Odom 2000; Amos et al. 2007; Cantaert et al. 2012; Jiang et al. 2013). Employment of patterned self-assembled monolayers varying in hydrophilicity enables effective templating of the mineral film (Kim et al. 2007). If the PILP film is generated beneath Langmuir monolayers, single-crystalline aragonite patches of films can be produced which are similar in morphology (a half micron in thickness and several microns in width) to molluscan nacreous tablets (Amos et al. 2007) and which in fact were successfully used for producing multilayered and nacre-like structures by multiple Langmuir–Schäfer transfers (Volkmer et al. 2005). The liquid-like colloidal intermediates can be accumulated and molded into an array of non-equilibrium morphologies by templating approaches. This has been exemplified by the retrosynthesis of nacre

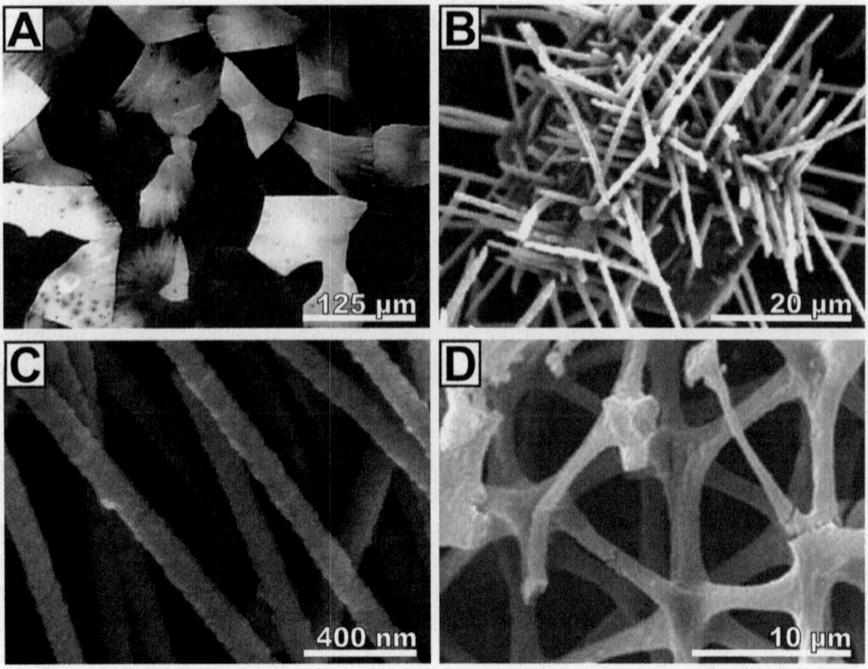

Fig. 3.1 Morphologies generated by means of the polymer-induced liquid-precursor (PILP) process: (**a**) a mosaic mesocrystalline film of calcium carbonate, (**b**) microfibrous calcite, (**c**) calcium carbonate fibres obtained by capillary infiltration into track-etch membranes and (**d**) bicontinuous calcium carbonate structure obtained by infiltration of a PHEMA hydrogel replica of a sea urchin spine. Subfigures are reprinted with permission from (**a**) Gower & Odom (Gower and Odom 2000), copyright 2000 Elsevier; (**b**) Olszta et al. (2004), copyright 2004 American Chemical Society; (**d**) Kim et al. (2011), copyright Wiley-VCH Verlag GmbH & Co. KGaA, Weinheim, 2011; (**e**) Cheng and Gower (2006), copyright 2006 American Chemical Society

or sea urchin spines (Volkmer et al. 2005; Gehrke et al. 2005), by infiltration of PILP phase in small volumes (Olszta et al. 2003a), coating of scaffolds (Sommerdijk et al. 2007) or exploitation of capillarity effects to produce, for example, nanowires (see Fig. 3.1) (Gower 2008; Kim et al. 2011). Fibrous mineral structures can be readily synthesized not only by infiltration but also in a self-organized fashion (see Fig. 3.1). This process was originally proposed to follow a solution–precursor–solid (SPS) mechanism starting from a seed crystal analogous to the established vapour–liquid–solid (VLS) and solution–liquid–solid (SLS) mechanisms (Olszta et al. 2004; Homeijer et al. 2010). Interestingly, Gower's group observed via time-lapse video that fibrous calcite can also be formed in the absence of seed crystals. In these experiments, amorphous fibres seem to be "extruded" from cracks in a gelatinous globule of precursor (Olszta et al. 2003b). Several hypotheses have been proposed to explain the formation of these fibrous morphologies, such as oriented attachment (although the fibres were stated as being formed from amorphous nanoparticles) or polymer coating of the sides of the fibre to allow for attachment of particles solely at the growth tip (Cantaert et al. 2013). The finding of Gower et al. that seed crystals are not required showed that fibres can also form by an apparently different mechanism. Therefore, the hypothesis was modified to include a mechanism by which an autocatalytic assembly of nanodroplets or nanoparticles at high energy surface protrusions could provide the "seed" (site-specific growth center) necessary for generating one-dimensional growth of fibers (Homeijer et al. 2010). The pronounced nanogranular texture of the generated fibres seems to indicate that the precursor particles are not always purely liquid-like, even though they coalesce fairly well to form single-crystalline fibres (if a singular nucleation event drives the crystallization across the accumulated amorphous phase). With regard to the variable stability of the precursor colloids, it seems clear that multiple formation mechanisms exist which all allow for initiating or collecting the colloids at a site-specific growth centre.

The PILP process was recently applied to calcium phosphates in order to establish a new route to biomimetic bone graft substitutes. In these studies, collagenous tissues were mineralized with calcium phosphate under PILP-generating conditions. The obtained fibres mineralized with hydroxyapatite show remarkable similarities to native bone as they feature intrafibrillar and uniaxially oriented crystals which are oriented roughly parallel to the fibril axis, generating diffraction patterns similar to those of native bone (Olszta et al. 2007; Jee et al. 2010a); such diffraction patterns cannot be generated by standard in vitro mineralization approaches. Olszta et al. proposed that the liquid mineral phase is drawn into the gap regions of collagen fibrils by capillary effects. A cryogenic transmission electron microscopy (cryoTEM) study by Nudelman et al. corroborated the existence of an amorphous precursor which selectively binds to and enters the fibres and proposed further that a net negative surface charge of the transient intermediate of the mineral may be crucial, as it may allow for the infiltration of the precursor complexes and their interaction with positively charged regions within the collagen fibril (Nudelman et al. 2010). In addition to these remarkable achievements, the PILP system has also been successfully employed to reconstitute partially demineralized human dentin resulting in a partial recovery of its mechanical properties (Burwell et al. 2012).

All of these studies exemplify the key feature of the PILP process: due to the presence of the polymer, the process of crystallization undergoes "conversion from a solution crystallization process to a solidification process" (Gower and Odom 2000). This is in sharp contrast to classical crystallization in which crystal morphology is dominated by crystal facets that result from the slowest growing crystal faces with lowest surface energy; crystallinity emerges concomitantly to the initial liquid/solid phase separation. In the PILP process, crystallization appears to proceed as a gel or solid/solid transformation, which is *after* phase separation and thus downstream to the formation of the transient liquid-amorphous intermediate and its solidification. This union of a solidification process with a final pseudomorphic transformation to crystallinity paves the way for the multitude of morphologies presented above.

Because the formation of the transient liquid-condensed mineral precursor phase is in contradiction with classical theories of nucleation and growth, its physicochemical basis was enigmatic from the beginning and still remains unclear.

Initially, it was merely attributed to the action of the polyanionic polymer additive and the assumption that the polymer induces and stabilizes the liquid precursor by sequestering ions and water and thereby triggering the unusual liquid/liquid phase separation. However, the formation of a liquid-amorphous calcium carbonate phase in the absence of additives – thus a liquid/liquid phase separation of a pure solution of a mineral (i.e. water/calcium carbonate) – was first proposed by Faatz et al. (2004). They speculatively explained the formation of spherical amorphous calcium carbonate (ACC) with monomodal particle size distribution by a binodal liquid/liquid phase separation leading to the formation of a transient liquid phase.[1] However, they based their speculation only on the temperature-dependent monodispersity of the ACC spheres and did not provide any proof for their phase diagram, let alone for the claimed liquid-like state of the early mineralization stages. A few years later, Rieger et al. provided in a cryoTEM study first indications for liquid/liquid phase separation (Rieger et al. 2007): 100 milliseconds after turbulent mixing of the two educts sodium carbonate and calcium chloride at $pH = 11$, they observed the formation of a hydrated liquid-like phase resembling the initial bicontinuous state of a spinodal phase. This experiment has been interpreted several times as an indication of spinodal decomposition which is reasonable for this particular experiment given the rapidity of ascending supersaturation. However, we believe caution should be exercised in interpreting these findings and extrapolating them to all liquid phase precursors. First, the educts were mixed under turbulent conditions which can lead easily to vortex formation and thus to an initial inhomogeneous mixing of educts. The droplet-like structures are then an artificial result of the solution's inhomogeneous supersaturation, the precipitation reaction starts heterogeneously at the interfaces of the intermingled educt solutions (Haberkorn et al. 2003; see also the Faraday Discussion transcripts 2007; Rieger et al. 2007).[2] Second, the cryogenic sampling was conducted with a guillotine-like device which shoots the TEM grid through a free jet of mother solution into the cryogen. The intermittent sheer stress during such a sampling can lead to coalescence of droplets and might feign by this a bicontinuous thus spinodal state. In the face of these problems, Wolf et al. devised a contact-free crystallization setup based on acoustical levitation, by which the formation of calcium carbonate could be studied under purely diffusion-controlled conditions. This approach provides reliable homogenous crystallization conditions so that artefacts arising from foreign interfaces (except the air/liquid interface) can be ruled out, be they due to reaction vessels or due to turbulent mixing. These experiments further substantiated the occurrence of a liquid-like mineral phase at the onset of the homogenous formation of calcium carbonate in

[1] Although often misinterpreted, the contribution of Faatz et al. does not claim spinodal decomposition. According to them, spinodal decomposition would lead to "ill-defined morphologies" whereas binodal decomposition to the observed monodisperse spheres. In subsequent contributions, they state their speculation about the binodal nature of this phase separation process much more clearly (Faatz 2005; Faatz et al. 2005).

[2] Rieger pointed out that one should not interpret his findings as a claim for a spinodal demixing (see Faraday Discussion transcripts 2007).

the absence of any additives (Wolf et al. 2008). The initial liquid-condensed phase (LCP) consisted of highly hydrated amorphous calcium carbonate droplets which were emulsion-like in appearance (Wolf et al. 2008; Wolf et al. 2011b). However, a direct evidence for their liquid state in solution was still pending. Later, Bewernitz et al. further corroborated the liquid state of this transient mineral phase by ^{13}C T_2 relaxation and ^{13}C PFG-STE diffusion NMR measurements (Bewernitz et al. 2012). However, there is still a missing link connecting the pure LCP mineral precursor and the PILP process; only the latter has thus far been shown to be suitable for morphosynthesis.

So far it is only clear that the polymer is not literally *inducing* the liquid-amorphous mineral phase, as it was evidenced that the LCP can also form in the absence of polymer. However, far more of this phase is produced in the presence of polymer (>16× the amount of bound calcium, see Bewernitz et al. 2012), so the polymer therefore enhances the stability of the LCP increasing the quantity formed. Without polymer, the concentration of the LCP was too low to be detected by conventional light scattering. In principle, this effect can be basically understood if one considers the generic effect of inhibiting additives: the additive raises the barrier(s) for nucleation and crystal growth, and thus the nucleation rate of a crystalline phase drops. Amorphous phases are usually thermodynamically unstable but have a lower surface energy due to their structural disorder or the presence of hydration waters. Because the nucleation rate is a sensitive function of the surface energy, amorphous phases thus typically have a higher rate of nucleation than the crystalline phases, especially in the presence of an inhibiting polymer. In later stages of ripening, amorphous phases have to compete against crystalline phases and dissolve in favour of the crystalline phases according to the Ostwald–Lussac rule of stages, feeding the growth of the crystalline phase due to the lower solubility product of the amorphous phases. The presence of inhibiting polymer abrogates or at least diminishes this "life-threatening" competition between the amorphous and crystalline phases and abets the formation of amorphous phases and prolongs their lifetime. These fundamental considerations already show that it is the polymer which takes the essential kinetic role in transforming an amorphous phase from a mere short-lived or unstable species to the principal agent of the precipitation. However, this is not a sufficient explanation for the emergence of a liquid-condensed phase instead of a solid amorphous phase and for the phenomenon of the PILP process and its capability to produce crystalline bodies with non-equilibrium morphologies. Throughout this chapter, we will provide several less expected actions of the polymeric additive. All of these actions are crucial for inducing a PILP process and will underline the fact that the polymer acts in the PILP system not only as an additive affecting the structure and morphology of the forming material, but it does so as a major process-directing agent (Olszta et al. 2007).

In further consideration of this perplexing system, we know various additives like magnesium, phosphate, silicate or citric acid were shown to do a similar job in favouring the formation of an amorphous phase (Aizenberg et al. 1996; Loste et al. 2003; Addadi et al. 2003; Gower 2008; Kellermeier et al. 2010; Tobler et al. 2015;

Kababya et al. 2015); but none of them was suitable for generating non-equilibrium morphologies like those that can be achieved with a PILP system – so what is the special and pivotal twist the polymer gives to the PILP process, turning it into a versatile route for morphosynthesis? Recent contributions of Wolf et al. gave a first insight into this and provided the missing link between the emergence of a pure liquid-condensed mineral phase and the PILP process. They give clear evidence for classical emulsion-like behaviour of the liquid-condensed phase at early stages of mineralization (see Fig. 3.2) (Wolf et al. 2011a, b, 2012). Therefore, we have to realize that the stage of mineralization which is critical for morphosynthesis and which follows directly the initial liquid/liquid phase separation is clearly governed by laws of colloid science (as is true for any other colloidal and dispersed system) and that all morphogenetic phenomena in the PILP process can only be thoroughly understood if the colloidal character of the key players, the emulsified droplets and their interaction with the different entities in the reaction volume – such as surfaces, crystal, ions or polymers – are fully taken into account. The contributions of Wolf et al. revealed that the emulsified state is electrostatically stabilized by the negative surface charge of the mineral droplets and is strongly affected by additives, in full accordance with the DLVO theory. As predicted by this classic theory, the addition of "innocent" spectator ions like sodium and chloride, which are intrinsically present in all standard metathesis precipitations (e.g. in Faatz et al. 2004; Pontoni et al. 2004), leads to aggregation and slight coalescence of the emulsion (see Fig. 3.2) (Wolf et al. 2011b). Consequently, polymers, be they basic or acidic, can have a pronounced effect on the emulsified state of the transient LCP mineral precursor. As we know from colloid science, the actual impact of an additive on dispersions can be diverse and may vary with the molecular charge, size and concentration of the employed additive; for a short summary of these interactions, please see the excursus at the end of this chapter. In the case of the PILP process, it was evidenced that addition of small amounts of a polyanionic polymer can lead to a collapse of the emulsion, whereas a larger amount of polyelectrolytes can lead to stabilization of the emulsion (Wolf et al. 2011a, see also the excursus at the end of this chapter; Wolf et al. 2012). This shows clearly that the response of the system (stabilization vs. demulsification) is a complex function of additive concentration but also of surface charge of the dispersed colloid, the interparticle distance and, therefore, of the particle number density of the dispersed phase (which is markedly enhanced with polymer). Applied to the PILP system, we can infer that coalescence and sedimentation of the PILP phase droplets is thus not triggered until a sufficient and critical amount of droplets are formed. Thus, one may conceptualize the PILP process to a fundamental yet Janus-faced action of the polymeric additive: first, stabilization and accumulation of an ion-enriched liquid-condensed phase (LCP) by inhibiting classical nucleation in the early stages of mineralization and, second, destabilization of the subsequent emulsified state, allowing for its final coalescence at later stages of the reaction (Wolf et al. 2011a, b).

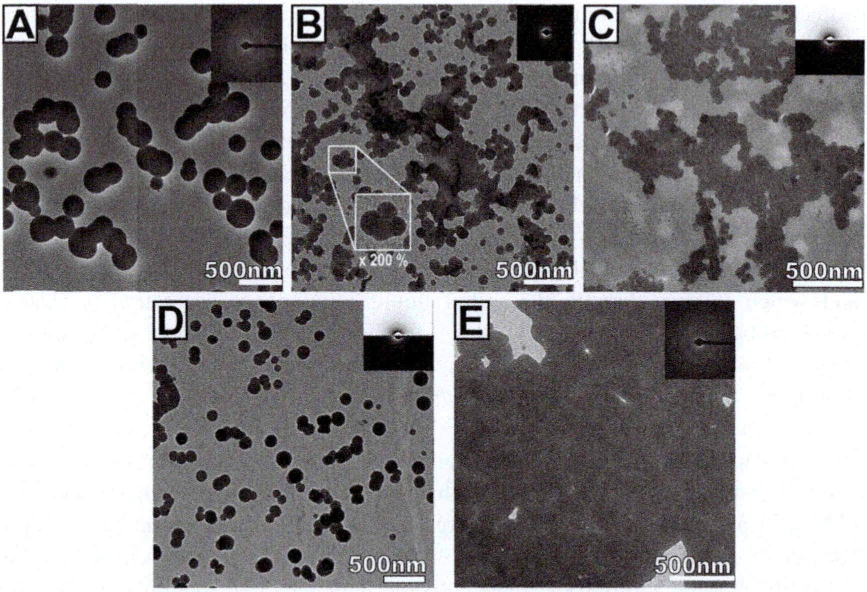

Fig. 3.2 Micrographs of liquid-condensed phases (LCPs) of calcium carbonate obtained under levitation; the insets show selected area electron diffractograms that indicate the amorphous state of the droplets. (**a**) LCP after 400 s of evaporation of a 4 μl droplet of saturated calcium bicarbonate solution. (**b**) LCP after 400 s, under addition of sodium chloride: aggregation and coalescence occurs due to screening of the particle surface charge. (**c**) LCP after 400 s obtained under influence of basic lysozyme (pI = 11.35): the droplets have coalesced and form a mineral film. (**d**) Under addition of the unusual acidic protein ovalbumin from the hen eggshell (pI = 4.6), the LCP is stabilized and still exists after 1000 s. (**e**) If only small amounts of an acidic protein is added (here 100 μg/mL of the aspartic-rich intracrystalline protein caspartin extracted the prismatic layer of *Pinna nobilis*), a well-developed mineral film is observed Subfigures (**a**) resp. (**c**) and (**d**) reprinted with permission from Wolf et al. (2008; resp. 2011a). Copyright 2008 resp. 2011 American Chemical Society. Subfigure (**b**) resp. (**e**) reproduced from Wolf et al. (2011b; resp. 2012) with permission from The Royal Society of Chemistry

3.1 Mechanistical Aspects

After this synopsis on our current knowledge of the PILP process, we will now dive deeper into various mechanistic aspects of the PILP process. First, we will start with a brief reprise of pertinent models of phase separation processes. Then, we will discuss each stage of the liquid/liquid phase separation process separately: pre-critical, critical – i.e. the phase separation itself – and the post-critical pseudomorphic phase transformation process. In each of these stages, we will discuss important aspects of how the polymer impacts on the crystallization process and how it abets the formation of a liquid-condensed mineral precursor phase. We will put a special focus on questions yet to be answered. Please note the distinction between liquid-condensed phase in the absence of additives (LCP) and the polymer-induced liquid-precursor (PILP) phase, which is stabilized and only formed in the presence of polymer. In the majority of studies which address or employ the PILP process, the mineral system of calcium carbonate served as the model system, mainly due to its high relevance in biomineralization systems and its apparent simplicity. Accordingly, we will focus in this review on the calcium carbonate system, although the PILP process has been expanded to non-calcium-based carbonates (Homeijer et al. 2008, 2010; Wolf et al. 2011b) and organic compounds like amino acids (Wohlrab et al. 2005; Jiang et al. 2011) and applied additionally to calcium phosphate (Jee et al. 2010b, 2011) and oxalate (Gower 2008), both of which are relevant to kidney stones (Amos et al. 2009).

3.2 Fundamental Concepts of Liquid/Liquid Phase Separations

Following the fundamental treatments of Gibbs (Gibbs 1877), for multicomponent systems to be stable against phase separation, they must resist two distinct scenarios: (α) they may not separate if a density fluctuation of high amplitude and finite size occurs; (β) they have to remain stable against small density fluctuations over a large volume. The first case (α) represents a typical metastable system whose curvature of the free energy of mixing is positive ($\partial^2 f/\partial c^2 > 0$). In the phase diagram, this region is delimited towards the stable region by the coexistence line, the so-called binodal (see Fig. 3.3a). By crossing this line, the system becomes metastable against spontaneous local density fluctuations, and separation *may* take place if these fluctuations are of sufficiently high amplitude. This case is subject to the classical nucleation theory (CNT) which describes the genesis of a new phase by the stochastic formation of nascent clusters (i.e. density fluctuations) which redissolve below a critical particle radius (Vekilov 2010). Once the cluster's surface energy is outbalanced by its bulk energy (i.e. the fluctuation reaches a sufficient amplitude), nucleation occurs, where the metastable and so-called critical nucleation cluster becomes stabilized by further growth that now lowers its free energy, and thus represents a new phase which – if the cluster does not redissolve – will grow in an ion-by-ion-mediated fashion as long as the supersaturation prevails, i.e. the system again becomes stable by reaching the binodal. The zone of metastability, which is governed by this first scenario, is separated from the zone of instability by the spinodal ($\partial^2 f/\partial c^2 < 0$). Beyond this line, the system becomes unstable even to infinitesimally small density fluctuations. This represents the second scenario (β), the thermodynamic background of which was elucidated by the pertinent works of Cahn and Hilliard (1959). In this regime of instability, phase separation is solely controlled by diffusion as there is no thermodynamic barrier against phase separation. Due to the curvature of the free energy of mixing, uphill diffusion against the concentration gradient takes place, leading to an increase in the concentration of the new phase with time, rather than size. In these stages, the density fluctuations

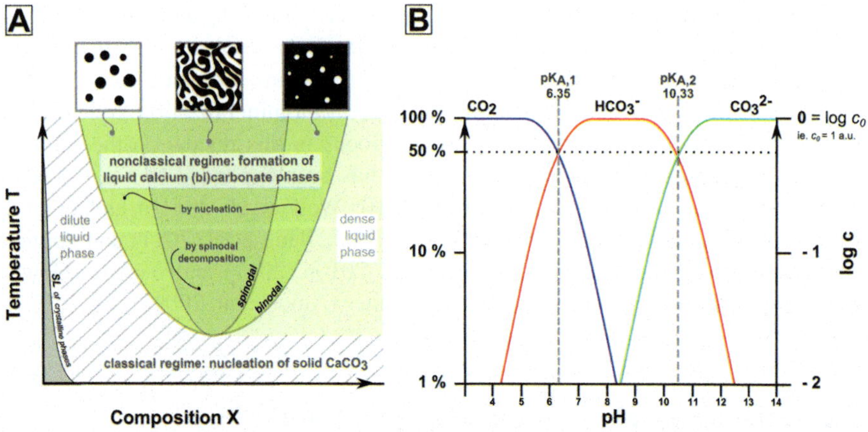

Fig. 3.3 (**a**) Hypothetical phase diagram of calcium (bi)carbonate solutions based on the works of Hilliard and Cahn. The *dashed region* represents the region of the phase diagram in which classical nucleation occurs. The *yellow* inset is the region in which a liquid-amorphous phase of calcium carbonate can form. If we assume a superposition of two phase diagrams of calcium carbonate and calcium bicarbonate, the SL line of the crystalline phases would represent the binodal line of the calcium carbonate system, and the *inset* would belong to the phase diagram of calcium bicarbonate. (**b**) Hägg diagram of a carbonate solution illustrating the ion speciation as a function of pH. Please note that in solution, virtually, no free carbonic acid exists but rapidly converts to solvated carbon dioxide. At $pH = pK_{A,2}$ 10.33, 50 % of the carbonate ions are protonated, i.e. HCO_3^-. At physiological relevant pH = 7, virtually no deprotonated carbonate is present: the solution contains 80 % bicarbonates, 20 % solvated carbon dioxide and only less than 0.4 ‰ deprotonated carbonates

gradually intensify and the solution's microstructure is distinctly fine and uniform. It may pass through a characteristic bicontinuous stage; this happens typically at a composition of 50:50 which is often where the spinodal curve is breached (see Fig. 3.3a). The system eventually ripens into a stage of spherical monomodal colloids and can resemble the morphology of colloids formed binodally.

3.3 The Pre-critical Stage in Real Systems

According to CNT, the formation of nuclei is a consequence of stochastic density fluctuations constrained to the metastable region. In the non-saturated regime, these clusters should be inherently unstable and thus dissolve and not be detectable. But in 2008, the existence of so-called prenucleation clusters (PNCs) was proclaimed by Gebauer et al. (2008). By means of analytical ultracentrifugation in combination with Ca^{2+}-activity measurements, they revealed the existence of calcium carbonate clusters of distinct size even in non-saturated solutions. PNCs are considered by some to be thermodynamically stable solutes; in the case of calcium carbonate, it was found that these PNCs are 7 k_BT more stable than the single, hydrated ions in solution. Furthermore, the stability of PNCs was shown to be dependent on the pH; their formation seems to be favoured at lower pH (Gebauer et al. 2008; Bewernitz et al. 2012). As of yet, detailed information regarding the structural composition at the molecular level of PNCs is not available. In order to fill this gap, computer simulations were undertaken in which PNCs were found to be hydrated and highly fluctuating oligomers of calcium carbonate ion pairs featuring (partially branched) chain- or ringlike motifs. The characteristic 1:1 ion ratio of this "dynamically ordered liquid-like oxyanion polymer" (DOLLOP) was corroborated by recent ESI-MS analysis (Wolf et al. 2011b; Demichelis et al. 2011). Other molecular dynamic simulations have also predicted the formation of a dense liquid phase through liquid/liquid separation within the concentration range in which clusters are observed (Wallace et al. 2013).

So far, the actual role and impact of PNCs on the mineralization process is still enigmatic and stirs disputes (see Faraday Discussion transcripts 2012). Do they provide a low-barrier pathway for nucleation? Or do they – due to their pronounced stability – pose an additional barrier for classical nucleation of a *crystalline* phase and allow, due to their structural disorder, the formation of a liquid-amorphous phase? Then the system's rate of nucleation is lowered so distinctly that the system may safely traverse the region of classical nucleation of the phase diagram and eventually enter the region of liquid/liquid phase separation (Wallace et al. 2013); or the prenucleation clusters start to segregate and consequently provide a low-energy barrier pathway for liquid/liquid phase separation (Wolf et al. 2011b; Gebauer et al. 2014). Their structural disorder and their high degree of hydration may then create lower surface energy than a crystalline nucleus, thereby favouring the formation of a liquid-amorphous phase. In the latter scenario, PNCs would represent the initial building blocks of the liquid-amorphous precursor phase. Initial validation for this assumption seems to be found in a recent report which showed that the distinct proto-crystalline features of ACC are related to different pH-dependent variants of PNCs (Gebauer et al. 2010). Further substantiation for this postulate is provided by the fact that in all existing systems which have been reported to exhibit liquid-condensed phases, prenucleation clusters have additionally been found: in a variety of bivalent carbonates (Wolf et al. 2011b), amino acids (Wohlrab et al. 2005; Jiang et al. 2011; Kellermeier et al. 2012) and calcium phosphate (Jee et al. 2010a). An extensive study of the early stages of calcium phosphate formation bringing together a set of different techniques suggested an ion-association process of prenucleation species, the Posner clusters, which assemble to polymeric entities of nanometre size. However a liquid state of the transient intermediate was not reported in this study (Habraken et al. 2013); the Posner cluster (aggregates) may be incapable of binding a sufficient amount of water to reach a liquid-like state. Yet in the presence of polymer, the precursor species are able to infiltrate the nanoscopic gap zones, while the conventional ACP phase does not; thus the polymer may provide additional hydration. In summary, while evidenced in a number of systems, further experimental proof is yet to be furnished demonstrating the potential role of PNCs as the structural unit of the liquid-amorphous phase.

3.4 The Critical Stage: Liquid/Liquid Phase Separation

It has often been rejected that the liquid/liquid phase separation in aqueous mineral systems is akin to that of polymer and protein systems (Petsev et al. 2003; see Faraday Discussion transcripts 2012). But now, the occurrence of a liquid/liquid phase separation in supersaturated solutions of minerals is an apparent and settled fact (Wolf et al. 2008; Bewernitz et al. 2012), albeit one in which the underlying mechanism is still enigmatic. The computational results suggesting that PNCs feature a polymer-like molecular structure underlines the fact that the calcium carbonate system may have more in common with polymer systems than initially assumed. Many thermodynamic features of polymers are dominated by entropic considerations, and the same is true of water in aqueous-based systems, so we believe close attention needs to be paid to the behaviour of ions in the presence of both.

Currently, we can only speculate under which conditions a binodal or spinodal decomposition may occur, and if a spinodal decomposition in mineral solutions is possible at all. We should keep in mind that, although often consulted, morphology of a ripe stage is rarely a good proxy for assessing a formation mechanism. A binodal process was put forward by Faatz et al. (2004), whereas the work of Rieger et al. (2007) was sporadically seen as a corroboration of a spinodal decomposition (see Faraday Discussion transcripts 2012; Gebauer et al. 2014). Also, the contribution of Wolf et al. (2008) was occasionally interpreted as evidence for binodal decomposition, but we clearly refrain from such speculation (as in Gebauer et al. 2014) as we did not probe the moment of phase separation, but a later stage. It is worth pointing out that the formation of monodispersed solid spheres were repeatedly interpreted as an indication of a spinodal *and* for a binodal mechanism (Pontoni et al. 2003; Faatz et al. 2004). However, such an observation points neither in the binodal nor in the spinodal direction; both pathways can yield monodispersed spheres. It is not even an indication for a liquid-like intermediate state because it can similarly be explained by the pertinent model of LaMer in full agreement with the notions of classical nucleation of a solid phase (LaMer and Dinegar 1950; Navrotsky 2004). Thus, without monitoring the initial stages of formation in real time, one should not employ morphological observations as an evidence of mechanism, only as an indicator.

Faatz et al. (2004) were the first to apply the generic model of Cahn and Hilliard (1959) to the calcium carbonate system. They proposed a schematic but nevertheless instructive phase diagram of the calcium carbonate system which was recently revisited by Wallace et al. (2013) in a slightly revised form (see Fig. 3.3). Beyond the solubility line(s) (SL) of solid calcium carbonate, formation of crystalline calcium carbonate may take place by the classical nucleation process, i.e. by binodal solid/liquid phase separation. The liquid/liquid coexistence line represents the border of the metastable region in which a liquid phase of calcium carbonate may form via a binodal route. Once the system crosses the coexistence line, we can expect that the rate of nucleation for a solid drops (Galkin and Vekilov 2000a).

Pushing the system even further beyond the liquid/liquid instability line, spinodal decomposition will take place. The schematic phase diagram, as given in Fig. 3.3, reflects the now established fact that a multitude of different pathways can occur in one system (please note that binodal solid/liquid demixing is still possible beyond the liquid/liquid coexistence line) and accounts for the formation of a liquid-condensed phase of calcium carbonate (Nielsen et al. 2014).

Another crucial question is still without reply. Do we know which kind of chemical species actually undergoes liquid/liquid phase separation? Commonly seen as a phase separation of the ionic constituents of calcium carbonate, we cannot rule out the option that only PNCs of calcium carbonate separate. If we review the pertinent literature, it seems even reasonable to predict that we are actually observing the liquid/liquid phase separation of a calcium *bi*carbonate(-biased) phase. Bicarbonate is virtually omnipresent in carbonate solutions – even at $pH = pK_{A,2} = 10.33$, half of the carbonates are protonated (in form of bicarbonate, HCO_3^-; see Fig. 3.3-b). At neutral pH, the solution contains virtually only bicarbonates and solvated carbon dioxide ($CO_{2\,(aq)}$); only a fraction (less than 0.4 ‰) are fully deprotonated carbonates (CO_3^{2-}). Thus, we have to realize that only minor amounts of free deprotonated carbonate are present under standard conditions; the majority of the carbonate ions are present in a protonated state. The idea that we observe the liquid/liquid phase separation of a calcium *bi*carbonate(-biased) phase finds corroboration in the fact that the formation/stability of a liquid-amorphous mineral phase is facilitated/enhanced at near neutral pH (Wolf et al. 2008, 2011b; Bewernitz et al. 2012). In situ NMR studies of Bewernitz et al. (2012) support this concept even further as the peak position indicated that the liquid-condensed phase is bicarbonate-biased. We could thus draw the conclusion that the complex behaviour of the calcium carbonate system is owed to an intrinsic competition between two different phase separation pathways of two closely related species in chemical equilibrium, i.e. the binodal solid/liquid phase separation of calcium carbonate and the liquid/liquid phase separation of calcium bicarbonate. Treating the different protonation states individually, we can see the phase diagram simply as a superposition of two independent phase diagrams, one of calcium carbonate overlaid with that of calcium bicarbonate. In other words, the nonclassical behaviour of the calcium carbonate system could then be traced back to the fundamental acid–base properties of the carbonate anion, i.e. its multiproticity paired with a relatively high $pK_{A,2}$. Consequently, liquid-amorphous intermediates are not only a feature of the extensively studied calcium carbonate system, but could be shown to be characteristic of a considerable set of different carbonate minerals (Wolf et al. 2011b).

Why should such a bicarbonate-biased system favour the formation of a liquid phase? First, the protonation can be seen as the simplest form of chemical protective groups (as used in organic synthesis), acting as protection against crystallization by raising the barrier of nucleation and crystal growth by introducing an additional step of deprotonation, along with dehydration. The ion-enriched phase would presumably be highly supersaturated if it contained mainly fully deprotonated carbonate species and thus tend to nucleate a crystalline phase. As shown by

Bewernitz et al. (2012), the liquid-condensed phase produced at pH 8.5 is enriched with bicarbonate ions. The presence of hydroxyl groups, i.e. the protonated $^{-}O_2C$–*OH* functional group in bicarbonate ions, would deliver an effective means for binding water via hydrogen bonding and thus facilitate the formation of a highly hydrated and thus liquid phase (Wolf et al. 2008). Therefore, the system of calcium carbonate at relatively low pH could be seen as a system which is – due to the high amount of protonated carbonate – frustrated with regard to crystallization and thus prone to the formation of highly hydrated and amorphous phases via liquid/liquid phase separation (Wolf et al. 2008).

The locus and shape of the liquid/liquid coexistence line would be an ideal parameter to characterize PILP systems, irrespective of the species which actually separate. It is well known from other systems that the presence of (even simple) additives can distinctly shift the locus and the shape of the coexistence line as well as the position or shape of the spinodal and the binodal (Galkin and Vekilov 2000b). In the case of the calcium carbonate system, it is reasonable to assume that the locus is strongly affected by pH, which may render the region of liquid/liquid phase separation more accessible at low, near neutral pH. Another important additive may be magnesium, which has been reported several times to facilitate the formation of a PILP phase (Kim et al. 2007; Cheng et al. 2007; Schenk et al. 2012; Cantaert et al. 2013). Polymeric additives may be able to shift the locus. Besides their pertinent inhibition effects on classical crystallizations which promote the formation of a (liquid) amorphous phase (see above), Bewernitz et al. (2012) showed recently that polyelectrolytes may change the pH profile of a precipitation process considerably. At early stages of a mineralization, this can diminish the pH slope which hampers "classical" nucleation of a crystalline phase by reducing supersaturation and, on the other hand, provides more bicarbonate ions for the formation of the bicarbonate-biased PILP phase. This behaviour is remarkable since the concentrations of polyelectrolytes in PILP reactions are typically "homeopathically" low (e.g. 20 μg ml^{-1}), yet can cause a pH shift of about 0.5 in late stages of precipitation (Bewernitz et al. 2012). However, the mechanism by which the additive so strongly affects pH evolution has not yet been revealed.

3.5 The Post-critical Stage, Part I: The Emulsified Stage of the Liquid-Condensed Phase

Once the liquid/liquid phase separation has taken place, the subsequent (additive-free) two-phased state behaves like a classical emulsion, as already described in the introductory synopsis. The droplets are electrostatically stabilized by a negative surface charge (Wolf et al. 2011a). Sheering of the emulsified liquid-condensed phase can lead to coalescence, and even the addition of the simplest additives may have a pronounced effect on the mineral emulsion (Wolf et al. 2011b). Sodium chloride leads to aggregation and slight coalescence of the emulsion (see Fig 3.2) due to screening of repulsive electrostatic forces by increasing the ionic strength of the solution (Wolf et al. 2011b). As outlined above, polymeric additives, be they basic or acidic, can also have a pronounced effect on emulsions (Wolf et al. 2011b). In the case of calcium carbonate, the actual impact depends on the molecular charge density and molecular weight, as well as initial concentration of the employed additive, and fits expectations based on the DLVO theory. Basic (bio)polymers lead to a demulsification by surface charge neutralization, whereas acidic polymers, if employed at sufficiently high concentrations, stabilize the emulsified state (see Fig 3.2) (Wolf et al. 2011a). But if the initial concentration of the acidic (bio)polymer is reduced considerably, the system enters the regime of depletion destabilization, and a prominent change from stabilization to destabilization occurs: the emulsified state of the PILP phase collapses and the droplets sediment and coalesce to a mineral film, the hallmark of the $CaCO_3$ PILP process (see Fig 3.2) (Wolf et al. 2012). To resume, *one* essential task of the polymer is that it allows for coalescence by inducing the collapse of the initial emulsified state of the liquid-condensed precursor phase (e.g. once a critical particle number density is reached). But this demulsification action is only one of the crucial steps in transforming a classical crystallization process into a pseudomorphic solidification process. Besides this colloido-chemical effect, the polymer additionally exerts numerous effects on the precipitation reaction which are essential for turning it into a PILP process. Initially, it allows for the accumulation of large amounts of LCP since it is able to sequester and to stabilize more ions, PNCs and/or LCP. By this, and by its preferential binding to active crystal growth sites, the polymer represses classical crystal nucleation and growth which leads, in turn, to the important effect that already formed amorphous material is not redissolved, as the Ostwald–Lussac rule of stages would suggest. The reverse case is impressively visualized by a recent liquid cell TEM study in which the formation of, what appeared to be, a liquid-condensed phase and aragonite was observed side by side. While the aragonite crystal grows, the liquid-condensed phase seems to be “slurped up”, thus simply serving as an ion supply as the abutting crystal grows into its characteristic twinned crystal habit (Nielsen et al. 2014). This system did not contain polymer, so even though a liquid-condensed phase may have been present, the classic PILP pseudomorphic transformation did not occur, consistent with many other studies that cannot achieve this without polymer. Ultimately, the second foundation of the PILP process thus relies on the polymer’s influence on the stability and transformation kinetics of the various mineral phases.

To this point, a considerable variety of acidic (bio)polymers have been shown to be capable of inducing PILP formation: polyaspartate (Gower and Tirrell 1998; Gower and Odom 2000; Jee et al. 2010a), polyacrylic acid (Balz et al. 2005; Harris et al. 2015), various block copolymers (Kim et al. 2009; Zhu et al. 2009), intracrystalline proteins from biominerals like ovalbumin from the hen eggshell (Wolf et al. 2011a) or caspartin and calprismin from the prismatic layer of *Pinna nobilis* (Wolf et al. 2012) and osteopontin from bovine milk (Rodriguez et al. 2014). Even short synthetic peptides (Sugawara et al. 2006; Dai et al. 2008a), DNA

(Sommerdijk et al. 2007) or polysaccharides (Hardikar and Matijevic 2001; Zhong and Chu 2009) have been shown to be employable in the PILP process. Additionally, Wolf et al. (2011a) were the first to show that basic polymers are capable of inducing a PILP phase: they evidenced that the PILP process can be induced by the addition of the basic protein lysozyme, which proceeds presumably by interconnection of the mineral precursor droplets. Recently, Cantaert et al. (2012; 2013) and Schenk et al. (2012; 2014) extended this observation to cationic polymers such as poly(allylamine) for the morphosynthesis of calcium carbonate films and fibres. This finding is of particular importance since polycationic polymers, like lysozyme or poly(allylamine), are not able to sequester cations from solution, in contrast to polyanionic polymers like polyaspartate (Wolf et al. 2011a); nevertheless, they were shown to be suitable for morphosynthesis (Schenk et al. 2012; Cantaert et al. 2012, 2013; Schenk et al. 2014). This clearly suggests that the impact of the polymer on the precipitation is much more complex than simply affecting calcium or carbonate ions and rather involves interactions with (pre)nucleation clusters or the liquid-condensed phase.

In the following we try to give a short account of the different, beneficial effects of PILP-active polymers. A polymer suitable for PILP may show all of these traits, like polyaspartate, or may lack some of them, like poly(allylamine). It is still not fully understood which traits are indispensable for a polymer to qualify as an additive for the PILP process.

1. At first glance, one may expect that the polymer reduces the overall solution's supersaturation by sequestering calcium ions (depending on the polymer's molecular charge), a behaviour demonstrated for acidic polyelectrolytes, e.g. proteins (Pipich et al. 2008), at higher concentration. However, some results show that calcium activity is *not* affected by the presence of acidic polyelectrolytes such as PAA if concentrations are used which are characteristic for PILP processes (Gebauer et al. 2009).
2. It inhibits nucleation and growth of a crystalline phase of calcium carbonate (Kim et al. 2009; Gebauer et al. 2009; Wolf et al. 2011a). None or only minor amounts of a crystalline and thermodynamically more stable phase of calcium carbonate is formed if the conditions are ideally stabilized. As stated above, this has two far-reaching and crucial consequences. First, the amorphous phases need not to compete for ions during growth, and thus more LCP can form until supersaturation ceases. Second, no Ostwald step ripening occurs and thus saturation with respect to the amorphous phases prevails. No crystalline phase grows in solution at the expense of the amorphous phases. This considerably increases the lifetime of the thermodynamically unstable amorphous phases.
3. It affects the pH profile of the mineralization since it acts as a typical buffering agent. Depending on the polymer concentration, the asymptote of the pH development may be attuned to the polymer's pK_A value, which can increase the supply of bicarbonate, which in turn can abet the formation of an amorphous phase or shift the solution's composition in favour of a bicarbonate-biased liquid phase (Pipich et al. 2008; Bewernitz et al. 2012).

4. It may change the actual composition of the droplets: the water content is increased, calcium is sequestered hyper-stoichiometrically in the early stages, and binding of bicarbonate may be favoured over carbonate ions (Dai et al. 2008b; Bewernitz et al. 2012).
5. It apparently influences the interactions of the droplets/phase with functionalized surfaces (e.g. patterned SAMs or collagen fibrils).

3.6 The Post-critical Stage, Part II: Pseudomorphic Transformation Amorphous to (Meso)crystalline Material

How does, as the final step, the material transform into a crystalline material? Apparently, there are several possible ways. One possibility may be the formation of a crystal by classical nucleation within one of the still liquid precursor droplets. Such a process has already been reported for protein crystallization in which the supersaturation of protein inside the droplets was shown to be higher than in the initial solution (Vekilov 2012). As a matter of fact, this mode has actually not yet been documented for liquid-condensed phases of minerals. In the absence of additives and under sufficient saturation, the liquid-condensed phase can densify, dehydrate and solidify over the course of time forming (assumingly monohydrated) ACC, as is the case in the Faatz method (Faatz et al. 2004). The final transformation to crystalline material proceeds typically via redissolution/recrystallization, i.e. Ostwald–Lussac ripening, and may occur well before solidification of the droplets – as impressively imaged by Nielsen et al. (2014) – or when the droplets reached a dry state (Wolf et al. 2008; Ihli et al. 2014). The presence of solid ACC seems to abet this transformation in solution by a template action of the ACC spheres (Zhang and Liu 2007; Wolf et al. 2008; Freeman 2014) and can result in interesting morphologies like polycrystalline hedgehog spheres (Shen et al. 2006; Loges et al. 2006). Under essentially dry conditions, ACC is remarkably stable owing to the fact that the first transformation to crystalline material can preferably occur via a dissolution/reprecipitation pathway. It is probable that this proceeds in localized, small domains, be it on the ACC surface or in small intergranular pockets entrapped within ACC (Ihli et al. 2014).

In the presence of polymeric additives, the mechanism described above gives way to a pseudomorphic transformation, i.e. the transformation of the amorphous material to a crystalline state with preservation of its morphology. Thus a combination of these critical stages provides a morphosynthetic route to non-equilibrium-shaped crystals by a mechanism we refer to as colloid assembly and transformation (CAT).

If we pool various facts and observations concerning ripening and crystallization of amorphous calcium carbonate intermediates obtained both from calcareous biominerals and biomimetic processes, we can draft a relatively detailed working model describing the process of crystallization in PILP-based systems (see Fig. 3.4):

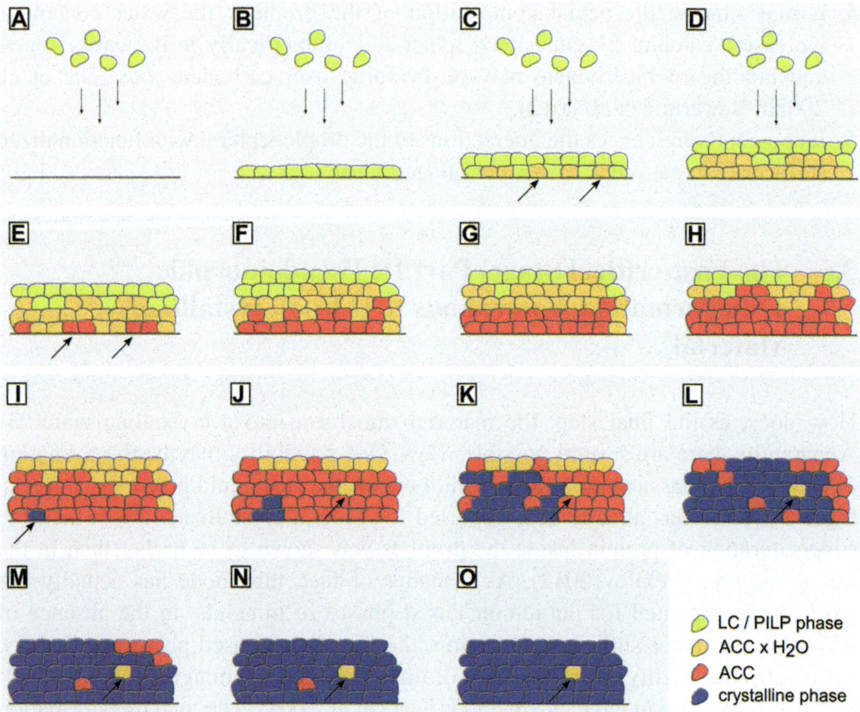

Fig. 3.4 Generic model of mineralization by colloid assembly and transformation (CAT), i.e. the formation of mesocrystals *via* a PILP intermediate; crucial details in each step are indicated by *diagonal arrows*. Stages **a**–**d**: Sedimentation of still liquid-like or viscous droplets. Stages **b**–**d**: The droplets accumulate in a space-filling fashion. Stage **c**: First steps of solidification, i.e. first dehydration from liquid-like ACC, respectively, PILP, to ACC × H_2O. Stages **e**–**h**: Second stage of dehydration, from ACC × H_2O to ACC. Stage **i**: Initial nucleation and solid–solid transformation to crystalline material. Stage **j**: Propagation of crystallinity throughout nearly the entire mineral volume via isoepitaxy across the interconnected and coalesced ACC particles. Stages **n**–**o**: Singular granules may remain amorphous throughout the process; their transformation may be impeded if the composition does not meet the requirements, e.g. it contains too much water, foreign ions like magnesium, polymeric impurities or improper stoichiometry

(a) The highly hydrated and thus still liquid-like or viscous ACC phase is formed. If the polymer inhibits classical crystal nucleation and growth in the neighbouring solution, the PILP phase is not redissolved for feeding the growth of a crystalline phase, as the Ostwald step rule would predict. At a critical point, e.g. when a critical droplet number density is reached, demulsification is induced by the polymeric additive via the mechanism of depletion destabilization; the initially stable emulsion collapses and the hydrated PILP phase starts to sediment (Fig 3.4, stages A to D) (Gower and Odom 2000; Wolf et al. 2011a).
(b) The droplets sediment on the substrate/mould. One of the key actions of the polymeric additive is to prolong the hydrated and thus liquid-like state of

Fig. 3.5 Atomic force microscopy micrographs of the nanogranular texture from (**a**) a transverse and mirror-polished section through a calcite prism from the shell of *Pinna nobilis* and (**b**) a mineral film generated by the PILP process. (**c**) A typical phase image of the biogenic calcium carbonate from the calcite prisms extracted from the shell of *Pinna nobilis*. The phase contrast indicates organic material which ensheathes the individual nanogranules. Subfigure (**b**) is reprinted with permission from Kim et al. (2007), copyright 2007 American Chemical Society. Subfigure (**c**) reproduced from Wolf et al. (2012) with permission from The Royal Society of Chemistry, copyright 2012

the droplets, so that the droplets accumulate in a space-filling fashion (Yang et al. 2011); each droplet adopts its shape to its individual mould/surrounding. Although they appear to coalesce fully, some "memory" of their individual colloidal shapes is retained, which is sometimes described as a nanocolloidal or nanogranular texture when observed in the final product by AFM measurements (see Fig. 3.5, A to C) (Kim et al. 2007). However, as hydration waters are also exiting the PILP phase with time (Dai et al. 2008b), the coalescence ability is reduced as the droplets become more gel or solid-like particles. For this reason, morphologies such as film formation are mainly found in reaction vessels with short transport distances. The dehydration kinetics is likely influenced by the polymer characteristics (such as molecular weight and charge density) and seems to be markedly influenced by interaction of the colloids with substrates and templates. For instance, it was recently exemplified that one can tune the wettability of the PILP phase for a given organic surface functionalization just by adjusting the magnesium content of the LCP mineral precursor (Berg et al. 2013).

(c) The forming mineral body initially remains amorphous since its chemical state is not yet ready for crystallization, which may require exclusion of the polymer, dehydration, deprotonation of bicarbonate or balancing of ion stoichiometry (Fig 3.4, stages A to H) (Politi et al. 2004; Dai et al. 2008b; Wolf et al. 2012; Gong et al. 2012; Ihli et al. 2014). If the chemical state for crystallization is not reached, singular granules may stay amorphous even if they are surrounded by crystallinity (as indicated by an arrow in stage J) (Wolf et al. 2012; Gong et al. 2012). Note that it has been suggested that only anhydrous ACC readily transforms to calcite (Politi et al. 2008; Gong et al. 2012; Ihli et al. 2014); this seems counterintuitive since hydration of water could provide more mobility to the ions during assembly.

(d) From the time when the droplets have formed, they start to dehydrate and subsequently solidify and form anhydrous ACC (e.g. stages C and E, transition from yellow via orange to red). This dehydration occurs concomitantly with an exclusion of the polymer (Dai et al. 2008b). The process of syneresis can exert a strain on the dehydrating bulk phase which sometimes leads to marked lattice strain, or jointing and crack formation; similar processes lead to prism formation in basalt or starch (Gower and Odom 2000; Toramaru and Matsumoto 2004; Gilman 2009). For PILP systems, photoelasticity could be observed (Leupold and Wolf 2015), and regular cracking was reported in large-scale PILP films of amino acids (Jiang et al. 2013).

(e) Nucleation occurs somewhere in the dehydrating amorphous bulk mineral, e.g. by heterogeneous nucleation on the surrounding matrix (stage I, the first crystallizing granule is indicated by an arrow, transition from red to blue) (Wolf et al. 2012).

(f) From this point on, the crystallinity percolates throughout the bulk material, sometimes following a stochastic and tortuous path; one dehydrated granule after the other crystallizes by solid/solid transformation induced by isoepitaxial nucleation (stages I to O) (Killian et al. 2009; Seto et al. 2011; Wolf et al. 2012; Gal et al. 2014). This process, because of the initial colloidal assembly of colloidal droplets/particles, yields a mesocrystalline material: the crystalline nanogranules are all highly co-oriented since the crystal orientation originates from a singular nucleation event, with neighbouring granule transformations inheriting the same orientation by isoepitaxy (Wolf et al. 2012; Gong et al. 2012; Gal et al. 2013, 2014). In some cases, crystallinity may proceed across the amorphous body following more "traditional" trails, such as dislocation defects that lead to spiral growth hillocks as the crystallinity slowly propagates within the amorphous film. This is evidenced by transition bars with crystallographic symmetry indicative of a lattice directed organization of species; the transition bars can be seen midway in the amorphous-to-crystalline transformation, either as faceted rhombs resulting from the calcite symmetry or radial patterns emanating from spherulitic growth (Dai et al. 2008a).

A recent study which reported the formation of thin mosaic and mesocrystalline thin films of DL-lysine by a PILP process provided initial evidence that the above model is essentially correct and valid as well for non-carbonate-based organic systems (Jiang et al. 2013).

3.7 Relevance to Biomineralization Systems

The idea that the PILP process might be of high relevance in biomineralization has been championed by Gower since its initial discovery in 1998 (Gower and Tirrell 1998). Interestingly, the discovery of the PILP phenomenon coincided with a paradigmatic change in our understanding of biomineralization processes. On the one side, key contributions in the analysis of biominerals from the Weizman institute showed that amorphous calcium carbonate plays a ubiquitous role in biogenic crystallization (Aizenberg et al. 1996; Weiss et al. 2002; Levi-Kalisman et al. 2002; Addadi et al. 2003; Politi et al. 2004, 2006, 2008). On the other side, the in vitro studies of the PILP process clearly demonstrated the immense impact minor amounts of polyanionic additives can have in transforming the mineralization to a precursor process. Polyaspartate, which was employed as an additive in the first experiments evidencing the PILP process, is a clear analogue to intracrystalline proteins found in biominerals. The PILP process evidenced for the first time that these biopolymers, which are typically unusually acidic due to a high fraction of aspartic acid in the primary sequence or post-translational modifications, are able to shift gears towards a crystallization pathway which proceeds in a particle-mediated route via an amorphous intermediate, akin to the processes observed in calcareous biominerals. Moreover, the strikingly similar nanogranular appearance of PILP materials and calcareous biominerals renders Gower's assumption very appealing: both calcareous biominerals, like mollusc shells (Li et al. 2011; Wolf et al. 2012) or sea urchin spines (Killian et al. 2009; Seto et al. 2011), and PILP-generated materials are composed from individual nanogranules which are about 100 nm in diameter and (partially) enwrapped in an organic ensheathment (Jacob et al. 2011; Wolf et al. 2012); see Fig. 3.5. This nanogranular texture occurs in most types of calcareous biominerals, be it in calcareous sponge spicules (Sethmann 2005), in nacre (Mutvei 1978; Rousseau et al. 2005; Jacob et al. 2011; Wolf et al. 2015), in prismatic and lamellar layers of bivalves (Dauphin 2008; Li et al. 2011; Wolf et al. 2012, 2015), in brachiopod shells (Dauphin 2008) or even in urchin spines (Seto et al. 2011) and coccoliths (Henriksen et al. 2004). Nanogranularity is also observed in calcareous fossils or non-calcareous biominerals (for an exhaustive review on its occurence, see Wolf et al. 2016). Since it strongly affects numerous properties, be it crack propagation, nanoplasticity or self-healing, it represents a fundamental and ubiquitous process-structure-property relationship of biominerals (Wolf et al. 2016). The nanogranular unity of biominerals and PILP-generated materials may portend that both form via the same pathway involving at least some form of liquid-like colloidal amorphous intermediate (Wolf et al. 2012). But this is not the only aspect indicating a PILP-like process in vivo. Besides the distinct nanogranular texture, calcareous biominerals are typically space-filling mesocrystals formed via a colloid-mediated crystallization pathway (Yang et al. 2011; Seto et al. 2011; Wolf et al. 2012; Hovden et al. 2015). Furthermore, the unexpected finding of intracrystalline preservation of nanoscale amorphous regions can be easily explained with the described ripening model simply by keeping in mind that individual nanogranules may not always crystallize (see Fig. 3.4, stages N to O) (Seto et al. 2011; Wolf et al. 2012; Gong et al. 2012), particularly when they contain too much water or polymer and have the wrong stoichiometry or when they have only insufficient contact points with the surrounding mineral so that isoepitaxial nucleation is not possible. Since dehydration is inherent to a PILP process, effects associated with syneresis should be also traceable in biominerals. The formation of prismatic calcite in mollusc shells could be a syneresis phenomenon similar to columnar jointing in basalt or starch (Toramaru and Matsumoto 2004; Gilman 2009). However this speculation still awaits its verification. Growth processes in vivo occur by a layer-wise deposition mode and proceed at a very slow speed so that syneresis effects may only occur in thick- or fast-growing biominerals, if at all.

It is obvious that all calcifying species have to cope with the same physicochemical "quirks" of the calcium carbonate system. As we have seen, this mineral system is not the simple and well-behaved model system as was initially thought. One quirk of the calcium carbonate system is clearly the occurrence of liquid-amorphous phases both in the presence of acidic (bio)polymers or even without, especially at near neutral/physiological conditions. Nature seems to suppress nucleation virtually at all costs, mainly by means of unusually acidic proteins. It seems that this was the basic requirement that promoted biomineral evolution in the first place, and recent studies revealed that such unusual acidic "inhibitory" proteins like caspartin extracted from *Pinna nobilis* in fact abet the formation of an LC/PILP phase (Wolf et al. 2012). Why shouldn't calcifying species exploit then the exact quirks that they have to face for their biogenic production of calcium carbonate when they are ultimately rewarded with a protective shell and other evolutionary advantages? Given this, it seems reasonable that different phylogenetic species developed individual but nevertheless comparable sets of molecular and physicochemical tweaks to deal with the calcium carbonate system. With millions of years of evolution, today these tweaks are highly sophisticated, employing elegantly self-organizing entities like liquid crystal formation of collagen and chitin templates, or aggregation-prone intrinsically disordered proteins (Evans 2013). Notably though, these more sophisticated features may be more related to the organic matrices and scaffolds, while the actual mineralization process can seemingly be accomplished with less sophisticated proteins (that can be effectively emulated with simple polyelectrolytes). Of course there may be minor tweaks in the proteins involved in controlling crystal phase, orientation and texture; but the overriding morphological control may be simpler than was long perceived. This may be pertinent with respect to phylogenetic and evolutionary analysis of various biomineralizing organisms. From an engineering perspective, these concepts may provide a great deal of inspiration for material synthesis via the PILP process.

3.8 Conclusions and Outlook

Still being rather a curiosity than an accepted morphosynthetic approach, the full potential of the PILP mechanism for morphosynthesis is yet to be unlocked. However, the myriad of morphologies generated by the PILP process to date and the path-breaking contributions exemplifying the applicability of the PILP process for biomimetic bone graft substitutes clearly demonstrates the potential of the PILP process. Aside from this, currently, it is still essentially confined to relatively mundane and mainly carbonate-based minerals. Only for organic pigments was the PILP process successfully employed in the generation and property tuning of a functional material (Ma et al. 2009). This, and our limited mechanistic insight into all stages of this nonclassical crystallization process, severely narrows the applicability and transferability of the PILP process. However, its power for generating non-equilibrium shapes is already proven, and its potential role in biomineralization only affirms our need for more intensive research. The PILP process is one of the most prominent examples of nonclassical crystallization, although it may be not as nonclassical as first perceived: classical colloido-chemical scenarios intimately entangled with liquid/liquid phase separations come together in a kinetically steered mineral precipitation pathway. This said, the nonclassical PILP crystallization process may be a much more classical assembly pathway than currently assumed and thus be transferable to a wide range of compounds. This perspective is an appealing one because as of yet, a broadly applicable method which allows for the generation of nanoparticle superstructures and nanocomposite materials under morphological control does not exist. The relatively facile PILP process is currently one of the strongest candidates to fill this gap and thus clearly requires in-depth studies into its underlying mechanisms.

Excursus: Electrostatically Stabilized Dispersions and Their Interaction with Polyelectrolytes

Electrostatic Stabilization In the absence of (surface-modifying) additives, dispersions of colloids are only stable if the colloids carry a (near-) surface charge. As expounded in the models of Gouy–Chapman, Stern and Grahame (see Grahame (1947) and references therein), counterions form a diffuse ionic layer around the charged colloids from which the effective Coulombic repulsion arises. The interaction potential between the colloids results from a superposition of the attractive van der Waals' forces and the repulsive electrostatic interaction and features a potential barrier which may save the colloids from coagulation. The electrostatic interaction is strongly dependent on the electrolyte concentration in the continuous medium. The Debye screening length $1/k$ is a measure for the size of the diffusive double layer and represents the distance at which the electric potential drops to $1/e$ of its initial value. It drops with increasing ionic strength of the solution; thus, in relatively pure water with low amounts of a single charged electrolyte ($c_e = 10$ mmol l^{-1}), the diffusive double layer is roughly 500 nm thick; at higher values ($c_e = 0.1$ mol l^{-1}) the double layer shrinks to about 5 nm. At a critical concentration of c_e, the diffusive layer is so thin that the particles can approach so closely that van der Waals' forces dominate and coagulation can occur. Under orthokinetic conditions, i.e. the colloids perform not only Brownian movement but experience as well sheer forces due to stirring or shaking, the coagulation speed can increase by a factor of 10^4. Then, the hydrodynamic forces can easily push the particles across the potential barrier.

(continued)

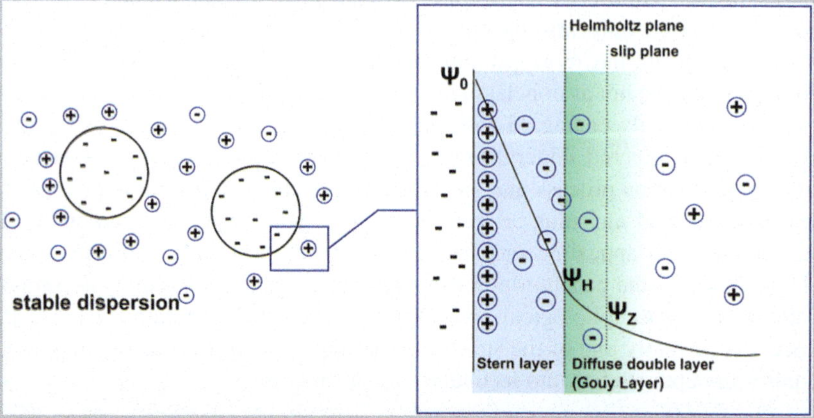

Effects of Heterocharged Polymers The effect of a charged polymeric additive on a dispersion depends on its interaction with the colloids. If the polyelectrolyte is heterocharged with respect to the colloid, the polymer can adsorb and lead to bridging flocculation by interconnecting the colloids. In the case of highly charged polymers, charge neutralization is the most common mechanism for bridging flocculation because these polyelectrolytes tend to adsorb in a flat configuration. If the colloids have a rather low surface charge density, adsorption of highly charged polyelectrolytes can provide an overall charge neutralization but lead to a heterogeneous surface charge distribution, which enhances coagulation by the so-called electrostatic patch effect. At very high concentrations of surface-bounded polymer, re-stabilization of the dispersion may occur by surface charge reversal.

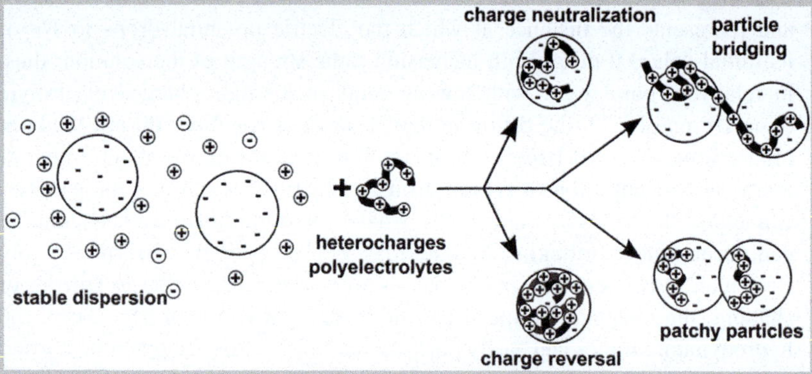

(continued)

Effects of Non-adsorbing Polymers Non-adsorbing polymers, e.g. homocharged polyelectrolytes, can cause so-called depletion effects which can occur in solutions in which the polymeric additive remains in the continuous medium and does not adsorb to the colloid's surface. The effect of depletion destabilization has been known since the early works of Traube in 1925 (Traube 1925); a theoretical explanation was first given by Asakura and Oosawa in 1958 (Asakura and Oosawa 1958). The driving force of depletion effects arises from the tendency of polymers to adopt the energetically most favoured conformation in solution, i.e. the minimization of the solution's free energy. If a free polymer experiences a confinement, e.g. by approaching closely an interface, a colloid or a second polymer, its conformational freedom is restrained which increases the free energy of the dispersion. Since this is unfavourable, each colloid is surrounded by a volume with a diameter δ from which non-adsorbing polymers are excluded, the so-called depletion layer. If the interparticle distance of two dispersed colloids becomes smaller than 2δ, – i.e. the depletion volumes overlap, e.g. due to an increase in the colloid's particle number density – the colloids force polymers to leave the interparticle space. The polymers diffuse out of the interparticle space and generate a polymer-excluded volume of pure solvent. This is energetically unfavourable: the solution's entropy is reduced by uphill diffusion, and the free energy of solvation arising from polymer-solvent interactions is diminished, especially if the liquid is a good solvent for the polymer. At low polymer concentration, polymer–polymer interactions are negligible, and the out-diffusion of the polymer and the concomitant gain in conformational freedom outbalance these costs. This results in a non-isotropic concentration gradient which creates a net osmotic pressure which pushes the colloids together and squeezes out the solvent: the dispersion experiences depletion destabilization and collapses. At high polymer concentrations, polymer–polymer interactions cannot be neglected anymore because the polymer's conformational freedom is already diminished by polymer–polymer interactions. The gain in conformational freedom of the polymer by out-diffusion cannot compensate for the cost of generating the polymer-excluded volume. Thus, the polymer stays in the interparticle space, enters the depletion volume and therefore exerts a counterforce on the colloidal particles which keeps them apart: the solution is stabilized by depletion stabilization.

(continued)

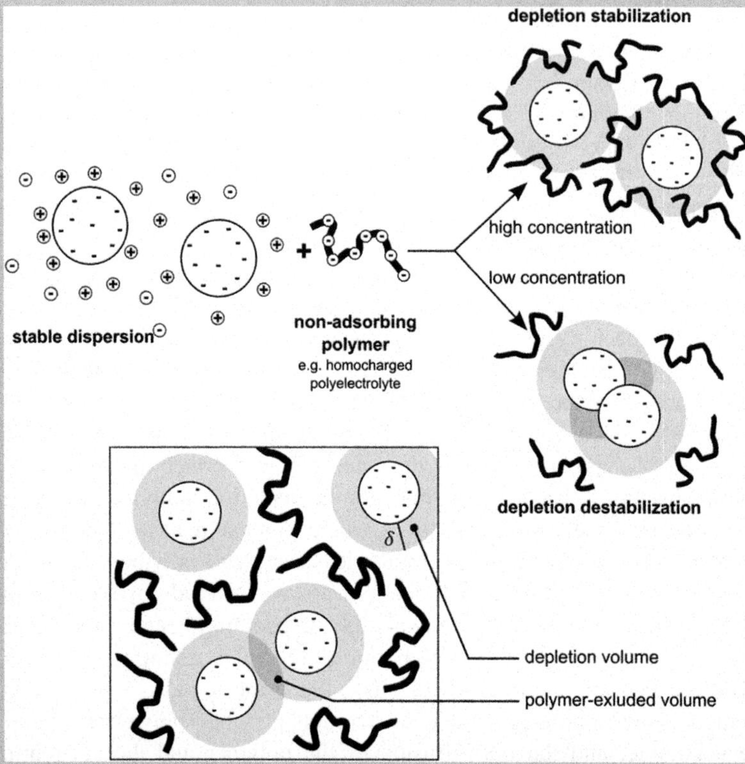

For further information, see Tadros (2013), Somasundaran (2006), Lagaly et al. (1997) or other pertinent monographies and reviews on colloido-chemistry and the DLVO theory.

Acknowledgements This material is based upon the work supported by the National Science Foundation (NSF) under Grant Number DMR-1309657 and by the National Institute of Diabetes and Digestive and Kidney Diseases (NIDDK) of the National Institutes of Health (NIH) under Award Number R01DK092311 and the National Institute of Dental and Craniofacial Research (NIDCR) Award Number 5R01DE016849-07. Any opinions, findings and conclusions or recommendations expressed in this material are those of the authors and do not necessarily reflect the views of the National Science Foundation or the National Institutes of Health.

SEW gratefully acknowledges financial support by an Emmy Noether research grant issued by the German Research Foundation (DFG, N° WO1712/3-1) and received further support by the Cluster of Excellence "Engineering of Advanced Materials—Hierarchical Structure Formation for Functional Devices" funded by the DFG (EXC 315).

References

Addadi L, Raz S, Weiner S (2003) Taking advantage of disorder: amorphous calcium carbonate and its roles in biomineralization. Adv Mater 15:959–970. doi:10.1002/adma.200300381

Aizenberg J, Lambert G, Addadi L, Weiner S (1996) Stabilization of amorphous calcium carbonate by specialized macromolecules in biological and synthetic precipitates. Adv Mater 8:222–226

Amos FF, Sharbaugh DM, Talham DR, Gower LB, Fricke M, Volkmer D (2007) Formation of single-crystalline aragonite tablets/films via an amorphous precursor. Langmuir 23:1988–1994. doi:10.1021/la061960n

Amos F, Dai L, Kumar R, Khan S, Gower LB (2009) Mechanism of formation of concentrically laminated spherules: implication to Randall's plaque and stone formation. Urol Res 37:11–17

Andreassen J-P, Lewis AE (2017) Classical and non-classical theories of crystal growth. In: Benning LG, Gebauer D, Kellermeier M, Van Driessche AES (ed) New perspectives on mineral nucleation and growth. Springer, Cham, pp 137–154

Asakura S, Oosawa F (1958) J Polym Sci 33:183–192

Balz M, Therese HA, Li J, Gutmann JS, Kappl M, Nasdala L, Hofmeister W, Butt H-J, Tremel W (2005) Crystallization of vaterite nanowires by the cooperative interaction of tailor-made nucleation surfaces and polyelectrolytes. Adv Funct Mater 15:683–688. doi:10.1002/adfm.200400333

Berg JK, Jordan T, Binder Y, Börner HG, Gebauer D (2013) Mg^{2+} tunes the wettability of liquid precursors of CaCO3: toward controlling mineralization sites in hybrid materials. J Am Chem Soc 135:12512–12515. doi:10.1021/ja404979z

Bergström L, Sturm (née Rosseeva) EV, Salazar-Alvarez G, Cölfen H (2015) Mesocrystals in biominerals and colloidal arrays. Acc Chem Res 150504141715009. doi: 10.1021/ar500440b

Bewernitz MA, Gebauer D, Long J, Cölfen H, Gower LB (2012) A metastable liquid precursor phase of calcium carbonate and its interactions with polyaspartate. Faraday Discuss 159:291. doi:10.1039/c2fd20080e

Burwell AK, Thula-Mata T, Gower LB, Habeliz S, Kurylo M, Ho SP, Chien Y-C, Cheng J, Cheng NF, Gansky SA, Marshall SJ, Marshall GW (2012) Functional remineralization of dentin lesions using polymer-induced liquid-precursor process. PLoS One 7:e38852. doi: 10.1371/journal.pone.0038852

Cahn JW, Hilliard JE (1959) Free energy of a nonuniform system. III. Nucleation in a two-component incompressible fluid. J Chem Phys 31:688

Cantaert B, Kim Y-Y, Ludwig H, Nudelman F, Sommerdijk NAJM, Meldrum FC (2012) Think positive: phase separation enables a positively charged additive to induce dramatic changes in calcium carbonate morphology. Adv Funct Mater 22:907–915. doi:10.1002/adfm.201102385

Cantaert B, Verch A, Kim Y-Y, Ludwig H, Paunov VN, Kröger R, Meldrum FC (2013) Formation and structure of calcium carbonate thin films and nanofibers precipitated in the presence of poly(allylamine hydrochloride) and magnesium ions. Chem Mater 25:4994–5003. doi:10.1021/cm403497g

Cheng X, Gower LB (2006) Molding mineral within microporous hydrogels by a polymer-induced liquid-precursor (PILP) process. Biotechnol Prog 22:141–149. doi:10.1021/bp050166+

Cheng X, Varona PL, Olszta MJ, Gower LB (2007) Biomimetic synthesis of calcite films by a polymer-induced liquid-precursor (PILP) process. J Cryst Growth 307:395–404. doi:10.1016/j.jcrysgro.2007.07.006

Cölfen H, Antonietti M (2008) Mesocrystals and nonclassical crystallization. Wiley-VCH, Chichester

Dai L, Cheng X, Gower LB (2008a) Transition bars during transformation of an amorphous calcium carbonate precursor. Chem Mater 20:6917–6928. doi:10.1021/cm800760p

Dai L, Douglas EP, Gower LB (2008b) Compositional analysis of a polymer-induced liquid-precursor (PILP) amorphous $CaCO_3$ phase. J Non-Cryst Solids 354:1845–1854. doi:10.1016/j.jnoncrysol.2007.10.022

Dauphin Y (2008) The nanostructural unity of Mollusc shells. Mineral Mag 72:243–246. doi:10.1180/minmag.2008.072.1.243
De Yoreo JJ, Sommerdijk N, Dove P (2017) Nucleation pathways in electrolyte solutions. In: Benning LG, Gebauer D, Kellermeier M, Van Driessche AES (ed) New perspectives on mineral nucleation and growth. Springer, Cham, pp 1–24
Demichelis R, Raiteri P, Gale JD, Quigley D, Gebauer D (2011) Stable prenucleation mineral clusters are liquid-like ionic polymers. Nat Commun 2:590. doi:10.1038/ncomms1604
Evans JS (2013) "Liquid-like" biomineralization protein assemblies: a key to the regulation of non-classical nucleation. CrystEngComm 15:8388. doi:10.1039/c3ce40803e
Faatz M (2005) Kontrollierte Fällung von amorphem Calciumcarbonat durch homogene Carbonatfreisetzung. Johannes-Gutenberg University of Mainz, Germany
Faatz M, Gröhn F, Wegner G (2004) Amorphous calcium carbonate: synthesis and potential intermediate in biomineralization. Adv Mater 16:996–1000. doi:10.1002/adma.200306565
Faatz M, Gröhn F, Wegner G (2005) Mineralization of calcium carbonate by controlled release of carbonate in aqueous solution. Mater Sci Eng C 25:153–159. doi:10.1016/j.msec.2005.01.005
Faraday Discussions Transcripts (2007) Transcript of the general discussion. In: Crystal growth and nucleation: Faraday Discussion N° 136. RSC Publishing
Faraday Discussions Transcripts (2012) Transcript of the general discussion. In: Crystallisation – a biological perspective: Faraday Discussion N° 159. RSC Publishing
Freeman CL (2014) Personal communication
Gal A, Habraken W, Gur D, Fratzl P, Weiner S, Addadi L (2013) Calcite crystal growth by a solid-state transformation of stabilized amorphous calcium carbonate nanospheres in a hydrogel. Angew Chem Int Ed 125:4967–4970. doi:10.1002/ange.201210329
Gal A, Kahil K, Vidavsky N, DeVol RT, Gilbert PUPA, Fratzl P, Weiner S, Addadi L (2014) Particle accretion mechanism underlies biological crystal growth from an amorphous precursor phase. Adv Funct Mater 24:5420–5426, 10.1002/adfm.201400676
Galkin O, Vekilov PG (2000a) Are nucleation kinetics of protein crystals similar to those of liquid droplets? J Am Chem Soc 122:156–163. doi:10.1021/ja9930869
Galkin O, Vekilov PG (2000b) Control of protein crystal nucleation around the metastable liquid-liquid phase boundary. Proc Natl Acad Sci U S A 97:6277–6281. doi:10.1073/pnas.110000497
Gebauer D, Völkel A, Cölfen H (2008) Stable prenucleation calcium carbonate clusters. Science 322:1819–1822. doi:10.1126/science.1164271
Gebauer D, Cölfen H, Verch A, Antonietti M (2009) The multiple roles of additives in $CaCO_3$ crystallization: a quantitative case study. Adv Mater 21:435–439. doi:10.1002/adma.200801614
Gebauer D, Gunawidjaja PN, Ko JYP, Bacsik Z, Aziz B, Liu L, Hu Y, Bergström L, Tai C-W, Sham T-K, Edén M, Hedin N (2010) Proto-calcite and proto-vaterite in amorphous calcium carbonates. Angew Chem Int Ed 122:9073–9075. doi:10.1002/ange.201003220
Gebauer D, Kellermeier M, Gale JD, Bergström L, Cölfen H (2014) Pre-nucleation clusters as solute precursors in crystallisation. Chem Soc Rev 43:2348–2371. doi:10.1039/c3cs60451a
Gehrke N, Nassif N, Pinna N, Antonietti M, Gupta HS, Cölfen H (2005) Retrosynthesis of nacre via amorphous precursor particles. Chem Mater 17:6514–6516. doi:10.1021/cm052150k
Gibbs JW (1877) On the equilibrium of heterogeneous substances. Trans Connecticut Acad Arts Sci 3:343–524
Gilman JJ (2009) Basalt columns: large scale constitutional supercooling? J Volcanol Geotherm Res 184:347–350. doi:10.1016/j.jvolgeores.2009.04.017
Gong YUT, Killian CE, Olson IC, Appathurai NP, Amasino AL, Martin MC, Holt LJ, Wilt FH, Gilbert PUPA (2012) Phase transitions in biogenic amorphous calcium carbonate. Proc Natl Acad Sci U S A 109:1–6. doi:10.1073/pnas.1118085109
Gower LA (1997) The influence of polyaspartate additive on the growth and morphology of calcium carbonate crystals. University of Massachusetts, Amherst
Gower LB (2008) Biomimetic model systems for investigating the amorphous precursor pathway and its role in biomineralization. Chem Rev 108:4551–4627. doi:10.1021/cr800443h
Gower LB, Odom D (2000) Deposition of calcium carbonate films by a polymer-induced liquid-precursor (PILP) process. J Cryst Growth 210:719–734. doi:10.1016/S0022-0248(99)00749-6

Gower LB, Tirrell DA (1998) Calcium carbonate films and helices grown in solutions of poly(aspartate). J Cryst Growth 191:153–160. doi:10.1016/S0022-0248(98)00002-5

Grahame DC (1947) The electrical double layer and the theory of electrocapillarity. Chem Rev 41:441–501. doi:10.1021/cr60130a002

Haberkorn H, Franke D, Frechen T, Goesele W, Rieger J (2003) Early stages of particle formation in precipitation reactions—quinacridone and boehmite as generic examples. J Colloid Interface Sci 259:112–126. doi:10.1016/S0021-9797(03)00024-9

Habraken W, Tao J, Brylka LJ, Friedrich H, Bertinetti L, Schenk AS, Verch A, Dmitrovic V, Bomans PHH, Frederik PM, Laven J, van der Schoot P, Aichmayer B, de With G, De Yoreo JJ, Sommerdijk NAJM (2013) Ion-association complexes unite classical and non-classical theories for the biomimetic nucleation of calcium phosphate. Nat Commun 4:1507. doi:10.1038/ncomms2490

Hardikar VV, Matijevic E (2001) Influence of ionic and nonionic dextrans on the formation of calcium hydroxide and calcium carbonate particles. Colloids Surf A Physicochem Eng Asp 186:23–31

Harris J, Mey I, Hajir M, Mondeshki M, Wolf SE (2015) Pseudomorphic transformation of amorphous calcium carbonate films follows spherulitic growth mechanisms and can give rise to crystal lattice tilting. CrystEngComm 17:6831–6837. doi:10.1039/C5CE00441A

Henriksen K, Young J, Bown P, Stipp S (2004) Coccolith biomineralisation studied with atomic force microscopy. Palaeontol 47:725–743. doi:10.1111/j.0031-0239.2004.00385.x

Homeijer SJ, Olszta MJ, Barrett RA, Gower LB (2008) Growth of nanofibrous barium carbonate on calcium carbonate seeds. J Cryst Growth 310:2938–2945. doi:10.1016/j.jcrysgro.2008.02.009

Homeijer SJ, Barrett RA, Gower LB (2010) Polymer-Induced Liquid-Precursor (PILP) process in the non-calcium based systems of barium and strontium carbonate. Cryst Growth Des 10:1040–1052. doi:10.1021/cg800918g

Hovden R, Wolf SE, Holtz ME, Marin F, Muller DA, Estroff LA (2015) Nanoscale assembly processes revealed in the nacroprismatic transition zone of *Pinna nobilis* mollusc shells. Nat Commun. doi:10.1038/ncomms10097

Ihli J, Wong WC, Noel EH, Kim Y-Y, Kulak AN, Christenson HK, Duer MJ, Meldrum FC (2014) Dehydration and crystallization of amorphous calcium carbonate in solution and in air. Nat Commun 5:3169. doi:10.1038/ncomms4169

Jacob DE, Wirth R, Soldati A, Wehrmeister U, Schreiber A (2011) Amorphous calcium carbonate in the shells of adult Unionoida. J Struct Biol 173:241–249. doi:10.1016/j.jsb.2010.09.011

Jee SS, Culver L, Li Y, Douglas EP, Gower LB (2010a) Biomimetic mineralization of collagen via an enzyme-aided PILP process. J Cryst Growth 312:1249–1256. doi:10.1016/j.jcrysgro.2009.11.010

Jee SS, Thula TT, Gower LB (2010b) Development of bone-like composites via the polymer-induced liquid-precursor (PILP) process. Part 1: influence of polymer molecular weight. Acta Biomater 6:3676–3686. doi:10.1016/j.actbio.2010.03.036

Jee SS, Kasinath RK, DiMasi E, Kim Y-Y, Gower LB (2011) Oriented hydroxyapatite in turkey tendon mineralized via the polymer-induced liquid-precursor (PILP) process. CrystEngComm 13:2077. doi:10.1039/c0ce00605j

Jiang Y, Gower LB, Volkmer D, Cölfen H (2011) Hierarchical DL-glutamic acid microspheres from polymer-induced liquid precursors. Cryst Growth Des 11:3243–3249. doi:10.1021/cg200504n

Jiang Y, Gong H, Grzywa M, Volkmer D, Gower LB, Cölfen H (2013) Microdomain transformations in mosaic mesocrystal thin films. Adv Funct Mater 23:1547–1555. doi:10.1002/adfm.201202294

Kababya S, Gal A, Kahil K, Weiner S, Addadi L, Schmidt A (2015) Phosphate–water interplay tunes amorphous calcium carbonate metastability: spontaneous phase separation and crystallization vs stabilization viewed by solid state NMR. J Am Chem Soc 137:990–998. doi:10.1021/ja511869g

Kellermeier M, Melero-García E, Glaab F, Klein R, Drechsler M, Rachel R, García-Ruiz JM, Kunz W (2010) Stabilization of amorphous calcium carbonate in inorganic silica-rich environments. J Am Chem Soc 132:17859–17866. doi:10.1021/ja106959p

Kellermeier M, Rosenberg R, Moise A, Anders U, Przybylski M, Cölfen H (2012) Amino acids form prenucleation clusters: ESI-MS as a fast detection method in comparison to analytical ultracentrifugation. Faraday Discuss 159:23. doi:10.1039/c2fd20060k

Killian CE, Metzler R, Gong YUT, Olson IC, Aizenberg J, Politi Y, Wilt FH, Scholl A, Young A, Doran A, Kunz M, Tamura N, Coppersmith SN, Gilbert PUPA (2009) Mechanism of calcite co-orientation in the sea urchin tooth. J Am Chem Soc 131:18404–18409. doi:10.1021/ja907063z

Kim Y-Y, Douglas EP, Gower LB (2007) Patterning inorganic $CaCO_3$ thin films via a polymer-induced liquid-precursor process. Langmuir 23:4862–4870

Kim Y-Y, Kulak A, Li Y, Batten T, Kuball M, Armes SP, Meldrum FC (2009) Substrate-directed formation of calcium carbonate fibres. J Mater Chem 19:387. doi:10.1039/b813101e

Kim Y-Y, Hetherington NBJ, Noel EH, Kröger R, Charnock JM, Christenson HK, Meldrum FC (2011) Capillarity creates single-crystal calcite nanowires from amorphous calcium carbonate. Angew Chem Int Ed 50:12572–12577. doi:10.1002/anie.201104407

Lagaly G, Schulz O, Ziemehl R (1997) Dispersionen und Emulsionen. Steinkopff, Darmstadt

LaMer VK, Dinegar RH (1950) Theory, production and mechanism of formation of monodispersed hydrosols. J Am Chem Soc 72:4847–4854

Leupold S, Wolf SE (2015) Unpublished Results.

Levi-Kalisman Y, Raz S, Weiner S, Addadi L, Sagi I (2002) Structural differences between biogenic amorphous calcium carbonate phases using X-ray absorption spectroscopy. Adv Funct Mater 12:43

Li H, Xin HL, Kunitake ME, Keene EC, Muller DA, Estroff LA, Muller A (2011) Calcite prisms from mollusk shells (Atrina rigida): swiss-cheese-like organic-inorganic single-crystal composites. Adv Funct Mater 21:2028–2034. doi:10.1002/adfm.201002709

Loges N, Graf K, Nasdala L, Tremel W (2006) Probing cooperative interactions of tailor-made nucleation surfaces and macromolecules: a bioinspired route to hollow micrometer-sized calcium carbonate particles. Langmuir 22:3073–3080. doi:10.1021/la0528596

Loste E, Wilson RM, Seshadri R, Meldrum FC (2003) The role of magnesium in stabilising amorphous calcium carbonate and controlling calcite morphologies. J Cryst Growth 254:206–218. doi:10.1016/S0022-0248(03)01153-9

Lutsko JF (2017) Novel paradigms in non-classical nucleation theory. In: Benning LG, Gebauer D, Kellermeier M, Van Driessche AES (ed) New perspectives on mineral nucleation and growth. Springer, Cham, pp 25–42

Ma Y, Mehltretter G, Plüg C, Rademacher N, Schmidt MU, Cölfen H (2009) PY181 pigment microspheres of nanoplates synthesized via polymer-induced liquid precursors. Adv Funct Mater 19:2095–2101. doi:10.1002/adfm.200900316

Mutvei H (1978) Ultrastructural characteristics of the nacre in some gastropods. Zool Scr 7:287–296

Navrotsky A (2004) Energetic clues to pathways to biomineralization: precursors, clusters, and nanoparticles. Proc Natl Acad Sci U S A 101:12096–12101. doi:10.1073/pnas.0404778101

Nielsen MH, Aloni S, De Yoreo JJ (2014) In situ TEM imaging of CaCO3 nucleation reveals coexistence of direct and indirect pathways. Science 345:1158–1162. doi: 10.1126/science.1254051

Nudelman F, Pieterse K, George A, Bomans PHH, Friedrich H, Brylka LJ, Hilbers PAJ, De With G, Sommerdijk NAJM (2010) The role of collagen in bone apatite formation in the presence of hydroxyapatite nucleation inhibitors. Nat Mater 9:9–14. doi:10.1038/NMAT2875

Olszta MJ, Douglas EP, Gower LB (2003a) Scanning electron microscopic analysis of the mineralization of type I collagen via a polymer-induced liquid-precursor (PILP) process. Calcif Tissue Int 72:583–591. doi:10.1007/s00223-002-1032-7

Olszta MJ, Odom D, Douglas EP, Gower LB (2003b) A new paradigm for biomineral formation: mineralization via an amorphous liquid-phase precursor. Connect Tissue Res 44:326–334. doi:10.1080/03008200390181852

Olszta MJ, Gajjeraman S, Kaufman M, Gower LB (2004) Nanofibrous calcite synthesized via a solution-precursor-solid mechanism. Chem Mater 16:2355–2362. doi:10.1021/cm035161r

Olszta MJ, Cheng X, Jee SS, Kumar R, Kim Y-Y, Kaufman MJ, Douglas EP, Gower LB (2007) Bone structure and formation: a new perspective. Mater Sci Eng R 58:77–116. doi:10.1016/j.mser.2007.05.001

Penn RL, Li D, Soltis JA (2017) A perspective on the particle-based crystal growth of ferric oxides, oxyhydroxides, and hydrous oxides. In: Benning LG, Gebauer D, Kellermeier M, Van Driessche AES (ed) New perspectives on mineral nucleation and growth. Springer, Cham

Petsev DN, Chen K, Gliko O, Vekilov PG (2003) Diffusion-limited kinetics of the solution-solid phase transition of molecular substances. Proc Natl Acad Sci U S A 100:792–796. doi:10.1073/pnas.0333065100

Pipich V, Balz M, Wolf SE, Tremel W, Schwahn D (2008) Nucleation and growth of $CaCO_3$ mediated by the Egg-white protein ovalbumin: a time-resolved in situ study using small-angle neutron scattering. J Am Chem Soc 130:6879–6892. doi:10.1021/ja801798h

Politi Y, Arad T, Klein E, Weiner S, Addadi L (2004) Sea urchin spine calcite forms via a transient amorphous calcium carbonate phase. Science 306:1161–1164. doi: 10.1126/science.1102289

Politi Y, Levi-Kalisman Y, Raz S, Wilt FH, Addadi L, Weiner S, Sagi I (2006) Structural characterization of the transient amorphous calcium carbonate precursor phase in sea urchin embryos. Adv Funct Mater 16:1289–1298. doi:10.1002/adfm.200600134

Politi Y, Metzler RA, Abrecht M, Gilbert B, Wilt FH, Sagi I, Addadi L, Weiner S, Gilbert PUPA (2008) Transformation mechanism of amorphous calcium carbonate into calcite in the sea urchin larval spicule. Proc Nat Acad Sci USA 105:17362–17366. doi:10.1073/pnas.0806604105

Pontoni D, Bolze J, Dingenouts N, Narayanan T, Ballauff M (2003) Crystallization of calcium carbonate observed in-situ by combined small- and wide-angle X-ray scattering. J Phys Chem B 107:5123–5125. doi:10.1021/jp0343640

Pontoni D, Bolze J, Ballauff M, Narayanan T, Cölfen H (2004) Time-resolved SAXS study of the effect of a double hydrophilic block-copolymer on the formation of $CaCO_3$ from a supersaturated salt solution. J Colloid Interface Sci 277:84–94. doi:10.1016/j.jcis.2004.04.029

Rao A, Cölfen H (2017) Mineralization schemes in the living world: mesocrystals. In: Benning LG, Gebauer D, Kellermeier M, Van Driessche AES (ed) New perspectives on mineral nucleation and growth. Springer, Cham

Rieger J, Frechen T, Cox G, Heckmann W, Schmidt C, Thieme J (2007) Precursor structures in the crystallization/precipitation processes of $CaCO_3$ and control of particle formation by polyelectrolytes. Faraday Discuss 136:265. doi:10.1039/b701450c

Rodriguez DE, Thula-Mata T, Toro EJ, Yeh Y-W, Holt C, Holliday LS, Gower LB (2014) Multifunctional role of osteopontin in directing intrafibrillar mineralization of collagen and activation of osteoclasts. Acta Biomater 10:494–507. doi:10.1016/j.actbio.2013.10.010

Rousseau M, Lopez E, Stempflé P, Brendlé M, Franke L, Guette A, Naslain R, Bourrat X (2005) Multiscale structure of sheet nacre. Biomaterials 26:6254–6262. doi:10.1016/j.biomaterials.2005.03.028

Schenk AS, Zope H, Kim Y-Y, Kros A, Sommerdijk NAJM, Meldrum FC (2012) Polymer-induced liquid precursor (PILP) phases of calcium carbonate formed in the presence of synthetic acidic polypeptides—relevance to biomineralization. Faraday Discuss 159:327. doi:10.1039/c2fd20063e

Schenk AS, Cantaert B, Kim Y-Y, Li Y, Read ES, Semsarilar M, Armes SP, Meldrum FC (2014) Systematic study of the effects of polyamines on calcium carbonate precipitation. Chem Mater 26:2703–2711. doi:10.1021/cm500523w

Sethmann I (2005) Observation of nano-clustered calcite growth via a transient phase mediated by organic polyanions: a close match for biomineralization. Am Mineral 90:1213–1217. doi:10.2138/am.2005.1833

Seto J, Ma Y, Davis SA, Meldrum FC, Gourrier A, Kim Y-Y, Schilde U, Sztucki M, Burghammer M, Maltsev S, Jäger C, Cölfen H (2011) Structure-property relationships of a biological mesocrystal in the adult sea urchin spine. Proc Natl Acad Sci U S A 109:1–6. doi:10.1073/pnas.1109243109

Shen Q, Wei H, Zhou Y, Huang Y, Yang H, Wang D, Xu D (2006) Properties of amorphous calcium carbonate and the template action of vaterite spheres. J Phys Chem B 110:2994–3000. doi:10.1021/jp055063o

Somasundaran P (ed) (2006) Encyclopedia of surface and colloid science, 2nd edn. CRC Press, Boca Raton

Sommerdijk NAJM, van Leeuwen ENM, Vos MRJ, Jansen JA (2007) Calcium carbonate thin films as biomaterial coatings using DNA as crystallization inhibitor. CrystEngComm 9:1209. doi:10.1039/b710277a

Sugawara A, Nishimura T, Yamamoto Y, Inoue H, Nagasawa H, Kato T (2006) Self-organization of oriented calcium carbonate/polymer composites: effects of a matrix peptide isolated from the exoskeleton of a crayfish. Angew Chem Int Ed 45:2876–2879. doi:10.1002/anie.200503800

Tadros T (ed) (2013) Encyclopedia of colloid and interface science. Springer Verlag, Heidelberg

Tobler DJ, Rodriguez-Blanco JD, Dideriksen K, Bovet N, Sand KK, Stipp SLS (2015) Citrate effects on Amorphous Calcium Carbonate (ACC) structure, stability, and crystallization. Adv Funct Mater 25:3081–3090. doi:10.1002/adfm.201500400

Toramaru A, Matsumoto T (2004) Columnar joint morphology and cooling rate□: a starch-water mixture experiment. 109:1–10. doi: 10.1029/2003JB002686

Traube J (1925) Gummi Ztg 39:434

Vekilov PG (2010) Nucleation. Cryst Growth Des 10:5007–5019. doi:10.1021/cg1011633

Vekilov PG (2012) Crystal nucleation: nucleus in a droplet. Nat Mater 11:838–840. doi:10.1038/nmat3441

Volkmer D, Harms M, Gower LB, Ziegler A (2005) Morphosynthesis of nacre-type laminated $CaCO_3$ thin films and coatings. Angew Chem Int Ed 44:639–644. doi:10.1002/anie.200461386

Wallace AF, Hedges LO, Fernandez-Martinez A, Raiteri P, Gale JD, Waychunas GA, Whitelam S, Banfield JF, De Yoreo JJ (2013) Microscopic evidence for liquid-liquid separation in supersaturated $CaCO_3$ solutions. Science 341:885–889. doi: 10.1126/science.1230915

Weiss IM, Tuross N, Addadi L, Weiner S (2002) Mollusc larval shell formation: amorphous calcium carbonate is a precursor phase for aragonite. J Exp Zool 293:478–491

Wohlrab S, Cölfen H, Antonietti M (2005) Crystalline, porous microspheres made from amino acids by using polymer-induced liquid precursor phases. Angew Chem Int Ed 44:4087–4092. doi:10.1002/anie.200462467

Wolf SE, Leiterer J, Kappl M, Emmerling F, Tremel W (2008) Early homogenous amorphous precursor stages of calcium carbonate and subsequent crystal growth in levitated droplets. J Am Chem Soc 130:12342–12347. doi:10.1021/ja800984y

Wolf SE, Leiterer J, Pipich V, Barrea R, Emmerling F, Tremel W (2011a) Strong stabilization of amorphous calcium carbonate emulsion by ovalbumin: gaining insight into the mechanism of "polymer-induced liquid precursor" processes. J Am Chem Soc 133:12642–12649. doi:10.1021/ja202622g

Wolf SE, Müller L, Barrea R, Kampf CJ, Leiterer J, Panne U, Hoffmann T, Emmerling F, Tremel W (2011b) Carbonate-coordinated metal complexes precede the formation of liquid amorphous mineral emulsions of divalent metal carbonates. Nanoscale 3:1158–1165. doi:10.1039/c0nr00761g

Wolf SE, Lieberwirth I, Natalio F, Bardeau J-F, Delorme N, Emmerling F, Barrea R, Kappl M, Marin F (2012) Merging models of biomineralisation with concepts of nonclassical crystallisation: is a liquid amorphous precursor involved in the formation of the prismatic layer of the Mediterranean Fan Mussel Pinna nobilis? Faraday Discuss 159:433–448. doi:10.1039/b000000x

Wolf SE, Böhm C, Harris J, Hajir M, Mondeshki M, Marin F (2015) Single nanogranules preserve intracrystalline amorphicity in biominerals. Key Eng Mater 672:47–59. doi:10.4028/www.scientific.net/KEM.672.47

Wolf SE, Böhm CF, Harris J, Demmert B, Jacob DE, Mondeshki M, Ruiz-Agudo E, Navarro CR (2016) Nonclassical crystallization in vivo et in vitro (I): process-structure-property relationships of nanogranular biominerals. J Struct Biol. doi:10.1016/j.jsb.2016.07.016

Yang H, ter Horst JH (2017) Crystal nucleation of small organic molecules. In: Benning LG, Gebauer D, Kellermeier M, Van Driessche AES (ed) New perspectives on mineral nucleation and growth. Springer, Cham

Yang L, Killian CE, Kunz M, Tamura N, Gilbert PUPA (2011) Biomineral nanoparticles are space-filling. Nanoscale 3:603–609. doi:10.1039/c0nr00697a

Zhang TH, Liu XY (2007) How does a transient amorphous precursor template crystallization. J Am Chem Soc 129:13520–13526

Zhong C, Chu CC (2009) Acid polysaccharide-induced Amorphous Calcium Carbonate (ACC) films: colloidal nanoparticle self-organization process. Langmuir 25:3045–3049. doi:10.1021/la803541m

Zhu J-H, Yu S-H, Xu A-W, Cölfen H (2009) The biomimetic mineralization of double-stranded and cylindrical helical BaCO(3) nanofibres. Chem Commun 1106–1168

Chapter 4
Structural Characteristics and the Occurrence of Polyamorphism in Amorphous Calcium Carbonate

Alejandro Fernandez-Martinez, Hugo Lopez-Martinez, and Dongbo Wang

4.1 Introduction: Why Are the Structures of Amorphous Precursors Important?

Engineering new materials to fulfill desired functions requires knowledge of their structure in order to reproduce the atomic mechanisms controlling the material's properties. Amorphous materials and liquids do not escape this requirement. A number of experimental and computational methods have been developed over the last century that allow for determination of local order in amorphous materials and liquids, under a variety of external constraints such as pressure, temperature, humidity, etc. Of special interest in the fields of liquids and amorphous materials is understanding how liquid or amorphous structures are related to the crystalline polymorphs of the same composition. Structural relationships established between the ordered structure of a crystalline polymorph and its corresponding liquid phase (formed upon melting of the former) often show that the average local order (the atomic arrangement around every atom in the structure) is maintained upon melting. An example is that of liquid cesium. A first-order phase transition is observed upon

A. Fernandez-Martinez (✉)
University Grenoble Alpes, CNRS, ISTerre, F-38000 Grenoble, France
e-mail: Alex.Fernandez-Martinez@univ-grenoble-alpes.fr

H. Lopez-Martinez
Université Grenoble Alpes, ISTerre, F-38041 Grenoble, France

Departamento de Física, Universidad de Oviedo, 33007 Oviedo, Spain
e-mail: correodehugo92@gmail.com

D. Wang
National Institute of Standards and Technology, Gaithersburg, MD 20899, USA
e-mail: dongbo.wang@nist.gov

A.E.S. Van Driessche et al. (eds.), *New Perspectives on Mineral Nucleation and Growth*,
DOI 10.1007/978-3-319-45669-0_4

compression at 3.9 GPa at 220 and 350 °C. This transition is associated with a change in the coordination number from 12 to 8, similar to what is observed in the crystalline system when passing from the Cs II (fcc) to the Cs IV phase (tetragonal) (Falconi et al. 2005). This kind of transitions in both liquid and solid systems has been reported for a variety of systems including water (Mishima et al. 1985), phosphorous (Katayama et al. 2000), or sodium (Raty et al. 2007) to name a few.

While these phenomena are clearly interesting, focusing on understanding the physical basis for liquid polymorphism has strong implications for the area of amorphous precursors in biomineralization and for development of new biomimetic materials. The amorphous precursor strategy is a process by which many organisms mold their skeletons and shells with a given shape, occupying space in the confining membranes, making use of the "flexible" and "moldable" hydrated amorphous structure, before crystallization takes place (Addadi et al. 2003). The existence of liquid polymorphism and of polyamorphism is relevant, as one could predict that the local order in the liquid or in the precursor amorphous structure could be maintained after crystallization. However, this whole process, although not well understood, is also thought to be controlled by the confining matrix, acting as a template for crystallization (Vekilov 2010; Giuffre et al. 2013; Hamm et al. 2014), and by the presence of different impurities (proteins, inorganic and organic ions), that exert controls over the kinetics of crystallization and over the polymorph selection (Rodriguez-Blanco et al. 2012; Ihli et al. 2013; Sun et al. 2015). Therefore, a first step toward understanding this crystallization process would be to robustly understand the structure of the amorphous precursor.

4.2 The Structure of Amorphous Calcium Carbonate

The structural and energetic characteristics of amorphous calcium carbonate (ACC) were described using a variety of analytical techniques: nuclear magnetic resonance (NMR) (Gebauer et al. 2010; Singer et al. 2012; Reeder et al. 2013), infrared and Raman spectroscopies (Addadi et al. 2003; Rodriguez Blanco et al. 2008; Politi et al. 2004; Wehrmeister et al. 2011; Wang et al. 2012), thermogravimetry and calorimetry analyses (Radha et al. 2010, 2012; Schmidt et al. 2014), X-ray absorption (Levi-Kalisman et al. 2002; Michel et al. 2008) and high-energy X-ray scattering combined with pair distribution function (PDF) analyses (Michel et al. 2008; Goodwin et al. 2010; Radha et al. 2012; Cobourne et al. 2014), and reverse Monte Carlo (RMC) modeling (Goodwin et al. 2010; Fernandez-Martinez et al. 2013). Results concerning the local order structure of ACC converge toward a picture where Ca^{2+} ions are in an average six- to eightfold coordination, and water has at least two components that can be classified as adsorbed and structural (restricted mobility and high energy of desorption). However, no studies have conclusively offered a picture useful for constructing function-structure relationships. This is due at least in part to the complexity of naturally occurring ACC, most of which is biological in origin. Such ACC has biomolecules incorporated in structural "sites" that are difficult to describe.

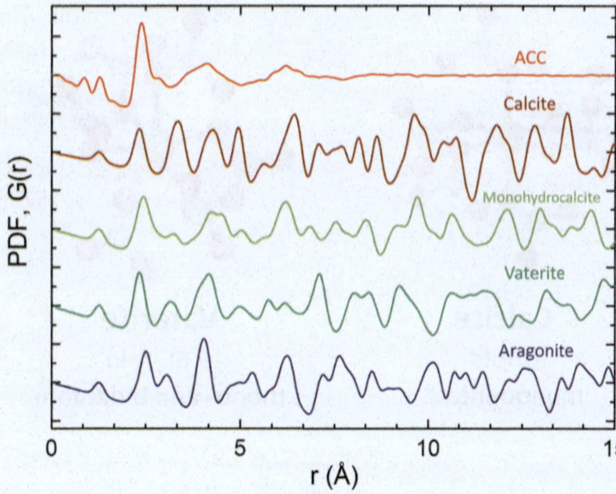

Fig. 4.1 Pair distribution functions of ACC and of some of the crystalline polymorphs of $CaCO_3{\cdot}nH_2O$ (n = 0, 1). The trigonal vaterite structure (space group $P3_221$), shown by Demichelis et al. (2012) to be the more stable, has been chosen

While crystalline materials are defined by the presence of periodicity in their atomic structures, which translates into the presence of diffraction peaks in their diffraction pattern, amorphous materials lack long-range order and thus peak in their diffraction pattern. Such materials only preserve an average local order that can be described through other techniques, such as the pair distribution function (PDF) technique (see Fig. 4.1). An amorphous material is at a metastable state formed by a large ensemble of configurational microstates. The PDF gives a map of the average interatomic distances over this ensemble (a radial distribution function). It is thus a useful way to describe the local order of amorphous precursors and to ascertain whether its structure is different from that of crystalline polymorphs of the same composition. An example is given in Fig. 4.1, where the PDFs of ACC, calcite, aragonite, vaterite, and monohydrocalcite are shown. It can be seen that the local order in ACC does not fit any of the crystalline polymorphs.

From the PDF data, it is not likely that ACC is a nanocrystalline "version" of crystalline calcium carbonate. However, other studies have pointed some similarities between ACC and some of the $CaCO_3$ polymorphs. For instance, some studies further examined the structural motifs in ACC using X-ray and neutron scattering, X-ray absorption spectroscopy (XAS), and other experimental probes sensitive to short-range order, such as NMR. Gebauer et al. (2010) showed using carbon-13 nuclear magnetic resonance (^{13}C-NMR) and XAS that the solution pH influences local order of the final ACC, with vaterite-like motifs present in ACC precipitated from more alkaline solutions and calcite-like ACC at less alkaline solutions (Gebauer et al. 2010). Fernandez-Martinez et al. (2013) noted also that the Ca-Ca partial PDF obtained from a RMC analysis of high-energy X-ray scattering data of high-pH synthetic ACC shows two broad oscillations at $\sim$4 and $\sim$6 Å, resembling those of vaterite except for a shift ($\sim$0.2 Å) to a longer distance for the second shell (Fernandez-Martinez et al. 2013). This study also shows that the Ca-O partial PDF of ACC shows a sharp peak at 2.48 Å that corresponds to the first shell of

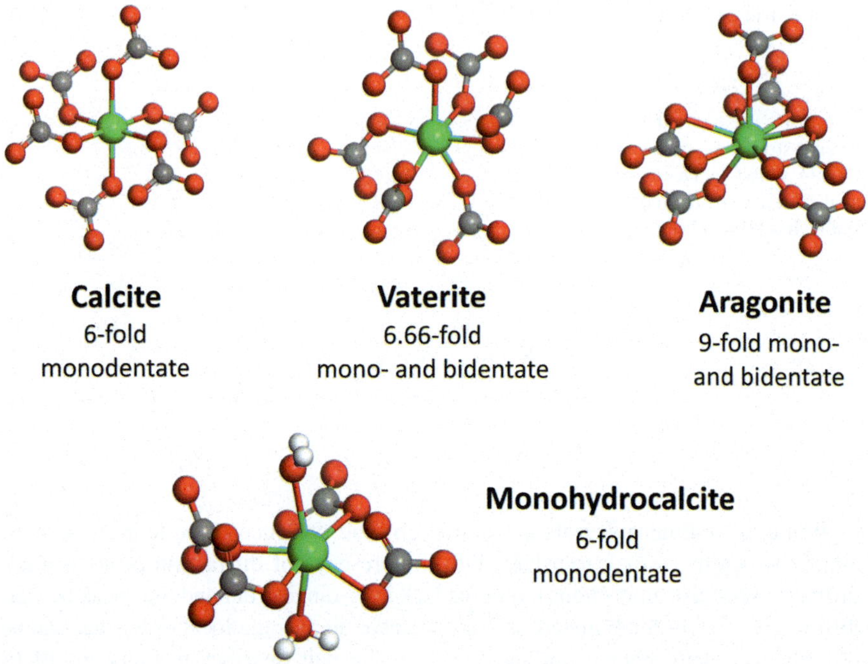

Fig. 4.2 Ball-and-stick representation of Ca^{2+} local environment in some of the more common calcium carbonate polymorphs. The planar carbonate molecule CO_3^{2-}, with D_{3h} symmetry, can be bound to a Ca^{2+} ion by sharing one or two oxygen atoms with the Ca^{2+} coordination sphere, i.e., through a monodentate or a bidentate ligand, respectively. The Ca^{2+} coordination number and the more common ligand type are shown. The coordination number given for vaterite is an average taken from the different structures that have been proposed

oxygen atoms coordinating Ca^{2+} ions, a distance that falls in between the distance for vaterite (2.36 Å) and aragonite (2.51 Å). Other two Ca-O distances at ~4 and ~6.2 Å indicate a disordered structure with no recognizable interatomic distances shared by the crystalline polymorphs (Fig. 4.2).

Summed together, the studies by Gebauer et al. (2010), Michel et al. (2008), and Goodwin et al. (2010) point to the idea of a disordered vaterite-like mixed with calcite-like local order in ACC. The existence of many different structures for vaterite, and the recent finding that the free energies of formation of some of the most stable structure types fall within the range of thermal energy at room temperature (Demichelis et al. 2012; Demichelis et al. 2017, Chap. 6), suggests that many possible bonding geometries are possible for the Ca-CO_3 system, which are possibly quenched during the formation of ACC from saturated solutions. The inclusion of water molecules in the structure possibly adds steric constraints resulting in disordered structures for ACC with no long-range order.

The PDF data clearly suggests that ACC is likely not a highly disordered crystalline calcium carbonate, though some authors have continued to argue that ACC is composed of an assemblage of 1 nm size calcite nanocrystals, with a high defect density (Rez et al. 2014). While this concept does not seem to be widely accepted in the literature (as noted above, results from other techniques indicate the opposite), the presence of small nanocrystalline domains in biological and synthetic ACC should not be completely ruled out (Wehrmeister et al. 2011). TEM imaging and electron diffraction are possibly the best suitable techniques to test this particular hypothesis (see De Yoreo et al. 2017, Chap. 1 and Nielsen and De Yoreo 2017, Chap. 18), but TEM data have to be examined with caution due to the high radiation doses received by the materials under study and the possible beam-induced crystallization.

4.3 Water in ACC

Water has an important role in (1) the structure, stability, and formation of ACC and (2) it has been implicated as critical to the ACC to crystalline calcium carbonate transformation.

4.3.1 Water as a Structural Component in ACC

Most of the studies dealing with the structure of amorphous calcium carbonates have reported characteristics related to the atomic structure of the Ca^{2+} and $CO_3{}^{2-}$ ions. Only a few studies have dealt with the structure and energetics of water, in spite of its abundance in ACC and other precursors, with values up to 1.4 H_2O per $CaCO_3$ in ACC (Radha et al. 2012). Water has been implicated as the key to the structural stability of ACC (Raiteri and Gale 2010) through formation of stable H bonds with carbonate molecules (Saharay et al. 2013). A recent NMR study reported that water in synthetic ACC has four discrete components: (1) translationally rigid structural H_2O, (2) restrictedly mobile H_2O, (3) fluidlike mobile H_2O that is decoupled from rigid H and C, and (4) hydroxyl (Schmidt et al. 2014). The same group reported that only the structural H_2O components (2) and (3) are present in biogenic ACC (Reeder et al. 2013). Molecular modeling work performed by Raiteri and Gale (2010) showed that ACC surface roughness is a key step allowing for the cluster aggregation. This mechanism involves the disruption of the surface layer of adsorbed water. They proposed that water content per formula unit increases proportionally with ACC cluster size. On the contrary, Wallace et al. (2013) and (Tribello et al. 2009) showed that ACC clusters dehydrate partially as they grow, with most water being trapped between clusters during aggregation. Until now, experimental evidence for the short-range structure and the dynamics of water in the ACC structure is still lacking.

Understanding the mesoscale structure of ACC (i.e., the topology of H_2O, Ca^{2+}, and CO_3^{2-} densities) is of interest to know how inorganic and organic impurities can be incorporated. Scanning electron microscopy (SEM) images revealed a lobular structure at the scale of hundreds of nanometers, with ACC forming spherical particles of 20–200 nm diameter size (Koga et al. 1998). This mesoscopic picture gives an idea about ACC's porosity, explaining the high content of labile adsorbed water readily identified in TGA studies (Radha et al. 2012; Schmidt et al. 2014). TGA and ^{1}H-NMR have also revealed the existence of water molecules with restricted mobility. From the structural point of view, the question remains about the internal distribution of H_2O vs. $CaCO_3$ inside the ACC spheres. At the nanoscale, a study by Goodwin et al. (2010) described the ACC structure as a "nanoporous charge-separated framework formed by Ca-rich hydrated regions and Ca-poor regions, where carbonate ions and water molecules are not bound to Ca ions." This description resulted from a reverse Monte Carlo (RMC) analysis of high-energy X-ray diffraction data. Further work by Singer et al. (2012) using molecular dynamics simulations showed that charge separation in that framework was not stable and that the structure reordered after a few simulation steps forming one that resulted more homogeneous in density and charge distribution. The simulation results by Wallace et al. (2013) showed similar homogeneous ACC particles. This result highlights the necessary caution in analyzing RMC simulations. The typical length of the "charge-separated pores" described by Goodwin et al. (2010) is about 1–2 nm. However, this range of distances is not typically represented in high-energy X-ray scattering data of amorphous materials (i.e., lacking diffraction peaks, or periodicity). This points to the need to include small-angle X-ray or neutron scattering data in the modeling of these structures, in order to obtain information about porosity or density fluctuations in the structure.

4.3.2 The Role of Water in the ACC to Crystalline Transformation

A decrease of the water content in ACC during its transformation to vaterite or calcite was reported. A transient ACC phase, often named "anhydrous ACC," due to its low water content as compared to synthetic or biogenic ACC (Raz et al. 2003), was shown to form prior to crystallization, with the crystallization kinetics depending upon the impurity content (Ihli et al. 2013). A recent study linked the size of ACC spheres with their solubility and with the polymorph selection: smaller particle sizes (~66 nm) were obtained at higher supersaturations that recrystallize into vaterite when put in aqueous solution (Zou et al. 2015). On the contrary, larger particle sizes (~200 nm) were obtained at lower supersaturations, with calcite being the final polymorph. This study reported also a value for the interfacial free energy of ACC of $\alpha_{ACC} = 0.33$ J/m^2. This value is higher than the interfacial free energy values usually reported for calcite, which are in the order of 0.1 J/m^2 (see (Fernandez-Martinez et al. 2014) and references therein).

Questions remain open as to what drives this dehydration in biological systems, if other control factor is present rather than simple thermodynamics. The water content seems to be inversely proportional to the thermodynamic stability of ACC (Radha et al. 2010). These observations, together with the results from Wallace et al. (2013) and Raiteri and Gale (2010), suggest a continuous evolution from a dense, liquid-like calcium carbonate system, in which polymeric units of $Ca\text{-}CO_3\text{-}H_2O$ (as those described by Raiteri and Gale 2010) evolve to more condensed atomic complexes or clusters (as those described by Wallace et al. 2013), with most of the water being expelled out of the clusters (as suggested by Tribello et al. 2009), while the final solid ACC is formed through a cluster aggregation process. This process is favorable thermodynamically, as shown by Wallace et al. (2013). The final ACC precipitate still contains a high water content, but most of it (up to 75 %) is water that is loosely bound, as shown by Schmidt et al. (2014). Questions remain open about the role of the remaining "structural" water in ACC's thermodynamic stability and kinetic persistence.

4.4 Ionic and Molecular Impurities in ACC

Impurities may have a strong influence over the structure and energetics of ACC. Magnesium impurities in calcite are probably one of the most studied ones. Magnesium is known to control the polymorph selection in crystalline calcium carbonates—calcite vs. aragonite—and to heavily influence the rate of calcite growth (Davis et al. 2000). Radha et al. (2012) reported that the inclusion of Mg^{2+} in ACC resulted in increased thermodynamic stability, in addition to the known kinetic persistence of Mg-bearing ACC. The complete compositional diagram, ranging from pure ACC to pure amorphous magnesium carbonate (AMC), was studied using PDF analyses, TGA coupled with evolved gas FTIR, and dissolution calorimetry. The results show that at low Mg^{2+} contents ($\sim$10 % at. Mg^{2+}), similar to those found in most biominerals, Mg^{2+}-ACC is more stable (see also Chap. 5 by Rodriguez Blanco et al. 2017, this volume). This behavior is similar to that observed for Mg^{2+} incorporation in calcite at low Mg^{2+} contents (see Fig. 4.3), suggesting that Mg^{2+} content in biominerals may be dictated by thermodynamics. Furthermore, the absolute maximum in stability is found at a concentration of $\sim$50 mol % Mg^{2+}, corresponding to the composition of the stable polymorph dolomite. Amorphous solids with Mg^{2+} contents over 50 % are composed of two phases, most likely AMC and the 50 % Mg-ACC. This study highlights the role that Mg^{2+}, a common impurity in biogenic ACC, has as stabilizer of the amorphous structure. It also highlights the ill-defined concept of "amorphous solid solution." By definition, a solid solution is a mixture in which one compound is distributed within the crystalline lattice of another compound. As shown by Radha et al. (2012),

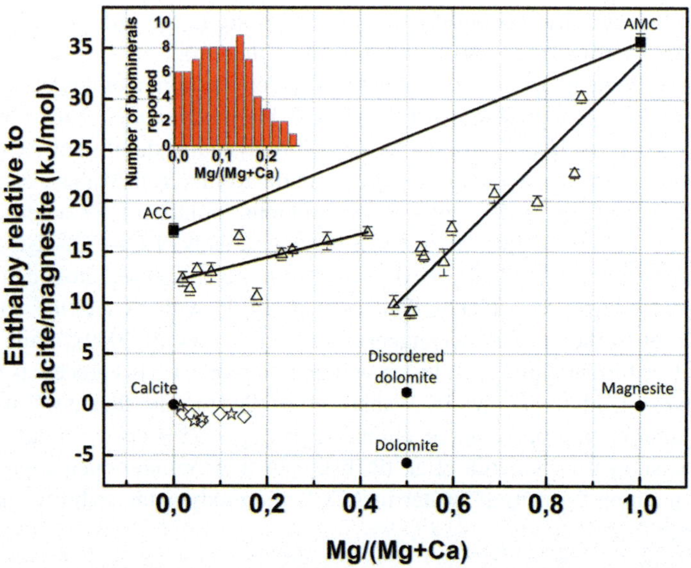

Fig. 4.3 Enthalpy of formation of amorphous and crystalline compounds with compositions $Ca_xMg_{1\text{-}x}CO_3{\cdot}nH_2O$. Mg^{2+} abundance in biogenic ACC is shown in the inset. Data for amorphous compounds is represented in *empty triangles* and *filled squares*. *Empty diamonds* and *stars* correspond to synthetic magnesian calcite data by (Bischoff 1998) and (Navrotsky and Capobianco 1987). Reprinted from Radha et al. (2012), with permission from Elsevier

the coherent domain sizes in Mg-bearing ACC are not longer than 1 nm. These authors also showed that the local order in Mg-ACC differs from that of pure ACC, with Mg-O bond lengths shorter than Ca-O ones, as expected (2.4 vs. 2.1 Å). A possible interpretation of this observation is that, at the atomic scale, the amorphous material is composed of Ca-rich regions or domains spatially separated from Ca domains, each of them of less than 1 nm in size making them by definition a physical mixture. However, the behavior of Mg-bearing ACC from the point of view of energetics (measured enthalpies of formation) describes a situation in which there is an energetic gain in stability associated with the addition of small percentages of Mg^{2+}, as observed in the $Mg_xCa_{1\text{-}x}CO_3$ solid solution (see Fig. 4.3; the enthalpy-composition curve of the amorphous system shows a similar behavior to that of the crystalline system). One possible rationalization for this observation is that, if we consider that the solid is composed of sub-nm Mg-rich and Ca-rich domains, the interfacial free energy contribution to the total free energy of formation would be high, possibly adding a negative term constituting the enthalpy of mixing. More research is needed not only to ascertain the structural arrangements at the atomic scale but also to correlate these structural descriptions with observation from other techniques, such as dissolution calorimetry, which gives information at the macroscale.

Ionic and molecular impurities affect polymorph selection mechanisms in different ways: (i) they can shift the thermodynamic equilibrium of one phase with respect to another; (ii) they can act on the crystallization kinetics, poisoning (or enhancing) the growth of one phase and thus favoring (or inhibiting) the formation of another one.

4.5 Polyamorphism in ACC

The phenomenon of polyamorphism, or the occurrence of different short-range order structures for the same substance, remains quite unexplored in the case of ACC, with only a handful of papers showing experimental results (see review by (Cartwright et al. 2012), where different experimental evidences pointing to the existence of ACC polyamorphism are shown). One of the first papers suggesting the presence of different atomic arrangements in ACC referred to it as: "an amorphous calcium carbonate phase in which the structure of aragonite is already preformed" (Hasse et al. 2000). However, studies dealing with the short-range order of biogenic ACC should be taken with caution due to the existence of impurities such as co-ions (Mg^{2+}, Na^{+} ...), organics such as proteins or carboxylic acids, polyphosphates, or even variable water contents, which have an influence on its local order (Politi et al. 2008; Sinha and Rez 2015). Along these lines, many studies on biogenic ACC do not report impurity or water contents, making it more difficult to catalog different ACC "polyamorphs" and to ascertain composition-structure-function relationships unequivocally. ACC polyamorphism and perhaps more importantly the physicochemical factors determining it remain therefore a quite unexplored field of research.

Defining polyamorphism is not a simple task due to the numerous effects that can create artifacts in the definition of a pure amorphous structure, such as the presence of nanocrystalline domains. Also, some criteria used to define ACC polyamorphism, such as the comparison of Ca-O bond lengths from EXAFS (see (Cartwright et al. 2012) and references therein), are not restrictive enough and do not suffice to declare a structure as a different calcium carbonate polyamorph. This is due to the large distribution of bond lengths in amorphous materials that are affected by structural and thermal disorder, as well as to bond anharmonicity. To our knowledge, only two studies have reported clear evidence for ACC polyamorphism from spectroscopic or scattering data: the study by Gebauer et al. (2010) using ^{13}C-NMR showed different local orders in ACC as a function of the solution pH value, with vaterite-like and calcite-like short-range order present at high and low pH, respectively. The second study by Fernandez-Martinez et al. (2013) used RMC modeling of X-ray total scattering data as a function of applied pressure to describe the formation of aragonite-like ACC at high pressures and a first-order polyamorphic transition with a critical pressure of $P_c \sim 10$ GPa (see Fig. 4.4).

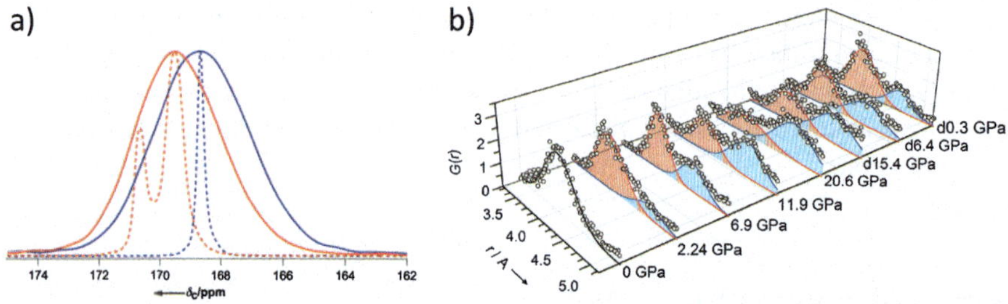

Fig. 4.4 Experimental proofs of polyamorphism. (**a**) ^{13}C solid-state NMR spectra of calcite (*dashed blue lines*), vaterite (*dashed red lines*), proto-calcite ACC (*blue line*) and proto-vaterite ACC (*red line*) (Gebauer et al. 2010); (**b**) Ca-Ca partial pair distribution function (pPDF) of ACC at different hydrostatic pressures. The double peak observed at high pressures matches the Ca-Ca pPDF of aragonite. Reprinted with permission of John Wiley and Sons from Gebauer et al. (2010) and Fernandez-Martinez et al. (2013)

A useful way to describe polyamorphism is to rationalize the structural arrangements of Ca and CO_3 molecules by determining the number of monodentate vs. bidentate ligands in Ca-CO_3 units. This was done in a couple of RMC-based studies (Goodwin et al. 2010; Fernandez-Martinez et al. 2013), where the results were compared to the bonding geometry of Ca^{2+} and $CO_3{}^{2-}$ in crystalline polymorphs. Calcite, the most stable crystalline $CaCO_3$ polymorph, is formed by $CaCO_3$ groups linked through monodentate ligands (see Fig. 4.2). Most of the structures that have been reported for vaterite show equivalent number of monodentate and bidentate ligands. RMC analyses by Goodwin et al. (2010) and Fernandez-Martinez et al. (2013) reported an approximate number of 50 % for each of these two molecular configurations. In addition, Fernandez-Martinez et al. (2013) showed that this proportion changes as a function of the external pressure, with the number of bidentate ligands increasing with pressure, reaching values close to those of aragonite (the $CaCO_3$ polymorph stable at high pressure). It is therefore possible that the changes in these proportions can be the underlying reason for the polymorph selection (e.g., more bidentate ligands favors aragonite over calcite). Further studies are necessary to systematically address the effects of external factors (pressure, temperature, humidity, solution pH value, etc.) on ACC polyamorphism (Fernandez-Martinez et al. 2013).

The significance of ACC polyamorphism is also under debate (Cartwright et al. 2012). The question of whether or not distinct structural motifs in the amorphous material are intimately related to the structure of the $CaCO_3$ crystalline polymorph formed ultimately remains an open question, with multiple ramifications. Underneath this question lies a more fundamental unknown about the mechanism of crystallization of the amorphous structure itself. Two mechanisms are usually considered: a dissolution-reprecipitation process (Nielsen et al. 2014) or a solid state transition (Politi et al. 2008; Gong et al. 2012; Ihli et al. 2014). The relevance of the

ACC polyamorphism seems thus to be directly dependent on the ACC crystallization mechanism. In a dissolution-reprecipitation crystallization mechanism, the structure of the amorphous precursor itself could become irrelevant. In contrast, the ACC local structure would be determinant if a solid state transition is the preferred crystallization pathway, due to the supposed similarities in the local order of the amorphous and the final crystalline structure. Of course, the answer to the question is that both mechanisms are relevant, each of them under different conditions. Recent in situ TEM investigations showed that dissolution-reprecipitation of ACC is the dominant mechanism of ACC crystallization in aqueous solution (Nielsen et al. 2014), a fact that seems logical given its high solubility (Kellermeier et al. 2014; see also Rodriguez-Blanco et al. 2017, Chap. 5). Solid state transitions at relevant temperatures have been less studied (Politi et al. 2008; Gong et al. 2012; Ihli et al. 2014). A different type of pathway that has until now only been tested by Ihli et al. (2014) (to the best of our knowledge) could be also relevant here: several experimental studies reported the existence of interface-mediated processes that result in coupled interfacial dissolution-reprecipitation of minerals at mineral-water interfaces (Ruiz-Agudo et al. 2014; Hellmann et al. 2015). In situ studies showed that these mechanisms lead to mineral replacement through precipitation processes occurring under conditions where the bulk fluid is undersaturated with respect to the precipitating phase (Ruiz-Agudo et al. 2014). The explanation for this behavior is based on the fact that ion concentrations at water-mineral interfaces differ from those at the bulk due to the existence of an electric double layer. The thermodynamic concept of bulk fluid supersaturation is therefore not useful here, and a precise theoretical account for these phenomena is still lacking due to the existing limitations to describe the structure and dynamics of the electric double layer. Given the hydrous nature of ACC, it could be hypothesized that similar phenomena could be at play at the ACC-water interface. Going beyond this picture of the dissolution-reprecipitation, one could hypothesize as well that if interfacial dissolution is only partial, structural motifs of the original ACC could persist during the reprecipitation process. This would make ACC polyamorphism a relevant phenomenon, given the “survival” of the structural features of the amorphous phase during the crystallization process.

Some recent studies on pre-nucleation clusters suggest that the local order in the amorphous precipitate is inherited from that of the clusters from which it is formed (Habraken et al. 2013; Wallace et al. 2013). This involves thus the presence of an aggregation-based process of pre-nucleation clusters, which were shown to be accompanied by cluster dehydration (Raiteri and Gale 2010; Wallace et al. 2013). Although, the structure of the pre-nucleation clusters has not been described yet, the different short-range ordering observed in the final ACC suggests that the clusters could have variable structures depending on the solution pH value from which they are formed (Gebauer et al. 2010).

A similar type of mechanism has been observed in a different family of amorphous precursors: amorphous calcium phosphate (ACP). Habraken et al. (2013) showed that ACP nucleation on collagen goes through a nonclassical

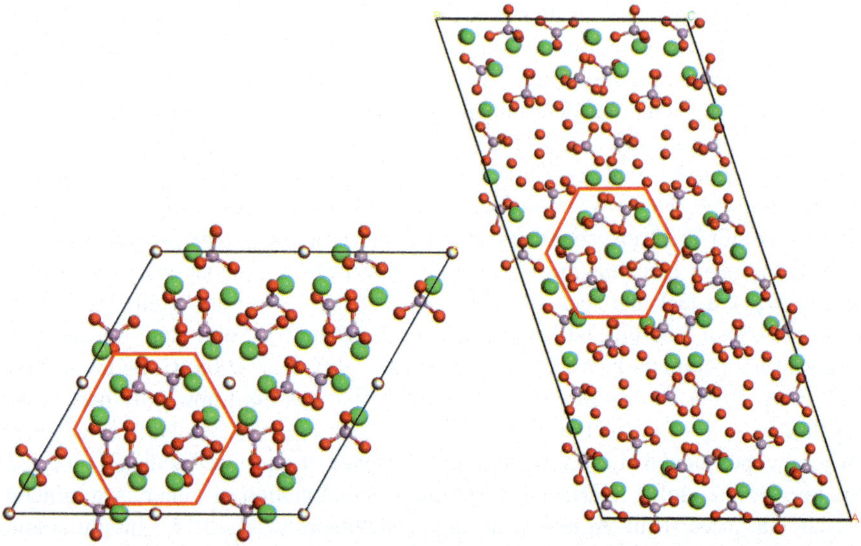

Fig. 4.5 *Left*: 2 × 2 × 1 supercell of hydroxyapatite (Hughes 1989). *Right*: 2 × 2 × 1 supercell of octacalcium phosphate (Brown 1962). The hexagons in both structures highlight the same cluster of atoms formed by a central Ca^{2+} and six neighboring $PO_4{}^{3-}$ anions. The main difference between both structures comes from the different packing of these atomic clusters

mechanism involving the formation of pre-nucleation clusters of $[Ca(HPO_4)_3]^{4-}$ that aggregate in solution forming different fractal arrangements and eventually precipitate forming ACP. The structure of these clusters was to be constant, with only changes in their aggregation geometry being responsible for the resulting crystalline polymorph formed. Actually, similar atomic arrangements at the local scale can be observed in two of the polymorphs in the calcium phosphate system, octacalcium phosphate and hydroxyapatite (see Fig. 4.5), so the fact that they are formed from the same type of clusters is not surprising.

As mentioned above, the presence of impurities in ACC and their effect on the local order can be an impediment to properly define polyamorphism. The existence of an "aragonitic" ACC was suggested by different authors in biogenic ACC (Hasse et al. 2000; Marxen et al. 2003). However, none of these studies reported the chemical compositions, making it more difficult to ascertain whether true ACC polyamorphism is present in those systems or not (most likely not, given the usual high concentrations of other cations such as Mg^{2+} in biominerals at about 10 % (Radha et al. 2012)). This is an important point, because the polymorphism and polyamorphism are circumscribed by definition to a single compositional phase. The question remains open as to whether structural variations in ACC are caused by external physicochemical factors that induce different local order in the $CaCO_3{\cdot}H_2O$ phase (i.e., true polyamorphism), or if the different structures are due to the presence of other ions or molecules.

4.6 Outlook

The existence of amorphous precursors to calcium carbonates and calcium phosphates has been known for some 40 or more years. But it has been only in the last 10 years that the community has made significant progress in understanding the structure-property relationships in amorphous precursors. This is partly due to the continuous development of advanced characterization tools that allow obtaining unprecedented molecular-scale information on atomic structures and processes. Some of these tools include in situ electron microscopy, modern X-ray scattering methods, 2D NMR on quadrupolar nuclei such as ^{43}Ca, and molecular dynamics simulations on very large systems. However, much work is still needed to be able to construct a detailed picture of the ACC system. Efforts need to be directed to the development of an ACC model that is testable and predictive of structure-function relationships. An implicit problem to the study of highly disordered phases is our difficulty to ascertain to what extent the amorphous structure contains structural motifs of its crystalline counterparts. This is especially important because we still do not know how polymorph selection is controlled in calcifying organisms. Related to this is the need of building an understanding of the role of impurities. These are responsible for ACC's kinetic behavior and therefore are of great interest for the development of applications such as the synthesis of functional biomimetic materials. More studies need to be addressed to the study of the dynamic behavior of ACC, with emphasis on the structure and dynamics of water. ACC is a highly hydrated compound, with a dynamic structural behavior. Open questions remaining include (1) how the presence of ionic or organic impurities affects atomistic dynamics and (2) what their effect is on the crystallization kinetics. These questions need to be addressed with the aid of advanced characterization techniques, both from an experimental and theoretical perspective, in multidisciplinary approaches, including the contributions from the biological sciences.

Acknowledgments The authors are very grateful to Gaston Garbarino (ESRF) for enriching discussions about pressure-induced polyamorphism. Funding was provided by the EC2CO (Biohefect) "NUCLEATION" program (INSU, CNRS).

References

Addadi L, Raz S, Weiner S (2003) Taking advantage of disorder: amorphous calcium carbonate and its roles in biomineralization. Adv Mater 15:959–970

Bischoff WD (1998) Dissolution enthalpies of magnesian calcites. Aquat Geochem 4:321–326

Brown WE (1962) Octacalcium phosphate and hydroxyapatite: Crystal structure of octacalcium phosphate. Nature 196:1048–1050

Cartwright JHE, Checa AG, Gale JD, Gebauer D, Sainz-Díaz CI (2012) Calcium carbonate polyamorphism and its role in biomineralization: how many amorphous calcium carbonates are there? Angew Chem Int Ed 51:11960–11970

Cobourne G, Mountjoy G, Rodriguez-Blanco JD, Benning LG, Hannon AC, Plaisier JR (2014) Neutron and X-ray diffraction and empirical potential structure refinement modelling of magnesium stabilised amorphous calcium carbonate. J Non-Cryst Solids 401:154–158

Davis KJ, Dove PM, De Yoreo JJ (2000) The role of Mg2+ as an impurity in calcite growth. Science 290:1134–1137

De Yoreo JJ, Sommerdijk NAJM, Dove PM (2017) Nucleation pathways in electrolyte solutions. In: Van Driessche AES, Kellermeier M, Benning LG, Gebauer D (eds) New perspectives on mineral nucleation and growth, Springer, Cham, pp 1–24

Demichelis R, Raiteri P, Gale JD, Dovesi R (2012) A new structural model for disorder in vaterite from first-principles calculations. CrystEngComm 14:44

Demichelis R, Raiteri P, Gale JD (2017) Ab Initio modelling of the structure and properties of crystalline calcium carbonate. In: Van Driessche AES, Kellermeier M, Benning LG, Gebauer D (eds) New perspectives on mineral nucleation and growth, Springer, Cham, pp 113–136

Falconi S, Lundegaard LF, Hejny C, McMahon MI (2005) X-ray diffraction study of liquid Cs up to 9.8 GPa. Phys Rev Lett 94:9–12

Fernandez-Martinez A, Kalkan B, Clark SJ, Waychunas GA (2013) Pressure-induced polyamorphism and formation of "aragonitic" amorphous calcium carbonate. Angew Chem Int Ed 52:8354–8357

Fernandez-Martinez A, Hu Y, Lee B, Jun Y-S, Waychunas GA (2014) In situ determination of interfacial energies between heterogeneously nucleated CaCO3 and quartz substrates: thermodynamics of CO2 mineral trapping. Environ Sci Technol 47:102–109

Gebauer D, Gunawidjaja PN, Ko JYP, Bacsik Z, Aziz B, Liu L, Hu Y, Bergström L, Tai C-W, Sham T-K et al (2010) Proto-calcite and proto-vaterite in amorphous calcium carbonates. Angew Chem Int Ed 49:8889–8891

Giuffre AJ, Hamm LM, Han N, De Yoreo JJ, Dove PM (2013) Polysaccharide chemistry regulates kinetics of calcite nucleation through competition of interfacial energies. Proc Natl Acad Sci U S A 110:9261–9266

Gong YUT, Killian CE, Olson IC, Appathurai NP, Amasino AL, Martin MC, Holt LJ, Wilt FH, Gilbert PUPA (2012) Phase transitions in biogenic amorphous calcium carbonate. Proc Natl Acad Sci U S A 109:6088–6093

Goodwin AL, Michel FM, Phillips BL, Keen DA, Dove MT, Reeder RJ (2010) Nanoporous structure and medium-range order in synthetic amorphous calcium carbonate. Chem Mater 22:3197–3205

Habraken WJEM, Tao J, Brylka LJ, Friedrich H, Bertinetti L, Schenk AS, Verch A, Dmitrovic V, Bomans PHH, Frederik PM et al (2013) Ion-association complexes unite classical and non-classical theories for the biomimetic nucleation of calcium phosphate. Nat Commun 4:1507

Hamm LM, Giuffre AJ, Han N, Tao J, Wang D, De Yoreo JJ, Dove PM (2014) Reconciling disparate views of template-directed nucleation through measurement of calcite nucleation kinetics and binding energies. Proc Natl Acad Sci U S A 111:1304–1309

Hasse B, Ehrenberg H, Marxen JC, Becker W, Epple M (2000) Calcium carbonate modifications in the mineralized shell of the freshwater snail Biomphalaria glabrata. Chem Eur J 6:3679–3685

Hellmann R, Cotte S, Cadel E, Malladi S, Karlsson LS, Lozano-Perez S, Cabié M, Seyeux A (2015) Nanometre-scale evidence for interfacial dissolution-reprecipitation control of silicate glass corrosion. Nat Mater 14:307–311

Hughes JM, Cameron M, Crowley KD (1989) Structural variations in natural F, OH, and Cl apatites. Am Mineral 74:870–876

Ihli J, Kim Y-Y, Noel EH, Meldrum FC (2013) The effect of additives on Amorphous Calcium Carbonate (ACC): Janus behavior in solution and the solid state. Adv Funct Mater 23:1575–1585

Ihli J, Wong WC, Noel EH, Kim Y-Y, Kulak AN, Christenson HK, Duer MJ, Meldrum FC (2014) Dehydration and crystallization of amorphous calcium carbonate in solution and in air. Nat Commun 5:3169

Katayama Y, Mizutani T, Utsumi W, Shimomura O, Yamakata M, Funakoshi K (2000) A first-order liquid-liquid phase transition in phosphorus. Nature 403:170–173

Kellermeier M, Picker A, Kempter A, Cölfen H, Gebauer D (2014) A straightforward treatment of activity in aqueous CaCO3 solutions and the consequences for nucleation theory. Adv Mater (Deerfield Beach, Fla.) 26:752–757

Koga N, Nakagoe YZ, Tanaka H (1998) Crystallization of amorphous calcium carbonate. Thermochim Acta 318:239–244

Levi-Kalisman Y, Raz S, Weiner S, Addadi L, Sagi I (2002) Structural differences between biogenic amorphous calcium carbonate phases using X-ray absorption spectroscopy. Adv Funct Mater 12:43–48

Marxen JC, Becker W, Finke D, Hasse B, Epple M (2003) Early mineralization in Biomphalaria glabrata: microscopic and structural results. J Molluscan Stud 69:113–121

Michel FM, MacDonald J, Feng J, Phillips BL, Ehm L, Tarabrella C, Parise JB, Reeder RJ (2008) Structural characteristics of synthetic amorphous calcium carbonate. Chem Mater 20:4720–4728

Mishima O, Calvert LD, Whalley E (1985) An apparently first-order transition between two amorphous phases of ice induced by pressure. Nature 314:76–78

Navrotsky A, Capobianco C (1987) Enthalpies of formation of dolomite and of magnesian calcites. Am Mineral 72:782–787

Nielsen MH, De Yoreo JJ (2017) Liquid phase TEM investigations of crystal nucleation, growth, and transformation. In: Van Driessche AES, Kellermeier M, Benning LG, Gebauer D (eds) New perspectives on mineral nucleation and growth, Springer, Cham, pp 353–371

Nielsen MH, Aloni S, De Yoreo JJ (2014) In situ TEM imaging of CaCO3 nucleation reveals coexistence of direct and indirect pathways. Science 345:1158–1162

Politi Y, Arad T, Klein E, Weiner S, Addadi L (2004) Sea urchin spine calcite forms via a transient amorphous calcium carbonate phase. Science 306:1161–1164

Politi Y, Metzler RA, Abrecht M, Gilbert B, Wilt FH, Sagi I, Addadi L, Weiner S, Gilbert P (2008) Transformation mechanism of amorphous calcium carbonate into calcite in the sea urchin larval spicule. Proc Natl Acad Sci U S A 105:17362–17366

Radha AV, Forbes TZ, Killian CE, Gilbert PUPA, Navrotsky A (2010) Transformation and crystallization energetics of synthetic and biogenic amorphous calcium carbonate. Proc Natl Acad Sci 107:16438–16443

Radha AV, Fernandez-Martinez A, Hu Y, Jun YS, Waychunas GA, Navrotsky A (2012) Energetic and structural studies of amorphous Ca1-xMgxCO3 nH2O ($0 \leq x \leq 1$). Geochim Cosmochim Acta 90:83–95

Raiteri P, Gale JD (2010) Water is the key to nonclassical nucleation of amorphous calcium carbonate. J Am Chem Soc 132:17623–17634

Raty J-Y, Schwegler E, Bonev SA (2007) Electronic and structural transitions in dense liquid sodium. Nature 449:448–451

Raz S, Hamilton PC, Wilt FH, Weiner S, Addadi L (2003) The transient phase of amorphous calcium carbonate in sea urchin larval spicules: the involvement of proteins and magnesium ions in its formation and stabilization. Adv Funct Mater 13:480–486

Reeder RJ, Tang Y, Schmidt MP, Kubista LM, Cowan DF, Phillips BL (2013) Characterization of structure in biogenic amorphous calcium carbonate: pair distribution function and nuclear magnetic resonance studies of lobster gastrolith. Cryst Growth Des 13:1905–1914

Rez P, Sinha S, Gal A (2014) Nanocrystallite model for amorphous calcium carbonate. J Appl Crystallogr 47:1651–1657

Rodriguez-Blanco JD, Shaw S, Benning LG (2008) How to make 'stable' ACC: protocol and preliminary structural characterization. Mineral Mag 72:283–286

Rodriguez-Blanco JD, Shaw S, Bots P, Roncal-Herrero T, Benning LG (2012) The role of pH and Mg on the stability and crystallization of amorphous calcium carbonate. J Alloys Compd 536:S477–S479

Rodriguez-Blanco JG, Sand KK, Benning LG (2017) ACC and vaterite as intermediates in the solution-based crystallization of $CaCO_3$. In: Van Driessche AES, Kellermeier M, Benning LG, Gebauer D (eds) New perspectives on mineral nucleation and growth, Springer, Cham, pp 93–112

Ruiz-Agudo E, Putnis CV, Putnis A (2014) Coupled dissolution and precipitation at mineral–fluid interfaces. Chem Geol 383:132–146

Saharay M, Yazaydin AO, Kirkpatrick RJ (2013) Dehydration-induced amorphous phases of calcium carbonate. J Phys Chem B 117:3328–3336

Schmidt MP, Ilott AJ, Phillips BL, Reeder RJ (2014) Structural changes upon dehydration of amorphous calcium carbonate. Cryst Growth Des 14:938–951

Singer JW, Yazaydin AÖ, Kirkpatrick RJ, Bowers GM (2012) Structure and transformation of amorphous calcium carbonate: a solid-state 43 Ca NMR and computational molecular dynamics investigation. Chem Mater 24:1828–1836

Sinha S, Rez P (2015) Distortions of the calcite and aragonite atomic structures from interstitial water. Mater Chem Phys 157:56–62

Sun W, Jayaraman S, Chen W, Persson KA, Ceder G (2015) Nucleation of metastable aragonite CaCO3 in seawater. Proc Natl Acad Sci U S A 112:3199–3204

Tribello GA, Bruneval F, Liew C, Parrinello M (2009) A molecular dynamics study of the early stages of calcium carbonate growth. J Phys Chem B 113:11680–11687

Vekilov PG (2010) The two-step mechanism of nucleation of crystals in solution. Nanoscale 2:2346–2357

Wallace AF, Hedges LO, Fernandez-Martinez A, Raiteri P, Gale JD, Waychunas GA, Whitelam S, Banfield JF, De Yoreo JJ (2013) Microscopic evidence for liquid-liquid separation in supersaturated CaCO3 solutions. Science 341:885–889

Wang D, Hamm LM, Bodnar RJ, Dove PM (2012) Raman spectroscopic characterization of the magnesium content in amorphous calcium carbonates. J Raman Spectrosc 43:543–548

Wehrmeister U, Jacob DE, Soldati AL, Loges N, Häger T, Hofmeister W (2011) Amorphous, nanocrystalline and crystalline calcium carbonates in biological materials. J Raman Spectrosc 42:926–935

Zou Z, Bertinetti L, Politi Y, Jensen ACS, Weiner S, Addadi L, Fratzl P, Habraken WJEM (2015) Opposite particle size effect on amorphous calcium carbonate crystallization in water and during heating in air. Chem Mater 27:4237–4246

Chapter 5
ACC and Vaterite as Intermediates in the Solution-Based Crystallization of $CaCO_3$

Juan Diego Rodriguez-Blanco, Karina K. Sand, and Liane G. Benning

5.1 Introduction

Calcium carbonate minerals are ubiquitous on Earth where they play a key role in many marine and terrestrial biomineralization processes (e.g., see also Chap. 9 by Falini and Fermani in this book). Their ubiquitous nature makes them key players in controlling a large part of the global carbon cycle. At ambient temperatures, they very often crystallize from solution via two intermediate phases: amorphous calcium carbonate (ACC) and vaterite.

ACC is a poorly ordered material (50–500 nm) (Rodriguez-Blanco et al. 2008) that in its pure form (no other cations but Ca^{2+} present at formation) consist of a Ca-rich framework with 1-nm diameter interconnected pores that contain water and carbonate ions (Goodwin et al. 2010). In recent years, ACC has been shown to

J.D. Rodriguez-Blanco (✉)
Department of Geology, Trinity College Dublin, Dublin 2, Ireland

NanoScience Center, Department of Chemistry, University of Copenhagen, Copenhagen, Denmark
e-mail: jblanco@nano.ku.dk; j.d.rodriguez-blanco@tcd.ie

K.K. Sand
NanoScience Center, Department of Chemistry, University of Copenhagen, Copenhagen, Denmark

Physical Sciences Division, Pacific Northwest National Laboratories, Richland, WA, USA

L.G. Benning
German Research Center for Geosciences, GFZ, Interface Geochemistry Section, 14473 Potsdam, Germany

Department of Earth Sciences, Free University of Berlin, 12249 Berlin, Germany

School of Earth and Environment, University of Leeds, Leeds LS2 9JT, UK

A.E.S. Van Driessche et al. (eds.), *New Perspectives on Mineral Nucleation and Growth*,
DOI 10.1007/978-3-319-45669-0_5

exhibit a complex chemical make up with variable hydration states linked to its "structure" at the short-range scale (Radha et al. 2010). Laboratory-synthesized ACC is often highly hydrated (<1.6 mols of water per mole of $CaCO_3$) and thermodynamically unstable. In the absence of additives, and when equilibrated with the solution, ACC transforms extremely fast (<2 min at 25° C; Rodriguez-Blanco et al. 2011) to crystalline $CaCO_3$ polymorphs. This transformation can take multiple, and sometimes complex, pathways that are dependent on the physico-chemical environment. Conversely, biogenic ACC is often initially hydrated, but during aging of ACC-rich biominerals (e.g., spine spicules; Gong et al. 2012), it slowly dehydrates concomitantly increasing its short-range order. In contrast to synthetic ACC, biogenic ACC can persist for days in the animal (Gong et al. 2012) and months when extracted and stored in dry conditions (Politi et al. 2008; Gong et al. 2012). Similarly carbonate concretions that are released by certain earthworm species (e.g., Brinza et al. 2014; Hodson et al. 2015) contain ACC stabilized for very long time periods, yet the reason of this stabilization is still unknown.

The main problem is the fact that the degree of ordering, hydration, particle size, and crystallization of ACC are largely controlled by its origin and conditions of formation. These include a number of factors like temperature, pH, and concentration of foreign ions in solution (e.g., Mg^{2+}, $SO_4{}^{2-}$, $PO_4{}^{3-}$, organics, etc.). For example, pure ACC usually transforms to calcite via a vaterite intermediate at low temperatures (<30 °C) and via aragonite at higher temperatures (>60 °C). However, small variations in initial pH of the aqueous solution (Tobler et al. 2016) or the presence of divalent ions (e.g., Mg^{2+}, Sr^{2+}) or organics (e.g., aspartic acid, glutamic acid, citric acid, etc.) can dramatically affect the crystallization rates and pathways of ACC, also resulting in the incorporation of variable amounts of these ions in the ACC structure (e.g., Wang et al. 2009; Rodriguez-Blanco et al. 2014; Tobler et al. 2015).

So far most studies focused on the formation and crystallization of ACC in abiotic systems (e.g., Ogino et al. 1987; Faatz et al. 2004; Radha et al. 2010; Rodriguez-Blanco et al. 2011). However, many biomineralization processes that ultimately lead to calcite or aragonite as the prime calcium carbonate phase occur within a biological membrane where (1) the activity of water is lower compared to a "pure" aqueous solution and (2) there is a high concentration of organic macromolecules, which fulfill a wide variety of functions (e.g., energy storage, structural protection, insulation). It is well known that the two most abundant $CaCO_3$ biominerals (i.e., calcite and aragonite) have remarkable morphologies, defined crystallographic orientations, and extraordinary functional properties (Dove et al. 2003). However, many biomineralizing organisms use the ACC to crystalline calcium carbonate pathway to control particle shape, crystallographic orientation, and crystalline polymorph during the formation of their shells, spines, etc. One of the best-known examples of ACC produced in biomineralization processes is found in sea urchins, which produce elongated single crystals of calcite with the c-axis aligned parallel to the length of the spicules. However, important for this chapter is the fact that they do this by controlling the deposition and crystallization of ACC within a biological membrane (Beniash et al. 1997). Similarly, ACC has also been found in the intestinal lumen of gilt-head sea bream *Sparus aurata* (Foran et al.

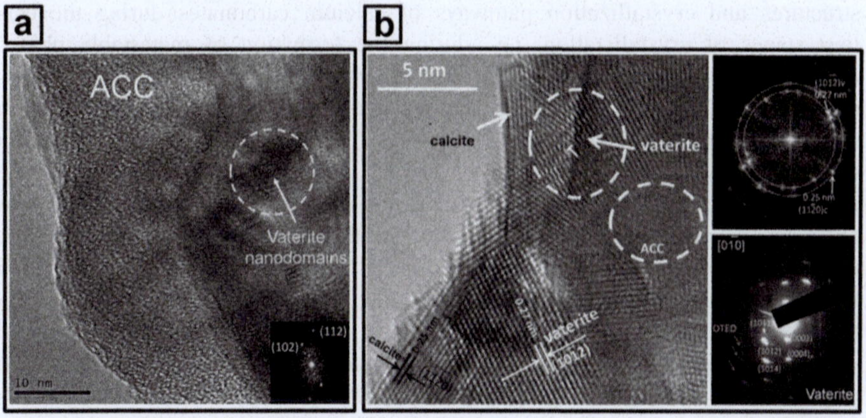

Fig. 5.1 $CaCO_3$ biomineralization produced by earthworms. (**a**) spherulitic deposit showing an ACC region with some minor contributions of crystalline phases in the form of small clusters (*inset circle*: selected area electron diffraction diagram of a vaterite single crystal represented along the [−2 0 1] zone axis). (**b**) Area of the spherulite showing vaterite nanocrystals partially transformed to calcite (calcite domains tend to be oriented with lattice planes parallel to the spherulite's edge and the regions between vaterite domains are separated by amorphous areas). The selected area electron diffraction diagram shown in the *inset* at the *bottom* of this figure corresponds to a single crystal of vaterite viewed along [0−1 0] showing these features. The *top inset* shows the fast Fourier transform analysis of the whole micrograph revealing a powderlike pattern (See more details in Gago-Duport et al. 2008)

2013), in some plants as the main component of cystoliths (Gal et al. 2012) and in earthworms' calciferous glands (Gago-Duport et al. 2008; Hodson et al. 2015). On the other hand, vaterite formed from ACC is often unstable. Nevertheless, although vaterite is rarely preserved in geologic settings (compared to calcite and aragonite), it is more common as a product of biomineralization. For example, vaterite has been found in earthworms calciferous glands (e.g., Fig. 5.1; Gago-Duport et al. 2008; Brinza et al. 2014), otoliths (Wehrmeister et al. 2011), lackluster pearls (Qiao et al. 2007), or spicules of tropical ascidian (Lowenstam and Abbott 1975; Kabalah-Amitai et al. 2013) or in aberrant growth of mollusk shells repaired after an injury (Isaure et al. 2010).

In recent years, these unusual biomineralization processes have attracted considerable interest because of their potential to be replicated, and controlled, in the laboratory. By using different methods (e.g., molecular templates, tailored self-assembly mechanisms), it could be possible to manufacture industrial $CaCO_3$ nanoparticles with specific sizes and shapes that would result in many applications for material sciences, biomedical research, food, agriculture, etc. For example, the crystallization processes of carbonates have great potential for incorporation of proteins or pharmaceutical compounds (e.g., insulin; Fujiwara et al. 2010), opening the door to simple and cost-effective methods for storage and targeted drug-delivery applications (e.g., vaterite nanocapsules; Parakhonskiy et al. 2012). All these applications require an understanding of the fundamental factors controlling the structures and crystallization pathways of calcium carbonates during the earliest stages of crystallization, i.e., during the formation of metastable phases like ACC and vaterite. However, the instability and rapid transformation tendency of ACC (e.g., in the pure system, seconds–minutes) can make an in depth characterization very difficult. Furthermore, to follow the rapid transformation reactions to crystalline carbonate, especially when they proceed via short-lived intermediate phases (e.g., vaterite, aragonite) is also not trivial. However, over the last few years, new experimental and characterization approaches have been developed, combining classical characterization techniques with methods that allow in situ and real-time monitoring of the reactions (e.g., time-resolved synchrotron-based scattering and diffraction, in situ liquid cell high-resolution microscopy, time-resolved spectrophotometry, in situ titrations, and ultracentrifugation; Rodriguez-Blanco et al. 2011; Nielsen et al. 2014; see also Chap. 18 by Nielsen and De Yoreo in this book; Burke et al. 2015; Tobler et al. 2015). These techniques have provided very useful data to attain the mechanisms and quantify the kinetics of ACC crystallization in abiotic systems. In addition, they provided a detailed understanding of how calcium carbonate phases form during biomineralization processes when the system moves from ACC toward thermodynamic equilibrium with calcite.

5.2 Mechanisms of Crystallization of Vaterite and Calcite from ACC

Since the 1970s, vaterite was assumed to be a possible precursor phase during the formation of $CaCO_3$ in sediments (Rowlands and Webster 1971). This assertion was confirmed by experimental observations showing that vaterite forms prior to calcite, at moderate or high supersaturations (Ogino et al. 1987). In addition, now it is well known that natural ACC crystallizes to vaterite in solution. Yet, the debate about the mechanisms of crystallization of vaterite from ACC, also linked to the complexity of determining its precise mineral structure (Kabalah-Amitai et al. 2013) and understanding its thermodynamic stability (Wang and Becker 2012), is in part still ongoing.

Just recently, cryo-TEM studies revealed that the transformation from ACC to vaterite can occur through a direct solid-state transformation. Crystalline domains have been observed after the aggregation of ACC nanoparticles into larger particles, and this was coupled with dehydration of the precursor phase. This combined aggregation and crystallization of ACC is considered to result in the development of randomly oriented crystalline domains that forms the typically observed spherical aggregates (Pouget et al. 2009; Pouget et al. 2010).

However, most studies consider that the ACC–vaterite transformation is a solution-mediated transformation mechanism that consists of ACC dissolution coupled to vaterite crystallization (e.g., Shen et al. 2006). Over the last decade, a number of nucleation and growth studies have taken advantage of the developments in synchrotron-based scattering and diffraction technologies that allow quantifying such reactions in situ, and real time, for more realistic conditions. Such methods have been previously tested on other minerals systems like amorphous silica (Tobler et al 2009; Tobler and Benning 2011), iron compounds (e.g., sulfides: Cahill et al. 2000; oxides: Davidson et al. 2008; Vu et al. 2010; Brinza et al. 2015) and sulfates (Ahmed et al. 2010). All these studies followed the formation and transformation of phases in situ and in real time based on changes in diffraction or scattering properties of the solids that were reacting in the aqueous solutions. Recently, this approach was also applied to the crystallization of ACC (Fig. 5.2), and this was possible at very high temporal resolutions (1 s/scan), over realistic time spans (secs-10's of secs), and for a wide range of supersaturation, temperature, pH, and additive concentrations. These studies provided new data on the nucleation and growth of $CaCO_3$ phases, i.e., the evolution of their particle sizes and shapes as well as the changes in short- or long-range order (crystallinity) of solids that are in continual contact with the reacting solutions. We crystallized (Bots et al. 2012; Rodriguez-Blanco et al. 2012) vaterite and calcite from solution via ACC from mixtures of highly concentrated (1 M) Na_2CO_3 and $CaCl_2$ solutions. They followed the reactions at 1–30 s time frames using small- and wide-angle X-ray scattering (SAXS/WAXS) and showed that the polymorph selection during the crystallization of ACC was highly pH-dependent but that other additives also changed the reactions dramatically. While a neutral (<7) starting pH during mixing promoted the direct transformation of ACC into calcite, when ACC formed from a solution with a basic initial pH (<11.5), the transformation to calcite occurred via metastable vaterite (Fig. 5.2). The early stages of crystallization of vaterite from ACC at basic initial pH (<11.5) and starting saturation index of $SI = 4.2$ started with the rapid breakdown of ACC and was followed by an equally rapid nucleation and spherulitic growth of vaterite (<2 min).

Spherulitic growth is a fast growth process that proceeds via the continuous nucleation of new nanoparticles on the surface of an already existing particle (core) via non-crystallographic branching. This nucleation-controlled growth process is usually termed “growth front nucleation” or “secondary nucleation” (Gránásy et al. 2005; Shtukenberg et al. 2012). In this process, there is no structural relationship between the newly nucleated particles and the already existing ones. This results in the formation of micrometer-sized spherulites that consist of aggregates of monodisperse nanoparticulate crystallites. Spherulitic morphologies have been categorized in two types: spheres (category 1) and dumbbells (category 2). This morphological variability is related to an orientation-dependent grain boundary energy change. Andreassen et al. (2010) experimentally demonstrated that spherulitic growth of vaterite directly from solution (i.e., not via an ACC precursor) requires a high crystallization driving force (i.e., $SI > 2–3$) to promote fast crystallization rates, leading to the formation of spherical nanocrystal aggregates. Spherical aggregates of vaterite crystallized via ACC have often been observed (Fig. 5.2), yet morphological evidence on its own is not satisfactory to provide definitive confirmation for spherulitic growth. Such evidence was derived from the change in saturation index of vaterite in the aqueous solutions at equilibrium with respect to ACC which was $SI = 1.4$ (Bots et al. 2012). However, SAXS/WAXS data revealed a

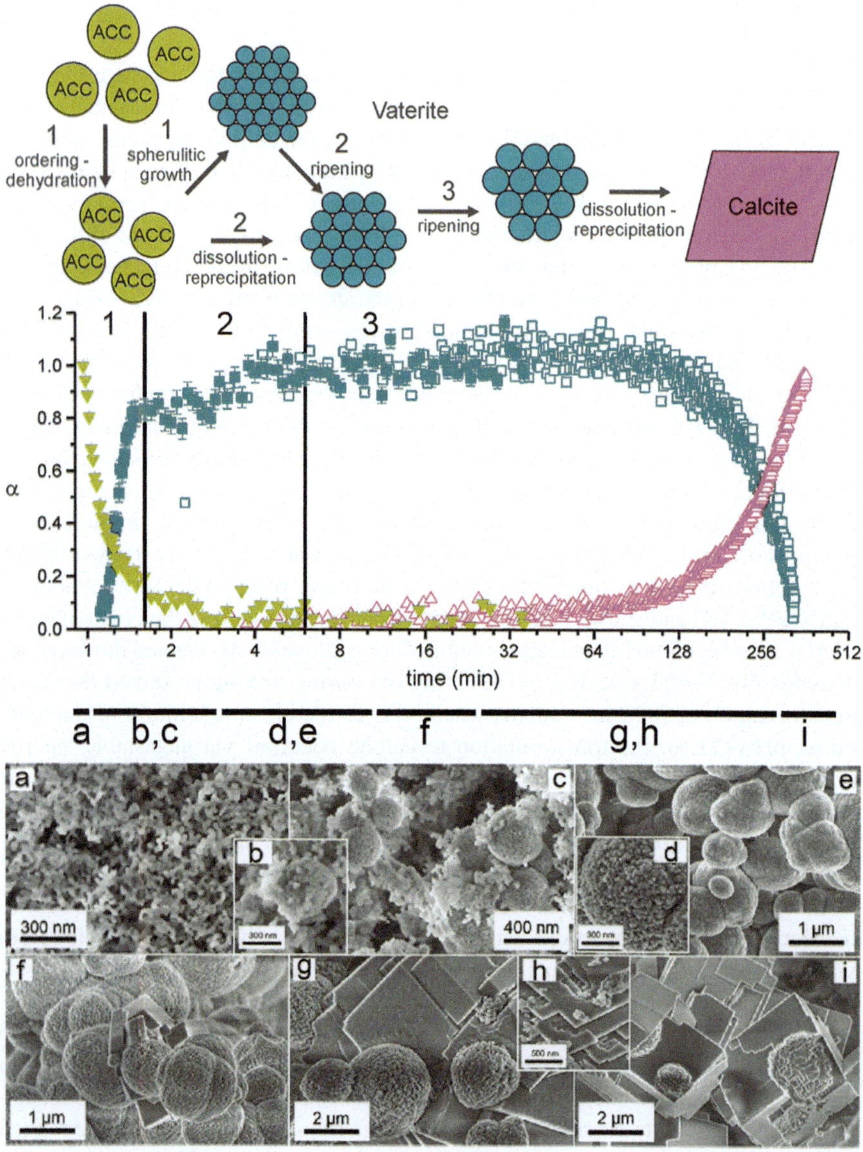

Fig. 5.2 *Top* diagram shows an schematic representation of the proposed multistage ACC → vaterite → calcite crystallization pathway (*top*) with the underlying combined reaction progress, α_{ACC}, $\alpha_{vaterite}$, and $\alpha_{calcite}$, for the full crystallization reaction in the pure ACC system (the *green triangles* and *full black squares* represent the ACC and vaterite from Bots et al. (2012) and the *open squares* and *red triangles* represent the vaterite and calcite from Rodriguez-Blanco et al. 2011); stages 1, 2, and 3 of the reaction mechanism are labeled on the figure. FEG-SEM images show the solid phases that form during these stages of the crystallization process. Pictures on the *top* correspond to the ACC → vaterite transformation: amorphous calcium carbonate (ACC) (**a**), ACC and vaterite nanoaggregates (**b**, **c**), and vaterite nanoaggregates (**d**, **e**). Pictures at the *bottom* correspond to vaterite → calcite transformation: vaterite nanoaggregates and first calcite crystals (**f**), calcite crystals attached to vaterite spheres with the development of growth steps on the calcite surface (**g**), calcite growth steps and vaterite nanoparticulate subunits, (**h**) and calcite crystals with vaterite casts (**i**) (after Rodriguez-Blanco et al. 2011)

fast breakdown of minor amounts of ACC prior to the crystallization of vaterite, resulting in a rapid increase in supersaturation, providing the high crystallization driving force needed for spherulitic growth. In addition, these data showed that the ACC–vaterite reaction proceeded via a continual dissolution of ACC and growth of vaterite. ACC that started off as ~35 nm particles was present until virtually all the vaterite had formed, keeping the supersaturation at a sufficiently high level to allow continuous vaterite nucleation. Furthermore, the SAXS/WAXS data clearly showed extremely rapid crystallization rates of vaterite, which are typical during a nucleation-controlled growth mechanism like spherulitic growth. Overall, the high level of supersaturation due to the breakdown of ACC promoted very rapid crystallization kinetics. The complete crystallization of vaterite occurred in <2 min and the average crystallite/particle sizes for the vaterite started off at ~10 nm, consistent with a nucleation dominated particle formation reaction with very little or no surface growth. At the end of the reaction, the particles reached a maximum size of ~40–60 nm, depending if sulfate was present in the aqueous solution or not.

Although spherulitic growth has been observed in other systems (i.e, polymers supercooled from a molten state or viscous magmas; Shtukenberg et al. 2012 and references therein), in carbonates it is promoted when the difference in solubility between the amorphous precursor phase and any crystalline phase is largest. Besides vaterite, other carbonates can also grow by a spherulitic mechanism: calcite (Rodriguez-Blanco et al. 2012), aragonite (Sand et al. 2012), monohydrocalcite (Rodriguez-Blanco et al. 2014), dolomite (Rodriguez-Blanco et al. 2015), or even REE-bearing carbonates like tengerite, kozoite, or hydroxylbastnasite (Vallina et al. 2014; Vallina et al. 2015) (Fig. 5.3). Interestingly, spherulitic morphologies are not just a vagary of experimental systems as they have also been observed in natural biogenic carbonates like corals that crystallize aragonite (Holcomb et al. 2009) or in microbial systems that produce proto-dolomite/dolomite (Warthmann et al. 2000; Sánchez-Román et al. 2011). Again, these reactions proceed when locally very high supersaturations are reached. The fact that Mg^{2+} and REE^{3+} ions play an important role in the process is primarily linked to their high dehydration energies (Fedorov et al. 2002; Di Tommaso and De Leeuw 2010). When they are present within the porous structure of the precursor phase, they retard its dehydration, thereby reducing the rate of dissolution and decreasing its overall solubility. As a result, in abiotic systems, high temperatures are sometimes required to promote the breakdown of the amorphous precursor phase and the fast increase in supersaturation levels during the crystallization process.

Above we primarily focused our discussions on how ACC transforms to vaterite. We also showed that once metastable vaterite has fully formed, it starts transforming into calcite. This process is temperature-dependent. We (Rodriguez-Blanco et al. 2011) derived the rates and energetics for the vaterite–calcite transformation reaction from 7.5 to 25° C, conditions which are relevant for both biomineralization and industrial calcium carbonate formation reactions. We found that the vaterite–calcite transformation is a surface-controlled process that occurs at much slower rates (e.g., 6 h at 25°°C) compared to the formation of vaterite via spherulitic growth (minutes). We also showed that the surface area of the newly forming calcite

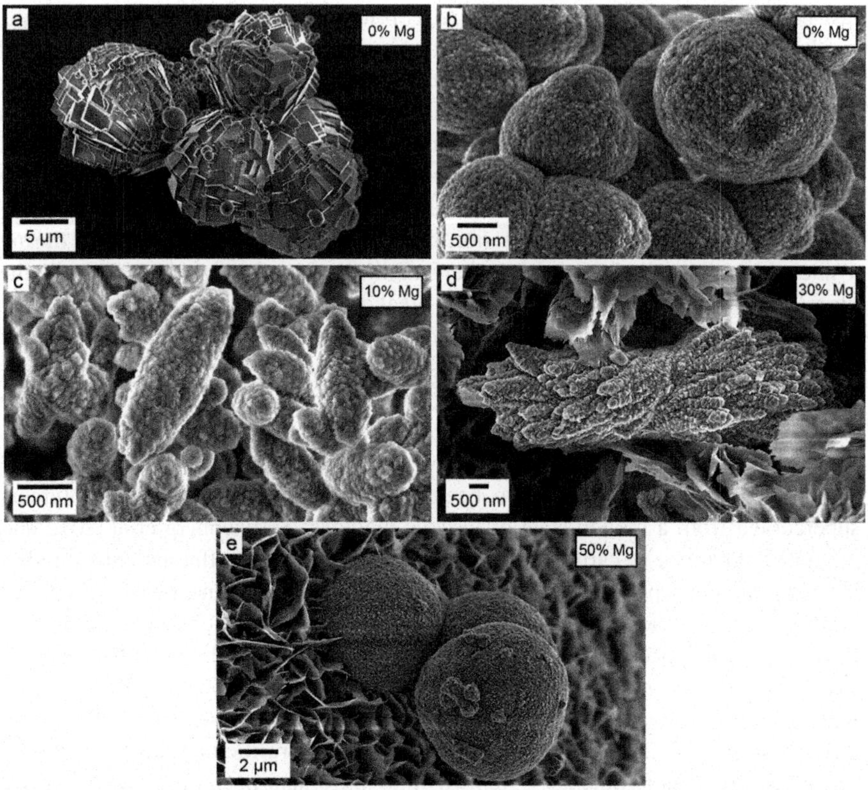

Fig. 5.3 Spherulitic morphologies of the first crystalline products obtained after the breakdown of pure or Mg-bearing ACC: (**a**) calcite, (**b**) vaterite, (**c**) Mg-calcite, (**d**) monohydrocalcite, (**e**) dolomite. Hydromagnesite ($Mg_5(CO_3)_4(OH)_2 \cdot 4H_2O$) is also a seen in the background of images (**d**) and (**e**) as it is usually a secondary product that forms during the crystallization of monohydrocalcite and dolomite. % Mg shown in each picture corresponds to $[Mg^{2+}]_{aq}/[Ca^{2+}]_{aq} \cdot 100$ in solution before the formation of ACC

controls the crystallization rate. This reaction proceeds via the slow dissolution of vaterite, followed by the gradual increase in its porosity and the release of calcium and carbonate ions into the solution. This leads to a surface-mediated dissolution–reprecipitation and growth of calcite crystals (Fig. 5.2f–h). This calcite consists of crystals with rhombohedral morphologies because during the vaterite–calcite transformation, the aqueous solution is in equilibrium with respect to vaterite, so the saturation index of calcite ($SI = 0.57$) is too low to favor spherulitic growth.

Although calcite is sometimes a product of the recrystallization of vaterite, the formation of ACC at neutral (~7) starting pH and at high supersaturations can also result in the primary crystallization of spherulitic calcite (Beck and Andreassen 2010; Rodriguez-Blanco et al. 2012). Again, this is driven by the large enough difference between the solubility of ACC and calcite to provide the high supersaturation ($SI > 2–3$) needed for spherulitic growth. Spherulitic calcite is not so common in nature compared to spherulitic vaterite, but there are examples that show that these morphologies may originate during biomineralization processes. Many species of earthworm secrete granules of calcium carbonate that are formed within the earthworm's calciferous glands. These granules are mainly made of calcite but also contain remnant ACC, vaterite, and aragonite (Canti and Piearce 2003; Briones et al. 2008), and their morphologies are similar to calcite crystals that form in laboratory experiments when ACC precipitates from highly supersaturated solutions and crystallizes to calcite via spherulitic growth (e.g., Fig. 5.3a).

Two main factors explain the different crystallization pathways of ACC to either primary vaterite or calcite. The first one relates the structure of ACC and its formation mechanisms and pathways. Gebauer et al. (2008) suggested that the binding strength of Ca^{2+} and $CO_3{}^{2-}$ ions within the ACC precursor clusters is pH-dependent and controls the structure, stability, and crystallization pathways of ACC. This way, ACC formed at a basic pH has a vaterite-type local structure and promotes the crystallization of vaterite over calcite, while ACC formed at a neutral pH has a calcite-type local structure and crystallizes directly to calcite. The second factor relates pH, solubility, and the dissolution rate of ACC (Kojima et al. 1993). The solubility (and dissolution rate) of ACC increase proportionally with formation pH, and thus, the saturation index for any anhydrous crystalline carbonate when an aqueous solution is in equilibrium with respect to ACC will be directly proportional to the pH at which ACC formed. Taking all this together, an ACC precursor formed at high pH does not only have a vaterite-type local structure but also promotes the crystallization of vaterite because of its higher solubility and dissolution rate. Conversely, ACC formed at a starting neutral pH has a calcite-type local structure and also a lower solubility so it tends to transform directly to calcite (Demichelis et al. 2011; Raiteri et al. 2012; Gebauer et al. 2014; also see Chap. 6 by Demichelis et al. 2017 in this book).

It is very important to take into account that ACC formation is extremely fast (<0.1 s), even at low supersaturation levels. This is critical for most experimental methods used for calcium carbonate crystallization and in particular when solution mixing-based methods are used (as opposed to slow diffusion methods). As ACC is forming during fast mixing, it is experimentally challenging to avoid local differences in the mixed solution and thus local ACC formation regardless how rapid or homogeneous mixing can be achieved. For example, when a $CaCl_2$ solution is added to a Na_2CO_3 solution, the first ACC nanoparticles form in a higher pH environment compared to those ones forming at the end of the mixing procedure. This means that during the mixing procedure, we have a changing pH that affects the short-range structure and chemical composition of any forming ACC (e.g., hydration, incorporation of OH^- ions). As a result, ACC nanoparticles with different short-range structures (vaterite and calcite type) can coexist in the same experiment/sample. This naturally downstream during the crystallization process can also result in a mixture of vaterite and calcite (Zou et al. 2015). There are other parameters that can also affect strongly the outcome of mixing experiments; these include temperature, concentrations/supersaturation levels of the starting solutions, aqueous $Ca^{2+}/CO_3{}^{2-}$ ratios, pH of the final solution once equilibrium with respect to ACC has been reached, stirring/shaking rate, additives, etc. In addition, the particle size of ACC is inversely proportional to the initial supersaturation, and this needs to be taken into account because the solubility of nanoparticles is inversely proportionally with the particle size (e.g., Hochella et al. 2008). Considering this large number of variables, it is not surprising to find experimental reproducibility problems during synthesis of crystalline $CaCO_3$ polymorphs at high supersaturations. In fact, some variables that seem to be negligible or difficult to control, like mixing procedure, stirring rate or small variations in the concentrations of the starting solutions, are actually significant enough to affect the short-range structure, dissolution kinetics, and crystallization pathways of ACC (e.g., Zou et al. 2015; Ševčík et al. 2015).

5.3 The Effect of (In)organics in the Formation of ACC and Vaterite

A wide range of industrial, medical, and pharmaceutical applications rely on simple protocols to produce $CaCO_3$ phases in a way that they can have a high degree of control over the resulting polymorph type and morphology. Organisms are experts in this field, and it is well known that the formation of calcium carbonates most often proceeds in the presence of, or through templating on, simple inorganic/complex organic molecules. This leads to elaborate morphologies and textures that could have innumerable applications (e.g., Trushina et al. 2014) in industry if we were able to reproduce them. Over the past decades, researchers have taken inspiration from biomineralizing organisms and employed a range of organic and inorganic compounds aiming to mimic their strategies and achieve control over morphology, textures, composition, and shapes of biomimetic $CaCO_3$ phases (e.g., Kim et al. 2011; Schenk et al. 2014a, b). This is not easy because the diversity of possible compounds that could be used as additives or templating agents span from simple ions such as Mg^{2+}, Sr^{2+}, REE^{3+}, SO_4^{2-}, PO_4^{3-}, and SiO_4^{2-} to a suite of organic molecules including (poly)carboxylates, cationic and anionic polyelectrolytes, amino acids, (block) copolymers, surfactants, proteins, polysaccharides, phosphate compounds (e.g., phosphoenolpyruvate), dendrimers, and alcohols (Meldrum and Cölfen 2008 and references therein).

The effects of (in)organic compounds on the formation of ACC and its crystallization to vaterite are primarily linked to variations in ACC lifetime, vaterite morphologies, and particle size. As explained above, these are usually coupled to variations in formation and crystallization mechanisms and pathways. Changes in vaterite morphology are quite common, both when it crystallizes as a single crystal (e.g., development of preferred crystal faces; Hu et al. 2012) or when it forms aggregates of nanoparticles (e.g., variation in angle of branching of spherulites, formation of hollow spheres; Cai et al. 2008; Beck and Andreassen 2010). The detailed outcome of such reactions varies greatly with the complexity and concentrations of additive, variations in starting and end solution compositions, and the experimental method used. The effects that (in)organic additives have on ACC to vaterite transformations also depend on density and chemistry of the functional groups in a polymer molecule but also a range of other factors such as their structure, chain length, hydrophobicity, or conformation. Provided a sufficient amount of an (in)organic additive is present, the lifetime of ACC and/or vaterite is usually prolonged, either by stabilization via a protective organic coating (Gal et al. 2013), inhibiting the formation of crystalline phases (e.g., Mg^{2+} unstabilizes the structure of vaterite; Rodriguez-Blanco et al. 2012) and/or incorporation of the additives which slow down the dehydration of ACC during its transformation (Rodriguez-Blanco et al. 2014; Rodriguez-Blanco et al. 2015).

Inorganic additives are usually incorporated into both ACC and vaterite. However, the incorporation of inorganic cations or anions into a crystalline $CaCO_3$ polymorph is favored or inhibited depending on its structure. For example, tetrahedral anions such as $SiO_4{}^{2-}$ and $PO_4{}^{3-}$ do not fit well into the calcite structure, and their presence results in the stabilization of ACC and vaterite but the inhibition of their transformation to calcite (Gal et al. 2010). Most of the studies about inorganic additives on carbonate formation pathways or stability have focused on the effects of Mg^{2+} and $SO_4{}^{2-}$ ions, as they are most abundant divalent ions in our oceans and because they are considered to control the primary inorganic marine calcium carbonate mineralogy throughout the Phanerozoic (Bots et al. 2011). On the other hand, sulfate slows down the rate of vaterite formation and particle growth (transformation of ACC to vaterite is $\sim 10\,\%$ slower than in the pure system), but most importantly, the presence of sulfate stabilizes vaterite (Bots et al. 2012). The effect of Mg^{2+} on the mechanism of ACC and vaterite formation is remarkable: Mg^{2+} gets incorporated in ACC and calcite, but the vaterite structure becomes unstable at higher Mg^{2+} concentrations (Bots et al. 2012). Mg^{2+} ions have a higher dehydration energy compared to Ca^{2+} (Di Tommaso and de Leeuw 2010), and hence, the presence of Mg^{2+} in ACC affects its dehydration prior to crystallization and reduces its dissolution rate. A slower rate of dissolution is translated into a slower increase of the solution supersaturation, influencing the crystallization pathways. In contrast to pure ACC, which directly crystallizes to vaterite or calcite (Rodriguez-Blanco et al. 2011), an increasing content of Mg^{2+} in ACC increases its stability and promotes the crystallization of Mg-calcite (10 % Mg^{2+}; Rodriguez-Blanco et al. 2012), monohydrocalcite (30 % Mg^{2+}; Rodriguez-Blanco et al. 2014), and dolomite (50 % Mg^{2+}; Rodriguez-Blanco et al. 2015). In particular, the crystallization of dolomite from Mg-bearing ACC requires high temperatures (> 60° C) because of the larger energy needed to dehydrate ACC with a high Mg content ($Ca_{0.606}Mg_{0.394}CO_3 \cdot 1.37H_2O$; Rodriguez-Blanco et al. 2015). After a temperature-dependent induction time, Mg-ACC partially dehydrates and orders prior to its rapid (<5 min) crystallization to nonstoichiometric proto-dolomite via spherulitic growth. Proto-dolomite has no ordering of Mg^{2+} and Ca^{2+} within its structure and transforms to highly crystalline and stoichiometric dolomite on a much longer timescale (hours to days), via an Ostwald-ripening mechanism (Malone et al. 1996). This crystallization route is completely different to the well-known dolomitization of calcite (secondary replacement of Mg^{2+} into previously precipitated calcite; e.g., Sibley et al. 1994) and is an excellent example of how inorganic ions can affect the crystallization pathways of carbonates and the stability and solubility of ACC.

In the case of organic additives, anionic groups have been proved important for ACC and vaterite stabilization in solution (Naka et al. 2006; Olderoy et al. 2009; Gal et al. 2010; Boyjoo et al. 2014; Tobler et al. 2014; Tobler et al. 2015). Some organics have given particularly interesting results, like polyacrylic acid (PAA; Xu et al. 2008) and polymerized dopamine (Wang and Xu 2013), stabilizing ACC up to half a month in solution and more than a year in dry conditions, respectively. In the case of vaterite, peptide block copolymers are able to inhibit the vaterite–calcite transformation at least a year (Kašparová et al. 2004). In most cases, the majority of the organic additives are adsorbed on the surface of the nanoparticles, so their lifetimes are proportional to the concentration of additive. This coating is thought to prevent particle aggregation and recrystallization by limiting diffusion (Gal et al. 2013) or by making them less soluble (Gal et al. 2014). Incorporation, and an associated change in the short-range structure of ACC particles, has been reported for citrate (Tobler et al. 2015). However, for amino acids such as glycine and aspartic acid, ACC was also stabilized, but incorporation was not observed at comparable concentrations (Tobler et al. 2014). Organic compounds have also been reported to change the crystallization mechanism of vaterite. The resulting change in solution chemistry with additive addition plays a large role for the fate of ACC and vaterite. Polymers with anionic groups can complex free calcium ions in solution, serving as nucleation points with decreased barriers for nucleation and also resulting into a change in supersaturation and promoting a change in particle size. A lower supersaturation results in lower nucleation density compared to a system with a higher supersaturation. Naka et al. (2006) studied cross sections of vaterite formed in the presence of PAA finding out that at high concentrations, the formation of vaterite occurred via an aggregation process. However, at low initial concentration, or when PAA was added later to the reaction, radial growth characteristic for spherulitic growth was observed. Olderoy et al. (2009) performed experiments at the same initial supersaturation and pH of $\sim$7 and found that alginate enhanced nucleation but inhibited growth and aggregation, resulting in a large number of small vaterite nanoparticles with no other significant change in morphology.

The vaterite morphology is particularly sensitive to changes in solution chemistry, and a wide range of shapes have been reported in the presence of additives. These shapes include dumbbells, disks, spheres, flowers, and flakes. A comprehensive overview is presented in Boyjoo et al. (2014) along with solution pH, method, additive, and concentration. They found that solution pH has a drastic effect on vaterite morphology but not on polymorphism. Addition of organic molecules can also cause a change in pH, which affects speciation, the degree of protonation of the added molecule, and the supersaturation levels. Decreased water activity also has a significant impact on stability, morphology, and polymorph selection of $CaCO_3$. By adding 10 and 50 % of short alcohol molecules, we (Sand et al. 2012) (Fig. 5.4) stabilized amorphous and crystalline phases and were able to change the

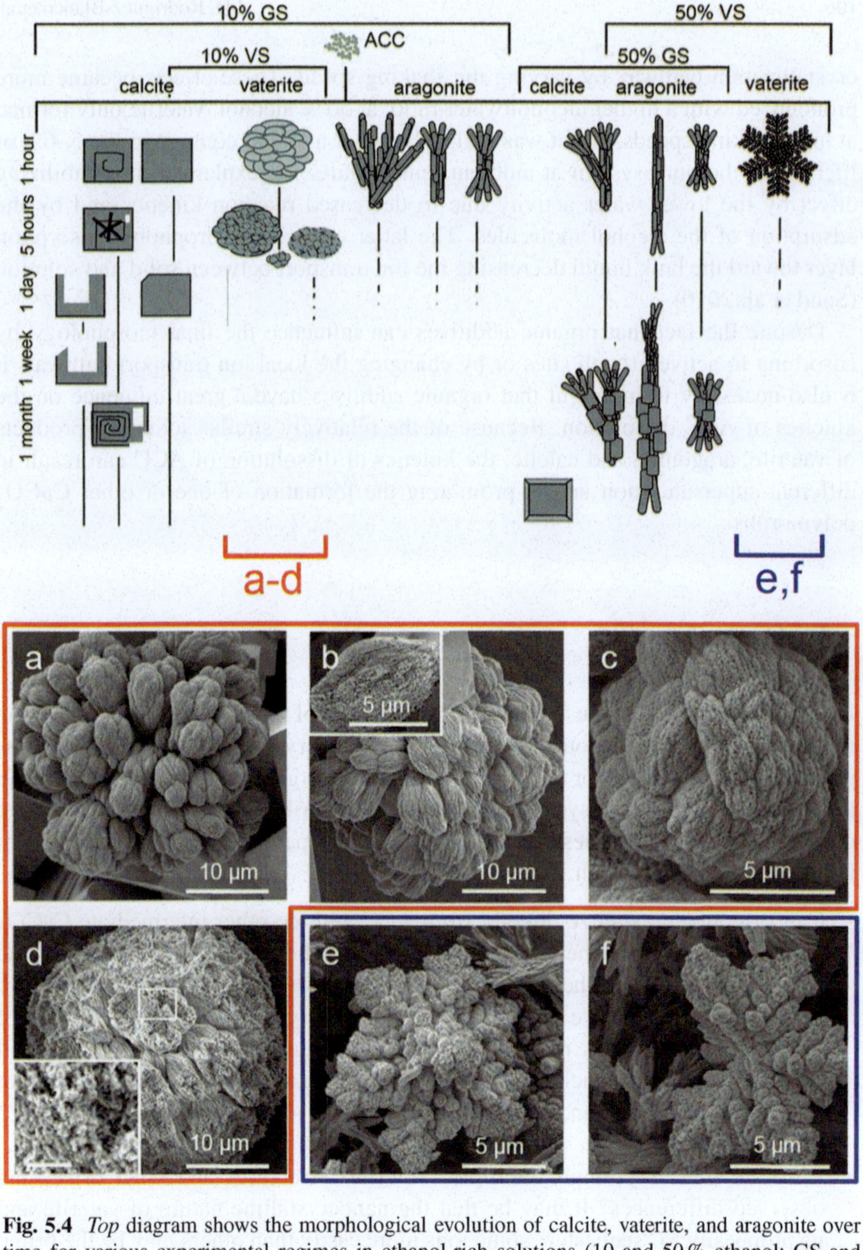

Fig. 5.4 *Top* diagram shows the morphological evolution of calcite, vaterite, and aragonite over time for various experimental regimes in ethanol-rich solutions (10 and 50 % ethanol; GS and VS refer to gentle shaking and vigorous shaking conditions, respectively). SEM images below represent the morphology for vaterite obtained from experiments with 10 and 50 % ethanol. With 10 % alcohol, the morphologies are independent of shaking speed (**a**–**d**): (**a**, **b**) cauliflower and layer cake (*inset* in **b**) crystals form in the GS regime after 1 h. Calcite crystals are often attached to the vaterite clusters, as seen in **a** and **b**; (**c**) VS regime after 1 h showing more porous vaterite cauliflower structures; (**d**) during the vaterite to calcite transformation, the cauliflower spheres become corroded (the *scale bar* in *inset* is 1 μm); (**e**, **f**) dendritic vaterite precipitated in the VS experiments with 50 % alcohol and sampled after 5 h; aragonite needles are present in both images (Adapted after Sand et al. 2012)

crystallization pathway by varying the shaking speed. These effects became more pronounced with a higher alcohol/water ratio: at 50 % alcohol, vaterite only formed at high shaking speeds, and it was still stable after a day in contrast to the 5–6 h of lifetime in the pure system at ambient temperature. We explained this stabilizing effect by the lower water activity due to decreased reaction kinetics and by the adsorption of the alcohol molecules. The latter creates a hydrophobic adsorption layer toward the bulk liquid decreasing the ion transport between solid and solution (Sand et al. 2010).

Despite the fact that organic additives can influence the final morphology by adsorbing to active growth sites or by changing the local ion transport pathway, it is also necessary to highlight that organic additives have a great influence on the kinetics of ACC dissolution. Because of the relatively similar solubility products of vaterite, aragonite, and calcite, the kinetics of dissolution of ACC can result in different supersaturation levels, promoting the formation of one or other $CaCO_3$ polymorphs.

5.4 Outlook

In the chapter above, we summarized the state of knowledge about how the transformation of amorphous calcium carbonate to crystalline $CaCO_3$ polymorphs proceeds in the presence, or absence, of important inorganic and organic additives in nature and in biomimetic systems. Although, as described, much research has been done to address these issues, there are still several open questions that we want to highlight as future research goals:

- Can aragonite crystallize directly from ACC with no other intermediate $CaCO_3$ polymorph affecting the reaction? Data about the pathways that lead to the formation of several other aragonite-like minerals (e.g., strontianite ($SrCO_3$) or whiterite ($BaCO_3$)) have suggested that a direct crystallization from solution of these carbonate phases may be feasible. However, the available data are still sketchy, and it is still unclear if in these systems, an amorphous precursor was not present or if there was an amorphous precursor present which was too short-lived to be detected.
- Why are there seemingly multiple "vaterite" structures and what drives the observed differences? It may be that the nanocrystalline nature of vaterite and its propensity to "sequester" some ions more easily than others may be the prime reason for the observed differences in structure. Yet, the fundamental mechanistic reasons for these differences are still unclear. The questions related to which (in)organic molecules affect the formation/aggregation/transformation of vaterite more than others and if, and how, such molecules become "incorporated" into the vaterite structure or if they just lead to defects is still debated and requires further research.

- Finally the bigger picture question! It is well known that the nature and stability (or not) of ACC and vaterite as precursors and/or intermediates to crystalline $CaCO_3$ polymorphs are crucial for many global biogeochemical cycles as well as many industrial processes. However, the complexity of the $CaCO_3$ system has so far prevented us from establishing a "unified" crystallization model. In natural settings, a chemically pure system does not exist, and the presence of (in)organic compounds, together with variations in physicochemical parameters, invariably affects the kinetics and mechanisms of crystallization. However, an elegant and simple model for ACC crystallization would improve, for example, our ability to design better biomimetic reactions and thus more efficient industrial processes. Equally, a "unified" model of $CaCO_3$ crystallization would also allow us to better predict the effects that, for example, changes in ocean chemistry due to climate change, have on the stability and formation of calcium carbonate shells in marine biota, which control to a large extent our global carbon cycle. Despite all the efforts done so far, it is clear that we are still at the beginning of an exciting research endeavor that will keep several generations of geochemists and chemists busy.

Acknowledgments J.D. Rodriguez-Blanco would like to acknowledge the EU-funded NanoCArB Marie Curie Intra-European Fellowship (IEF) under contract PIEF-GA-2013-624016. K.K. Sand is grateful for funding from the Danish Council for Independent Research on their Individual Post Docs (0602-02915B) and Sapere Aude program (0602-02654B) and support from the US Department of Energy, Office of Basic Energy Sciences, Division of Chemical Sciences, Geosciences, and Biosciences at Pacific Northwest National Laboratory, which is operated by Battelle for the US Department of Energy under Contract DE-AC05-76RL01830.

References

Ahmed IAM, Benning LG, Kakonyi G, Sumoondur A, Terrill N, Shaw S (2010) The formation of green rust sulfate: in situ and time-resolved scattering and electrochemistry. Langmuir 26:6593–6603

Andreassen JP, Flaten EM, Beck R, Lewis AE (2010) Investigations of spherulitic growth in industrial crystallization. Chem Eng Res Des 88:1163–1168

Beck R, Andreassen JP (2010) Spherulitic growth of calcium carbonate. Cryst Growth Des 10:2934–2947

Beniash E, Aizenberg J, Addadi L, Weiner S (1997) Amorphous calcium carbonate transforms into calcite during sea urchin larval spicule growth. Proc R Soc Lond Ser B 264:461–465

Bots P, Benning LG, Rickaby REM, Shaw S (2011) The role of SO4 in the switch from calcite to aragonite seas. Geology 39:331–334

Bots P, Benning LG, Rodriguez-Blanco JD, Roncal-Herrero T, Shaw W (2012) Mechanistic insights into the crystallization of amorphous calcium carbonate (ACC). Cryst Growth Des 12:3806–3814

Boyjoo Y, Pareek VK, Liu J (2014) Synthesis of micro and nano-sized calcium carbonate particles and their applications. J Mater Chem A 2:14270–14288

Brinza L, Schofield PF, Hodson ME, Weller S, Ignatyev K, Geraki K, Quinn PD, Mosselmans JFW (2014) Combining μXANES and μXRD mapping to analyse the heterogeneity in calcium carbonate granules excreted by the earthworm Lumbricus terrestris. J Synchrotron Radiat 21:235–241

Brinza L, Vu HP, Shaw S, Mosselmans FJ, Benning LG (2015) Effect of Mo and V on the hydrothermal crystallization of hematite from ferrihydrite: an in situ energy dispersive X-Ray diffraction and X-Ray absorption spectroscopy study. Cryst Growth Des 15:768–4780

Briones MJI, López E, Méndez J, Rodríguez JB, Gago-Duport L (2008) Biological control over the formation and storage of amorphous calcium carbonate by earthworms. Mineral Mag 72:227–231

Burke IT, Mosselmans FW, Shaw S, Peacock CL, Benning LG, Cocker VS (2015) Impact of the diamond light source on research in earth and environmental sciences: current work and future perspectives. In: Materlik G, Stuart D, Rayment T (eds) Research achieved with diamond and future perspectives. Philos Trans A 373:20130151

Cahill CL, Benning LG, Barnes HL, Parise JB (2000) In situ time-resolved X-ray diffraction of iron sulfides during hydrothermal pyrite growth. Chem Geol 167:53–63

Cai A, Xu X, Pan H, Tao J, Liu R, Tang R, Cho K (2008) Direct synthesis of hollow vaterite nanospheres from amorphous calcium carbonate nanoparticles via phase transformation. J Phys Chem C 112:11324–11330

Canti MG, Piearce TG (2003) Morphology and dynamics of calcium carbonate granules produced by different earthworm species. Pedobiologia 47:511–521

Davidson LE, Shaw S, Benning LG (2008) The kinetics and mechanisms of schwertmannite transformation to goethite and hematite under alkaline conditions. Am Mineral 93:1326–1337

Demichelis R, Raiteri P, Gale JD, Quigley D, Gebauer D (2011) Stable prenucleation mineral clusters are liquid-like ionic polymers. Nat Commun 2:590

Demichelis R, Raiteri P, Gale JD (2017) Ab Initio modelling of the structure and properties of crystalline calcium carbonate. In: Van Driessche AES, Kellermeier M, Benning LG, Gebauer D (eds) New perspectives on mineral nucleation and growth, Springer, Cham, pp 113–136

Di Tommaso D, De Leeuw NH (2010) Structure and dynamics of the hydrated magnesium ion and of the solvated magnesium carbonates: insights from first principles simulations. Phys Chem Chem Phys 12:894–901

Dove PM, De Yoreo JJ, Weiner S (eds) (2003) Reviews in mineralogy and geochemistry, vol 154. Mineralogical Society of America, Chantilly

Faatz M, Gröhn F, Wegner G (2004) Amorphous calcium carbonate: synthesis and potential intermediate in biomineralization. Adv Mater 16:996–1000

Fedorov PP, Nazarkin MV, Zakalyukin RM (2002) On polymorphism and morphotropism of rare earth sesquioxides. Crystallogr Rep 47:281–286

Foran E, Weiner S, Fine M (2013) Biogenic fish-gut calcium carbonate is a stable amorphous phase in the gilt-head seabream sparus aurata. Sci Rep 3:1700

Fujiwara M, Shiokawa K, Araki M, Ashitaka N, Morigaki K, Kubota T, Nakahara Y (2010) Encapsulation of proteins into $CaCO_3$ by phase transition from vaterite to calcite. Cryst Growth Des 10:4030–4037

Gago-Duport L, Briones MJ, Rodríguez JB, Covelo B (2008) Amorphous calcium carbonate biomineralization in the earthworm's calciferous gland: pathways to the formation of crystalline phases. J Struct Biol 162:422–435

Gal A, Weiner S, Addadi L (2010) The stabilizing effect of silicate on biogenic and synthetic amorphous calcium carbonate. J Am Chem Soc 132:13208–13211

Gal A, Hirsch A, Siegel S, Li C, Aichmayer B, Politi Y, Fratzl P, Weiner S, Addadi L (2012) Plant cystoliths: a complex functional biocomposite of four distinct silica and amorphous calcium carbonate phases. Chemistry 18:10262–10270

Gal A, Habraken W, Gur D, Fratzl P, Weiner S, Addadi L (2013) Calcite crystal growth by a solid-state transformation of stabilized amorphous calcium carbonate nanospheres in a hydrogel. Angew Chem Int Ed 52:4867–4870

Gal A, Kahil K, Vidavsky N, DeVol RT, Gilbert PUPA, Fratzl P, Weiner S, Addadi L (2014) Particle accretion mechanism underlies biological crystal growth from an amorphous precursor phase. Adv Funct Mater 24:5420–5426

Gebauer D, Völkel A, Cölfen H (2008) Stable prenucleation calcium carbonate clusters. Science 322:819–1822

Gebauer D, Kellermeier M, Gale JD, Bergström L, Cölfen H (2014) Pre-nucleation clusters as solute precursors in crystallisation. Chem Soc Rev 43:2348–2371
Gong YUT, Killian CE, Olson IC, Appathurai NP, Amasino AL, Martin MC, Holt LJ, Wilt FH, Gilbert PUPA (2012) Phase transitions in biogenic amorphous calcium carbonate. Proc Natl Acad Sci U S A 109:6088–6093
Goodwin AL, Michel FM, Phillips BL, Keen DA, Dove MT, Reeder RJ (2010) Nanoporous structure and medium-range order in synthetic amorphous calcium carbonate. Chem Mater 22:3197–3205
Gránásy L, Pusztai T, Tegze G, Warren JA, Douglas JF (2005) Growth and form of spherulites. Phys Rev E 72:011605
Hochella MF, Lower SK, Maurice PA, Penn RL, Sahai N, Sparks DL, Twining BS (2008) Nanominerals mineral nanoparticles and earth systems. Science 319:1631–1635
Hodson ME, Benning LG, Demarchi B, Penkman KEH, Rodriguez-Blanco JD, Schofield PF, Versteegh EAA (2015) Biomineralisation by earthworms – an investigation into the stability and distribution of amorphous calcium carbonate. Geochem Trans 16:4
Holcomb M, Cohen AL, Gabitov RI, Hutter JL (2009) Compositional and morphological features of aragonite precipitated experimentally from seawater and biogenically by corals. Geochim Cosmochim Acta 73:4166–4179
Hu Q, Zhang J, Teng H, Becker U (2012) Growth process and crystallographic properties of ammonia-induced vaterite. Am Mineral 97:1437–1445
Isaure MP, Sarret G, Harada E, Choi YE, Marcus MA, Fakra SC, Geoffroy N, Pairis S, Susini J, Clemens S, Manceau A (2010) Calcium promotes cadmium elimination as vaterite grains by tobacco trichomes. Geochim Cosmochim Acta 74:5817
Kabalah-Amitai L, Mayzel B, Kauffmann Y, Fitch AN, Bloch L, Gilbert PUPA, Pokroy B (2013) Vaterite crystals contain two interspersed crystal structures. Science 340:454–457
Kašparová P, Antonietti M, Cölfen H (2004) Double hydrophilic block copolymers with switchable secondary structure as additives for crystallization control. Colloids Surf A Physicochem Eng Asp 250:153–162
Kim YY, Hetherington NB, Noel EH, Kröger R, Charnock JM, Christenson HK, Meldrum FC (2011) Capillarity creates single-crystal calcite nanowires from amorphous calcium carbonate. Angew Chem Int Ed Engl 50:12572–12577
Kojima Y, Kawanobe A, Yasue T, Arai Y (1993) Synthesis of amorphous calcium carbonate and its crystallization. J Ceram Soc Jpn 101:1145–1152
Lowenstam HA, Abbott DP (1975) Vaterite: a mineralization product of the hard tissues of a marine organism (Ascidiacea). Science 188:363–365
Malone JM, Baker PA, Burns SJ (1996) Recrystallization of dolomite: an experimental study from 50–200°C. Geochim Cosmochim Acta 60:2189–2207
Meldrum FC, Cölfen H (2008) Controlling mineral morphologies and structures in biological and synthetic systems. Chem Rev 108:4332–4432
Naka K, Huang Y, Chujo Y (2006) Formation of stable vaterite with poly(acrylic acid) by the delayed addition method. Langmuir 22:7760–7767
Nielsen MH, Aloni S, De Yoreo JJ (2014) In situ TEM imaging of $CaCO_3$ nucleation reveals coexistence of direct and indirect pathways. Science 345:1158–1162
Ogino T, Suzuki T, Sawada K (1987) The formation and transformation mechanism of calcium carbonate in water. Geochim Cosmochim Acta 51:2757–2767
Olderoy MO, Xie ML, Strand BL, Flaten EM, Sikorski P, Andreassen JP (2009) Growth and nucleation of calcium carbonate vaterite crystals in presence of alginate. Cryst Growth Des 9:5176–5183
Parakhonskiy BV, Haase A, Antolini R (2012) Sub-micrometer vaterite containers: synthesis substance loading and release. Angew Chem Int 51:1195–1197
Politi Y, Metzler RA, Abrecht M, Gilbert B, Wilt FH, Sagi I, Addadi L, Weiner S, Gilbert PUPA (2008) Transformation mechanism of amorphous calcium carbonate into calcite in the Sea urchin larval spicule. Proc Natl Acad Sci U S A 105:20045–20045

Pouget EM, Bomans PHH, Goos JACM, Frederik PM, With G, Sommerdijk NAJM (2009) The initial stages of template-controlled CaCO3 formation revealed by cryo-TEM. Science 323:1455–1458

Pouget EM, Bomans PHH, Dey A, Frederik PM, de With G, Sommerdijk NAJM (2010) The development of morphology and structure in hexagonal vaterite. J Am Chem Soc 132:11560–11565

Qiao L, Feng QL, Li Z (2007) Special vaterite found in freshwater lackluster pearls. Cryst Growth Des 7:275–279

Radha AV, Forbes TZ, Killian CE, Gilbert PUPA, Navrotsky A (2010) Transformation and crystallization energetics of synthetic and biogenic amorphous calcium carbonate. Proc Natl Acad Sci U S A 107:16438–16443

Raiteri P, Demichelis R, Gale JD, Kellermeier M, Gebauer D, Quigley D, Wright LB, Walsh TR (2012) Exploring the influence of organic species on pre-and post-nucleation calcium carbonate. Faraday Discuss 1:61–85

Rodriguez-Blanco JD, Shaw S, Benning LG (2008) How to make ‘stable’ ACC: protocol and preliminary structural characterization. Mineral Mag 72:283–286

Rodriguez-Blanco JD, Shaw S, Benning LG (2011) The kinetics and mechanisms of amorphous calcium carbonate (ACC) crystallization to calcite via vaterite. Nanoscale 3:265–271

Rodriguez-Blanco JD, Shaw S, Bots P, Roncal-Herrero T, Benning LG (2012) The role of pH and Mg on the stability and crystallization of amorphous calcium carbonate. J Alloys Compd 536:S477–S479

Rodriguez-Blanco JD, Shaw S, Bots P, Roncal-Herrero T, Benning LG (2014) The role of Mg in the crystallisation of monohydrocalcite. Geochim Cosmochim Acta 127:204–220

Rodriguez-Blanco JD, Shaw S, Benning LG (2015) A route for the direct crystallization of dolomite. Am Mineral 100:1172–1181

Rowlands DLG, Webster RK (1971) Precipitation of vaterite in lake water. Nat Phys Sci 229:158

Sánchez-Román M, McKenzie JA, Wagener ALR, Romanek CS, Sánchez-Navas A, Vasconcelos C (2011) Experimentally determined biomediated Sr partition coefficient for dolomite: significance and implication for natural dolomite. Geochim Cosmochim Acta 75:887–904

Sand KK, Yang M, Makovicky E, Cooke DJ, Hassenkam T, Bechgaard K, Stipp SLS (2010) Binding of ethanol on calcite: the role of the OH bond and its relevance to biomineralization. Langmuir 26:15239–15247

Sand KK, Rodriquez-Blanco JD, Makovicky E, Benning LG, Stipp SLS (2012) Crystallization of CaCO3 in water-alcohol mixtures: spherulitic growth polymorph stabilization and morphology change. Cryst Growth Des 12:845–853

Schenk AS, Albarracin EJ, Kim YY, Ihli J, Meldrum FC (2014a) Confinement stabilises single crystal vaterite rods. Chem Commun 50:4729–4732

Schenk AS, Cantaert B, Kim YY, Li Y, Read ES, Semsarilar M, Armes SP, Meldrum FC (2014b) Systematic study of the effects of polyamines on calcium carbonate precipitation. Chem Mater 26:2703–2711

Ševčík R, Pérez-Estébanez M, Viani A, Šašek P, Mácová P (2015) Characterization of vaterite synthesized at various temperatures and stirring velocities without use of additives. Powder Technol 284:265–271

Shen Q, Wei H, Zhou Y, Huang Y, Yang H, Wang D, Xu D (2006) Properties of amorphous calcium carbonate and the template action of vaterite spheres. J Phys Chem B 110:2994–3000

Shtukenberg AG, Punin YO, Gunn E, Kahr B (2012) Spherulites. Chem Rev 112:1805–1838

Sibley DF, Nordeng SH, Borkowski ML (1994) Dolomitization kinetics in hydrothermal bombs and natural settings. J Sediment Res 64:630–637

Tobler DJ, Benning LG (2011) The microbial diversity in Icelandic hot springs: temperature salinity pH and sinter growth rate effects. Extremophiles 15:473–485

Tobler DJ, Shaw S, Benning LG (2009) Quantification of initial steps of nucleation and growth of silica nanoparticles: an in-situ SAXS and DLS study. Geochim Cosmochim Acta 73:5377–5393

Tobler DJ, Rodriguez-Blanco JD, Dideriksen K, Sand KK, Bovet N, Benning LG, Stipp SLS (2014) Effect of aspartic acid and glycine on amorphous calcium carbonate (ACC) structure stability and crystallisation. Procedia Earth Planet Sci 10:143–148

Tobler DJ, Rodriguez-Blanco JD, Dideriksen K, Sand KK, Bovet N, Benning LG, Stipp SLS (2015) Citrate effects on amorphous calcium carbonate (ACC) structure stability and crystallization. Adv Funct Mater 25:3081–3090

Tobler DJ, Rodriguez-Blanco JD, Sørensen HO, Stipp SLS, Dideriksen K (2016) Effect of pH on amorphous calcium carbonate structure and transformation. Cryst Growth Des 16:4500–4508

Trushina D, Bukreeva TV, Kovalchuk MV, Antipina MN (2014) $CaCO_3$ vaterite microparticles for biomedical and personal care applications. Mater Sci Eng C 45:644–65

Vallina B, Rodriguez-Blanco JD, Blanco JA, Benning LG (2014) The effect of heating on the morphology of crystalline neodymium hydroxycarbonate NdCO3OH. Mineral Mag 78:1391–1397

Vallina B, Rodriguez-Blanco JD, Brown AP, Blanco JA, Benning LG (2015) The role of amorphous precursors in the crystallization of La and Nd carbonates. Nanoscale 7:12166–12179

Vu HP, Shaw S, Brinza L, Benning LG (2010) Crystallization of hematite (alpha-Fe_2O_3) under alkaline condition: the effects of Pb. Cryst Growth Des 10:1544–1551

Wang J, Becker U (2012) Energetics and kinetics of carbonate orientational ordering in vaterite calcium carbonate. Am Mineral 97:1427–1436

Wang SS, Xu AW (2013) Amorphous calcium carbonate stabilized by a flexible biomimetic polymer inspired by marine mussels. Cryst Growth Des 13:1937–1942

Wang D, Wallace AF, De Yoreo JJ, Dove PM (2009) Carboxylated molecules regulate magnesium content of amorphous calcium carbonates during calcification. Proc Natl Acad Sci U S A 106:21511–21516

Warthmann R, van Lith Y, Vasconcelos C, McKenzie JA, Karpoff AM (2000) Bacterially induced dolomite precipitation in anoxic culture experiments. Geology 28:1091–1094

Wehrmeister U, Jacob DE, Soldati AL, Loges N, Häger T, Hofmeister W (2011) Amorphous nanocrystalline and crystalline calcium carbonates in biological materials. J Raman Spectrosc 42:926–935

Xu XR, Cai AH, Liu R, Pan HH, Tang RK, Cho KW (2008) The roles of water and polyelectrolytes in the phase transformation of amorphous calcium carbonate. J Cryst Growth 310:3779–3787

Zou Z, Bertinetti L, Politi Y, Jensen ACS, Weiner S, Addadi L, Fratzl P, Habraken WJEM (2015) Opposite particle size effect on amorphous calcium carbonate crystallization in water and during heating in air. Chem Mater 27:4237–4246

Chapter 6
Ab Initio Modelling of the Structure and Properties of Crystalline Calcium Carbonate

Raffaella Demichelis, Paolo Raiteri, and Julian D. Gale

6.1 Introduction

During the past decade, the nucleation and crystal growth of biominerals, and calcium carbonate in particular, have attracted the attention and the imagination of the scientific community, due to the discovery of pathways, mechanisms, species and phases that differ from those expected through strictly classical nucleation theory (Gebauer et al. 2008; Demichelis et al. 2011; Wallace et al. 2013).

A full description of what is now called "alternative" or "non-classical" nucleation and crystal growth is presented in the previous chapters of this book, and at present it can be qualitatively summarised as follows:

(i) ions interact in water solution (see De Yoreo et al. 2017, Chap. 1; Lutsko 2017, Chap. 2; Penn et al. 2017, Chap. 13; Reichel and Faivre 2017, Chap. 14; Tobler et al. 2017, Chap. 15);
(ii) stable pre-nucleation clusters (PNCs) form (see De Yoreo et al. 2017, Chap. 1; Lutsko 2017, Chap. 2; Andreassen and Lewis 2017, Chap. 7);
(iii) as the concentration increases these clusters aggregate with each other giving rise to a liquid-liquid phase separation (De Yoreo et al. 2017, Chap. 1; Wolf and Gower 2017, Chap. 3);
(iv) dehydration and structural rearrangement lead to the formation of amorphous nanoparticles (De Yoreo et al. 2017, Chap. 1; Wolf and Gower 2017, Chap. 3; Fernandez-Martinez et al. 2017, Chap. 4; Rodriguez-Blanco et al. 2017, Chap. 5; Birkedal 2017, Chap. 10);

R. Demichelis (✉) • P. Raiteri • J.D. Gale
Curtin Institute for Computation, The Institute for Geoscience Research, and Department of Chemistry, Curtin University, PO Box U1987, Perth, WA 6845, Australia
e-mail: raffaella.demichelis@curtin.edu.au

A.E.S. Van Driessche et al. (eds.), *New Perspectives on Mineral Nucleation and Growth*,
DOI 10.1007/978-3-319-45669-0_6

(v) crystalline phases form following aggregation and structural rearrangement of amorphousnanoparticles (De Yoreo et al. 2017, Chap. 1; Fernandez-Martinez et al. 2017, Chap. 4; Rodriguez-Blanco et al. 2017, Chap. 5; Birkedal 2017, Chap. 10; Penn et al. 2017, Chap. 13; Tobler et al. 2017, Chap. 15);

(vi) the final crystalline polymorph grows, in some cases after transformation of one crystalline form to another (De Yoreo et al. 2017, Chap. 1; Andreassen and Lewis 2017, Chap. 7; Rao and Cölfen 2017, Chap. 8; Penn et al. 2017, Chap. 13; Van Driessche et al. 2017, Chap. 12; Tobler et al. 2017, Chap. 15).

In the specific context of calcium carbonate, much recent research has been focused on investigating the details of the early stages of mineral nucleation, leading to the formation of stable pre-nucleation clusters and of amorphous nanoparticles, including their polyamorphism and size-dependent stability (Cartwright et al. 2012). However, in this chapter we will focus on the subsequent steps involving the appearance of intermediate crystalline phases and their eventual transformation into more stable mineral phases (Munemoto and Fukushi 2008; Tang et al. 2009; Pouget et al. 2009; Rodriguez-Blanco et al. 2011), which are equally challenging and important to understand.

Calcium carbonate is arguably the most studied biomineral (Falini and Fermani 2017, Chap. 9). There are five known crystalline phases that are stable at room pressure; three of these correspond to the anhydrous $CaCO_3$ polymorphs calcite, aragonite and vaterite, while the remaining two are the metastable hydrated phases ikaite ($CaCO_3{\cdot}6H_2O$) and monohydrocalcite ($CaCO_3{\cdot}H_2O$). Recently, nanosized crystals of two high-pressure calcium carbonate phases, namely, calcite-III and calcite-IIIb, have been detected for the first time in natural geological samples at ambient pressure (Schaebitz et al. 2015). If we consider magnesium and other divalent cations along with calcium, we obtain a variety of hydrated and hydroxylated species, many of which have structures and properties that are still undetermined. Many of these compounds are difficult to obtain as pure phases and are generally unstable unless in extreme conditions (e.g. high salinity, lower or higher temperatures than standard conditions, Swainson and Hammond 2003; Swainson 2008; Munemoto and Fukushi 2008) or under particular biogenic conditions (Skinner et al. 1977; Mikkelsen et al. 1999; Kabalah-Amitai et al. 2013). Nonetheless, they are important since they are likely to represent a link between the early stages of mineralisation and the formation and growth of the stable polymorphs. In particular, knowing the reasons why these intermediates form in the first instance, their formation mechanisms and the details of their structures and properties may provide insights that are relevant to the evolution of species during nucleation, as well as regarding the polymorphism of the final stable minerals.

In this chapter we will focus on the calcium carbonate crystalline phases that are most frequently observed in biomineralisation processes, i.e. calcite, aragonite, vaterite, monohydrocalcite and ikaite. In particular, we will devote most of our attention to vaterite, whose complex, disordered structure has been recently, at least partially, explained through applying ab initio methods. This represents evidence that computational tools can provide accurate results that are complimentary to those obtained through experimental techniques and therefore have the potential to play a significant role in advancing this field. In fact, ab initio methods allow the exploration of the atomic and electronic structure of a given compound, which are indeed responsible for defining its properties and reactivity. Moreover, computational techniques, in general, have almost unlimited freedom to explore conditions that may be difficult to access via experiment, such as high pressures and temperatures, or specific isotopic compositions, as well as hypothetical structures and transition states.

Despite the focus of this chapter being on calcium carbonate, the methods and the analysis shown here can serve as a guide for undertaking similar studies on the many other crystalline phases that can appear in biomineralisation processes.

6.2 Methodology

The aim of this chapter is to illustrate how ab initio techniques can be applied to the investigation of the structure and properties of biominerals. We begin by providing a brief summary of the theoretical framework underpinning such techniques. A more exhaustive description of these methods can be found in Sherman (2016), Erba and Dovesi (2016), De La Pierre et al. (2016), and references therein.

6.2.1 The Potential Energy Surface

The energy of a system can be defined as a function of its atomic coordinates. This function is called the *potential energy surface* (PES);

$$PES = E(\mathbf{x}, \mathbf{H}) = E(x_1, x_2, \ldots, x_{3N}, H_{11}, H_{12}, \ldots, H_{33}) \quad (6.1)$$

where $\mathbf{x}$ represents the $3N$ atomic coordinates of an N-atom system and $\mathbf{H}$ is the matrix containing the Cartesian components of the three lattice vectors. More generally, at finite temperature one could include the momenta of the atoms to arrive at the free energy surface (FES), but for simplicity we will start by taking the approach of lattice dynamics in which the PES is explored first and then the effect of temperature is subsequently accounted for through consideration of the lattice vibrations (i.e. phonons).

Different conformations, configurations, polymorphs and transition states of a given system will therefore correspond to distinct points on the PES. In particular, stationary points of the PES (i.e. points where all the components of the gradient—the vector whose elements correspond to the partial first derivatives of the PES with respect to the atomic coordinates and lattice parameters—are zero) have a physical meaning: minima correspond to equilibrium geometries (minimum energy structures), whereas saddle points and maxima correspond to transition states.

The search for stationary points on the PES is known as *geometry optimisation*. Most of the optimisation techniques are able to locate the stationary point nearest to the starting geometry, through algorithms that minimise the energy and the components of the gradient. The search for the other stationary points requires the use of advanced optimisation tools, whose details are beyond the aim of this chapter. For most minerals, a reasonable guess for the atomic positions and for the lattice parameters is often available. As a consequence, the optimised structure is expected to correspond to either a realistic minimum energy structure or to a transition state close to it.

Once the stationary point has been found, it is of crucial importance to understand whether it corresponds to a minimum or to a saddle point. In order to do this, second derivatives must be calculated. In particular, for a $3N$ dimensional function, a $3N \times 3N$ matrix can be defined, the *Hessian matrix*, that contains all the partial second derivatives of the energy with respect to the atomic coordinates. Through diagonalising this matrix, the eigenvalues are obtained, and if they are all positive, it means that the structure is a minimum energy structure. If one or more eigenvalues are negative, then the structure is a transition state.

The physical quantities directly linked to the second derivatives of the energy with respect to the atomic coordinates are the vibrational frequencies. In particular, the eigenvalues of the mass-weighted Hessian matrix, as defined in the previous paragraph, are related to the square of the vibrational frequencies, whereas the eigenvectors correspond to the atomic motions associated with the various modes.

Transition states are characterised by the presence of *imaginary* frequencies. In the presence of such modes, a successful scheme that allows the minimum energy structure to be reached is highlighted here. First, the nature of the imaginary mode can be examined through its corresponding eigenvector. Then, the system is allowed to relax following the direction of this eigenvector. This may require the removal of one or more symmetry constraints. Finally, vibrational frequencies (second derivatives) are computed for the new structure. This procedure is repeated until the final structure is free from the presence of imaginary modes. The application of this scheme has been used extensively in the search for valid possible structures of vaterite, as will be described in Sect. 6.3.

Aside from the vibrational frequencies, there are many physical properties that can be calculated through computing the derivatives of the PES. For example, elastic constants are related to the second derivatives of the PES with respect to strain being applied to the lattice parameters; the dipole moment (and then also the intensities of infrared active modes) corresponds to the first derivative of the PES with respect to an applied electric field and so on. Many thermodynamic properties can also be derived from vibrational frequencies as well. It is then clear the importance of ensuring that the system we are dealing with has been accurately optimised and that its final geometry corresponds to a minimum energy structure.

6.2.2 Ab Initio Methods: A Quick Summary

In the previous section, we have stated that the energy of a system is a function of its geometry (Eq. 6.1). However, we have not yet defined the functional form of the PES. In fact, one of the main differences between the many possible computational approaches is the definition of such a functional form.

The basic assumption of ab initio (Latin: "from the beginning") or first-principles methods is that a system can be fully defined through its wave function, $\psi(\mathbf{x})$.[1] The energy of such system can be obtained through solving Schrödinger's equation:

$$\widehat{H}\psi(\mathbf{x}) = E\psi(\mathbf{x}) \tag{6.2}$$

where $\widehat{H}$ is the Hamiltonian operator and E the energy. In particular, we say that $\psi(\mathbf{x})$ is an eigenfunction of the Hamiltonian operator and E is its eigenvalue.

The main difference between the many types of ab initio techniques is in the definition of $\psi(\mathbf{x})$ (usually approximated through a linear combination of a set of either Gaussian-type or plane wave functions, the so-called basis set) and of $\widehat{H}$. In general, the Hamiltonian is expressed as a sum of the kinetic (T) and the potential (V) energy operators:

$$\widehat{H} = \widehat{T} + \widehat{V} \tag{6.3}$$

that in terms of the nuclei (n) and electrons (e) can be expressed as

$$\widehat{H} = \widehat{T}_n + \widehat{T}_e + \widehat{V_{nn}} + \widehat{V_{ne}} + \widehat{V_{ee}} \tag{6.4}$$

where we see the kinetic contributions of all nuclei ($\widehat{T}_n$, that within the Born-Oppenheimer approximation are separable and can be treated classically), the kinetic contribution $\widehat{T}_e$ of all electrons, the interactions between nuclei ($\widehat{V_{nn}}$), nuclei and electrons ($\widehat{V_{ne}}$), and electrons and electrons ($\widehat{V_{ee}}$).[2]

While the interaction potentials between nuclei-nuclei and nuclei-electrons can be fully described in terms of classical Coulomb interactions, $\widehat{V_{ee}}$ also has non-classical components, known as exchange and correlation. The exact formulation of the latter is unknown, though it plays a crucial role in defining the energy of a system, especially where weak interactions and van der Waals forces dominate.

In the Hartree-Fock (HF) approach, once the wave function has been defined, all terms apart from electron correlation can be calculated exactly and their accuracy depends only the quality of the basis set and numerical factors. Post HF methods

[1] We do not consider time dependence here.

[2] Additional terms that are generally null or negligible should be considered in certain circumstances, like when in the presence of an external applied field (e.g. if we want to explore dielectric properties) or if we are interested in computing magnetic properties.

are available to evaluate the electron correlation, which can be achieved with high accuracy using methods such as configuration interaction. However, such techniques are computationally expensive, and, up to very recently, implementations were only available for molecular cases. At present, these methods are rarely applied to solid-state system.

An alternative and popular method is density functional theory (DFT), which is based on the assumption that the energy of a system is a functional of its electron density, $\rho(x, y, z)$, which corresponds to the product of the wavefunction with its complex conjugate (or just the square of the wave function if it is real) integrated over the electron coordinates. Within this approach, all non-classical terms are grouped into one contribution, namely, the exchange-correlation functional (V_{xc}), whose exact formulation is unknown. Several kinds of approximations are available, the most widely adopted for minerals being the local density approximation (LDA), the generalised gradient approximation (GGA) and the hybrid DFT/HF approach (a percentage of exact HF exchange is included into V_{xc}). Despite being quite extensively used in the past (e.g. Yu et al. 2010 and Li et al. 2007), predictions made at the LDA level often lack accuracy, especially when dealing with systems that contain hydrogen or that are dominated by weak interactions.

In the following sections, we will make use of a Gaussian-type basis set specifically optimised for calcite (Valenzano et al. 2006) and of GGA and hybrid functionals, as implemented in the CRYSTAL code (Dovesi et al. 2014a). In particular, the PBE (Perdew et al. 1996) series of functionals will be considered, which comprises the pure GGA, PBE itself; its corresponding formulation devised for solids, PBEsol (Perdew et al. 2008); the 25 % hybrid, PBE0 (Adamo and Barone 1999); and PBE and PBE0 with inclusion of a long-range empirical correction terms, PBE-D2 (Grimme 2006) and PBE0-DC (Demichelis et al. 2013a), the latter with a dispersion contribution specifically fitted for carbonates. The popular hybrid functional B3LYP (Becke 1993) and the corresponding dispersion-corrected B3LYP-D2 (Grimme 2006) will also be used for further comparison. Specific details regarding the calculation set-up, parameters and algorithms adopted can be found in Demichelis et al. (2013a,b) and in De La Pierre et al. (2014b).

6.2.3 *Calcite and Aragonite: Reference Systems to Assess the Accuracy of Methods and Parameters*

One of the most delicate tasks, even for ab initio methods, is in ensuring that the model that we want to adopt is predictive for the system and properties that we would like to investigate. Therefore, we dedicate this section to validating the computational approach that we propose as a tool for the investigation of more complex calcium carbonate crystalline phases. Calcite and aragonite are relatively well-known and can be confidently used as reference systems to assess the accuracy

Table 6.1 Relative % difference between calculated and experimental structures of calcite at 298 K (unit cell volume, *Vol* [Å^3]; hexagonal lattice parameters, *a* and *c* [Å]; and bond lengths [Å]). Here the experimental structure is taken from Antao et al. (2009), Antao and Hassan (2010) and the values are given on the first line

	Vol	*a*	*c*	Ca-O	C-O
Exp	367.45	4.9877	17.0558	2.355	1.288
PBEsol	−0.1	+0.2	−0.6	−0.1	+0.3
PBE	+3.9	+1.2	+1.5	+1.4	+0.7
PBE-D2	+0.7	+0.7	−0.6	+0.2	+0.6
PBE0	+1.7	+0.4	+0.9	+0.8	−0.4
PBE0-DC	0.0	+0.1	−0.2	+0.1	−0.4
B3LYP	+3.8	+1.1	+1.6	+1.6	0.0
B3LYP-D2	−0.6	+0.4	−1.4	0.0	−0.3

of our methods. In particular, results obtained through our calculations will be compared to experimental data for a set of properties of one or both these two phases.

We will focus on testing and defining only those properties that we will use to characterise vaterite, without entering into the specific theoretical details of how they are computed. For a more comprehensive review of the many crystal properties that can be explored through ab initio methods and for further insights into the properties introduced in this section, we refer readers to Nye (1985), Dovesi et al. (2014b), Erba and Dovesi (2016) and De La Pierre et al. (2016).

6.2.3.1 Structural and Elastic Properties

The computed structure of calcite and some of its properties related to the elastic tensor are reported in Tables 6.1 and 6.2, where it is also compared against the available experimental data. Similar trends have been obtained for aragonite, but they are not reported here. The structure is obtained through minimising the energy with respect to the unit cell parameters and the atomic coordinates. Elastic properties can be calculated through evaluating the second derivatives of the energy with respect to the strain tensor components after the equilibrium structure is obtained (Nye 1985; Dovesi et al. 2014b).

All DFT methods considered here provide a good description of the lattice parameters and of the atomic positions, reported in terms of bond lengths (Demichelis et al. 2013a). It is a known feature that PBE and B3LYP tend to overestimate the volume by 3–4 %, but the inclusion of empirical dispersion improves the results.

Table 6.2 is an example of how ab initio methods can be applied to realistically predict quantities that can be directly related to macroscopic properties of materials, such as elastic constants, bulk, shear and Young's moduli and Poisson ratios. The

Table 6.2 Bulk (K), shear (G) and Young's (E) moduli, Poisson's ratio ν, and symmetry independent components of the elastic tensor (C) [GPa] for calcite. Subscripts of V and R indicate whether the definitions of Voigt or Reuss were used, respectively (Experimental data are reproduced from Chen et al. 2001)

	K_V	K_R	G_V	G_R	E	ν	C_{11}	C_{33}	C_{44}	C_{12}	C_{13}	C_{14}
Exp	76.1		32.8		–	–	149.4	85.2	34.1	57.9	53.5	−20.0
PBEsol	88.3	79.7	38.0	29.6	89.4	0.32	161.2	88.4	35.0	66.9	63.5	−18.7
PBE	80.3	73.8	36.6	30.0	87.3	0.31	149.9	85.5	32.6	58.8	56.0	−15.5
PBE-D2	86.4	76.8	37.8	28.3	87.4	0.32	160.2	84.1	34.5	65.5	61.5	−19.8
PBE0	85.4	78.3	39.6	31.9	93.6	0.31	159.9	90.3	36.0	62.0	58.9	−17.9
PBE0-DC	88.7	80.3	40.1	31.0	93.4	0.32	164.1	89.6	37.6	65.5	62.7	−20.1
B3LYP	82.2	75.7	39.6	32.2	93.5	0.30	156.8	88.8	35.2	58.0	55.2	−17.2
B3LYP-D2	91.3	80.0	41.8	29.8	94.2	0.32	172.8	86.7	38.6	52.6	63.6	−23.8

table shows that relatively good agreement between experimental and calculated properties is achieved for calcite with the adopted methods.

6.2.3.2 Thermodynamic Properties

The energy that is obtained through solving Schrödinger's equation (Eq. 6.1) corresponds to the *electronic energy*, U_{el}, and strongly depends on the formulation of the exchange-correlation functional, on the basis set and on the accuracy of other computational parameters (convergence thresholds, selected algorithms and approximations). As anticipated in a previous paragraph, the basis set adopted here was specifically optimised for calcite and therefore already shown to offer excellent predictions for several calcium carbonate crystal properties (Valenzano et al. 2006, 2007, Demichelis et al. 2012, 2013a,b, 2014, Carteret et al. 2013, De La Pierre et al. 2014a,b). The accuracy of the computational parameters was increased to a point that a further increase did not significantly affect the crystal properties. Therefore, in this chapter, any inaccuracy in predicting ΔU_{el} between two phases can be attributed to the approximation of the exchange-correlation functional.

To obtain quantities that can be compared to experimental values, i.e. the enthalpy (ΔH) and the free energy (ΔG) differences at a given temperature T, the *vibrational contributions* must also be computed. These are the zero-point energy, $ZPE = \frac{1}{2} h \sum_{\mathbf{k},i} \nu_{\mathbf{k},i}$ (where $\nu_{\mathbf{k},i}$ is the ith vibrational frequency at point $\mathbf{k}$ of the reciprocal space and h is Planck's constant); the temperature-dependent constant-pressure specific heat, $C_P(T)$; and the temperature-dependent entropy, $S(T)$. In particular, the quantities that should be evaluated are $H(T) = U_{el}(T) + PV(0) + ZPE + \int_0^T C_P(T)dT$ and $G(T) = H(T) - \int_0^T S(T)dT$, where $V(T)$ is the volume at temperature T and P is the pressure. Here, our model is based on a few main assumptions that are considered valid for many ionic, covalent and semi-ionic crystalline systems at room temperature. Firstly, vibrations are assumed to

Table 6.3 Values of the thermodynamic functions associated with the process aragonite → calcite at 298 K (S in J $K^{-1}mol^{-1}$, the rest in kJ/mol). Vibrational contributions were not calculated for B3LYP and PBE since they are expected to be of the order of fractions of kJ/mol and therefore would not be sufficient to lead to correct values of ΔH and ΔG (Experimental values from Wolf et al. 1996 and Königsberger et al. 1999)

	ΔU_{el}	$\Delta H(298)$	$\Delta S(298)$	$\Delta G(298)$
PBEsol	+1.3	+1.3	+3.7	+0.2
PBE[a]	−6.7	–	–	–
PBE-D2	+0.7	+0.8	+3.1	−0.1
PBE0	−4.8	−5.2	+4.1	−6.4
PBE0-DC[b]	+0.5	+0.4	+4.0	−0.8
B3LYP[a]	−12.5	–	–	–
B3LYP-D2	−3.1	−2.9	+0.2	−2.9
Exp	–	+0.44±0.05	+4.3±0.2	−0.8±0.2
Exp	–	+0.19	+3.72	−0.92

be harmonic. Secondly, thermal expansion is disregarded, so that $V(0) \simeq V(T)$ and $C_P(T) \simeq C_V(T)$. Finally, only vibrational entropy is considered in the entropic term.

Computing the difference between these thermodynamic functions obtained at the DFT level for two polymorphs would, in principle, predict their relative stability. However, there are cases where this approach fails. In particular, the variation in relative stabilities associated with different DFT functionals is known to be larger than the actual values for phases that exhibit major structural differences and small energy differences. This is due to short- and long-range dispersion interactions not being properly accounted for Casassa and Demichelis (2012). Calcite and aragonite represent one of those cases: their structures are very different and their energetics very challenging, with ΔH and ΔG having opposite signs and being on the order of fractions of kJ/mol. On the contrary, for polymorphs having similar densities, such as calcite and vaterite, this error is fortuitously cancelled.

The relative energy differences between calcite and aragonite computed with different DFT functionals are compared to the experimental values in Table 6.3. A detailed description of the reasons behind the apparent discrepancies for many of the functionals is given in Demichelis et al. (2013a) and references therein, together with a discussion of the more complex case of hydrated polymorphs and water incorporation in the solid phase. Data reported in Table 6.3 show that DFT schemes that are able to provide an accurate description of many crystal properties (e.g. B3LYP, PBE0) are not necessarily able to reproduce the thermodynamics between polymorphs that are too different in structure and density. It is also shown that the choice of more appropriate dispersion coefficients (PBE0-DC) can at least partly redress this lack of accuracy.

The full energetics of vaterite with respect to calcite will be discussed in detail in Sect. 6.3. Table 6.4 reports only the values of the electronic energy differences between calcite and one of the stable phases of vaterite (viz. *Vh* in the table) obtained

Table 6.4 Differences in electronic energy [kJ/mol per fu] between hexagonal vaterite (*Vh*, $P3_221$) and calcite (*C*) and between hexagonal and monoclinic vaterite (*Vm*, *C*2). For the different structures of vaterite, see Sect. 6.3

	$\Delta U_{el}(Vh - C)$	$\Delta U_{el}(Vh - Vm)$
PBEsol	+3.1	−0.2
PBE	+4.5	−0.1
PBE0	+3.9	−0.1
PBE0-DC	+4.1	−0.1
B3LYP	+4.7	0.0

with a subset of DFT functionals, including those that give the largest deviations from experiment in Table 6.3. The differences in electronic energy between two forms of vaterite, namely, *Vh* and *Vm*, are also reported. Due to the more similar densities of the considered polymorphs, all DFT methods are able to reproduce the right stability order with small variations in the electronic energy difference.

6.2.3.3 Vibrational Spectroscopy

Raman and IR spectra of calcite, aragonite and related phases have been extensively investigated from both a computational and an experimental point of view (e.g. Donoghue et al. 1971, Frech et al. 1980, Gillet et al. 1996, Alía et al. 1997, Prencipe et al. 2004, Valenzano et al. 2006, 2007, Wehrmeister et al. 2010, 2011, Carteret et al. 2013, De La Pierre et al. 2014a,b). The method that we are applying here is the same as the one adopted in many of the aforementioned publications (see, e.g. De La Pierre et al. 2014a,b), with a demonstrated ability to provide accurate predictions of both vibrational frequencies and intensities. Here we simply highlight that the accuracy in computing vibrational properties is important for three main reasons. Firstly, a predictive method, like the one we apply here, can be successfully used to investigate phases whose properties and structure are still a matter of debate. Secondly, as mentioned above, many thermodynamic functions can be obtained *via* computing the vibrational frequencies of the system; the more accurate the vibrational frequencies, the more accurate the quantities that depend on them. Thirdly, since computed vibrational frequencies contain information about the topography of the PES, they allow us to discriminate between stable structures and transition states. Crucially, the third of these points has been used to investigate the crystallography of vaterite and provide a model for its complex structure.

Table 6.5 shows that the B3LYP hybrid functional is, in principle, one of the best choices for reproducing vibrational properties.[3] The maximum ($|\Delta_{max}|$) and the mean absolute difference ($|\bar{\Delta}|$) between computed and experimental frequencies

[3] Aragonite is used in this example, and only Raman active frequencies have been considered, but similar comments hold also for other minerals and for IR active modes.

Table 6.5 Statistical analysis of the differences between calculated and experimental Raman active modes of aragonite. $|\Delta_{max}|$ and $|\bar{\Delta}|$ are the maximum and the mean absolute difference, respectively, and $\bar{\Delta}$ is the mean differences. All data are in cm^{-1} (Experimental data from De La Pierre et al. 2014a and Carteret et al. 2013)

	PBEsol	PBE0	PBE0-DC	B3LYP
$\lvert\bar{\Delta}\rvert$	+10.9	+19.5	+16.1	+7.8
$\bar{\Delta}$	−7.4	+19.2	+16.1	+0.7
$\lvert\bar{\Delta}_{max}\rvert$	+57.9	+69.0	+65.7	+23.3

are the lowest, and the mean difference ($\bar{\Delta}$) is around 0 cm^{-1}, meaning that there is no systematic shift towards higher or lower values with respect to experiment. However, Table 6.3 shows that B3LYP is not accurate in predicting the electronic energy difference between different phases. As ΔU_{el} is the contribution that dominates the energetics, we should look for a compromise since ideally we want a model that is able to reasonably reproduce vibrational spectra and thermodynamics simultaneously. PBEsol can be considered a good compromise, as it provides reasonable results for carbonate properties, including vibrational spectra, and for their energetics. Despite the data in Table 6.5 suggesting that PBEsol is significantly worse than B3LYP for predicting vibrational frequencies, if we consider the five main regions of the spectrum separately (i.e. lattice modes, L; symmetric and asymmetric stretching of CO_3^{2-}, ν_1 and ν_3; in-plane and out-of-plane bending of CO_3^{2-}, ν_4 and ν_2), all frequencies within a single region are shifted with respect to the actual value by nearly the same quantity ($\Delta L = +3.1$; $\Delta\nu_1 = -11.4$; $\Delta\nu_2 = -44.4$; $\Delta\nu_3 = -7.8$; $\Delta\nu_4 = -24.3$; all in cm^{-1}). To obtain a more accurate prediction, it is therefore sufficient to subtract this quantity from each frequency of the region (De La Pierre et al. 2014b).

6.3 The Multiple Structures of Vaterite

As mentioned in Sect. 6.1, vaterite is one of the three crystalline polymorphs of anhydrous calcium carbonate. Though it is metastable with respect to calcite and aragonite, it is found in nature as a result of biomineralisation (Cartwright et al. 2012; Hasse et al. 2000; Qiao and Feng 2007). For example, vaterite is crystallised in bivalve organisms and mussels to repair damage in their hard tissues (Wilbur and Watabe 1963; Spann et al. 2010). Vaterite is also often observed as an intermediate phase during the transformation that leads from amorphous calcium carbonate nanoparticles (ACC) to calcite and aragonite (Cartwright et al. 2012; Gebauer et al. 2010; Rodriguez-Blanco et al. 2011). For these reasons, vaterite is now attracting much interest as further insight into its features could reveal new details about non-classical nucleation pathways, polymorphism and crystallisation under biogenic conditions.

Table 6.6 Structures proposed for vaterite up to 2012: space group (SG) and lattice parameters (a, b, c [Å]). Cell angles from Mugnaioli et al. (2012): β=118.94° ($C2/c$); α=γ=90.0°, β=99.22° ($\overline{C}1$); α =82.1°, β=97.9°, γ=119.3° ($\overline{P}1$, in parentheses, corresponds to the primitive unit cell of the structure with $\overline{C}1$ symmetry). Cell angles from Wang and Becker (2012): β=115.9° ($C2/c$)

Reference	SG	a	b	c
Meyer (1959)	$Pbnm$	4.13	7.15	8.48
McConnell (1960), Bradley et al. (1966)	$P6_322$	7.135	7.135	8.524
Kahmi (1963), Sato and Matsuda (1969)	$P6_3/mmc$	4.13	4.13	8.49
Meyer (1969), Gabrielli et al. (2000)	$P6_3/mmc$	7.15	7.15	16.96
Dupont et al. (1997)	$P6_3/mmc$	7.169	7.169	16.98
Le Bail et al. (2011)	$Ama2$	8.7422	7.1576	4.1265
Medeiros et al. (2007)	$Pbnm$[a]	4.531	6.640	8.480
Wang and Becker (2009)	$P6_522$[a]	7.290	7.290	25.302
Wang and Becker (2012)	$C2/c$[b]	12.62	7.29	9.37
Mugnaioli et al. (2012)	$C2/c$	12.408	7.1372	9.4067
Mugnaioli et al. (2012)	$\overline{C}1$	12.17	7.12	25.32
	($\overline{P}1$)	7.05	7.05	25.32

[a] First-principles study with no calculation of the second derivatives
[b] Force-field simulation

The structure of vaterite has been debated for over 50 years, due to what is often referred to as "disorder" of the carbonate anion sites. Table 6.6 presents a summary of the most relevant studies from 1959 to 2012 aimed at resolving the structure of vaterite, showing a clear disagreement as to the crystal system (orthorhombic, hexagonal, monoclinic and triclinic), the crystal symmetry and site occupancy and even as to the unit cell size. We refer the readers to the original literature for more details about these studies. Here we show how the approach described in this chapter has contributed a new perspective through proposing a model which is in agreement with most of the experimental findings.

In Demichelis et al. (2012, 2013b), electronic structure calculations were performed for the orthorhombic models ($Pbnm$, $Ama2$), for the hexagonal model ($P6_522$) that was obtained by Wang and Becker (2009) as a result of re-examining previously proposed models through DFT and force-field-based calculations and for the monoclinic and triclinic models ($C2/c$ and $\overline{C}1$) proposed by Mugnaioli et al. (2012) as a result of combined automated diffraction tomography (ADT) and precession electron diffraction (PED). Note that Wang and Becker (2012) predicted the presence of a monoclinic $C2/c$ basin slightly less stable than their hexagonal $P6_522$ one through applying molecular dynamics techniques.

The calculation of second derivatives, which was not performed in previous computational studies, showed that all of these structures but one ($Ama2$) correspond to transition states rather than to minimum energy structures and therefore cannot be a true representation of vaterite.

As mentioned in Sect. 6.2, transition states are characterised by the presence of one or more imaginary modes (i.e. the Hessian matrix has one or more negative eigenvalues). The structure with *Pbnm* symmetry exhibits one imaginary mode, the $P6_522$ hexagonal model four and the monoclinic ($C2/c$) and triclinic ($\overline{C}1$) structures two. Following the displacement along the eigenvectors related to the imaginary modes, it was found that they all correspond to symmetry forbidden rotations of carbonate anions. After removing the corresponding symmetry constraints, the true minimum energy structures were found.

As a result of this procedure, it was found that the structure with *Pbnm* symmetry relaxes to a $P2_12_12_1$ configuration that is 1–2 kJ/mol per formula unit more stable than the original one, depending on the DFT functional. The $P6_522$ arrangement relaxes towards three minimum energy structures, $P3_221$ (−3.5 kJ/mol), $P6_5$ (−2.5 kJ/mol) and $P112_1$ (−3.0 kJ/mol),[4] and one transition state with two imaginary modes. The latter was not fully analysed, but through removing the symmetry constraint related to the imaginary modes and reoptimising, $P3_221$ and a $P112_1$ structures were obtained. The $C2/c$ structure assumes either a $C2$ or a Cc configuration, which are 1.4 and 0.5 kJ/mol more stable, whereas the $\overline{C}1$ structure falls into two very similar structures, both with symmetry $C1$ and about 0.1 and 0.2 kJ/mol more stable.[5]

A more accurate analysis of the structure and relative energies shows that both of the orthorhombic models considered (*Ama*2 and $P2_12_12_1$) do not fit with the experimental expectations. Apart from being too dense, due to the optimised b lattice parameter being about 10 % shorter than that suggested by Meyer (1959) and Le Bail et al. (2011), $P2_12_12_1$ and *Ama*2 are about 1 kJ/mol and 18 kJ/mol less stable, respectively, than the lowest energy structure derived from Wang and Becker's model ($P3_221$). A second point against the most stable orthorhombic model comes from a study performed by Balan et al. (2014) through simulation of the mixing energy of sulphate into vaterite and reanalysis of the anomalous data presented by Fernández-Díaz et al. (2010). Notably, this study also ruled out the hexagonal $P6_3/mmc$ structure in favour of the $P3_221$ model proposed by Demichelis et al. (2012), whereas the monoclinic and the triclinic models were not considered.

The seven structures that were obtained through the computational procedure highlighted in this chapter and that have structural and energetic features that are in agreement with the experimental observations were divided into three groups: six-layer hexagonal, referring to the structures derived from the hexagonal model that has six layers of carbonate anions in the unit cell ($P3_221$, $P6_5$, $P112_1$); two-layer monoclinic, referring to those derived from the monoclinic structure that has two layers of carbonate anions in the unit cell ($C2$, Cc); and six-layer triclinic, referring to those obtained from Mugnaioli's triclinic structure ($C1$, $C1$). Figure 6.1 shows that these groups of structures can be considered different polytypes of vaterite, differing only in the sequence of the carbonate layers stacked along the z Cartesian

[4]Unless stated otherwise, all energies given in the text were calculated with the PBEsol functional; for vaterite similar values are obtained also with other functionals.

[5]The conventional choice for this space group would be $P1$; however, $C1$ is used to indicate that it has been obtained from the $\overline{C}1$ structure originally proposed by Mugnaioli et al. (2012). Despite this structure being triclinic, its conventional cell has $\alpha \simeq \gamma \simeq 90°$, and this is why in a previous publication we referred to this model as quasi-monoclinic or "6-layer" monoclinic.

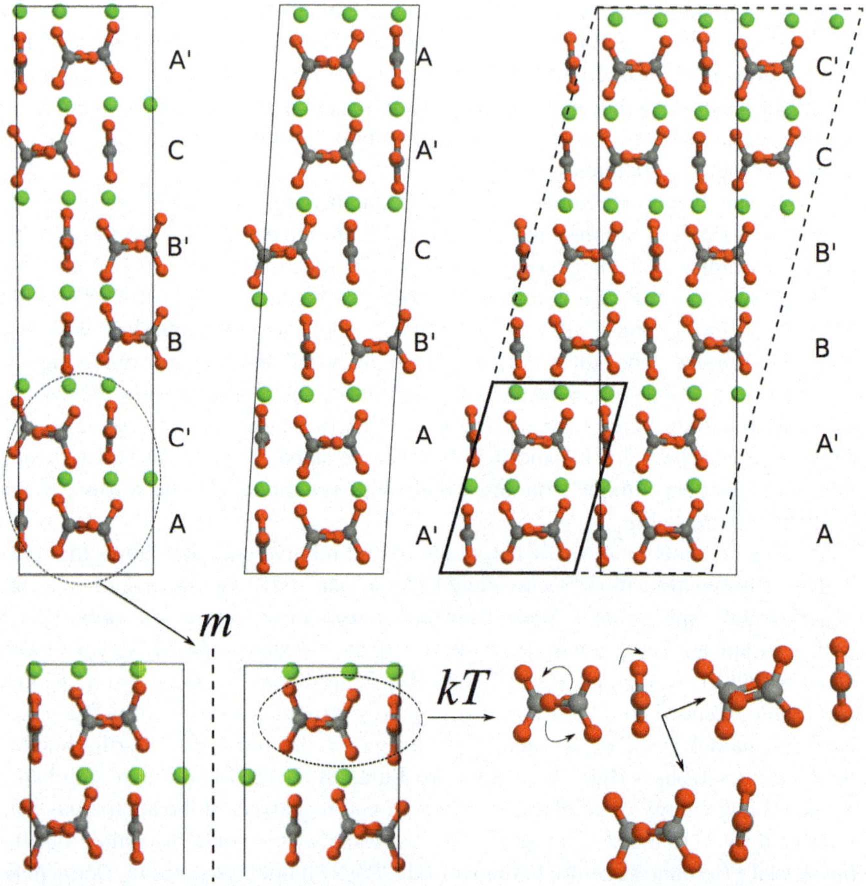

Fig. 6.1 *Top*: *Ball* and *stick* representation of the structures for vaterite belonging to the three different families of polytypes, viewed along the *a* lattice vector. O, Ca and C are represented in *red*, *green* and *grey* colours, respectively. From *left* to *right*: the six-layer hexagonal model, the six-layer triclinic model and the two-layer monoclinic model. The latter has a smaller unit cell (*bold black line*), so that a six-layer supercell (*thin black line*) is built to show the structural similarities and differences with respect to the other models. The stacking sequence of CO_3^{2-} layers refers to the orientation of the CO_3^{2-} units that have a C-O bond parallel to *a*: planes with the same shift with respect to *b* are labelled with the same letter (A for 0; B for 1/3; C for 2/3); planes with the C-O bond pointing towards $+a$ and $-a$ are labelled with and without a prime, respectively. *Bottom*: The chiral image of a portion of the hexagonal structure is represented (*left*), as well as the meaning of rotational disorder of CO_3^{2-}, which is allowed at room temperature (Reprinted with permission from Demichelis, Raiteri, Gale, Dovesi, The Multiple Structures of Vaterite, Cryst. Growth Des., 13 (6), 2247–2251. Copyright 2013 American Chemical Society)

axis. In particular, one can go from the monoclinic to the hexagonal model by switching two planes, whereas the triclinic model has two missing planes with respect to the other two models, causing its γ lattice parameter to deviate from

90°. Starting from the hexagonal model, by performing all the possible physically allowed permutations of the six different planes along the *c* axis, the only structures that are found correspond to one of those in the hexagonal or monoclinic groups (including their chiral images).

Notably, the hypothesis of stacking faults in vaterite was originally suggested by Meyer (1969), and twin stacking faults were detected by Qiao and Feng (2007) using HRTEM and SAED techniques on natural samples of vaterite. Wehrmeister et al. (2010) suggested the existence of different polytypes of vaterite, though there was not enough evidence in their data to fully support the claim. On the contrary, chirality was never suggested prior to the studies of Demichelis et al. (2013b). At this point, it is reasonable to hypothesise that a certain number of other structures may exist, having missing planes and different sequences with respect to the three groups of structures reported here.

Symmetry and thermodynamic data obtained for the aforementioned hexagonal, monoclinic and triclinic models are summarised in Table 6.7. All structures present similar free energies and are equally likely to represent a good model for vaterite, with $C2$ and $P3_221$ models being the most stable. Data in the table also show that switching two carbonate layers with each other has no or little cost in energy. Figure 6.2 shows a plot of ΔU_{el} between a given structure and $P3_221$. While the data presented here are unable to predict whether it is possible to transform a structure belonging to one basin into a structure belonging to another one, the energy barrier for rotating CO_3^{2-} anions is so low that within the same basin we can suppose that structures can interconvert between each other at 298 K.

From the above discussion, we can conclude that from a theoretical point of view, the hypothesis of vaterite existing in several forms rather than assuming one particular structure is very likely to be correct. In particular, our analysis shows that

Table 6.7 Symmetry and stability of the seven structures of vaterite found by Demichelis et al. (2012, 2013b). For each structure, the corresponding chiral image, crystal system and number of independent carbonate anions in the unit cell (N_{CO_3}) are indicated. The thermodynamic quantities (PBEsol) associated with the process vaterite→calcite at 298 K are calculated (all in kJ/mol except for ΔS in J K^{-1} mol^{-1}). The Boltzmann relative populations p at 298 K are also calculated for each structure. The Boltzmann-weighted average of the thermodynamic quantities and the corresponding experimental data from Königsberger et al. (1999) are shown

		Chiral image	Crystal system	N_{CO_3}	ΔU_{el}	ΔH	ΔS	ΔG	p
Hexagonal	$P3_221$	$P3_121$	Trigonal	4	+3.1	+3.3	+0.8	+3.1	0.090
(6-layer)	$P6_5$	$P6_1$	Hexagonal	3	+3.8	+3.8	−1.1	+4.1	0.060
	$P112_1$	$P112_1$	Monoclinic	9	+3.6	+3.9	+1.2	+3.6	0.074
Monoclinic	Cc	–	Monoclinic	3	+4.0	+4.0	−0.5	+4.2	0.058
(2-layer)	$C2$	$C2$	Monoclinic	4	+3.2	+3.2	−1.0	+3.5	0.077
Triclinic	$C1$	$C1$	Triclinic	18	+3.5	+3.7	+1.7	+3.2	0.087
(6-layer)	$C1$	$C1$	Triclinic	18	+3.6	+3.9	+2.0	+3.3	0.083
Average					+3.5	+3.6	+0.6	+3.5	
Exp.						+3.60	+1.30	+3.21	

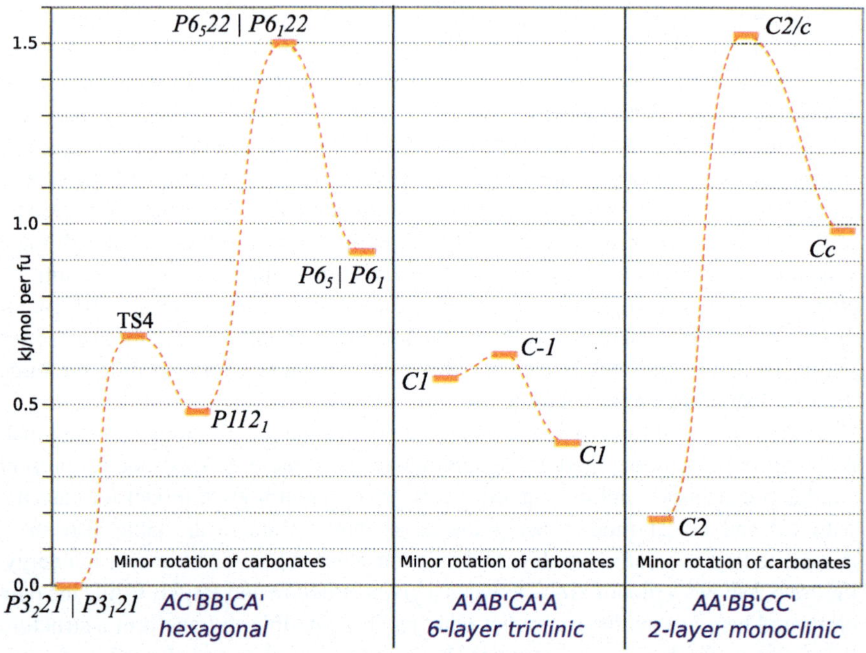

Fig. 6.2 Representation of ΔU_{el} for the three polytypes of vaterite, separated by vertical lines. Space groups having a chiral image that belong to another space group are indicated. The structure labelled as *TS*4 is a transition state obtained after the analysis of one of the four imaginary modes found for $P6_522$; see text for more details (Reprinted with permission from Demichelis, Raiteri, Gale, Dovesi, The Multiple Structures of Vaterite, Cryst. Growth Des., 13 (6), 2247–2251. Copyright 2013 American Chemical Society)

there are three levels of complexity in the structure of vaterite: the intrinsic rotational disorder of carbonate anions, the possibility of having different carbonate layer sequences and the presence of chirality. The claim of vaterite assuming multiple structures was confirmed for the first time in the study by Kabalah-Amitai et al. (2013), where HRTEM images show that there are at least two interdispersed crystal structures in a vaterite sample. However, this work does not investigate the details of the structural features of the two different phases. It was found that their main difference is related to carbonate anions, whereas calcium layers are nearly the same. While one of the phases is claimed to be hexagonal, the structure of the other one could not be determined.

As a further attempt to definitively solve the problem of the structure of vaterite, two independent studies have been recently undertaken, using experimental and computed NMR, XRD and Raman spectra, all leading to the same result.

One of these works De La Pierre et al. (2014b) re-examined existing and new experimental Raman data in the light of the new structural information from Demichelis et al. (2013b) and Kabalah-Amitai et al. (2013). The rich collection of Raman spectra recorded on a variety of vaterite samples of geological, biological and synthetic origin from Wehrmeister and co-workers (2010; 2011) was used. New samples considered in this study included *Herdmania momus* spicules from the Great Barrier Reef (Queensland, Australia), the same kind of samples where Kabalah-Amitai et al. (2013) identified two distinct structures.

One of the most significant points arising in this study was Wehrmeister et al.'s (2010) observation, based on the analysis of their Raman spectra, that vaterite has at least three independent carbonate units. Notably, all structures proposed in the literature prior to Demichelis et al. (2012, 2013b) have no more than two such groups. This means that either the new structures are a better choice or that multiple structures within the same sample are present or both.

The Raman spectra for the seven structures proposed here as possible candidates were computed by De La Pierre et al. (2014b) and compared to the experimental data. Overall, they all look very similar. However, all the computed spectra but two have peaks that do not appear in the most intense region of the experimental spectra, namely, ν_1 (symmetric stretching of CO_3^{2-}, 1020–1160 cm^{-1}) and ν_4 (in-plane bending of CO_3^{2-}, 660–750 cm^{-1}). Figure 6.3 shows that *C*2 and *P*$3_2$21, the most stable two-layer monoclinic and six-layer hexagonal structures, respectively, have Raman spectra that match with those experimentally recorded. The latter structure is in agreement with Kabalah-Amitai et al. (2013), who found that one of the structures in their sample has a hexagonal unit cell.

On the basis of Raman spectroscopy alone, it is not possible to tell whether the first, the second or both structures are present in the experimental samples. Also, the remaining five theoretical structures seem to be ruled out according to this study. These results then seem to provide evidence that opposes the room temperature intra-basin structural interconversion via carbonate rotations suggested by Demichelis et al. (2012). This would result in different domains being present in the same crystal and thus in a spectra that is a Boltzmann-weighted average of all seven spectra. A possibility is that the system becomes kinetically trapped in a subset of the possible stable structures. To verify this hypothesis, further investigation is required in order to determine the activation barriers for the interconversion between all of the different energy minima.

A second study, by Burgess and Bryce (2015), has recently shown that only *C*2 and *P*$3_2$21 structures are compatible with ^{43}Ca NMR spectra recorded on samples of vaterite, after considering most of the structures that have been proposed in the literature. *C*2 and *P*$3_2$21 were also shown to be compatible with the measured XRD diffraction pattern. As with the Raman spectroscopy, also ^{43}Ca NMR seems to be unable to provide a definitive answer on whether the former, the latter or both structures are present in the real sample.

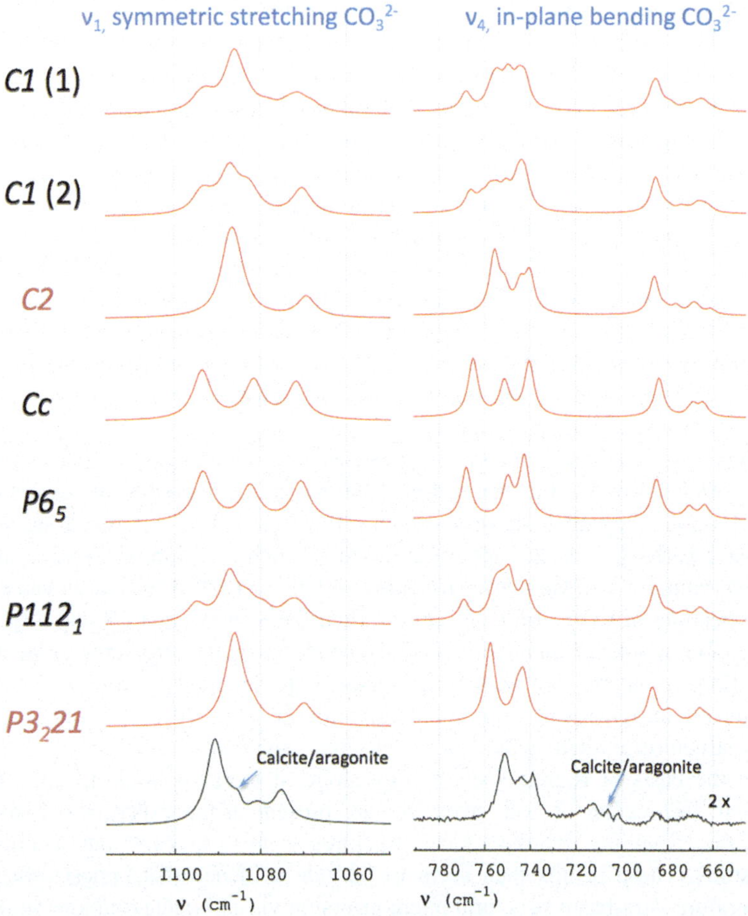

Fig. 6.3 Calculated Raman spectra for the seven structures of vaterite (*red*) and comparison with one experimental spectrum recorded on a geological sample from Wehrmeister et al. (2010) (*black*). Other samples exhibit very similar features. In the latter, signals resulting from calcite and aragonite impurities are highlighted

6.4 Conclusions and Future Challenges

The theoretical approach highlighted in this chapter is highly predictive in determining the structures and properties of crystalline phases. These tools are therefore particularly useful when studying phases and processes that are challenging to investigate with experimental techniques. The case of vaterite represents one of the best examples of how ab initio methods can be applied to unravel the complexities of metastable biomineral phases and more generally of minerals that are not easily accessible as large single crystals for diffraction. It also represents a successful example of how computational schemes can be combined with experimental techniques to investigate complex mineral structures.

Future challenges in this field are related to the development of algorithms able to address the non-generality of the many approximations on which ab initio methods rely. In the particular case of biominerals, one of the most important features that would open the way for an appropriate investigation of the composition, the stability and the stoichiometry of the many hydrated and hydroxylated phases would be the ability to properly account for weak interactions in a fast and accurate way. Methods that are currently available are either too approximate or not available in solid-state codes or computationally too expensive for systems that contain tens of atoms in their unit cell (Pisani et al. 2008; Del Ben et al. 2012).

In the particular case of vaterite, future challenges are mostly related to finding or developing experimental techniques that are able to distinguish the different structures and confirm their features. Moreover, the average size of a vaterite single crystal (tens of nm in the best case scenario) is too small to undertake any investigation as to its chirality with available optical techniques (Bilotti et al. 2002). From a biomineralisation perspective, chirality would be an extremely interesting property to investigate, since vaterite often grows under biogenic conditions, therefore in the presence of biological molecules that are often optically active. In this context, also other biominerals may exhibit optical activities, such as monohydrocalcite, which like vaterite belong to one of those space groups that are intrinsically chiral (Demichelis et al. 2014).

Acknowledgements The research described in this chapter was supported by the Australian Research Council through Discovery Grant DP0986999 and Future Fellowship FT130100463 and by Curtin University through the Curtin Research Fellowship scheme. Both the Pawsey Centre and the Australian National Computational Infrastructure facilities are acknowledged for the provision of computer time.

We are also grateful to Dorrit E. Jacob (Macquarie University), Ursula Wehrmeister (Universität Mainz), Marco De La Pierre (Curtin University), David Quigley (University of Warwick) and Pupa Gilbert (University of Wisconsin-Madison) for their collaboration in investigating the structure of vaterite and to Roberto Dovesi and Roberto Orlando (Università degli Studi di Torino) for providing us with access to the CRYSTAL code.

References

Adamo C, Barone V (1999) Toward reliable density functional methods without adjustable parameters: the PBE0 model. J Chem Phys 110:6158–6170

Alía M, Díaz de Mera Y, Edwards HGM, González Martín P, López Andrés S (1997) FT-Raman and infrared spectroscopic study of aragonite-strontianite ($Ca_xSr_{1-x}CO_3$) solid solution. Spectrochim Acta A 53:2347–2362

Andreassen JP, Lewis AE (2017) Classical and nonclassical theories of crystal growth. In: Van Driessche AES, Kellermeier M, Benning LG, Gebauer D (eds) New perspectives on mineral nucleation and growth, Springer, Cham, pp 137–154

Antao SM, Hassan I (2010) Temperature dependence of the structural parameters in the transformation of aragonite to calcite, as determined from in situ synchrotron X-ray diffraction data. Can Miner 48:1225–1236

Antao SM, Hassan I, Mulder WH, Lee PL, Toby BH (2009) In situ study of the $R\bar{3}c \rightarrow R\bar{3}m$ orientational disorder on calcite. Phys Chem Miner 36:159–169

Balan E, Blanchard M, Pinilla C, Lazzeri M (2014) First-principles modeling of sulfate incorporation and $^{34}S/^{32}S$ isotopic fractionation in different calcium carbonates. Chem Geol 374–375:84–91

Becke AD (1993) Density functional theochemistry. III The role of exact exchange. J Chem Phys 98:5648–5652

Bilotti I, Biscarini P, Castiglioni E, Ferranti F, Kuroda A (2002) Reflectance circular dichroism of solid-state chiral coordination compounds. Chirality 14:750–756

Birkedal H (2017) Phase transformations in calcium phosphate crystallization. In: Van Driessche AES, Kellermeier M, Benning LG, Gebauer D (eds) New perspectives on mineral nucleation and growth, Springer, Cham, pp 199–210

Bradley WF, Graf DL, Roth RS (1966) The vaterite-type ABO_3 rare-earth borates. Acta Crystallogr 20:283–287

Burgess KNM, Bryce DL (2015) On the crystal structure of the vaterite polymorph of $CaCO_3$ & on. A calcium-43 solid-state NMR and computational assessment. Solid State Nucl Magn Reson 65:75–83

Carteret C, De La Pierre M, Dussot M, Pascale F, Erba A, Dovesi R (2013) The vibrational spectrum of $CaCO_3$ aragonite: a combined experimental and quantum-mechanical investigation. J Chem Phys 138:014201

Cartwright JHE, Checa AG, Gale JD, Gebauer D, Sainz-Díaz CI (2012) Calcium carbonate polyamorphism and its role in biomineralization: how many amorphous calcium carbonates are there? Angew Chem Int Ed 51:11960–11970

Casassa S, Demichelis R (2012) Relative energy of aluminum hydroxides: the role of electron correlation. J Phys Chem C 116:13313–13321

Chen C-C, Lin C-C, Liu L-G, Sinogeikin SV, Bass JD (2001) Elasticity of single-crystal calcite and rhodochrosite by Brillouin spectroscopy. Am Miner 86:1525–1529

De La Pierre M, Carteret C, Maschio L, André E, Orlando R, Dovesi R (2014a) The Raman spectrum of $CaCO_3$ polymorphs calcite and aragonite. A combined experimental and computational study. J Chem Phys 140:164509

De La Pierre M, Demichelis R, Wehrmeister U, Jacob DE, Raiteri P, Gale JD, Orlando R (2014b) Probing the multiple structures of vaterite through combined computational and experimental Raman spectroscopy. J Phys Chem C 118:27493–27501

De Yoreo JJ, Sommerdijk NAJM, Dove PM (2017) Nucleation pathways in electrolyte solutions. In: Van Driessche AES, Kellermeier M, Benning LG, Gebauer D (eds) New perspectives on mineral nucleation and growth, Springer, Cham, pp 1–24

De La Pierre M, Demichelis R, Dovesi R (2016) Molecular modeling of geochemical reactions: an introduction, chapter 10. In: Vibrational spectroscopy of minerals through ab initio methods. Wiley, Chichester, pp 341–374

Del Ben M, Hutter J, VandeVondele J (2012) Second-order Møller Plesset perturbation theory in the condensed phase: An efficient and massively parallel Gaussian and plane waves approach. J Chem Theory Comput 8:4177–4188

Demichelis R, Raiteri P, Gale JD (2014) Structure of hydrated calcium carbonates: a first-principles study. J Cryst Growth 401:33–37

Demichelis R, Raiteri P, Gale JD, Quigley D, Gebauer D (2011) Stable prenucleation mineral clusters are liquid-like ionic polymers. Nat Commun 2:590

Demichelis R, Raiteri P, Gale JD, Dovesi R (2012) A new structural model for disorder in vaterite from first-principles calculations. CrystEngComm 14:44–47

Demichelis R, Raiteri P, Gale JD, Dovesi R (2013a) Examining the accuracy of density functional theory for predicting the thermodynamics of water incorporation into minerals: the hydrates of calcium carbonate. J Phys Chem C 117:17814–17823
Demichelis R, Raiteri P, Gale JD, Dovesi R (2013b) The multiples structure of vaterite. Cryst Growth Des 13:2247–2251
Donoghue M, Hepburn PH, Ross SD (1971) Factors affecting the infrared spectra of planar anions with D_{3h}, symmetry—V: the origin of the splitting of the out-of-plane bending mode in carbonates and nitrates. Spectrochim Acta A 27:1065–1072
Dovesi R, Orlando R, Erba A, Zicovich-Wilson CM, Civalleri B, Casassa S, Maschio L, Ferrabone M, De La Pierre M, D'Arco P, Noël Y, Causà M, Rérat M, Kirtman B (2014a) CRYSTAL14: a program for the ab initio investigation of crystalline solids. Int J Quantum Chem 114: 1287–1317
Dovesi R, Saunders VR, Roetti C, Orlando R, Zicovich-Wilson CM, Pascale F, Civalleri B, Doll K, Harrison NM, Bush IJ, D'Arco Ph, Llunell M, Causà M, Noël Y (2014b) CRYSTAL 2014 user's manual
Dupont L, Portemer F, Figlarz M (1997) Synthesis and study of a well crystallized $CaCO_3$ vaterite showing a new habitus. J Mater Chem 7:797–800
Erba A, Dovesi R (2016) Molecular modeling of geochemical reactions: an introduction, chapter 3. In: Quantum-mechanical modeling of minerals. Wiley, Chichester, pp 77–105
Falini G, Fermani S (2017) Nucleation and growth from a biomineralization perspective. In: Van Driessche AES, Kellermeier M, Benning LG, Gebauer D (eds) New perspectives on mineral nucleation and growth, Springer, Cham, pp 185–198
Fernández-Díaz L, Fernández-González A, Prieto M (2010) The role of sulfate groups in controlling $CaCO_3$ polymorphism. Geochim Cosmochim Acta 74:6064–6076
Fernandez-Martinez A, Lopez-Martinez H, Wang D (2017) Structural characteristics and the occurrence of polyamorphism in amorphous calcium carbonate. In: Van Driessche AES, Kellermeier M, Benning LG, Gebauer D (eds) New perspectives on mineral nucleation and growth, Springer, Cham, pp 77–92
Frech R, Wang EC, Bates JB (1980) The IR and Raman spectra of $CaCO_3$ (aragonite). Spectrochim Acta A 36:915–919
Gabrielli C, Jaouhari R, Joiret S, Maurin G (2000) In situ Raman spectroscopy applied to electrochemical scaling. Determination of the structure of vaterite. J Raman Spectrosc 31:497–501
Gebauer D, Völkel A, Cölfen H (2008) Stable prenucleation calcium carbonate clusters. Science 322:1819–1822
Gebauer D, Gunawidjaja PN, Ko JYP, Bacsik Z, Aziz B, Liu LJ, Hu YF, Bergström L, Tai CW, Sham TK, Eden M, Hedin N (2010) Proto-calcite and proto-vaterite in amorphous calcium carbonates. Angew Chem Int Ed 49:8889–8891
Gillet P, McMillan P, Schott J, Badro J, Grzechnik A (1996) Thermodynamic properties and isotopic fractionation of calcite from vibrational spectroscopy of ^{18}O-substituted calcite. Geochim Cosmochim Acta 60:3471–3485
Grimme S (2006) Semiempirical GGA-type density functional constructed with a long-range dispersion contribution. J Comput Chem 27:1787–1799
Hasse B, Ehrenberg H, Marxen JC, Becker W, Epple M (2000) Calcium carbonate modifications in the mineralized shell of the freshwater snail *Biomphalariaglabrata*. Chem Eur J 6:3679–3685
Kabalah-Amitai L, Mayzel B, Kauffmann Y, Fitch AN, Bloch L, Gilbert PUPA, Pokroy B (2013) Vaterite crystals contain two interspersed crystal structures. Science 340:454–456
Kahmi SR (1963) On the structure of vaterite, $CaCO_3$. Acta Crystallogr 16:770–772
Königsberger E, Königsberger LC, Gamsjäger H (1999) Low-themperature thermodynamic model for the system Na_2CO_3-$MgCO_3$-$CaCO_3$-H_2O. Geochim Cosmochim Acta 63:3105–3199
Le Bail A, Ouheni S, Chateigner D (2011) Microtwinning hypothesis for a more ordered vaterite model. Powder Diffr 26:16–21
Li L, Wentzcovitch RM, Weidner DJ, Da Silva CRS (2007) Vibrational and thermodynamic properties of forsterite at mantle conditions. J Geophys Res 112:B05206

Lutsko JF (2017) Novel paradigms in nonclassical nucleation theory. In: Van Driessche AES, Kellermeier M, Benning LG, Gebauer D (eds) New perspectives on mineral nucleation and growth, Springer, Cham, pp 25–42

McConnell JDC (1960) Vaterite from Ballycraigy, Larne, Nothern Ireland. Mineral Mag 32:535–544

Medeiros SK, Albuquerque EL, Maia Jr FF, Caetano EWS, Freire VN (2007) First-principles calculations of structural, electronic, and optical absorption properties of $CaCO_3$ Vaterite. Chem Phys Lett 435:59–64

Meyer HJ (1959) Über vaterit und seine struktur. Angew Chem 71:678–679

Meyer HJ (1969) Struktur und Fehlordnung des Vateritis. Z Kristallogr 128:183–212

Mikkelsen A, Andersen AB, Engelsen SB, Hansen HCB, Larsen O, Skibsted LH (1999) Presence and dehydration of ikaite, calcium carbonate hexahydrate, in frozen shrimp shell. J Agric Food Chem 47:911–917

Mugnaioli E, Andrusenko I, Schüler T, Loges N, Dinnebier RE, Panthöfer M, Tremel W, Kolb U (2012) Ab initio structure determination of vaterite by automated electron diffraction. Angew Chem Int Ed 51:1–7

Munemoto T, Fukushi K (2008) Transformation kinetics of monohydrocalcite to aragonite in acqueous solutions. J Miner Petrol Sci 103:345–349

Nye JF (1985) Physical properties of crystals. Oxford University Press, Oxford

Penn RL, Li D, Soltis JA (2017) A perspective on the particle-based crystal growth of ferric oxides, oxyhydroxides, and hydrous oxides. In: Van Driessche AES, Kellermeier M, Benning LG, Gebauer D (eds) New perspectives on mineral nucleation and growth, Springer, Cham, pp 257–274

Perdew JP, Ruzsinsky A, Csonka GI, Vydrov OA, Scuseria GE, Constantin LA, Zhou X, Burke K (2008) Restoring the density-gradient expansion for exchange in solids and surfaces. Phys Rev Lett 100:136406

Perdew JP, Burke K, Ernzerhof M (1996) Generalized gradient approximation made simple. Phys Rev Lett 77:3865–3868

Pisani C, Maschio L, Casassa S, Halo M, Schütz M, Usvyat D (2008) Periodic local MP2 method for the study of electronic correlation in crystals: theory and preliminary applications. J Comput Chem 29:2113–2124

Pouget EM, Bomans PHH, Goos JACM, Frederik PM, de With G, Sommerdijk NAJM (2009) The initial stages of template-controlled $CaCO_3$ Formation Revealed by cryo-TEM. Science 323:1455–1458

Prencipe M, Pascale F, Zicovich-Wilson CM, Saunders VR, Orlando R, Dovesi R (2004) The vibrational spectrum of calcite ($CaCO_3$): an ab initio quantum-mechanical calculation. Phys Chem Miner 31:559–564

Qiao L, Feng QL (2007) Study on twin stacking faults in vaterite tablets of freshwater lacklustre pearls. J Cryst Growth 304:253–256

Reichel V, Faivre D (2017) Magnetite nucleation and growth. In: Van Driessche AES, Kellermeier M, Benning LG, Gebauer D (eds) New perspectives on mineral nucleation and growth, Springer, Cham, pp 275–292

Rao A, Cölfen H (2017) Mineralization schemes in the living world: mesocrystals. In: Van Driessche AES, Kellermeier M, Benning LG, Gebauer D (eds) New perspectives on mineral nucleation and growth, Springer, Cham, pp 155–184

Rodriguez-Blanco JD, Shaw S, Benning LG (2011) The kinetics and mechanisms of amorphous calcium carbonate (ACC) crystallization to calcite, via vaterite. Nanoscale 3:265–271

Rodriguez-Blanco JG, Sand KK, Benning LG (2017) ACC and vaterite as intermediates in the solution-based crystallization of $CaCO_3$. In: Van Driessche AES, Kellermeier M, Benning LG, Gebauer D (eds) New perspectives on mineral nucleation and growth, Springer, Cham, pp 93–112

Sato M, Matsuda S (1969) Structure of vaterite and infrared spectra. Z Kristallogr 129:405–410

Schaebitz M, Wirth R, Janssen C, Dresen G (2015, in press) First evidence of $CaCO_3$-iii and $CaCO_3$-iiib – high-pressure polymorphs of calcite – authigenically formed in near surface sediments. Am Mineral

Sherman D (2016) Molecular modeling of geochemical reactions: an introduction, chapter 1. In: Introduction to the theory and methods of computational chemistry. Wiley, Chichester, pp 1–31

Skinner HCW, Osbaldiston GW, Wilner AN (1977) Monohydrocalcite in a guinea pig bladder stone, a novel occurrence. Am Mineral 62:273–277

Spann N, Harper EM, Aldridge DC (2010) The unusual mineral vaterite in shells of the freshwater bivalve corbicula fluminea from the UK. Naturwissenschaften 97:743–751

Swainson IP (2008) The structure of monohydrocalcite and the phase composition of the beachrock deposits of Lake Butler and Lake Fellmongery, South Australia. Am Mineral 93:1014–1018

Swainson IP, Hammond RP (2003) Hydrogen bonding in ikaite, $CaCO_3 \cdot 6\ H_2O$. Mineral Mag 67:555–562

Tang CC, Thompson SP, Parker JE, Lennie AR, Azoughc F, Katod K (2009) The ikaite-to-vaterite transformation: new evidence from diffraction and imaging. J Appl Cryst 42:225–233

Tobler DJ, Stawski TM, and Benning LG (2017) Silica and alumina nanophases: natural processes and industrial applications. In: Van Driessche AES, Kellermeier M, Benning LG, Gebauer D (eds) New perspectives on mineral nucleation and growth, Springer, Cham, pp 293–316

Valenzano L, Torres FJ, Doll K, Pascale F, Zicovich-Wilson CM, Dovesi R (2006) Ab initio study of the vibrational spectrum and related properties of crystalline compounds; the case of $CaCO_3$ calcite. Z Phys Chem 220:893–912

Valenzano L, Noël Y, Orlando R, Zicovich-Wilson CM, Ferrero M, Dovesi R (2007) Ab initio vibrational spectra and dielectric properties of carbonates: magnesite, calcite and dolomite. Theor Chem Acc 117:991–1000

Van Driessche AES, Stawski TM, Benning LG, Kellermeier M (2017) Calcium sulfate precipitation throughout its phase diagram. In: Van Driessche AES, Kellermeier M, Benning LG, Gebauer D (eds) New perspectives on mineral nucleation and growth, Springer, Cham, pp 227–256

Wallace AF, Hedges LO, Fernandez-Martinez A, Raiteri P, Gale JD, Waychunas GA, Whitelam S, Banfield JF, De Yoreo J (2013) Microscopic evidence for liquid-liquid separation in supersaturated $CaCO_3$ solutions. Science 341:885–889

Wang J, Becker U (2009) Structure and carbonate orientation of vaterite ($CaCO_3$). Am Mineral 94:380–386

Wang J, Becker U (2012) Energetics and kinetics of carbonate orientational ordering in vaterite calcium carbonate. Am Mineral 97:1427–1436

Wehrmeister U, Jacob DE, Soldati AL, Loges N, Häger T, Hofmeister W (2011) Amorphous, nanocrystalline and crystalline calcium carbonates in biological materials. J Raman Spectrosc 42:926–935

Wehrmeister U, Soldati L, Jacob DE, Häger T, Hofmeister W (2010) Raman spectroscopy of synthetic, geological and biological vaterite: a Raman spectroscopic study. J Raman Spectrosc 41:193–201

Wilbur KM, Watabe N (1963) Experimental studies on calcification in mollusc and the alga coccolithus Huxleyi. Ann N Y Acad Sci 109:82–112

Wolf SE, Gower LB (2017) Challenges and perspectives of the polymer-induced liquid-precursor process: the pathway from liquid-condensed mineral precursors to mesocrystalline products. In: Van Driessche AES, Kellermeier M, Benning LG, Gebauer D (eds) New perspectives on mineral nucleation and growth, Springer, Cham, pp 43–76

Wolf G, Lerchner J, Schmidt H, Gamsjäger H, Königsberger E, Schmidt P (1996) Thermodynamics of $CaCO_3$ phase transitions. J Therm Anal 46:353–359

Yu YG, Wentzcovitch RM, Angel RJ (2010) First principles study of thermodynamics and phase transition in low-pressure ($P2_1/c$) and high-pressure ($C2/c$) clinoenstatite $MgSiO_3$. J Geophys Res 115:B02201

Chapter 7
Classical and Nonclassical Theories of Crystal Growth

Jens-Petter Andreassen and Alison Emslie Lewis

In this chapter, we discuss classical and nonclassical concepts of crystal growth that coexist in the literature as explanations for the formation of both mono- and polycrystalline particles, often of the same substances. Crystalline particles with intraparticle nanosized subunits, nanoparticulate surface features, and complex morphologies have led to the development of new nonclassical theories of crystal growth based on the aggregation of nanocrystals in solution. At the same time, similar morphologies are explained by monomer incorporation at conditions of stress incorporation, which results in nucleation at the growth front and accompanying branching at the nanoscale. The two mechanisms are differently affected by important process variables like supersaturation, temperature, or additives and are analyzed with respect to their capability of predicting crystal growth rates. A quantitative description of the formation kinetics of the solid phases is essential for the design and operation of industrial precipitation and crystallization processes and for the understanding of fundamental principles in material design and biomineralization processes. In this chapter, we emphasize the importance of supersaturation in order to account for the extensive nanoparticle formation required to build micron-sized particles by nano-aggregative growth, as well as the accompanying change in the population density.

J.-P. Andreassen (✉)
Department of chemical engineering, Norwegian university of science and technology (NTNU), Trondheim, Norway
e-mail: Jens-Petter.Andreassen@ntnu.no

A.E. Lewis
Department of chemical engineering, University of Cape Town, Cape Town, South Africa

A.E.S. Van Driessche et al. (eds.), *New Perspectives on Mineral Nucleation and Growth*,
DOI 10.1007/978-3-319-45669-0_7

7.1 Introduction

During the last decades, various nonclassical concepts of crystal growth have emerged in the scientific literature (Matijevic 1993; Cölfen and Antonietti 2008). They depart from the classical growth theories (Chernov 1984) by proposing that crystals, both mono- and polycrystalline, are produced by aggregation or assembly of nucleated nanocrystals, as opposed to integration of solute species into the crystal lattice surface. These novel concepts have received considerable attention and are now used extensively in the literature to analyze crystallization and precipitation phenomena in both nature and industry.

Classical mechanistic explanations of crystal growth have traditionally focused on explaining the microscopic features of the crystal surfaces and the integration of growth units leading to the development of faceted crystals. Derivation of the corresponding rate expressions, accompanied by experimental verification, has provided industrial practitioners with a tool to model and predict crystal morphology and size. Emerging fields in material technology, nanoparticle science, and biomineralization have introduced new experimental conditions and protocols, resulting in particle morphologies that are very often quite different from the expected equilibrium morphologies. Precipitation occurring at higher supersaturation and temperatures, often in combination with additives and templates, results in various examples of complex shapes and surface feature expressions that are not in accordance with the classical morphology predictions from an energy minimization viewpoint. The advancement of characterization techniques has opened up the ability to carry out more detailed studies of processes going on during nucleation, growth, and aggregation. New nonclassical concepts of crystal growth have been proposed as a consequence of this. Phase change kinetics by these alternative routes will be very different from what the classical theories predict. In this chapter, we present some of the dominant concepts of crystal growth and discuss the main differences between them, since the many different and often conflicting explanations presented in the literature make studying and analyzing the phenomena of precipitation a challenge.

The aim is to identify which investigations are required to resolve the seemingly conflicting descriptions in literature when it comes to determination and prediction of growth kinetics during precipitation. This is motivated by the need for a consistent description of crystal enlargement processes and the effect of the main operating parameters during industrial manufacture and separation of crystalline materials from solution.

7.2 Driving Force and Size Enlargement

Enlargement of crystalline particles in solution is traditionally described by two processes: crystal growth and agglomeration due to encounters between the growing crystals. In this classical picture, crystal growth is considered to progress by attachment of monomeric growth units from solution. These solute species are the ions, molecules, or atoms which correspond to the primary constituents of the crystal lattice. The driving force for solute incorporation is the difference in their activity as solution species, a, and as incorporated in the solid, represented by the activity of solution in equilibrium with the solid, a_{eq}. The corresponding chemical potential difference, $\Delta\mu$, for a non-dissociating compound can be expressed as

$$\frac{\Delta\mu}{RT} = \ln\frac{a}{a_{\text{eq}}} \tag{7.1}$$

leading to definition of the supersaturation ratio, S_a

$$S_{\text{a}} = \frac{a}{a_{\text{eq}}} \tag{7.2}$$

For minerals and electrolyte crystals where each formula unit consists of a total number of ν ions, being the sum of ν_+ cations and ν_- anions, the expression becomes

$$\frac{\Delta\mu}{RT} = \nu\ln\frac{a_\pm}{a_{\pm,\text{eq}}} \tag{7.3}$$

where $a_\pm$ is the mean activity of the ionic species, resulting from the correction of the free concentration of ions, c, by the mean activity coefficient, $\gamma_\pm$. For crystals composed of two divalent ions, like $CaCO_3$, the corresponding activity-based supersaturation is thus

$$S_{\text{a}} = \frac{a_\pm}{a_{\pm,\text{eq}}} = \sqrt{\frac{c_{\text{Ca}^{2+}} \cdot c_{\text{CO}_3{}^{2-}} \cdot \gamma_\pm^2}{K_{\text{sp,CaCO}_3}}} \tag{7.4}$$

In this classical picture, agglomeration is a rather rare event that will occur if the forces acting on the growing crystalline particles in solution cause them to approach each other closely enough so that a crystalline bridge develops resulting in a new stable particle. Hence in order for agglomeration to take place, crystal growth must occur during the time of contact, which also implies that the system must be supersaturated with respect to the crystallizing compound (Andreassen and Hounslow 2004). Investigations of agglomeration kinetics have concentrated on the contact brought about by shear forces in solution. In this context, agglomeration will depend on the collision rate of the growing crystals and an efficiency parameter that depends on the shape and contact point geometry and which also takes into account the disruptive hydrodynamic forces. In order for particles to be brought into contact by shear forces, their size needs to be in the micron meter range.

The recent new nonclassical theories challenge these ideas by claiming that also crystal growth is an aggregation process that can take place by assembly of nanosized crystalline particles by both oriented and non-oriented attachment, operating at much smaller length scales. The attachment of particles must in this case be governed by very different forces (De Yoreo et al. 2015). The outcome of these assembly processes can be both poly- and monocrystalline, which in the latter case makes the end product indistinguishable from crystals growing according to a classical mechanism. For systems like calcium carbonate and barium sulfate, and many more, both classical and nonclassical explanations of crystal growth coexist in the recent literature. We will thus start by describing classical crystal growth and thereafter look into the newer theories and compare how the different conceptions explain prediction of morphology and the kinetics of the enlargement process.

7.3 Classical Crystal Growth by Monomers

7.3.1 At Low and Intermediate Supersaturation

The incorporation of growth units from solution into the crystal lattice is a complicated process of attachment and detachment at active sites on the crystal surface. The attachment process involves desolvation of the monomers, adsorption onto the surface, and diffusion along the surface toward the active sites. Attachment at the active sites leads to advancement of steps over the surface. For the crystals to continue growing, new steps must be generated constantly. The dominant source of step formation varies with the supersaturation in the system. At driving forces lower than the critical value for surface nucleation, inherent dislocations in the crystal lattice are responsible for the presence of steps. Stacking faults lead to screw dislocations that represents a nonterminating step source since growth preserves the initial stacking fault of the lattice. Attachment along the screw dislocation step results in spiral growth and growth hillocks that eventually merge and cause the crystal face to propagate outwards. At a sufficient level of supersaturation, new steps are formed around two-dimensional islands that nucleate on the surface. The nucleation rate increases with supersaturation leading to a rough surface where integration of new units can potentially take place anywhere. The microscopic features of the surfaces growing by these two mechanisms of step generation can be observed by in situ AFM experiments as shown for the two examples of $BaSO_4$ and $CaCO_3$ in Fig. 7.1.

7.3.2 At High Supersaturation

Despite that the surface features of spiral growth and 2-D nucleation growth can be observed at the microscopic level, crystal faces resulting from these mechanisms are macroscopically smooth and the crystals are dense. However, when further increasing the supersaturation, the interface becomes unstable due to a change from integration-controlled growth to mass-transfer-controlled growth, by diffusion from

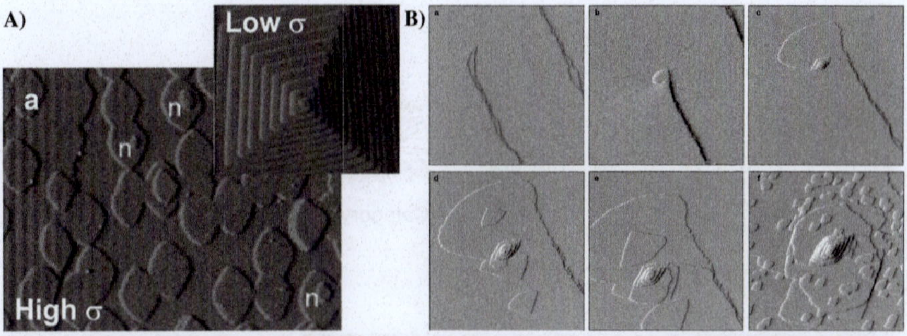

Fig. 7.1 AFM images of crystal faces showing examples of spiral growth and growth hillocks at lower supersaturation and two-dimensional (2-D) nucleation islands at higher supersaturation for (**a**) calcite, $CaCO_3$ at two levels of supersaturation (σ corresponds to the left-hand side of Eq. 7.1) (Reproduced from De Yoreo and Vekilov (2003) with permission from the Mineral Society of America), and (**b**) barite, $BaSO_4$ – first five pictures in the panel is a growth sequence over 60 min for a supersaturation of $S_a^2 = 12$ (when S_a is according to the definition in Eq. 7.4). In the last picture, the supersaturation has been increased, corresponding to $S_a^2 = 26$ (Reproduced from (Pina et al. 1998) with permission from Nature Publishing Group)

the surrounding bulk liquid. This transition (Fig. 7.2a) comes about when the surface turns rough and highly kinked due to extensive surface nucleation. Integration of new units is not rate limiting, and depending on the diffusion field around the crystal and the crystal morphology, certain edges and corners of the crystal will get access to solution of higher supersaturation. The consumption of supersaturation at these locations prevents growth of the central part of the crystal faces, and so-called "hopper" crystals develop. This elevation of the growth rate for certain parts of the crystal is self-enforcing, since the same edges, corners, or surface perturbations will access yet higher supersaturation levels and as a consequence dendritic growth starts to dominate. The crystal branches off in directions of higher supersaturation in an interplay dictated by the nature of the crystal lattice and the changing supersaturation profile from the surface into the bulk of the surrounding liquid. This branching is crystallographic, leading to a monocrystalline and usually symmetric object of complex shape. The shift in growth morphologies is a consequence of the change in growth regime and not the supersaturation itself. The supersaturation is merely responsible for introducing the diffusion limitation by increasing the monomer integration rate. This has been illustrated by similar morphology shifts due to diffusion limitations imposed by the surrounding medium, by performing mineral precipitation in hydrogels, relevant for biomineralization processes (Asenath-Smith et al. 2012).

The overall growth rate of crystals in solution is determined by the chemical affinity, the chemical potential difference between the solution and the crystal. The power law relationship shown in Eq. 7.5 is frequently used to relate the overall growth rate, G [ms^{-1}], to the activity-based supersaturation ratio, S_a (Eq. 7.4), and the solubility- and temperature-dependent growth rate constant, k_g. The growth order, g, signifies the mechanism of crystal growth. At low supersaturation, and

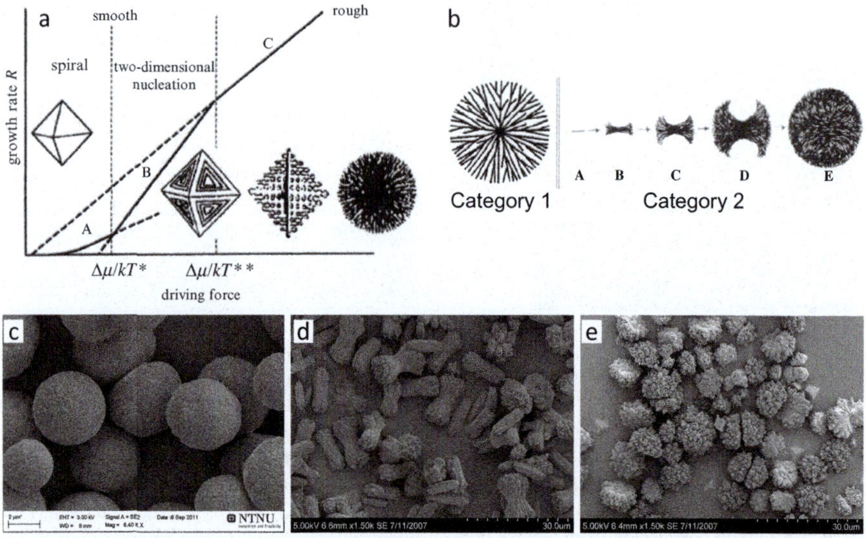

Fig. 7.2 (**a**) Growth rate, crystal growth mechanisms (where A is spiral growth, B is growth by two-dimensional nucleation and C is rough growth) and corresponding morphology changes as a function of driving force for a hypothetical crystal bounded by {111} faces (Reproduced from Sunagawa (2005) with permission from Cambridge University Press). (**b**) Two categories of spherulitic growth: Category 1 describes multidirectional growth from a central nucleus, and category 2 describes growth front nucleation and resulting branching on the fast-growing tips of an elongated precursor crystal (A), leading to intermediate dumbbell morphologies (B-D) and potentially to a polycrystalline sphere (E) (Reproduced from Granasy et al. (2005) with permission from Cambridge University Press). (**c**) Category 1 vaterite spherulites grown in water at an initial supersaturation of $S_a = 5.7$ (with respect to vaterite) (Reproduced from Andreassen et al. (2012) with permission from RSC) (**d**) Intermediate-shaped category 2 spherulites grown at an initial S_a (vaterite) $= 3.8$ in a mixture of ethylene glycol and water (90/10 %) and (**e**) nearly fully developed spherulites after 45 min when the initial supersaturation was increased to $S_a = 7.0$; the spherulites tend to break in two halves along the resulting equator (Reproduced from Andreassen et al. (2010) with permission from Elsevier)

provided that growth is dominated by simple, single sourced dislocation spirals, the value of g is 2, resulting in the so-called parabolic growth law. When supersaturation is increased and 2-D nucleation starts to dominate, g takes on higher numbers, evident of the exponential nature of the nucleation process. When the surface becomes rough and the growth rate is mass-transfer controlled, g is equal to 1. Although the power law model is a simplified and averaged approach to the detailed crystal growth processes and composite mechanisms taking place on the surface of individual crystal faces (Teng et al. 2000), it has had success in the operation of industrial crystallization processes, by being used to adjust liquid phase parameters in order to control solid phase characteristics.

$$G = k_g(S_a - 1)^g \tag{7.5}$$

In contrast to the crystallographic branching that characterizes dendritic growth, a new branching mechanism starts to become operative at yet higher driving forces. Surface nucleation is no longer crystallographic and, as a result, the particle now becomes polycrystalline. This growth front nucleation will, as opposed to dendritic growth, often produce spherical space-filling structures, hence the term spherulitic growth. It has been a topic of numerous studies for well over a century, but is still not properly recognized in the precipitation literature. Spherulitic growth has been reported for various systems independent of the nature of the crystallizing compound. Molecular, atomic, and ionic crystals all grow by spherulitic growth, and although many of the studies are based on polymer crystallization and crystallization from viscous melts, it has been shown that neither large molecules, high viscosity, nor impurities are crucial for this mechanism (Beck and Andreassen 2010; Shtukenberg et al. 2012). Also crystals growing from pure aqueous solutions will produce spherulites when the conditions for internal stress accumulation, like high supersaturation, result in nucleation of new sub-individuals on the surface for the relaxation of this stress. The morphology will depend, firstly, on the progress of growth (supersaturation and branching frequency) and, secondly, on which of the two mechanisms of spherulitic growth in Fig. 7.2 that operates, which has been elegantly demonstrated by phase field modeling (Granasy et al. 2005). When growth front nucleation starts from a central nucleus, it results in isotropic type 1 spherulites, while type 2 spherulites are formed by branching on the two fast-growing faces of an elongated precursor, ultimately leading to polycrystalline spheres.

Both categories of spherulitic growth offer explanations of spherical polycrystalline particles without invoking an aggregation mechanism. Like for the other classical crystal growth mechanisms, the driving force $\Delta\mu$, is also a controlling factor for non-crystallographic branching, but the criteria for which type of spherulite that develops is not well known. As an example, both category 1 and 2 spherulites of the vaterite polymorph of calcium carbonate can be produced in solution at sufficiently high supersaturation (Fig. 7.1c–e). Category 1 spherulites grew by transformation from initially formed amorphous calcium carbonate (Andreassen 2005) in water, whereas category 2 spherulites developed in mixtures of ethylene glycol and water where the degree of branching was dependent on the initial supersaturation (Andreassen et al. 2010). Glycol reduces the growth rate of vaterite and allowed for a time-resolved observation of morphology development, but was not considered a prerequisite for this category of spherulitic growth. Category 2 spherulites of aragonite have been produced in water/ethanol mixtures (Sand et al. 2012) but also in aqueous solutions without additional solvents (Andreassen et al. 2012). The category 2 mechanism can offer an explanation for many other literature observations of sheaf of wheat and dumbbell morphologies that are described for crystalline mineral particles, like fluorapatite (Busch et al. 1999), $BaSO_4$ (Qi et al. 2000), and hematite (Sugimoto et al. 1993) but also in crystallization of higher-solubility compounds and from melts (Shtukenberg et al. 2012). The kinetic expression for spherulitic growth should reflect that this is an interface-controlled process driven by nucleation at the growth front and not diffusion controlled.

However, it has frequently been modeled as a diffusion problem, but this is related to the special conditions in many of the systems where spherulites are observed, i.e., high-viscosity metal or polymer melts where heat and mass transfer is limiting (Shtukenberg et al. 2012).

The mechanistic shifts and corresponding morphology predictions in classical crystal growth, from faceted monocrystals to surface-instability driven polycrystals, are explained with reference to the increasing supersaturation. Although there are some unresolved questions related to the phenomenon of non-crystallographic branching and the accompanying kinetic dependency on supersaturation, especially for growth in impurity-free low-viscosity media, the success of the classical monomer incorporation models has been their corresponding rate expressions that have been of great use to model industrial crystallization processes. The current alternative nonclassical concepts do not propose corresponding rate equations and do not necessarily obey the classical criterion related to the supersaturation dependence. They rather base their explanations on the colloidal stability of initially nucleated nanocrystals and how they may aggregate to create both mono- and polycrystalline structures.

7.4 Growth by Assembly of Precursor Crystals

A recent review (De Yoreo et al. 2015; see also Chap. 1 de Yoreo et al. 2017, this volume) presents a general concept of crystallization by particle attachment (CPA) by discussing attachment of a wider range of precursor particles, like oligomers, droplets, amorphous particles, or fully developed nanocrystals. We limit the following analysis to attachment and aggregation of prenucleated nanosized crystals.

Although polycrystallinity by growth of monomeric units can be explained by growth front nucleation, hardly any of the papers advocating an aggregation-based theory refer to such growth phenomena, indicating that it has been largely unknown in the field of mineral precipitation. However, it is of vital importance to understand whether growth is facilitated by particles nucleated in solution and transported to the crystal or if the resulting subunits nucleate and grow on the advancing crystal surface. In the former case, a high 3-D nucleation rate and hence a high supersaturation are required, as well as an efficient assembly or ordering principle, whereas in the latter case the degree of epitaxy and non-crystallographic branching during growth front nucleation is the determining factor for the level of polycrystallinity.

7.4.1 Spontaneous Precipitation Systems: Mesocrystals and Polycrystalline Particles

Egon Matijevic (Matijevic 1993) was a pioneer in establishing an alternative interpretation of particle growth and proposed aggregation of nanosized crystals as the mechanism responsible for the formation of polycrystalline micron-sized particles of various minerals. The underlying assumption was that an object that is spherical and polycrystalline can only be explained by assembly of already nucleated crystals, but other morphologies in the vast collection of "monodispersed" colloidal systems were also explained by a coagulation mechanism of nanoparticles. Ocaña et al. (Ocana et al. 1995) summarized some of the systems undergoing nano-aggregation and categorized the possible mechanisms as either undirectional or directional aggregation. Undirectional aggregation results in spherical particles, whereas ellipsoids, platelets, prisms, spheres, and rods could all be produced by directional aggregation, often assisted by additives.

Oriented attachment (OA) was demonstrated for a smaller assembly of nanoparticles some years later (Penn and Banfield 1999; see also Chap. 13 Penn et al. 2017, this volume) by hydrothermal treatment of 5 nm titania crystals (Fig. 7.3a). The particle size increased both by monomeric crystal growth and by assembly of up to ten individual crystals, a process that took place over several hours. When this oriented attachment process is non-perfect, it introduces dislocations in the resulting crystal (Penn and Banfield 1998), explaining how dislocations necessary for subsequent crystal growth can appear even though the initial nuclei are dislocation-free. Recent advances have allowed for direct observation of nanoparticle coalescence by high-resolution transmission electron microscopy using a fluid cell. In situ observations of platinum nanocrystals (Zheng et al. 2009) show that they grow by both monomer addition and particle-particle attachment events in the same order of numbers as that of the initial particles, illustrating that a few nanocrystals combine to make the new particles. This is an important growth

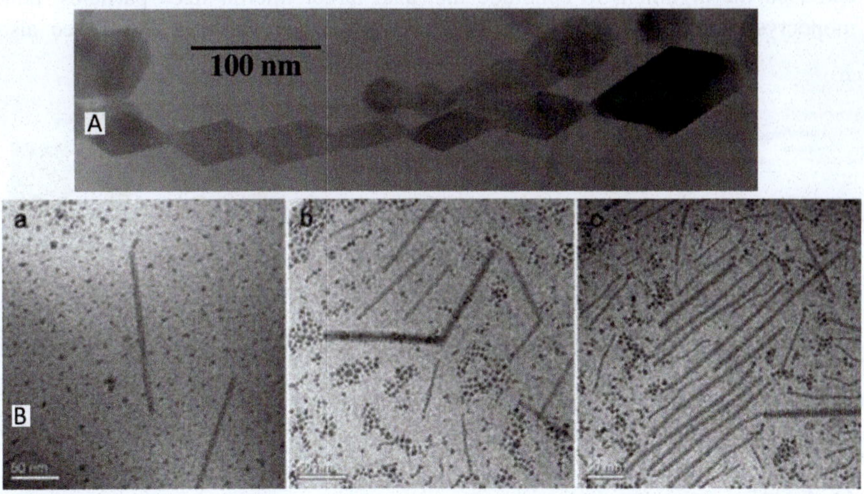

Fig. 7.3 (**a**) TEM micrograph of a single crystal of anatase demonstrating oriented attachment of titania nanocrystals in 0.001 M HCl (Reproduced from Penn and Banfield (1999) with permission from Elsevier). (**b**) Cryo-TEM images of goethite mesocrystals developing by aging of a suspension of ferrihydrite nanocrystals at 80 °C, after (*a*) 5 days, (*b*) 10 days, and (*c*) 24 days (Reproduced from Yuwono et al. (2010) after permission from American Chemical Society)

trajectory to explain the nearly monodisperse particles resulting from an initially broad size distribution. The oriented attachment processes of iron oxyhydroxide nanoparticles observed in fluid cell TEM demonstrate how the particles rotate and interact until they find a perfect lattice match, and the attachment is followed by atom-by-atom addition. The translational and rotational accelerations show that direction-specific interactions drive the attachment, and the electrostatic field can promote these oriented attachment events (Li et al. 2012; Zhang et al. 2014; Nielsen et al. 2014).

The development of larger particles involving oriented attachment of a higher number of nanoparticles, in the orders of ten to hundreds, has been shown (Yuwono et al. 2010) by cryo-TEM monitoring of a suspension of ferrihydrite (Fig. 7.3b; see also Chap. 13 Penn et al. 2017, this volume). Over the course of several days, long thin mesocrystal assemblies of oriented goethite nanocrystals are developed by (1) self-assembly of primary nanocrystals, (2) crystallographic reorganization within the self-assemblies, and (3) conversion to oriented aggregates, which are new secondary crystals.

Directional aggregation has also been found to take place in the early process of gypsum precipitation (Van Driessche et al. 2012). The process started by formation of nanoparticles of the hemihydrate bassanite which after growth to nanorods of some 100 nm in length aggregated to needle-shaped particles which later transformed to the dihydrate gypsum. This oriented aggregation of bassanite was explained by an increase in the enthalpy with decreasing surface area, favoring aggregation over crystal growth.

Along with, and often based on, these demonstrations of crystal growth by oriented attachment, a significant number of studies in the recent literature have extended the argument to conclude that also larger micron-sized particles, both monocrystalline (Fig. 7.4) mesocrystals (Cölfen and Antonietti 2005; see also

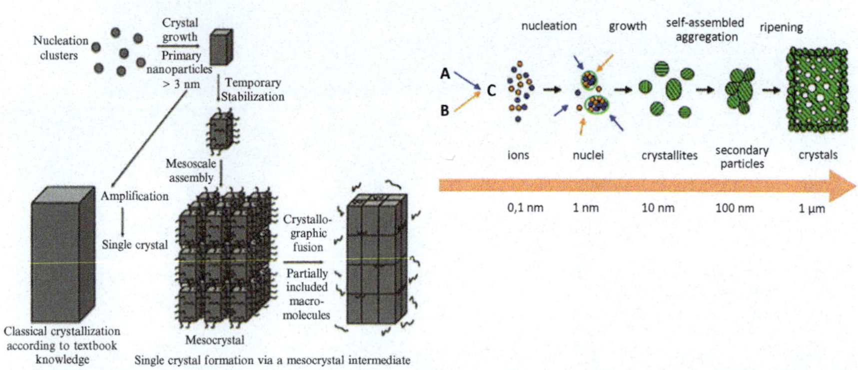

Fig. 7.4 Aggregation schemes for monocrystalline particles in solution, *left*, assisted by macromolecules resulting in mesocrystals (Reproduced from Cölfen and Antonietti (2005) after permission from John Wiley and Sons) and, to the *right*, self-assembled aggregation in the absence of additives leading to porosity in the final crystals (Reprinted from Judat and Kind (2004) after permission from Elsevier)

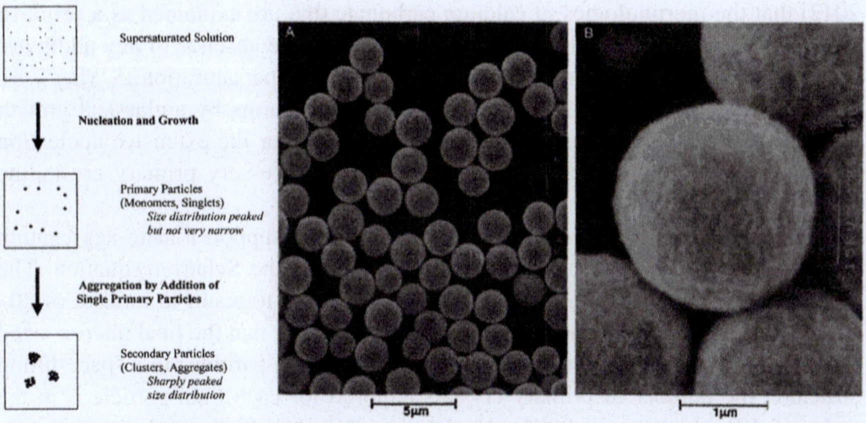

Fig. 7.5 Aggregation scheme (to the *left*) illustrating how growth of monodisperse polycrystalline secondary particles of gold (by reduction of auric acid) in solution (micrograph *A* and *B*) occurs by addition of primary particles to secondary particles, to ensure a sharply peaked particle size distribution (Reproduced from Park et al. (2001) with permission from American Chemical Society)

Chap. 8 Rao and Cölfen 2017, this volume) and polycrystalline particles (Fig. 7.5), are formed by assembly mechanisms, suggesting that aggregation of nanocrystals is a universal mechanism for growth of crystalline particles. However, many orders of magnitude, higher number of primary nanocrystals and a highly efficient assembly mechanism, are required to account for these rapidly formed large micron-sized particles. Kind et al. (Judat and Kind 2004) used cryo-TEM to study shock-frozen samples from the rapid precipitation of $BaSO_4$ in a T-mixer arrangement and concluded that the internal porosity of the particles was a consequence of highly ordered aggregation of nanoparticles formed in the beginning of the process (Fig. 7.4). A similar evaluation of copper oxalate precipitation (Soare et al. 2006) discussed how the high ionic strength due to high concentration during nucleation would lead to aggregation of the nuclei, by suppression of the electric double layer and accompanying low colloidal stability. This initial random aggregation resulted in a core of randomly oriented primary particles whereas the outer aligned crystallites were explained by directional aggregation due to lower ionic strength at lower supersaturation later on in the growth process.

Calcium carbonate precipitation of both mono- and polycrystalline particles, both in the absence and presence of additives, is frequently explained as a result of non-classical crystal growth based on observations of intraparticle subunit structure and nanoparticle surface features. The argument of aggregation of nanosized crystals has been used to explain both polycrystalline spherical particles and monocrystalline hexagonal particles of vaterite. In the case of polycrystalline particles, aggregation is not oriented, but for the formation of hexagonal plates (Xu et al. 2006), additives in the system are said to be necessary to aid the assembly process, following a similar principle as described in Fig. 7.4. By contrast, it has been shown (Andreassen et al.

2012) that the morphologies of calcium carbonate that are explained as a result of modulating additives can in fact also be produced in the absence of any additives, by simply controlling the value of the activity-based supersaturation, S_a (Eq. 7.4). It was concluded that calcium carbonate in this case forms by a classical growth mechanisms, since the supersaturation, S_a, is too low for the extensive nucleation required to supply an attachment process with the necessary primary crystalline particles.

Peak broadening in Powder XRD is widely used to support a nano-aggregation mechanism and the nanoparticle size is estimated by the Scherrer equation. The resulting size determination of the nano-domains tends to result in a range of 20–50 nm, irrespective of the mineral in question. Assuming that the final micron-sized particles would be constructed by aggregation of 20 nm particles in a space-filling structure, the number of primary crystals required for each final particle is in the order of 10^8 (Andreassen 2005). Nucleation of such a high number of primary crystals, or building blocks, would require a substantial supersaturation, and their presence should be easily detectable in solution by in situ or cryo-TEM or even in dry samples separated from the solution during the 5-min growth period. In this case, the vaterite spheres were growing at a moderate supersaturation dictated by the transformation of amorphous calcium carbonate, and as a result, it was concluded that spherulitic growth was the operating mechanism and not aggregation of nucleated precursor crystals. When polycrystalline spheres of vaterite were seeded to systems of constant supersaturation well below the nucleation threshold, the particles grew by incorporation of ions from solution, constantly creating surface units that varied in size with the applied level of supersaturation (Andreassen et al. 2012), as predicted by the dependency of branching on the thermodynamic driving force (Shtukenberg et al. 2012). In a recent critical analysis of monocrystalline calcite mesocrystals precipitated in the presence of polymers (Kim et al. 2014), it was shown that the use of the Scherrer equation to infer information about primary building blocks is not valid and that the main source of the peak broadening was substantial strain in the crystal lattice which occurred when the crystals were grown in the presence of additives. In addition, the high surface area due to the nanoparticulate surface features could be fully explained as a consequence of the additive action on crystal growth in a classical sense. In a recent investigation, a polycrystalline vaterite biomineral (Pokroy et al. 2015) was also suggested to form from solution by ion-by-ion spherulitic growth, as opposed to nano-aggregation, due to 0 and 30° angle spreads and no interfacial organic layers between adjacent crystals. Other biomineralization examples where spherulitic growth has been shown is in the formation of earthworm granules (Hodson et al. 2015) and to infer information about microbial activity responsible for the formation of dolomite rock where some organisms make spherulitic Ca-Mg carbonate by increasing the supersaturation levels locally (McKenzie and Vasconcelos 2009).

In the current literature that uses the argument of aggregation of nanosized crystals, the total number of nanoparticles required and the reduction in numbers due to aggregation is normally not part of the analysis when attachment processes are discussed. However, in the classical industrial crystallization literature, it is essential to show that there has been a reduction in the number of crystals in order to demonstrate that aggregation has taken place. A population balance analysis offers a consistent accounting tool to distinguish between particle enlargement by growth and by aggregation, and this has traditionally been applied for the agglomeration of larger crystals growing by classical crystal growth (Andreassen and Hounslow 2004; Costodes et al. 2006).

When it comes to the nonclassical growth mechanisms, only a few attempts have been made to quantify the kinetics of growth. Early attempts (Dirksen et al. 1990) to use a population balance approach to analyze aggregative growth of copper oxalate were performed, motivated by the morphology of the final particles. However, in order to explain the fairly monodisperse population of final particles from smaller crystals, it had to be assumed that aggregation takes place only by allowing nuclei to combine with larger particles, not by nuclei-nuclei or aggregate-aggregate events. A similar assumption was made to explain the rapid formation (10 s) of spherical micron-sized polycrystalline gold particles shown in Fig. 7.5 (Park et al. 2001). In order to account for the size selection leading to the final particles, the proposed aggregation units of 40 nm were assumed to only attach to secondary particles, which is not in accordance with later observations of single-single particle aggregation in liquid cell TEM (Zheng et al. 2009). However, the average secondary particle size predicted was smaller than the experimental values and the distribution was too narrow. In a later refined model (Libert et al. 2003) based on the synthesis of uniform cadmium sulfide particles, a new adjustable parameter was introduced. The number of primary particles in a typical secondary particle was estimated to be $2 \cdot 10^6$, but instead of allowing only for attachment of singlets to secondary particles, the experimental results were better explained by allowing cluster-to-cluster aggregation of secondary particles containing up to approximately 25 primary particles. These findings have not been verified by experiments. But this should be relatively uncomplicated, since such high numbers of particles in different states of aggregation should be clearly visible in both SEM and TEM.

7.4.2 Superparticles by Controlled Assembly of Nanosized Seed Crystals

Another class of superstructures reported over the last decade could possibly shed some light on the previous discussion. Superparticles are based on the assembly of pre-made nanoparticles, made possible by the recent advances in nanoparticle manufacture. Superparticles can be grown from solution, where nanocrystals that are initially well dispersed can be brought together in larger particles resulting in a well-defined superlattice structure. As opposed to the spontaneous formation of meso- and polycrystals, precursor nanoparticles of well-controlled size and shape are fabricated in a separate step before the assembly principle is initiated. Bai et al. (Bai et al. 2007) produced colloidal spheres based on oil-in-water emulsions where the precursor crystals are functionalized with a surfactant and dispersed in the oil phase, which is subsequently removed by evaporation. Hence, the particle size is a result of the nanocrystal concentration in the droplets as well as the emulsification process. For the assembly of 6.9 nm $BaCrO_4$ nanocrystals, the center-to-center distance inside the superparticles is 9.1 nm, which quantifies the TEM-observed interparticle gap to be 2.2 nm, caused by the presence of ligand molecules. This means that the nanocrystals kept their individual character and did not sinter into larger units despite prolonged heat treatment. The same was observed for nanoparticle micelles (Zhuang et al. 2008) that were dispersible in aqueous solution due to hydrophobic van der Waals interactions between the Fe_3O_4 nanoparticle ligands (oleic acid) and hydrocarbon chain of the surfactant (DTAB). By adding the nanoparticle-micelle solution to glycol, the micelles decomposed due to the loss of DTAB, and the Fe_3O_4 nanoparticles were then assembled due to solvophobic interactions between the nanoparticle ligands and glycol. The success and rate of the assembly was controlled by this solvophobic interaction by varying the amount of the surfactant, whereas CTAB prevented the assembly from taking place due to the lower solubility in glycol thereby preventing the solvophobic assembly. With DTAB, the rate of superparticle formation was generally fast and led initially to amorphous assemblies that only crystallized upon annealing. However, the individual nanocrystals were still separated within the superparticle crystal.

Assembly principles can also be applied for nanoparticles of opposite charge where the driving force for the aggregation process is governed by electrostatic forces. Kalsin et al. (Kalsin et al. 2006) prepared silver and gold NPs carrying positively and negatively alkane thiols, respectively. Both NP populations were stable separately, but when brought into contact, they eventually combined to superparticles, once the size distributions of the individual populations were appropriately tuned. Micron-sized supracrystals displaying the individual NP "ions" in a sphalerite structure were produced, but the assembly was occasionally unsuccessful producing a population of the individual NPs, or resulting in amorphous structures if the spread of the seed size distribution was less than optimal.

Based on the particle engineering principles described above, it seems clear that nanoparticles will assemble only once the conditions in terms of destabilization or association are realized at the right time and rate. This relies on separating the stages of nanoparticle formation and the subsequent formation of superparticle structures. In this respect, it differs significantly from the proposed formation mechanism of micron-sized mesocrystals and polycrystals which rely on nucleation of the sufficient number of nanocrystal building blocks in situ, a highly efficient assembly principle and mechanisms for removal of interparticle polymeric additives.

7.5 Outlook

The kinetics of crystal growth has a pivotal role in running processes for the design of particle products or for the prevention and inhibition of crystalline deposits. For that purpose, classical crystal growth has been successful in relating the chemical potential and temperature of the crystallizing medium to the rate of solid formation. The alternative aggregative growth concepts do not provide the same predictive capability. This outlook summarizes the arguments of the above discussion and aims at recommending investigations that are required to resolve the current situation of two coexisting growth theories.

It has been demonstrated by in situ investigations in liquid cell TEM that a smaller number of nanocrystals will aggregate after nucleation and that this is important for the development of the nanoparticle size distribution and for development of dislocations. However, there is a large gap between these observations and the numbers required for the proposed extensive aggregation of nanocrystals responsible for the construction of larger micron-sized mesocrystals and polycrystalline particles, which can be in the order of 10^6–10^8 primary crystals in each resulting secondary particle. The presence of such a high number of primary particles during an assembly process that goes on over a time period of seconds to minutes should also be easily identified in time-resolved TEM or cryo-TEM investigations, since the size of the building blocks is well within the detection limit of the microscopes.

The material balance is a classical tool that accounts for all species entering and leaving a system. The population balance method uses a number balancing approach in a similar way, to make sure that all particles in a process are accounted for and to ensure that number continuity is preserved. This is essential in order to prove that aggregation is taking place and to establish rate expressions for aggregation processes. Unlike classical crystal growth, only a few attempts have been performed to quantify precipitation rates in the case of nano-aggregative growth based on a population balance approach with varying and contradicting assumptions regarding aggregation events between primary particles and already formed assemblies. For formation of micron-sized particles based on nanocrystal assembly, the reduction in numbers should be accounted for in order to establish the mechanisms and kinetics of these growth processes.

Although variations in the free energy landscape and the surface energy can contribute to lower the critical free energy and hence change the pathway of the crystallization process, a significant reduction in the free energy due to increased supersaturation is still required to give the high nucleation rate of crystals necessary to provide the building blocks needed for the particle-based growth process. Hence, assembly-based growth of micron-sized particles from nanocrystals should only be possible in systems that are characterized by high supersaturation levels, unless a high nucleation rate is justified by other explanations. Future studies should be performed at quantified and varying levels of the supersaturation to establish that the nucleation rate is sufficient for the required number of nanocrystals. The effect of supersaturation on the aggregation mechanism must also be analyzed in order for nonclassical crystal growth to offer the same predictions as classical crystal growth when it comes to quantifying particle enlargement rates.

Due to the limited number of in situ observations at sufficiently high resolution, most of the evidence of nanocrystal aggregation is relying on observations of particles after completion of the formation pathway. However, many of the same features used to support an aggregation mechanism, like branched morphologies and nanosized surface units, can also be explained by far from equilibrium dendritic and spherulitic growth. It has been shown that diffraction line broadening is not evident of growth by aggregation, since lattice strain during classical growth can give the same result. Classical growth by fast monomer incorporation can lead to strain development which leads to growth front nucleation and ultimately to non-crystallographic branching by increasing the supersaturation. Future investigations should thus analyze how classical growth mechanisms influenced by additives or running at high driving force can explain porosity and polycrystallinity. Dendritic and spherulitic growth can explain monocrystalline nonequilibrium particle morphologies and polycrystalline particles. However, spherulitic growth from solution is not well understood in terms of the kinetics of growth front nucleation, especially regarding the effect of temperature and driving force on the branching process. In the same way, as external morphology observations cannot be used to support a mechanism of nano-aggregation, it is also not a sufficient evidence to conclude on dendritic or spherulitic growth, and future mineralization studies should thus analyze the results in the framework of both concepts.

Attachment by rearrangement and lattice matching by rotation has been demonstrated for smaller ensembles of nanoparticles. Are these assembly mechanisms efficient enough for the aggregation of millions of nanocrystals on a timescale of seconds or minutes? The forces involved are not well understood, but, from the deliberate aggregation studies to design superparticles based on prefabricated nanoparticles, it becomes evident that aggregation is not trivial and association principles need to be finely tuned in order to result in assembly. Destabilization of the electric double layer and accompanying collisions due to Brownian motion is frequently used to explain the efficiency of aggregation. This hypothesis can be tested by manipulating the ionic strength of initially dispersed nanoparticles.

Since the assembly mechanisms involve highly specific aggregation of similarly shaped nanocrystals, especially in the case of directional growth, it should also be possible to prevent this assembly by steric hindrance with appropriate additives. This could provide proof that the driving force is sufficient for extensive nanoparticle generation, and at the same time, it can provide an efficient route to produce nanoparticles of various materials.

In order to resolve the current situation of two coexisting growth theories and the accompanying uncertainties when it comes to the kinetic predictions, more research is required. These future investigations should consider both nonclassical and classical mechanisms of crystal growth and in a systematic way compare evidence required to discriminate between them. Analyses of industrial precipitation processes depend on it, but the same fundamental questions also apply for minerals precipitating in natural processes.

References

Andreassen JP (2005) Formation mechanism and morphology in precipitation of vaterite – nano aggregation or crystal growth? J Cryst Growth 274:256–264

Andreassen JP, Hounslow MJ (2004) Growth and aggregation of vaterite in seeded-batch experiments. Aiche J 50:2772–2782

Andreassen JP, Flaten EM, Beck R, Lewis AE (2010) Investigations of spherulitic growth in industrial crystallization. Chem Eng Res Des 88:1163–1168

Andreassen JP, Beck R, Nergaard M (2012) Biomimetic type morphologies of calcium carbonate grown in absence of additives. Faraday Discuss 159:247–261
Asenath-Smith E, Li H, Keene EC, Seh ZW, Estroff LA (2012) Crystal growth of calcium carbonate in hydrogels as a model of biomineralization. Adv Funct Mater 22:2891–2914
Bai F, Wang D, Huo Z, Chen W, Liu L, Liang X, Chen C, Wang X, Peng Q, Li Y (2007) A versatile bottom-up assembly approach to colloidal spheres from nanocrystals. Angew Chem Int Ed 46:6650–6653
Beck R, Andreassen JP (2010) Spherulitic growth of calcium carbonate. Cryst Growth Des 10:2934–2947
Busch S, Dolhaine H, Duchesne A, Heinz S, Hochrein O, Laeri F, Podebrad O, Vietze U, Weiland T, Kniep R (1999) Biomimetic morphogenesis of fluorapatite-gelatin composites: fractal growth, the question of intrinsic electric fields, core/shell assemblies, hollow spheres and reorganization of denatured collagen. Eur J Inorg Chem 1999:1643–1653
Chernov AA (1984) Modern crystallography III: crystal growth. Springer, Berlin
Cölfen H, Antonietti M (2005) Mesocrystals: inorganic superstructures made by highly parallel crystallization and controlled alignment. Angew Chem Int Ed 44:5576–5591
Cölfen H, Antonietti M (2008) Mesocrystals and nonclassical crystallization. Wiley, Chichester
Costodes VCT, Mausse CF, Molala K, Lewis AE (2006) A simple approach for determining particle size enlargement mechanisms in nickel reduction. Int J Miner Process 78:93–100
De Yoreo JJ, Vekilov PG (2003) Principles of crystal nucleation and growth. Rev Mineral Geochem 54:57–93
De Yoreo JJ, Gilbert PUPA, Sommerdijk NAJM, Penn RL, Whitelam S, Joester D, Zhang H, Rimer JD, Navrotsky A, Banfield JF, Wallace AF, Michel FM, Meldrum FC, Cölfen H, Dove PM (2015) Crystallization by particle attachment in synthetic, biogenic, and geologic environments. Science 349, aaa6760
De Yoreo JJ, Sommerdijk NAJM, Dove PM (2017) Nucleation pathways in electrolyte solutions. In: Van Driessche AES, Kellermeier M, Benning LG, Gebauer D (eds) New perspectives on mineral nucleation and growth, Springer, Cham, pp 1–24
Dirksen JA, Benjelloun S, Ring TA (1990) Modelling the precipitation of copper oxalate aggregates. Colloid Polym Sci 268:864–876
Granasy L, Pusztai T, Tegze G, Warren JA, Douglas JF (2005) Growth and form of spherulites. Phys Rev E 72:011605
Hodson ME, Benning LG, Demarchi B, Penkman KEH, Rodriguez-Blanco JD, Schofield PF, Versteegh EAA (2015) Biomineralisation by earthworms – an investigation into the stability and distribution of amorphous calcium carbonate. Geochem Trans 16:1–16
Judat B, Kind M (2004) Morphology and internal structure of barium sulfate – Derivation of a new growth mechanism. J Colloid Interface Sci 269:341–353
Kalsin AM, Fialkowski M, Paszewski M, Smoukov SK, Bishop KJM, Grzybowski BA (2006) Electrostatic self-assembly of binary nanoparticle crystals with a diamond-like lattice. Science 312:420–424
Kim YY, Schenk AS, Ihli J, Kulak AN, Hetherington NBJ, Tang CC, Schmahl WW, Griesshaber E, Hyett G, Meldrum FC (2014) A critical analysis of calcium carbonate mesocrystals. Nat Commun 5:4341
Li D, Nielsen MH, Lee JRI, Frandsen C, Banfield JF, De Yoreo JJ (2012) Direction-specific interactions control crystal growth by oriented attachment. Science 336:1014–1018
Libert S, Gorshkov V, Goia D, Matijević E, Privman V (2003) Model of controlled synthesis of uniform colloid particles: cadmium sulfide. Langmuir 19:10679–10683
Matijevic E (1993) Preparation and properties of uniform size colloids. Chem Mater 5:412–426
Mckenzie JA, Vasconcelos C (2009) Dolomite mountains and the origin of the dolomite rock of which they mainly consist: historical developments and new perspectives. Sedimentology 56:205–219
Nielsen MH, Li D, Zhang H, Aloni S, HanT FC, Seto J, Banfield JF, Cölfen H, De Yoreo JJ (2014) Investigating processes of nanocrystal formation and transformation via liquid cell TEM. Microsc Microanal 20:425–436

Ocana M, Rodriguezclemente R, Serna CJ (1995) Uniform colloidal particles in solution – formation mechanisms. Adv Mater 7:212–216

Park J, Privman V, Matijević E (2001) Model of formation of monodispersed colloids. J Phys Chem B 105:11630–11635

Penn RL, Banfield JF (1998) Imperfect oriented attachment: dislocation generation in defect-free nanocrystals. Science 281:969–971

Penn RL, Banfield JF (1999) Morphology development and crystal growth in nanocrystalline aggregates under hydrothermal conditions: insights from titania. Geochimica Et Cosmochimica Acta 63:1549–1557

Penn RL, Li D, Soltis JA (2017) A perspective on the particle-based crystal growth of ferric oxides, oxyhydroxides, and hydrous oxides. In: Van Driessche AES, Kellermeier M, Benning LG, Gebauer D (eds) New perspectives on mineral nucleation and growth, Springer, Cham, pp 257–274

Pina CM, Becker U, Risthaus P, Bosbach D, Putnis A (1998) Molecular-scale mechanisms of crystal growth in barite. Nature 395:483–486

Pokroy B, Kabalah-Amitai L, Polishchuk I, Devol RT, Blonsky AZ, Sun CY, Marcus MA, Scholl A, Gilbert PUPA (2015) Narrowly distributed crystal orientation in biomineral vaterite. Chem Mater 27:6516–6523

Qi L, Cölfen H, Antonietti M (2000) Crystal design of barium sulfate using double-hydrophilic block copolymers. Angew Chem Int Ed 39:604–607

Rao A, Cölfen H (2017) Mineralization schemes in the living world: mesocrystals. In: Van Driessche AES, Kellermeier M, Benning LG, Gebauer D (eds) New perspectives on mineral nucleation and growth, Springer, Cham, pp 155–184

Sand KK, Rodriguez-Blanco JD, Makovicky E, Benning LG, Stipp SLS (2012) Crystallization of $CaCO_3$ in water-alcohol mixtures: spherulitic growth, polymorph stabilization, and morphology change. Cryst Growth Des 12:842–853

Shtukenberg AG, Punin YO, Gunn E, Kahr B (2012) Spherulites. Chem Rev 112:1805–1838

Soare LC, Bowen P, Lemaitre J, Hofmann H (2006) Precipitation of nanostructured copper oxalate: substructure and growth mechanism. J Phys Chem B 110:17763–17771

Sugimoto T, Khan MM, Muramatsu A (1993) Preparation of monodisperse peanut-type α-Fe2O3 particles from condensed ferric hydroxide gel. Colloids Surf A Physicochem Eng Asp 70: 167–169

Sunagawa I (2005) Crystals: growth, morphology and perfection. Cambridge University Press, Cambridge

Teng HH, Dove PM, De Yoreo JJ (2000) Kinetics of calcite growth: surface processes and relationships to macroscopic rate laws. Geochimica Et Cosmochimica Acta 64:2255–2266

Van Driessche AES, Benning LG, Rodriguez-Blanco JD, Ossorio M, Bots P, García-Ruiz JM (2012) The role and implications of bassanite as a stable precursor phase to gypsum precipitation. Science 335:69–72

Xu AW, Antonietti M, Cölfen H, Fang YP (2006) Uniform hexagonal plates of vaterite $CaCO_3$ mesocrystals formed by biomimetic mineralization. Adv Funct Mater 16:903–908

Yuwono VM, Burrows ND, Soltis JA, Penn RL (2010) Oriented aggregation: formation and transformation of mesocrystal intermediates revealed. J Am Chem Soc 132:2163–2165

Zhang H, De Yoreo JJ, Banfield JF (2014) A unified description of attachment-based crystal growth. ACS Nano 8:6526–6530

Zheng H, Smith RK, Jun YW, Kisielowski C, Dahmen U, Paul Alivisatos A (2009) Observation of single colloidal platinum nanocrystal growth trajectories. Science 324:1309–1312

Zhuang J, Wu H, Yang Y, Cao YC (2008) Controlling colloidal superparticle growth through solvophobic interactions. Angew Chem Int Ed 47:2208–2212

Chapter 8
Mineralization Schemes in the Living World: Mesocrystals

Ashit Rao and Helmut Cölfen

In Aristotle's parable, the house is there that men may live in it; but it is also there because the builders have laid one stone upon another . . . – On Growth and Form (Thompson 1942)

8.1 Introduction

Biomaterials have been utilized by humankind since the emergence of complex cognition. Archeological records and aboriginal arts provide several instances for mineralized hard tissue being used for different purposes such as instruments, ornaments, and weapons based on their availability, macrostructural beauty, and durability (Fig. 8.1). With advances in mathematics as well as optical microscopy, our understanding of hard biological materials progressed with the sophisticated morphologies of biominerals described using concepts from geometry and physics (Thompson 1942). At present, investigations are driven by (1) the fascinating properties that emerge from structure–property–function relationships (Mann 1995; Aizenberg et al. 2004), (2) the need for material synthesis via green routes (Mann et al. 1993; Meldrum and Cölfen 2008), as well as (3) the inspiration for new practical materials based on designs optimized by evolution (Lee et al. 2012; Natalio et al. 2013; Xu et al. 2007; Niederberger and Cölfen 2006; Song and Cölfen 2010). The body of knowledge on the structure–property relations and formation of biominerals thus is rapidly growing, supported by concomitant advances in high-resolution analytical techniques such as electron microscopy (Nielsen and De Yoreo 2017, Chap. 18), solid-state NMR, and ultracentrifugation as well as

A. Rao (✉) • H. Cölfen (✉)
Physical Chemistry, Department of Chemistry, University of Konstanz, Universitätsstr. 10, D-78457 Konstanz, Germany
e-mail: Ashit.Rao@frias.uni-freiburg.de; Helmut.Coelfen@uni-konstanz.de

A.E.S. Van Driessche et al. (eds.), *New Perspectives on Mineral Nucleation and Growth*, DOI 10.1007/978-3-319-45669-0_8

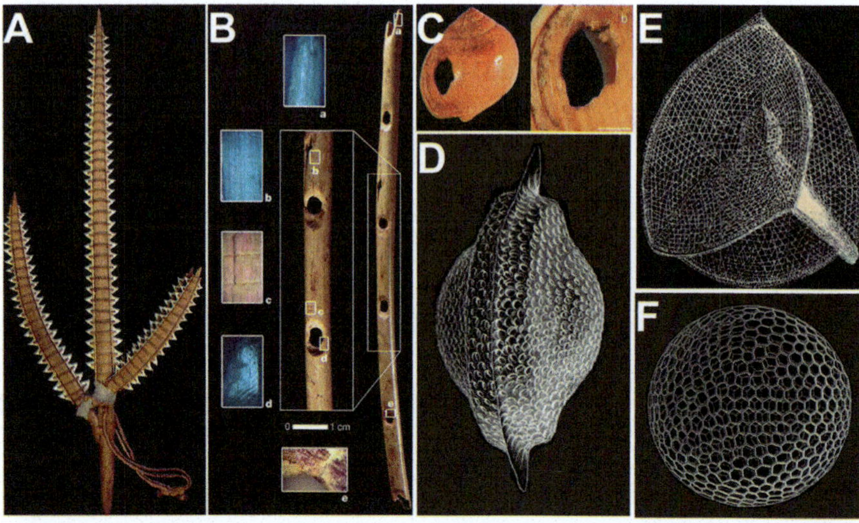

Fig. 8.1 Early applications and studies of biominerals. (**a**) Shark tooth weapon from Gilbert Islands, (**b**) paleolithic flutes made from the radius of a griffon vulture excavated at Hohle Fels, (**c**) decorative gastropod shell beads from the middle stone age. Depictions of exquisite skeletons from Radiolaria (**d**) *Perizona pterygota*, (**e**) *Callimitra agnesae*, and (**f**) an *Aulonia* sp. [Images reproduced from (**a**) Drew J. et al. *PLoS ONE 2013*, 8(4)], (**b**) Conard N.J. et al. *Nature 2009*, 460(737) with permission from Nature publishing group, (**c**) d'Errico F. et al. *J. Human Evol.* 2005, 48(3) with permission from Elsevier, (**d**, **e**, **f**) reproduced from the classical works of Ernst Häckel

breakthroughs in sequencing technologies and high-throughput tools for proteomics analysis. Fueled by the participation of groups from diverse disciplines, there has been a recent development of intriguing concepts pertaining to biomineralization (Falini and Fermani 2017) such as the hierarchical structures of certain biominerals (Seto et al. 2014; Currey 2005; Cölfen and Antonietti 2008; Aizenberg et al. 2005); the early stages of biomineralization involving pre-nucleation clusters, amorphous precursors, and mineral poly(a)morphism (Cartwright et al. 2012; Gebauer and Cölfen 2011; Gower and Odom 2000; Gong et al. 2012; De Yoreo et al. 2017, Chap. 1; etc.); as well as the role of water and biomolecules in the mineralization processes (Demichelis et al. 2011; Raiteri and Gale 2010; Evans 2008). Thus, our relation with biomaterials has journeyed about few million years ago, from harnessing mechanically robust, esthetic biomaterials to present times in which we use state-of-the-art techniques to discover how these organisms *have laid stone upon another*.

Of the recent discoveries in the field of biomineralization, the concept of mesocrystallinity is particularly interesting due to its elucidation for interesting physical properties of biominerals such the sea urchin spine (Seto et al. 2012; Bergström et al. 2015). Moreover this material type has an enormous potential for practical applications (Zhou and O'Brien 2012). Therefore the aim of this chapter is to present mineralization pathways as well as their contribution to mesocrystallinity and other biomineral structures.

8.2 Biogenic Minerals: Formation and Structure

Minerals formed by organisms vary significantly in terms of composition, shape, size, and orientation. Biological control over these material properties often leads to interesting physical properties. For instance, the design properties of nacreous layers from certain mollusks and gastropods lead to mechanical properties superior to the bulk inorganic constituent, i.e., aragonite (Sun and Bhushan 2012). The resulting fracture resistance of nacre material is 3000 times that of geological aragonite (Smith et al. 1999). The siliceous basal spicules of some sea sponges have a unique hierarchical organization with concentric layers composed of silica nanoparticles. This results in a flexible material with photonic properties (Aizenberg et al. 2004). On account of such fascinating structure–property relations, it is crucial to understand associated formation mechanisms. This is not an easy task due to minute length scales at which growth phenomena occur, involvement of transient in- or metastable precursors, as well as experimental challenges with addressing dynamic organic–inorganic interactions. However concepts from nonclassical nucleation and crystallization (De Yoreo et al. 2017, Chap. 1; etc.) can be effectively applied to address such problems (Fig. 8.2). Some important aspects of mineral nucleation and growth are presented here.

8.2.1 Pre-nucleation Clusters

Pre-nucleation clusters (PNCs) are small, thermodynamically stable, structurally dynamic solute precursors that nucleate to form mineral products via phase separation (Gebauer and Cölfen 2011; Gebauer et al. 2014). Direct in vivo characterization of PNCs during biomineralization is experimentally challenging due to the spatiotemporally dynamic nature of mineralization processes. Nevertheless the existence of PNCs has been well validated by several techniques such as analytical ultracentrifugation, mass spectrometry, molecular simulations, potentiometric titration, and transmission electron microscopy (Gebauer et al. 2008; Demichelis et al. 2011; Kellermeier et al. 2012a, b; Raiteri and Gale 2010; Pouget et al. 2009). The relevance of these species in biomineralization is recognized by the relation of the thermodynamic stabilities of PNCs and the solubility products of the phases formed post-nucleation (Cartwright et al. 2012). For instance, the stabilities of PNCs altered by a pH change during the pre-nucleation regime the solubility of

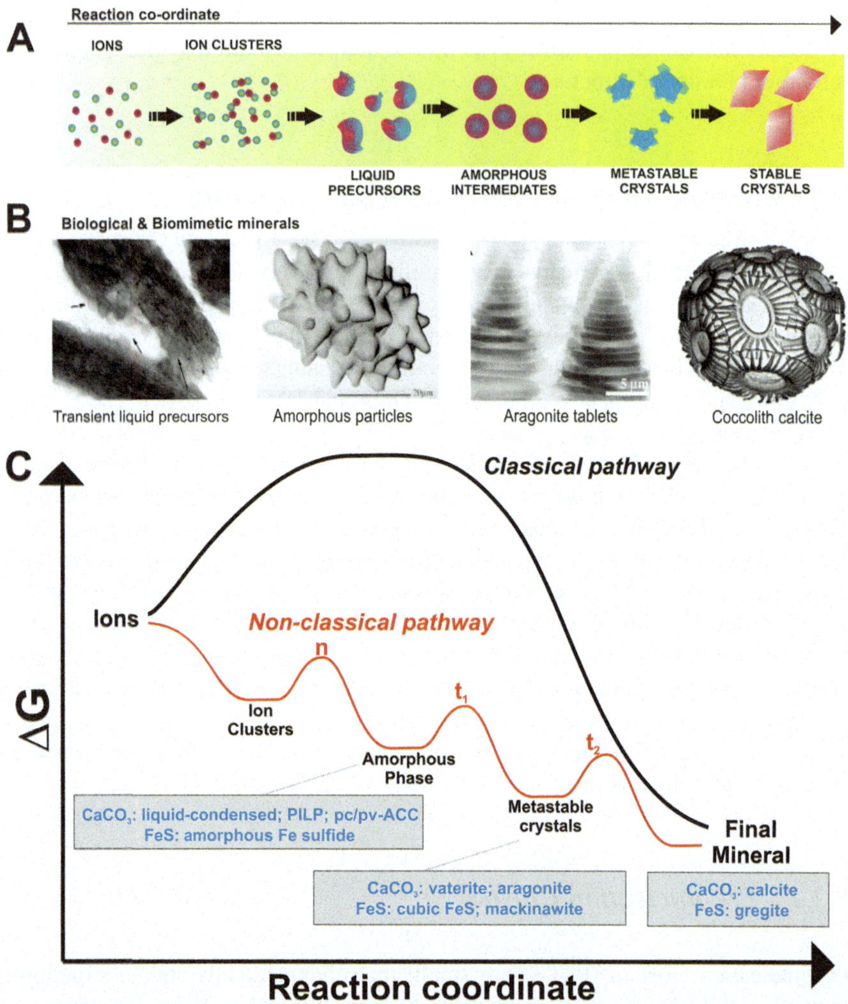

Fig. 8.2 (**a**) During nucleation, pre-nucleation clusters lead to mineral precursors including liquid and solid amorphous materials that subsequently form crystalline phases. (**b**) Kinetic modulation of the energy barrier for nucleation and phase transformation leads to transient or stable phases such as amorphous calcium phosphate droplets associated with the collagen fibrils, stable ACC in form of cystoliths from the leaves of *Ficus microcarpa*, nacreous tablets composed of aragonite, a metastable form of crystalline $CaCO_3$, and a calcite coccosphere from *Emiliania huxleyi* [*left* to *right*: reproduced from Olszta et al. *Mat. Sci. Eng. R 2007*, 58(77) with permission from Elsevier; Addadi et al. *Adv. Mat. 2003*, 15(959) with permission from Wiley; Blank S. et al *J. Microscopy 2003*, 212(280) with permission from Wiley; Young J. et al. *J. Struct. Biol. 1999*, 126(195) with permission from Elsevier]. (**c**) Free energy profile for mineralization described by classical and nonclassical pathways illustrated with calcium carbonate and iron sulfide. From the nonclassical standpoint, formation of ion clusters has a negligible energy barrier compared to nucleation (n) of an amorphous phase. The formation and stability of subsequent amorphous/crystalline particles are determined by energy barriers associated with phase transformation events (t_1, t_2). Textbook descriptions of the classical pathway however describe mineralization as a one-step pathway to the mineral formation. [Adapted from Cölfen and Mann, *Angew. Chemie Int. Ed. 2003* 42(2350) with permission from Wiley]

the corresponding nucleated phase (Gebauer et al. 2008, 2010). In the presence of synthetic and biological additives, such effects are complex and have a significant bearing on the mineralization outcome (Gebauer et al. 2009; Rao et al. 2014, 2016). Hence the fate of mineralization can be determined well before the nucleation event via optimal interactions between mineral precursors and additives in a chemically fine-tuned environment (Fig. 8.2).

8.2.2 *Amorphous Phases*

Mineral growth processes involving transient amorphous phases are prevalent in biological systems (Weiner et al. 2005). This is exemplified by the association of amorphous phases with biominerals such as enamel, fins of certain fish, mollusk shells, and sea urchin skeletal elements (Beniash et al. 2009; Mahamid et al. 2008; Nassif et al. 2005; Weiss et al. 2002; Beniash et al. 1997; Politi et al. 2006; Raz et al. 2002). Amorphous materials are devoid of long-range atomic periodicity and hence appear malleable to complex shapes that are typical of biominerals. More due to their high solubility, there is an inherent ease of mobility and distribution in confined spaces. In extension to their role as mineral precursors, amorphous phases also serve as temporary reserves of calcium in certain crustaceans (Raz et al. 2002). The biological mechanisms underlying the stabilization of amorphous phases are being explored because such phases are highly metastable and transform to crystalline forms. Studies show that poly(a)morphism is present in biogenic and synthetic mineral precursors (Cartwright et al. 2012; Weiss et al. 2002; Gebauer et al. 2010; Fernandez-Martinez et al. 2017, Chap. 4; Rodriguez-Blanco et al. 2017, Chap. 5). Although the exact functions of proto-structuring in mineral precursors are not understood, a possibility of biomolecular recognition of distinct proto-structures requires elucidation. The specificity of such interactions might not be surprising when one considers that the fairly analogous behavior of biomolecules that lack defined configurations (intrinsically disordered proteins) can interact with their binding partners (Tompa and Fuxreiter 2008). The occurrence of biogenic amorphous precursors indicates that in addition to the formation of a solid phase from PNCs, the phase transformation of amorphous to crystalline materials may represent a second regulative checkpoint. The kinetic modulation of energy barriers associated with nucleation and phase transformation by biomolecules is a key aspect of biomineralization (Fig. 8.2). This results in the vast polymorphic diversity of biominerals evident in Nature.

8.2.3 *Regulation of Crystal Orientation*

An important aspect of biogenic minerals is the consistency of crystallographic orientations across different specimens from the same species. Such remarkable control over the deposition and organization of mineral precursors as well as the consistent alignment at lower length scales is yet to be achieved in synthetic materials. The biological regulation of crystal orientation in biominerals is clearly evident from the calcitic triradiate spicule from sea urchin larvae (*Strongylocentrotus purpuratus*). The three spicule arms elongate in the negative direction along the trigonal axes of rhombohedral calcite, and subsequently two arms change their direction of growth toward the c-axis (Okazaki and Inoué 1976). Considering biominerals with larger subunits, the exoskeleton of the coccolithophore *Rhabdosphaera clavigera* is composed of rodlike calcite particles about $100 \times 300 \times 50$ nm in size regularly arranged in a chiral fivefold symmetric structure (van de Locht et al. 2014). Thus, precise regulation over crystallographic orientation across several length scales is evident in biominerals.

8.2.4 Structural Hierarchy

Compositionally diverse materials such as bone tissue, butterfly wings, shells, and wood exhibit hierarchical organization at multiple length scales (Seto et al. 2014). Such superstructures lead to the emergence of properties such as enhanced mechanics, remarkable photonics, as well as iridescent structural coloring (Aizenberg et al. 2004; Kinoshita and Yoshioka 2005; Barthelat et al. 2007). For biominerals, superstructural organization is a property of skeletal elements associated especially with multicellular organisms such as corals, echinoderms, gastropods, mammals, mollusks, and sponges. On the other hand, mineral elements formed by single-celled prokaryotes such as magnetotactic bacteria are typically monocrystalline (Bazylinski et al. 1994). Coccolithophores are unicellular eukaryotes which produce modular calcite elements that form intricate larger structures like micron-sized Lego® blocks (van de Locht et al. 2014). The structural complexity of coccoliths can thus be considered to be intermediate with respect to the single crystals formed by prokaryotes (possibly archaea as well) and the hierarchical composites in multicellular organisms. This trend can be explained by the suitability of composite biominerals (e.g., bone) toward bearing mechanical stress in larger organisms in comparison to brittle inorganic single crystals. Furthermore, repair of composite biominerals is efficient because mineral precursors and nanocrystalline particles can effectively fill in material fractures, in a process that retains macrostructural integrity (Cölfen 2010; Olszta et al. 2003a).

Such an immense diversity in the composition, orientation, and structure of biogenic minerals motivates the characterization of underlying growth mechanisms. Other aspects of biomineral growth such as transport of mineral precursors, kinetic control over crystallization, and control of polymorph nucleation/stabilization are also important (Meldrum and Cölfen 2008). However, for brevity, we address the concept of biogenic mesocrystallinity in context of the features discussed above. Of the recent discoveries in the field of biomineralization, mesocrystal, a class of novel materials, is particularly interesting due to the explanation it provides for the unique properties of biominerals such as the sea urchin spicule (Seto et al. 2012). The following sections present mineralization schemes as well as emergent structures in context of biogenic minerals, specifically addressing mesocrystals.

8.3 Biogenic Mesocrystals

Mesocrystals are 3-D superstructures composed of crystallographically co-oriented nanoparticles (Cölfen and Antonietti 2005, 2008). This concept was primarily inspired by the examination of earlier reports describing unusual architectures of certain synthetic crystals, in addition to the existence of non-facetted crystal morphologies in biominerals (Cölfen and Antonietti 2008). On account of the oriented subunits, mesocrystals exhibit scattering patterns or birefringence in polarization light similar to that of single crystals of the corresponding bulk material. Mesocrystals are therefore identified by examining material nanoscale structure and emergent properties. Well-studied examples of mesocrystalline biominerals are the sea urchin spine and nacre tablets (Seto et al. 2012; Li and Huang 2009).

8.3.1 Sea Urchin Spine

Among organisms that exhibit intricate mineral architectures, the sea urchin (*S. purpuratus*) spine has been a model system to understand mechanisms underlying biomineralization (Wilt 2005). Reconciling the apparently divergent material properties of the sea urchin skeletal elements had been a long-standing challenge. Techniques such as polarization microscopy, X-ray diffraction, and EBSD indicate that the spine is a single crystal (Moureaux et al. 2010; Su et al. 2000). However, typical of amorphous materials, the spine also displays porous and curved surfaces as well as conchoidal fractures after cleavage (Lowenstam and Weiner 1989). Studies probing the nanostructure of the sea urchin spine have provided answers to this conundrum. At the macroscale, the sea urchin spine exhibits a complex porous construction with a complete absence of crystal faces. However this structure is composed of mesocrystalline Mg-calcite nanoparticles integrated with a cement of Mg-stabilized amorphous calcium carbonate (ACC) and biomolecules (Seto et al. 2012; Raz et al. 2003). This endows the spine with unique properties such as single crystal-like diffraction and birefringence, conchoidal fracture behavior, as well as better flexibility and toughness in comparison to pure calcite. Under mechanical stress, a fracture would likely follow a tortuous path through the cement of ACC and biomolecules and not along the low energy {104} cleavage planes of the crystallites. In addition, this cement appears to inhibit the fusion of the calcite nanoparticles, thus help retaining their size regime (Seto et al. 2012). For brittle materials, the presence of nanoparticulate building units can impart superior mechanical strength with regard to the bulk material (Knudsen 1959). Thus the sea urchin spine is an exquisite material formed by biological processes.

The mesocrystalline superstructure in the sea urchin spine is demonstrated by different techniques. Observations of the fractured spine surface show the presence of conchoidal surfaces characteristic of amorphous materials as well as planar areas that appear facetted (Seto et al. 2012). At higher magnification, the substructure consists of calcite nanoparticles (100–300 nm) ordered in a space-filling manner (Yang et al. 2011). Single crystal-like diffraction patterns indicate the co-orientation of these subunits. Synchrotron X-ray diffraction measurements also support these findings reporting coherence lengths of 50–200 nm for the calcitic subunits (Aizenberg et al. 1997). Further evidence for the mesocrystallinity of the spine comes from the incorporation of organic compounds such as rhodamine B, eosin Y, polypyrrole, pyrene, and 9-vinylcarbazole in the intercrystalline nanospaces of the spine (Oaki and Imai 2006; Munekawa et al. 2015; Oaki et al. 2011).

8.3.2 *Nacre Tablets*

Nacre (or "mother of pearl") is the iridescent inner layer in certain mollusk shells. Composed of aragonite (95 wt%) and organic material (5 wt%), this composite biomaterial is a model system to understand the intricate relationships between hierarchical structures and emergent mechanical properties (Jackson et al. 1988). The mesocrystalline nature of nacre arises from the nanograins comprising aragonite tablets in nacre material (Li et al. 2004; Li and Huang 2009). This accounts for the tablet's ductility and plastic deformation, which is in contrast to the expected brittleness of aragonite. The existence of nanoparticles is evident from the grain rotation and deformation under tensile strain investigated using atomic force microscopy (Li et al. 2006). Structural integrity is retained by a glue-like hydrated organic framework and associated amorphous material surrounding the nanograins (Li et al. 2006). In the Japanese pearl oyster (*Pinctada fucata*), the aragonite tablets are composed of nano-building blocks in the range of 20–180 nm (Oaki and Imai 2005). Diffraction analyses show that the subunits in the adjacent tablet are preferentially oriented with the c-axes perpendicular to the shell surface (Yokoo et al. 2011). These results evidence a composite material formed of crystallographically co-oriented nanoparticles and shows that nacre material is mesocrystalline.

At the micron length scale, nacre is composed of columnar or sheetlike architectures composed of aragonite tablets. Due to this multilevel structural organization, the nacre material exhibits high fracture toughness, greater than that of bulk aragonite (Smith et al. 1999). On account of its remarkable mechanics, nacre material is an inspiration for designing synthetic materials engineered at different hierarchical levels.

8.3.3 *Other Examples*

Biologically controlled mineralization predominantly proceeds via the deposition of complexes of mineral precursors and biomolecules. This is reflected by the nanostructures of calcareous and siliceous spicules and associated organic contents (Sethmann et al. 2006; Aizenberg et al. 2004; Mann et al. 2008) as well as the involvement of vesicular structures during coccolith, magnetosome, and bone formation (Young and Henriksen 2003; Komeili et al. 2004; Mahamid et al. 2011). Triradiate spicules from the sponge *Pericharax heteroraphis* exhibit a "brick-and-mortar"-like arrangement with calcite nanoparticles supported with a proteinaceous organic matrix (Sethmann et al. 2006). Moreover, the spicule exhibits single crystal-like diffraction properties. The fracture behavior of this material is conchoidal similar to that of the sea urchin spine. This is indicative of a mesocrystal with an interparticle matrix composed of ACC and organic molecules. Zones in calcified tissue from corals such as *Corallium rubrum* and *Porites* sp. also exhibit mesocrystalline properties (Benzerara et al. 2011). These consist of aragonite nanoparticles aligned in a radial manner across the micrometer scale. Mesocrystalline structures are also present in calcified biomaterials from certain foramifers and avian eggshells (Oaki et al. 2006). Prismatic columns in certain bivalve shells consist of co-aligned calcite nanoparticles presenting mesocrystallinity (Kijima et al. 2011). In the presence of an acidic additive, in vitro repair of this biomineral proceeds by the deposition of nanocrystals in a specific crystallographic direction (Kijima et al. 2011).

Mineralization experiments in the presence of certain biomineral-derived additives also produce mesocrystalline products. For instance, the self-assembly properties of α-silicatein, a protein from sponge spicules, are used to synthesize flexible rodlike mesocrystals of calcite (Gebauer 2013; Natalio et al. 2013). These synthetic mesocrystals are about 10–300 μm in length and composed of crystallographically aligned nanoparticles. On account of the nano-sized building units and the high organic content (about 16 %), these structures exhibit a high stiffness as well as elasticity and are not fractured under conditions of extreme deformation. Mesocrystalline architecture is also exhibited by enamel material in which hydroxyapatite nano-rods are organized in the *c*-axis direction at the smallest level of hierarchy. During enamel repair in a solution of simulated body fluid supplemented with glutamate, apatite nanoparticles (~20 nm) serve as seed crystals for the growth of co-oriented nano-rods (Li et al. 2011). Another example of biomimetic mesocrystallinity is calcium carbonate films grown on chitin layers in the presence of an acidic peptide, CAP-1 derived from the crayfish exoskeleton (Yamamoto et al. 2008; Sugawara et al. 2006). These films are composed of smaller calcitic nanoparticles, which are crystallographically co-oriented in a uniaxial fashion. These results suggest that CAP-1 has multiple effects on mineralization including (1) binding to the chitin substrate, (2) forming an interface between the substrate and ACC particles, and (3) inhibition of growth and lattice fusion on account of an acidic phosphoserine residue. Hematite mesocrystals formed in the presence of silk fibroin also demonstrate that additives affect different aspects of mesocrystals such as composition and shape (Fei et al. 2014). Interestingly, silk fibroin possesses low-complexity regions also represented in certain mollusk shell-associated proteins (Nakahara 1983; Nakahara et al. 1991; Evans 2008). Another potential mesocrystalline material is derived from the topotactic transition of large dicalcium phosphate–gelatin tablets to a superstructure of oriented hydroxyapatite nanostructures (Furuichi et al. 2006). This structure is relevant to biominerals such as bone and teeth tissue. Such studies illustrate that the influence of additives on early nucleation stages, phase transformations, and subsequent stages is crucial in determining the fate of crystallization products in terms of composition, structure, and orientation.

8.4 Biological Control Over Mineral Structure

Conventional studies have typically focused on the growth of single crystals for purposes such as enantiomer purification, crystallography, and optical applications. However, biogenic mineral composites are decorated with "voids and impurities" such as intra-/intercrystal spaces and biomolecular additives. These features are crucial for the robust architecture and properties of the biomineral as well as, in certain cases, the emergence of mesocrystallinity. From a different perspective, the unique properties of natural composite materials arise on account of the biologically engineered "voids and impurities" and thus serve as motivation for their emulation in synthetic materials.

In nature, the growth of biominerals is achieved by spatiotemporally regulated processes such as transportation of ions/PNCs, nucleation, stabilization of intermediate amorphous mineral precursors, and subsequent phase transformation. This is exemplified by sea urchin spiculogenesis during which mineral precursors transform to a mesocrystalline composite. The initial stages of spicule formation involve vesicle-enclosed amorphous particles possibly kinetically stabilized by proteins and Mg^{2+} ions or by confinement mechanisms (Wilt 2002). These are transported to a syncytium, the site of spicule growth, where the amorphous phase transforms into calcite (Beniash et al. 1997). In addition, a mixture of stabilized amorphous phase and organic matter cements the calcite nanoparticles together (Seto et al. 2012). Studies on the growth of the sea urchin larval spicules using X-ray photoelectron emission spectromicroscopy show the presence of three closely associated mineral phases: a biogenic calcite, an initially hydrated ACC phase, and an intermediate transient ACC phase (Politi et al. 2008). The consistency of particle size distributions associated with the precursor phases and mineral particles suggested that phase transformation events follow a tortuous route in biomineral growth (Politi et al. 2008). Such processes appear to be under biochemical regulation because certain mineral-associated proteins appear to inhibit the dehydration of amorphous precursors, thereby stabilizing mineralization precursors (Gong et al. 2012). Phase transformations of synthetic ACC in the presence of sea urchin spicule extracts also suggest the role of local hydration and particle aggregation in the intercalation of biomolecules in biominerals and emergence of textured crystal surfaces, respectively (Gal et al. 2014). The aspects of nucleation and crystal growth operating here are not yet completely understood due to challenges associated with in situ observation of biomineral formation and growth. However the formation of biological minerals including mesocrystals can be effectively addressed in context of nonclassical pathways of crystallization (Gebauer and Cölfen 2011). According to the classical nucleation theory (CNT), a crystal nucleus can form or disintegrate based on the interplay between energy associated with formation of a new surface and the bulk energy gained from a crystal lattice. This theory does not provide a complete understanding for the early stages of biomineralization because precursors of important biominerals such as carbonates and phosphates of calcium do not completely dissociate to free ions, but form clusters that are stable with respect to the free ions in the solution state analogous to a “weak electrolyte” (Gebauer and Cölfen 2011).With the existence of PNCs in mineralization solutions including body fluids, approaches alternative to the ion-by-ion growth models envisaged by CNT are required for most biogenic minerals (Dey et al. 2010). Therefore, to elucidate the off-equilibrium morphologies and formation mechanisms of biominerals, nonclassical nucleation and nonclassical crystallization models are adopted here (Fig. 8.2).

8.4.1 Significance of the Pre-nucleation Regime

With regard to biominerals, the interactions of additives with mineral precursors and concomitant physiochemical conditions such as pH and ionic strength are crucial (Fig. 8.2). Changes in the dynamics of PNCs lead to internal condensation and the clusters become nanodroplets, which minimize their interfacial surface area by aggregation or coalescence (Gebauer et al. 2014). Additives that affect the structure and stability of PNCs also alter the course of mineralization, i.e., pre-nucleation events can significantly affect post-nucleation products in terms of composition and structure (Cartwright et al. 2012). In biological systems, there are several players regulating mineralization and this leads to a complex scenario wherein each additive modulates pre- and post-nucleation stages of mineralization (Fig. 8.3). The emergent biomineral is a consequence of a compositionally transient inorganic phase interacting with multiple biomolecules. A simple illustration is the formation of hetero-structures composed of calcite and vaterite in the presence of the polymer poly(sodium 4-styrenesulfonate) and folic acid (Wang et al. 2013b). Titration experiments show that these additives, complexed with Ca^{2+} ions, inhibit nucleation of solid particles as well as form selective post-nucleation products. Another important aspect of the pre-nucleation stage is that additives can mediate polymorph selection by modulating the stability of cluster species. For instance, D-glucose does not induce significant changes during the nascent stages of calcium carbonate mineralization wherein the typical post-nucleation product is ACC. In contrast, its enantiomer L-glucose leads to an enhancement in the stability of PNCs and the subsequent formation of vaterite, a crystalline form of calcium carbonate (possibly via a short-lived amorphous phase), after nucleation (Rao et al. 2014). These instances demonstrate that the pre-nucleation regime is of profound significance to the polymorphic composition of biominerals.

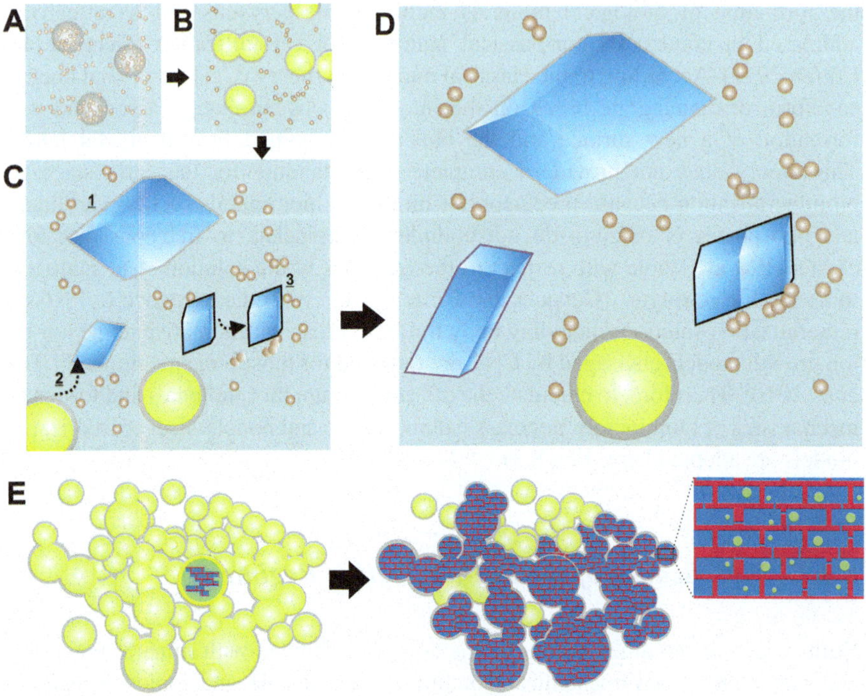

Fig. 8.3 Schematic representation of additive-mediated control of mineralization. (**a**) During the pre-nucleation regime, certain additives (*gray*) can affect mineralization by ion complexation, stabilization of PNCs, and the time required for nucleation of solid phase. (**b**) Nucleated amorphous phases (*yellow*) can be bound to and stabilized by additives. Such additives can also affect particle sizes by controlling the growth and coalescence of nucleated products. (**c, d**) (1) Certain crystalline faces are stabilized arresting particle growth. (2) Additives can control the transformation of amorphous to crystalline particles (*blue*). (3) Crystalline nanoparticles can attract by van der Waals/dipolar forces and then co-align to the lowest energy configuration and fuse by oriented attachment. Depending on the nature of the organic–inorganic junction at the solid–liquid interface, the step of coalescence may not be complete. (**e**) During biomineral growth, an interplay of these pathways can lead to tortuous growth paths and features such as mineral bridges, mesocrystalline structure, and nonequilibrium crystal morphologies. In addition to the mineralization environment, these processes are regulated by additives affecting phase transformation such as occluded intra- (*green*) and inter- (*pink*)crystalline additives

8.4.2 Role of the Amorphous Phase

Amorphous precursor phases are prevalent in biomineralization processes including formation of nacre and spicular materials (Zhang and Xu 2013; Beniash et al. 1997). Stabilized ACC can exhibit a $CaCO_3{\cdot}H_2O$ stoichiometry; however, the water content can also vary from 0.4 to 1.4 mol per $CaCO_3$ unit, with more stable forms having lower water contents (Michel et al. 2008; Radha et al. 2010; Rodriguez-Navarro et al. 2015). In an energetically favorable manner, crystallization proceeds by the dehydration of ACC to an anhydrous form (Radha et al. 2010; Cartwright et al. 2012). In nature, stabilization of such transient phases can be achieved by acidic biomolecules, confinement mechanisms, and the presence of Mg^{2+} ions (Kellermeier et al. 2010; Bentov et al. 2010). Although the mechanisms of biomolecule-mediated phase stabilization are not understood yet, Mg^{2+} is known to inhibit phase transformation due to its effective hydration layer and also by inducing distortions in the mineral structure (Lin et al. 2015; Politi et al. 2010). Recent simulation efforts also show that below a size of 4 nm, the stability of ACC is greater than calcite (Raiteri and Gale 2010). The consequence of such size-dependent phase stabilization (possibly involving confinement) over the size of the crystalline building units in biominerals requires elucidation. Localized in intercrystalline spaces, amorphous phases also have important roles as a cement material that provides structural integrity to a multi-crystalline assembly. This is evident from studies on the nanostructure of the sea urchin spine, nacre, and synthetic hydroxyapatite particles (Seto et al. 2012; Jäger et al. 2006; Nassif et al. 2005).

8.4.3 *Influence of Additives on Crystal Architecture*

In biomineralization, the role of additives on nucleation and phase transformation events is an important determinant of mono- or multi-crystallinity and also crystallinity itself (Fig. 8.3). Studies indicate a scheme of classifying mineralization additives based on their location in the final product, i.e., intracrystalline (incorporated/occluded) or intercrystalline (expelled) molecules (Fig. 8.4). The eventual spatial distributions of such additives affect crystal growth in terms of grain coalescence, particle size regimes, and structural integrity (Chen et al. 2007) and are susceptible to the kinetics of solidification (see Wolf and Gower 2017, Chap. 3). Several reports address biomolecules that are incorporated within crystals such as certain sea urchin and nacre proteins as well as agarose (Li et al. 2009; Berman et al. 1990). These molecules possibly exhibit weak interactions with the amorphous phase or the hydrated biochemical environment, but are structurally optimal for occlusion within crystals. This can be estimated from crystal face-specific incorporation energies for an additive (Clydesdale et al. 2003). An associated increase in the stiffness of inorganic crystals that are intrinsically brittle is an advantage of additive incorporation.

On the other hand, certain additives such as ions and biomolecules are expelled from the crystallization front during phase transformation similar to metallurgical zone melting (Nassif et al. 2005; Aizenberg et al. 2003). Some of these molecules can potentially induce stabilization of amorphous phases in a concentration-dependent manner. Such additives help elucidate incomplete phase transformations represented by the proximity of amorphous particles to crystals in certain biominerals. A high local concentration of such additives on the particle surface can result in the persistence of the intrinsically instable amorphous phases (Chen et al. 2007).

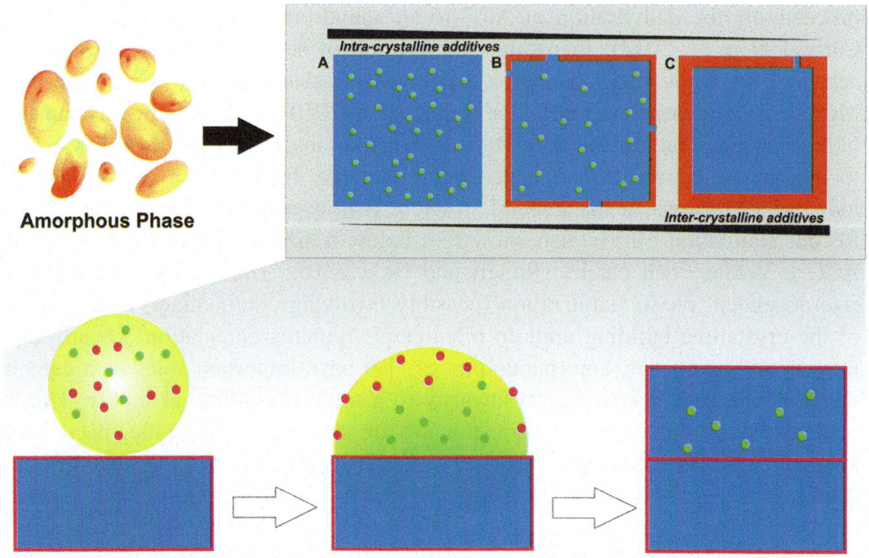

Fig. 8.4 Effect of additives (*red*) on phase transformation and emergent biomineral architecture: (**a**) intracrystalline additives lead to a lattice structure (*blue*) with defects due to additive (*red*) incorporation, (**b**) intra- and intercrystalline additives lead to a semicontinuous nature of crystallinity such as prevalence of mineral bridges, and (**c**) intercrystalline additives are expelled from the growing crystal front accumulating on the particle surface. In certain cases, such additives can stabilize particle surfaces and inhibit coalescence. The accompanying cartoon illustrates such events with respect to "wetting" of initial crystals with amorphous precursors (Homeijer et al. 2010; Berg et al. 2013) and phase transformation in the presence of intra- (*green*) and inter- (*pink*) crystalline additives. The partitioning of additives can be affected by the kinetics of solidification of the liquid precursors (see Wolf and Gower 2017, Chap. 3)

This is also demonstrated by the initial formation of highly concentrated amorphous droplets of mineral precursors in the presence of poly(acrylic) acid, which assemble into a "flux droplet" for formation of fibers, followed by crystallization of the entire assembly and formation of mesocrystalline fibers containing occluded additives (Homeijer et al. 2010; Gower and Odom 2000) (see Wolf and Gower 2017, Chap. 3). Thus, in special cases, the intercrystalline occurrence of such additives and stabilized amorphous material can inhibit coalescence and lattice fusion of the crystalline products (Fig. 8.3). For example, in the sea urchin spine, a higher content of biomolecules and Mg^{2+} ions can inhibit the dehydration and phase transformation of proximal ACC particles in the intercrystalline spaces (Seto et al. 2012). This may result in a decrease of scattering-derived coherence length in comparison to bulk crystals and reflect the presence of constituent individual particles (Seto et al. 2012; Aizenberg et al. 1997). Such effects are also indicated from products of gas diffusion experiments performed in the presence of recombinant domains of SM50 and n16.3 (Rao et al. 2013; Perovic et al. 2014; Keene et al. 2010).

The distinct effects of these additive classes are evident from structural comparison of the sea urchin spine, nacre material, and coccoliths from *Rhabdosphaera clavigera* (Fig. 8.5). The coccolith spine surface has a prominent homogenous organic layer (~25 nm thick), which is believed to inhibit growth in the {104} direction (van de Locht et al. 2014). Since these biomolecules are associated with mineral growth, their deposition is most likely driven by their expulsion during crystal growth. A lack of nanostructure reflects the inability of these additives to stabilize amorphous calcium carbonate (Didymus et al. 1993). During coccolith formation, the pre-nucleation regime is critical for spatially tuning mineralization during which a precursor phase composed of calcium-loaded macromolecules attach to a scaffold by molecular recognition (Gal et al. 2016). A deficit of potent ACC-stabilizing and nanoparticle-stabilizing biomolecules thus leads the coccolith units to having overall morphologies very similar to the calcite rhombohedra. The sea urchin spine however deviates from the equilibrium shape of calcite at multiple length scales (Seto et al. 2012). Biochemical and physiological studies show that biomolecules from the sea urchin spine can be either occluded or extruded from the calcite crystals (Berman et al. 1990; Rao et al. 2013). In the spine, a fine-tuned combination of these additives leads to mesocrystal formation wherein nanoparticles retain their size regime due to the intercrystalline cement mainly composed of Mg^{2+}-stabilized ACC and biomolecules (Seto et al. 2012). The proximity of amorphous inorganic phases to crystalline subunits of composite biomineral indicates a mechanism of stabilization by a locally high content of biomolecules and ions expelled from the newly formed crystal lattice. Such an environment promotes the stability of particle interfaces by limiting grain coalescence (Chen et al. 2007) and may also play an early role in phase selection as demonstrated by the formation of aragonite in the presence of mollusk shell proteins, respectively (Falini et al. 1996).

The effects of these additive classes are also illustrated by electron microscopy and diffraction-based observations of biogenic calcitic prisms (Okumura et al. 2012). In oyster shells, the distribution of biomolecules is inhomogeneous and splits the biogenic calcite into co-oriented nanograins. In contrast, in pen shells, a homogenous distribution of biomolecules does not lead to a significant substructure.

8.4.4 Crystallographic Co-orientation

Having addressed the influence of additives during phase transformation, we now look at mechanisms of co-orientation based on evidence from biological mesocrystals. Cells in multicellular organisms secrete copious amounts of extracellular matrices (ECM). This matrix primarily consists of self-assembling biomolecules that form a gel-like network to aid cellular sustenance (McDonald 1988). Such gel matrices are suitable for mesocrystal formation because diffusion-limited growth promotes high nucleation rate and offers control over particle size. Gels with ion/PNC complexation properties also affect local supersaturation and thereby provide regulation over polymorph selection (Wang et al. 2005; Grassmann et al. 2002; Helminger et al. 2014). In certain cases, the periodicity associated with

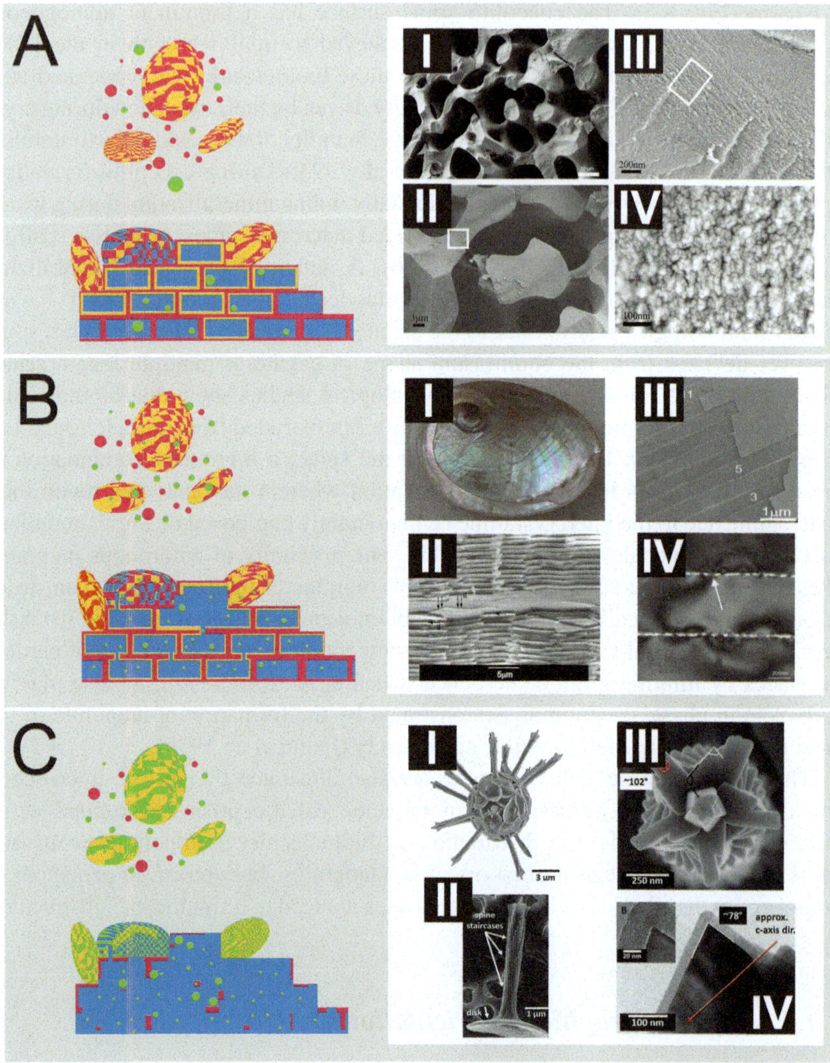

Fig. 8.5 Schematic representation illustrating the outcomes of phase transformations of amorphous (*yellow*) to crystalline (*blue*) phases under control of inter-(*pink*) and intra-(*green*) crystalline additives. (**a**) Mineral growth predominantly influenced by intercrystalline additives that inhibit particle coalescence and growth. Extrusion of additives that stabilize ACC from the crystallization front can favor crystal surfaces covered by a composite layer. Under viscoelastic conditions as well as mineral bridges that favor co-orientation, a mesocrystal is formed such as the sea urchin spine. (**b**) With intermediate contents of inter- and intracrystalline additives that promote limited grain coalescence, spatially regulated particle coalescence might lead to distinct mineral bridges depicted by those between tablets in nacre material. (**c**) For biominerals such as the coccolith calcite, additives allow for effective coalescence of the precursor phase to form large crystals. The thick amorphous organic layer surrounding coccolith calcite and continuous crystal lattice is evident of particle fusion and expulsion of mineralization additives. [Insets reproduced from (**a**) Seto et al. 2012 with permissions from *Proc. Natl. Acad. Sci. U.S.A.* (**b**) Bezares et al. *J. Struct. Biol. 2012* 170(484); Gries et al. *Ultramicroscopy 2009* 109(230) with permissions from Elsevier, and (**c**) van de Locht et al. 2014 with permission from American Chemical Society]

ECM biomolecules is transcribed into the arrangement of the mineral subunits by co-localized crystal nucleation and growth. For example, in bone tissue, hydroxyapatite nano-plates (45–20 nm) have a consistent crystallographic alignment wherein the *a*- and *c*-axes (corresponding to [110] and [001]) are aligned toward the groove and axial direction of collagen fibrils, respectively (Landis et al. 1996). Thus self-assembling additives that nucleate or stabilize specific crystal faces are crucial in the formation of biominerals with properties of crystallographic co-orientation. Although mineralization of collagen fibrils has been realized in vitro (Olszta et al. 2003a, b; Nudelman et al. 2010), the role of the mineral–organic interface being principally composed of polysaccharide and the impact of non-collagenous protein on macrostructural integrity need to be addressed (Wise et al. 2007; Hang et al. 2014). As discussed for nanograins in nacre tablets, the physical environment during phase transformation also has an important bearing on the biomineral architecture (Li et al. 2006). In nature, mesocrystal formation necessitates a viscoelastic or amorphous medium that not only confers certain spatial dynamism to nanoparticles but also inhibits lattice fusion. In the absence of effective nanoparticle stabilization, van der Waals and dipolar interaction-driven oriented attachment may lead to transient mesocrystals that gradually transform to a single crystal (Niederberger and Cölfen 2006; Schwahn et al. 2007). A physicochemical route for crystallographical alignment can also originate from the charges present on organic surfaces in biominerals. Oppositely charged polar inorganic surfaces (such as the (001) aragonite face in nacre) can possibly be nucleated on such organic layers. This circumvents the requirement for precise epitaxial alignment between interfaces of soft and hard materials (Nassif et al. 2005). Thus the cumulative effects of biomolecular properties such as ion complexation, self-assembly, viscoelasticity, and charge complemented with an appropriate crystallization system lead to the emergence of mesocrystallinity in nature.

The nonclassical pathways of crystallization further elucidate crystallographic orientation and mesocrystal formation in biomimetic systems. Central to this topic are the phenomena of oriented attachment in which particles spontaneously self-align, attach, and consequently share a common crystallographic orientation (Penn and Banfield 1998; Niederberger and Cölfen 2006). This process is particularly important for biominerals that have nanocrystalline subunits exhibiting high surface energies. These subunits can fuse to a structure with a relatively lower energy configuration and also lead to unusual crystal morphologies. In the presence of certain additives, this process can be controlled (Niederberger and Cölfen 2006). For instance, hematite mesocrystals can form in the presence of silk fibroin by processes that provide control over size and morphology (Fei et al. 2014). Steps in this mechanism involve (a) complexation of ions/ion clusters to the protein matrix, (b) nucleation of metastable goethite, and (c) dehydration of goethite to hematite nanoparticles (Fei et al. 2014). Mesocrystal formation is attributed to the lower surface energy of hematite nanoparticles stabilized by silk fibroin and also to the macromolecular self-assembly. Such studies show that a high prevalence of low-complexity regions in biomineralization proteins (suggestive of repetitive structural motifs) can lead to assembly of associated particles. "Particle accretion" is a recently proposed mechanism for crystal growth in order to explain phase transformation at the nanoscale and associated crystallographic co-orientation (Gal et al. 2014). The steps proposed are (a) partial dissolution of ACC particle to yield solute ions, (b) inhibition of step (a) and ion-by-ion crystallization by additives, (c) diffusion and surface tension-driven attachment of ACC to kink and steps on crystal surfaces, and (d) crystal growth by phase transformation of ACC and ion-by-ion attachment. This route is very similar to earlier described pathways of nonclassical nucleation and crystallization involving oriented attachment of nanoparticles to form larger crystal (Niederberger and Cölfen 2006); however it fundamentally differs by proposing ion-by-ion growth for ionic minerals such as calcium carbonate (Cölfen and Antonietti 2008; Gebauer and Cölfen 2011; Kim et al. 2008). In this study, the low coherence length of the synthesized crystals ($\sim$150 nm) reflects certain mesocrystallinity that might emerge from the process of oriented attachment in a viscoelastic gel medium (Gal et al. 2014; Niederberger and Cölfen 2006). This is validated for crystallization in hydrogels such gelatin that promote particle assembly in viscous environments and in some cases also favor particle anisotropy by stabilization of specific crystal faces (Tseng et al. 2009; Grassmann et al. 2002; Fei et al. 2014).

8.5 Characterization Techniques

We briefly introduce the reader to certain methods for investigating biomineral structure as well as associated precautions. To establish or disprove mesocrystallinity, we recommend the application of complementary techniques to investigate material substructures at appropriate experimental time scales because a mesocrystal is more a kinetically, metastable intermediate than a thermodynamically stable product (Cölfen and Antonietti 2005).

8.5.1 Electron Microscopy

Electron microscopy is a powerful method for structural characterization at the nanoscale (Nielsen and De Yoreo 2017, Chap. 18). It is also used to determine chemical and electronic material properties. Material processing in order to view the mesocrystalline interior might lead to artifacts. For example, after sectioning of the sea urchin spine, the cuts exposed to ethylene glycol and water lead to different observations (Seto et al. 2012). This is attributed to the insolubility of calcium carbonate in ethylene glycol that inhibits the transformation/dissolution of amorphous calcium carbonate. The user should also be aware of beam damage effects such as the instability of amorphous phases at high electron dosage (Weiss et al. 2002; Rodriguez-Blanco et al. 2008). Time-resolved studies also help identify the mechanisms involved in mesocrystal formation (Yuwono et al. 2010).

8.5.2 *Light Microscopy*

Limited by optical resolution, light microscopy may not be the most suitable technique for validation of mesocrystallinity. However, special imaging modes can be used to investigate iso-orientation as well as formation mechanisms at the micron length scale. For instance, quantitative polarization microscopy (Abrio®) provides distributions of structural orientation for optically resolved length scales and also can circumvent the formation of drying artifacts (Tritschler et al. 2013; Jiang et al. 2013). Unlike in standard electron, time-resolved imaging is possible under ambient conditions.

8.5.3 *Atomic Force Microscopy*

This is a useful tool to map sample topography, reconstruct 3-D surfaces, and probe mechanical properties at the nanoscale. As shown for nacre material, changes in surface topography under mechanical load can hint toward the presence of a substructure (Li et al. 2004, 2006). The nature of the intercrystalline spaces can also be probed by the phase contrast during AFM (Seto et al. 2012). Although information about crystallographic co-orientation is not provided by AFM, this is an established method to investigate material nanostructure.

8.5.4 *BET*

Due to high internal structure and porosity, synthetic mesocrystals can exhibit high Brunauer–Emmett–Teller (BET) surface area. For example, calcite mesocrystals formed in the presence of polystyrene sulfonate have surface areas larger than 260 m^2/g (Wang et al. 2005). However, in biogenic mesocrystals, this quantity can be significantly lower because of the space-filling nature of the nanoparticles as well as biomolecules and amorphous inorganic material occupying intercrystalline zones (Yang et al. 2011). Such analyses should be considered with caution due to ripening effects that can significantly decrease the associated surface area and porosity (Wang et al. 2009; Cölfen and Antonietti 2005).

8.5.5 *Other Techniques*

Small-angle X-ray scattering (SAXS) can be applied to identify phases differing in electron density as well as investigate the size distributions of primary particles and their domain spreads (Seto et al. 2012). An indirect approach for detection of mesocrystals is the infiltration of a putative crystal with monomeric units followed by a polymerization reaction and subsequent dissolution of the initial crystal. A residual porous template is obtained for a mesocrystalline material (Oaki et al. 2011). Organic compounds such as colored dyes can also permeate through the intercrystalline spaces of a mesocrystal (Munekawa et al. 2015; Oaki and Imai 2006). Different techniques for investigating growth composition and structure of the mesocrystals are comprehensively listed in previous literature (Cölfen and Antonietti 2008).

8.6 Conclusions and Outlook

There are several routes presented in literature for the formation of biogenic and synthetic mesocrystalline materials. Since material formation in nature solely relies on bottom-up approaches, we can describe biologically derived mesocrystallinity in terms of activation energy barriers associated with the aggregation, attachment, coalescence, and growth of nanoparticles after nucleation (Fig. 8.6).

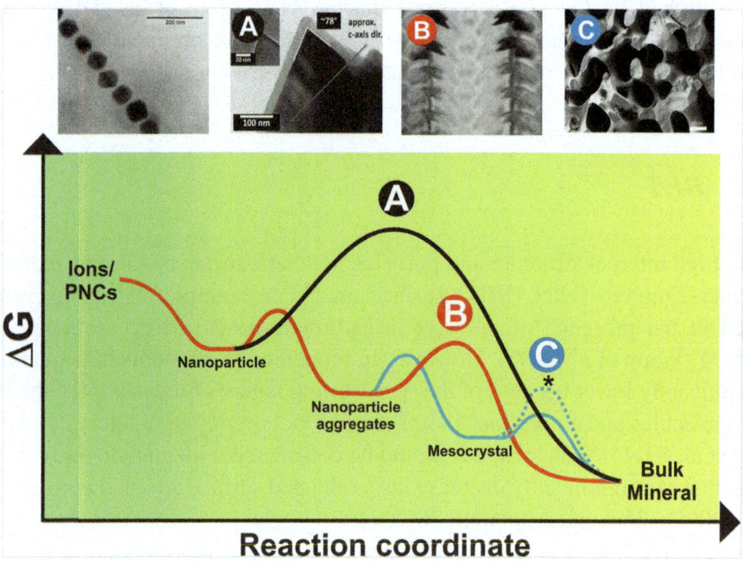

Fig. 8.6 Schematic free energy profiles for nonclassical crystallization pathways illustrating the nucleation stage (*green zone*) followed by schemes for the emergence of diverse biomineral structures (*yellow zone*) [see text for description]. Transient (*continuous blue*) or kinetically stable (*dotted blue*) nature of mesocrystals depends on the free energy of activation for particle coalescence [*top left* to *right*: Images reproduced from Bazylinski et al. 1994 with permissions from John Wiley and Sons; van de Locht et al. 2014 with permission from American Chemical Society; Wang et al. 2013a with permissions from John Wiley and Sons and Seto et al. 2012 with permissions from Proc. Natl. Acad. Sci. U.S.A.]

From the discussions presented above, the fate of the mineral structure is primarily determined by the effects of additives as well as the nature of inorganic constituents among other factors. Supported by literature evidence, the effect of these two factors on nonclassical crystallization mechanisms leads to the emergence of diverse mineral architectures such as stable nanoparticles, crystals close to equilibrium morphologies, polycrystalline composites, and mesocrystallinity. These are exemplified with greigite nanoparticles in magnetotactic bacteria (Bazylinski et al. 1994), single crystalline calcite rhombohedra from *R. clavigera* (van de Locht et al. 2014), polycrystalline magnetite in chiton teeth (Wang et al. 2013a), and mesocrystalline calcite in sea urchin spines (Seto et al. 2012). As depicted in Fig. 8.6, pathway (A) represents mineral formation from a single seed crystal in an environment that favors attachment and coalescence of precursors to the initial seed. Additives that stabilize particle surface and inhibit aggregation, alignment, and coalescence lead to stable nanoparticles such as those in magnetosomes. Pathway (B) represents a state for nanoparticles that attach forming aggregates. These are either amorphous/polycrystalline aggregates or nanoparticles co-oriented at small length scales. These nanoparticles can "seed" further crystal growth. This pathway associates well with the maturation of radular chiton teeth wherein magnetite particles grow to form parallel rods (Wang et al. 2013a). With inhibitors of particle coalescence/lattice fusion, a mesocrystal may form via pathway (C). Depending on the activation energy barrier associated with particle coalescence*, the mesocrystal may be transient or kinetically stable. Note that the prevalence of multiple pathways is possible during mineralization. Although this description does not directly address pre-nucleation stages, this regime is also of utmost significance to the control over nucleation as well as polymorph selection (Fig. 8.2). Nevertheless, we show that additive-induced kinetic modulation of activation energy barriers associated with inter-nanoparticle interactions can account for mineral architectures. This cannot be addressed by classical models of nucleation and growth.

Understanding formation mechanisms of mesocrystalline and other biominerals is critical for their practical use. The present applications of synthetic mesocrystalline materials have been toward photocatalyst, sensor, and electrode materials (Zhou and O'Brien 2012; Ma and Cölfen 2014). Synthetic mesocrystals generally require precise control over steps such as nanoparticle synthesis, evaporation-driven assembly, and hydrothermal treatment. However, in nature, mesocrystals are formed by the integration of organic and inorganic components through a bottom-up strategy under physiological conditions. Understanding these biological means for mesocrystal formation can allow for new strategies for material synthesis. However this entails investigating the structure, formation, and properties of biominerals in terms of biophysical and biochemical interactions operating at different length scales. This is a challenging task on account of the spatiotemporally dynamic interface between organic additives and inorganic precursors/particles (Fig. 8.7). Furthermore, with the emerging role of PNCs, biomineralization appears to comprise pre-nucleation as well as post-nucleation processes under regulation of biomolecules. Significant progress has been made toward understanding mineral-

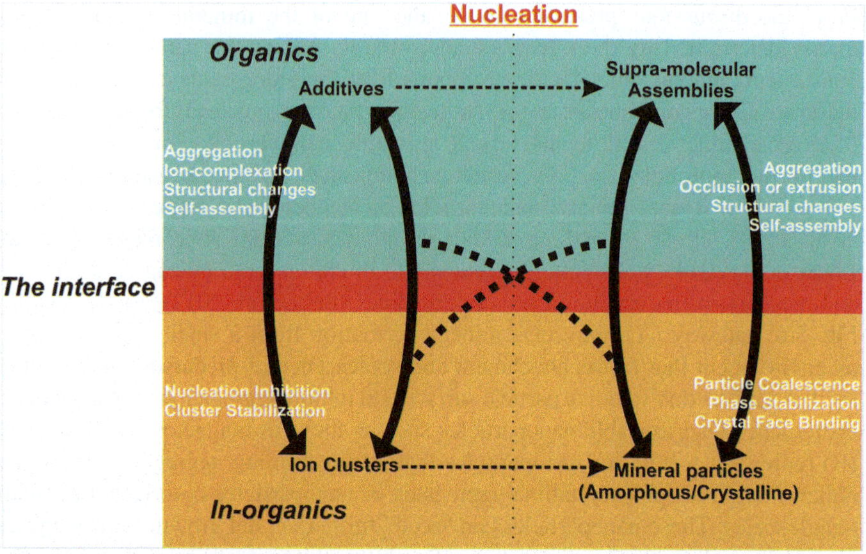

Fig. 8.7 During biomineralization, a spatiotemporally dynamic interface between organic additives and mineral precursors/particles spans the pre- and post-nucleation regimes. Interactions between inorganic and organic components are two ways leading to consequences for inorganic species as well as additives

ization processes by nonclassical nucleation and crystallization pathways. Further investigations into such exciting research problems will aid further elucidation of biomineralization as well as inspire routes for fascinating new materials.

References

Aizenberg J, Hanson J, Koetzle T, Weiner S, Addadi L (1997) Control of macromolecule distribution within synthetic and biogenic single calcite crystals. J Am Chem Soc 119: 881–886

Aizenberg J, Muller DA, Grazul JL, Hamann DR (2003) Direct fabrication of large micropatterned single crystals. Science 299:1205–1208

Aizenberg J, Sundar VC, Yablon AD, Weaver JC, Chen G (2004) Biological glass fibers: correlation between optical and structural properties. Proc Natl Acad Sci U S A 101: 3358–3363

Aizenberg J, Weaver JC, Thanawala MS, Sundar VC, Morse DE, Fratzl P (2005) Skeleton of Euplectella sp.: structural hierarchy from the nanoscale to the macroscale. Science 309: 275–278

Andreassen J-P, Lewis AE (2017) Classical and non-classical theories of crystal growth. In: Van Driessche AES, Kellermeier M, Benning LG, Gebauer D (eds) New perspectives on mineral nucleation and growth. Springer, Cham, pp 137–154

Barthelat F, Tang H, Zavattieri P, Li C-M, Espinosa H (2007) On the mechanics of mother-of-pearl: a key feature in the material hierarchical structure. J Mech Phys Solids 55:306–337

Bazylinski DA, Garratt-Reed AJ, Frankel RB (1994) Electron microscopic studies of magnetosomes in magnetotactic bacteria. Microsc Res Tech 27:389–401
Beniash E, Aizenberg J, Addadi L, Weiner S (1997) Amorphous calcium carbonate transforms into calcite during sea urchin larval spicule growth. Proc R Soc Lond Ser B Biol Sci 264:461–465
Beniash E, Metzler RA, Lam RS, Gilbert P (2009) Transient amorphous calcium phosphate in forming enamel. J Struct Biol 166:133–143
Bentov S, Weil S, Glazer L, Sagi A, Berman A (2010) Stabilization of amorphous calcium carbonate by phosphate rich organic matrix proteins and by single phosphoamino acids. J Struct Biol 171:207–215
Benzerara K, Menguy N, Obst M, Stolarski J, Mazur M, Tylisczak T, Brown GE Jr, Meibom A (2011) Study of the crystallographic architecture of corals at the nanoscale by scanning transmission X-ray microscopy and transmission electron microscopy. Ultramicroscopy 111:1268–1275
Berg JK, Jordan T, Binder Y, BöRner HG, Gebauer D (2013) Mg^{2+} tunes the wettability of liquid precursors of $CaCO_3$: toward controlling mineralization sites in hybrid materials. J Am Chem Soc 135:12512–12515
Bergström L, Sturm EV, Salazar-Alvarez G, Cölfen H (2015) Mesocrystals in biominerals and colloidal arrays. Acc Chem Res 48:1391–1402
Berman A, Addadi L, Kvick Å, Leiserowitz L, Nelson M, WEINER S (1990) Intercalation of sea urchin proteins in calcite: study of a crystalline composite material. Science 250: 664–667
Cartwright JH, Checa AG, Gale JD, Gebauer D, Sainz-Díaz CI (2012) Calcium carbonate polyamorphism and its role in biomineralization: how many amorphous calcium carbonates are there? Angew Chem Int Ed 51:11960–11970
Chen B, Zhang H, Gilbert B, Banfield JF (2007) Mechanism of inhibition of nanoparticle growth and phase transformation by surface impurities. Phys Rev Lett 98:106103
Clydesdale G, Hammond RB, Roberts KJ (2003) Molecular modeling of bulk impurity segregation and impurity-mediated crystal habit modification of naphthalene and phenanthrene in the presence of heteroimpurity species. J Phys Chem B 107:4826–4833
Cölfen H (2010) Biomineralization: a crystal-clear view. Nat Mater 9:960–961
Cölfen H, Antonietti M (2005) Mesocrystals: inorganic superstructures made by highly parallel crystallization and controlled alignment. Angew Chem Int Ed 44:5576–5591
Cölfen H, Antonietti M (2008) Mesocrystals and nonclassical crystallization. Wiley, Chichester
Currey JD (2005) Hierarchies in biomineral structures. Science 309:253–254
De Yoreo JJ, Sommerdijk N, Dove P (2017) Nucleation pathways in electrolyte solutions. In: Van Driessche AES, Kellermeier M, Benning LG, Gebauer D (eds) New perspectives on mineral nucleation and growth. Springer, Cham, pp 1–24
Demichelis R, Raiteri P, Gale JD, Quigley D, Gebauer D (2011) Stable prenucleation mineral clusters are liquid-like ionic polymers. Nat Commun 2:590
Dey A, Bomans PH, Müller FA, Will J, Frederik PM, De With G, Sommerdijk NA (2010) The role of prenucleation clusters in surface-induced calcium phosphate crystallization. Nat Mater 9:1010–1014
Didymus JM, Oliver P, Mann S, Devries AL, Hauschka PV, Westbroek P (1993) Influence of low-molecular-weight and macromolecular organic additives on the morphology of calcium carbonate. J Chem Soc Faraday Trans 89:2891–2900
Evans JS (2008) "Tuning in" to mollusk shell nacre- and prismatic-associated protein terminal sequences. Implications for biomineralization and the construction of high performance inorganic – organic composites. Chem Rev 108:4455–4462
Falini G, Fermani S (2017) Nucleation and growth from a biomineralization perspective. In: Van Driessche AES, Kellermeier M, Benning LG, Gebauer D (eds) New perspectives on mineral nucleation and growth. Springer, Cham, pp 185–198
Falini G, Albeck S, Weiner S, Addadi L (1996) Control of aragonite or calcite polymorphism by mollusk shell macromolecules. Science 271:67–69

Fei X, Li W, Shao Z, Seeger S, Zhao D, Chen X (2014) Protein biomineralized nanoporous inorganic mesocrystals with tunable hierarchical nanostructures. J Am Chem Soc 136:15781–15786

Fernandez-Martinez A, Lopez-Martinez H, Wang D (2017) Structural characteristics and the occurrence of polyamorphism in amorphous calcium carbonate. In: Van Driessche AES, Kellermeier M, Benning LG, Gebauer D (eds) New perspectives on mineral nucleation and growth, Springer. Cham, pp 77–92

Furuichi K, Oaki Y, Imai H (2006) Preparation of nanotextured and nanofibrous hydroxyapatite through dicalcium phosphate with gelatin. Chem Mater 18:229–234

Gal A, Kahil K, Vidavsky N, Devol RT, Gilbert PUPA, Fratzl P, Weiner S, Addadi L (2014) Particle accretion mechanism underlies biological crystal growth from an amorphous precursor phase. Adv Funct Mater 24:5420–5426

Gal A, Wirth R, Kopka J, Fratzl P, Faivre D, Scheffel A (2016) Macromolecular recognition directs calcium ions to coccolith mineralization sites. Science 353:590–593

Gebauer D (2013) Bio-inspired materials science at its best—flexible mesocrystals of calcite. Angew Chem Int Ed 52:8208–8209

Gebauer D, Cölfen H (2011) Prenucleation clusters and non-classical nucleation. Nano Today 6:564–584

Gebauer D, Völkel A, Cölfen H (2008) Stable prenucleation calcium carbonate clusters. Science 322:1819–1822

Gebauer D, Cölfen H, Verch A, Antonietti M (2009) The multiple roles of additives in $CaCO_3$ crystallization: a quantitative case study. Adv Mater 21:435–439

Gebauer D, Gunawidjaja PN, Ko J, Bacsik Z, Aziz B, Liu L, Hu Y, Bergström L, Tai CW, Sham TK (2010) Proto-calcite and proto-vaterite in amorphous calcium carbonates. Angew Chem 122:9073–9075

Gebauer D, Kellermeier M, Gale JD, Bergström L, Cölfen H (2014) Pre-nucleation clusters as solute precursors in crystallisation. Chem Soc Rev 43:2348–2371

Gong YU, Killian CE, Olson IC, Appathurai NP, Amasino AL, Martin MC, Holt LJ, Wilt FH, Gilbert P (2012) Phase transitions in biogenic amorphous calcium carbonate. Proc Natl Acad Sci 109:6088–6093

Gower LB, Odom DJ (2000) Deposition of calcium carbonate films by a polymer-induced liquid-precursor (PILP) process. J Cryst Growth 210:719–734

Grassmann O, Müller G, Löbmann P (2002) Organic – inorganic hybrid structure of calcite crystalline assemblies grown in a gelatin hydrogel matrix: relevance to biomineralization. Chem Mater 14:4530–4535

Hang F, Gupta HS, Barber AH (2014) Nanointerfacial strength between non-collagenous protein and collagen fibrils in antler bone. J R Soc Interface 11:20130993

Helminger M, Wu B, Kollmann T, Benke D, Schwahn D, Pipich V, Faivre D, Zahn D, Cölfen H (2014) Synthesis and characterization of gelatin-based magnetic hydrogels. Adv Funct Mater 24:3187–3196

Homeijer SJ, Barrett RA, Gower LB (2010) Polymer-Induced Liquid-Precursor (PILP) process in the non-calcium based systems of barium and strontium carbonate. Cryst Growth Des 10:1040–1052

Jackson A, Vincent J, Turner R (1988) The mechanical design of nacre. Proc R Soc Lond B Biol Sci 234:415–440

Jäger C, Welzel T, Meyer-Zaika W, Epple M (2006) A solid-state NMR investigation of the structure of nanocrystalline hydroxyapatite. Magn Reson Chem 44:573–580

Jiang Y, Gong H, Grzywa M, Volkmer D, Gower L, Cölfen H (2013) Thin films: microdomain transformations in mosaic mesocrystal thin films (Adv. Funct. Mater. 12/2013). Adv Funct Mater 23:1604

Keene EC, Evans JS, Estroff LA (2010) Silk fibroin hydrogels coupled with the n16N – β-chitin complex: an in vitro organic matrix for controlling calcium carbonate mineralization. Cryst Growth Des 10:5169–5175

Kellermeier M, Melero-Garcia E, Glaab F, Klein R, Drechsler M, Rachel R, García-Ruiz JM, Kunz W (2010) Stabilization of amorphous calcium carbonate in inorganic silica-rich environments. J Am Chem Soc 132:17859–17866

Kellermeier M, Gebauer D, Melero-García E, Drechsler M, Talmon Y, Kienle L, Cölfen H, García-Ruiz JM, Kunz W (2012a) Colloidal stabilization of calcium carbonate prenucleation clusters with silica. Adv Funct Mater 22:4301–4311

Kellermeier M, Rosenberg R, Moise A, Anders U, Przybylski M, Cölfen H (2012b) Amino acids form prenucleation clusters: ESI-MS as a fast detection method in comparison to analytical ultracentrifugation. Faraday Discuss 159:23–45

Kijima M, Oaki Y, Imai H (2011) In vitro repair of a biomineral with a mesocrystal structure. Chem Eur J 17:2828–2832

Kim IW, Giocondi JL, Orme C, Collino S, Spencer Evans J (2008) Morphological and kinetic transformation of calcite crystal growth by prismatic-associated asprich sequences. Cryst Growth Des 8:1154–1160

Kinoshita S, Yoshioka S (2005) Structural colors in nature: the role of regularity and irregularity in the structure. ChemPhysChem 6:1442–1459

Knudsen FP (1959) Dependence of mechanical strength of brittle polycrystalline specimens on porosity and grain size. J Am Ceram Soc 42:376–387

Komeili A, Vali H, Beveridge TJ, Newman DK (2004) Magnetosome vesicles are present before magnetite formation, and MamA is required for their activation. Proc Natl Acad Sci U S A 101:3839–3844

Landis WJ, Hodgens KJ, Arena J, Song MJ, Mcewen BF (1996) Structural relations between collagen and mineral in bone as determined by high voltage electron microscopic tomography. Microsc Res Tech 33:192–202

Lee K, Wagermaier W, Masic A, Kommareddy KP, Bennet M, Manjubala I, Lee S-W, Park SB, Cölfen H, Fratzl P (2012) Self-assembly of amorphous calcium carbonate microlens arrays. Nat Commun 3:725

Li X, Huang Z (2009) Unveiling the formation mechanism of pseudo-single-crystal aragonite platelets in nacre. Phys Rev Lett 102:075502

Li X, Chang W-C, Chao YJ, Wang R, Chang M (2004) Nanoscale structural and mechanical characterization of a natural nanocomposite material: the shell of red abalone. Nano Lett 4: 613–617

Li X, Xu Z-H, Wang R (2006) In situ observation of nanograin rotation and deformation in nacre. Nano Lett 6:2301–2304

Li H, Xin HL, Muller DA, Estroff LA (2009) Visualizing the 3D internal structure of calcite single crystals grown in agarose hydrogels. Science 326:1244–1247

Li L, Mao C, Wang J, Xu X, Pan H, Deng Y, Gu X, Tang R (2011) Bio-inspired enamel repair via glu-directed assembly of apatite nanoparticles: an approach to biomaterials with optimal characteristics. Adv Mater 23:4695–4701

Lin C-J, Yang S-Y, Huang S-J, Chan JCC (2015) Structural characterization of Mg-stabilized amorphous calcium carbonate by Mg-25 solid-state NMR spectroscopy. J Phys Chem C 119:7225–7233

Lowenstam HA, Weiner S (1989) On biomineralization. Oxford University Press, New York

Lutsko JF (2017) Novel paradigms in non-classical nucleation theory. In: Van Driessche AES, Kellermeier M, Benning LG, Gebauer D (eds) New perspectives on mineral nucleation and growth. Springer, Cham, pp 25–42

Ma M-G, Cölfen H (2014) Mesocrystals—applications and potential. Curr Opin Colloid Interface Sci 19:56–65

Mahamid J, Sharir A, Addadi L, WEINER S (2008) Amorphous calcium phosphate is a major component of the forming fin bones of zebrafish: indications for an amorphous precursor phase. Proc Natl Acad Sci 105:12748–12753

Mahamid J, Sharir A, Gur D, Zelzer E, Addadi L, Weiner S (2011) Bone mineralization proceeds through intracellular calcium phosphate loaded vesicles: a cryo-electron microscopy study. J Struct Biol 174:527–535

Mann S (1995) Biomineralization and biomimetic materials chemistry. J Mater Chem 5:935–946

Mann S, Archibald DD, Didymus JM, Douglas T, Heywood BR, Meldrum FC, Reeves NJ (1993) Crystallization at inorganic-organic interfaces: biominerals and biomimetic synthesis. Science 261:1286–1292

Mann K, Poustka AJ, Mann M (2008) The sea urchin (strongylocentrotus purpuratus) test and spine proteomes. Proteome Sci 6:22

Mcdonald JA (1988) Extracellular matrix assembly. Annu Rev Cell Biol 4:183–207

Meldrum FC, Cölfen H (2008) Controlling mineral morphologies and structures in biological and synthetic systems. Chem Rev 108:4332–4432

Michel M, Macdonald J, Feng J, Phillips BL, Ehm L, Tarabrella C, Parise JB, Reeder RJ (2008) Structural characteristics of synthetic amorphous calcium carbonate. Chem Mater 20:4720–4723

Moureaux C, Pérez-Huerta A, Compère P, Zhu W, Leloup T, Cusack M, Dubois P (2010) Structure, composition and mechanical relations to function in sea urchin spine. J Struct Biol 170:41–49

Munekawa Y, Oaki Y, Sato K, Imai H (2015) Incorporation of organic crystals into the interspace of oriented nanocrystals: morphologies and properties. Nanoscale 7:3466–3473

Nakahara H (1983) Calcification of gastropod nacre. In: Biomineralization and biological metal accumulation. Springer, Berlin

Nakahara H, Suga S, Kaigi NG (1991) Mechanisms and phylogeny of mineralization in biological systems. Springer, Tokyo

Nassif N, Pinna N, Gehrke N, Antonietti M, Jäger C, Cölfen H (2005) Amorphous layer around aragonite platelets in nacre. Proc Natl Acad Sci U S A 102:12653–12655

Natalio F, Corrales TP, Panthöfer M, Schollmeyer D, Lieberwirth I, Müller WE, Kappl M, Butt H-J, Tremel W (2013) Flexible minerals: self-assembled calcite spicules with extreme bending strength. Science 339:1298–1302

Niederberger M, Cölfen H (2006) Oriented attachment and mesocrystals: non-classical crystallization mechanisms based on nanoparticle assembly. Phys Chem Chem Phys 8:3271–3287

Nielsen MH, De Yoreo JJ (2017) Liquid phase TEM investigations of crystal nucleation, growth and transformation. In: Van Driessche AES, Kellermeier M, Benning LG, Gebauer D (eds) New perspectives on mineral nucleation and growth. Springer, Cham, pp 353–374

Nudelman F, Pieterse K, George A, Bomans PH, Friedrich H, Brylka LJ, Hilbers PA, de With G, Sommerdijk NA (2010) The role of collagen in bone apatite formation in the presence of hydroxyapatite nucleation inhibitors. Nat Mater 9:1004–1009

Oaki Y, Imai H (2005) The hierarchical architecture of nacre and its mimetic material. Angew Chem Int Ed 44:6571–6575

Oaki Y, Imai H (2006) Nanoengineering in echinoderms: the emergence of morphology from nanobricks. Small 2:66–70

Oaki Y, Kotachi A, Miura T, Imai H (2006) Bridged nanocrystals in biominerals and their biomimetics: classical yet modern crystal growth on the nanoscale. Adv Funct Mater 16: 1633–1639

Oaki Y, Kijima M, Imai H (2011) Synthesis and morphogenesis of organic polymer materials with hierarchical structures in biominerals. J Am Chem Soc 133:8594–8599

Okazaki K, Inoué S (1976) Crystal property of the larval sea urchin spicule*. Develop Growth Differ 18:413–434

Okumura T, Suzuki M, Nagasawa H, Kogure T (2012) Microstructural variation of biogenic calcite with intracrystalline organic macromolecules. Cryst Growth Des 12:224–230

Olszta M, Douglas E, Gower L (2003a) Scanning electron microscopic analysis of the mineralization of type I collagen via a polymer-induced liquid-precursor (PILP) process. Calcif Tissue Int 72:583–591

Olszta MJ, Odom DJ, Douglas EP, Gower LB (2003b) A new paradigm for biomineral formation: mineralization via an amorphous liquid-phase precursor. Connect Tissue Res 44:326–334

Penn RL, Banfield JF (1998) Imperfect oriented attachment: dislocation generation in defect-free nanocrystals. Science 281:969–971

Penn RL, Li D, Soltis JA (2017) A perspective on the particle-based crystal growth of ferric oxides, oxyhydroxides, and hydrous oxides. In: Van Driessche AES, Kellermeier M, Benning LG, Gebauer D (eds) New perspectives on mineral nucleation and growth. Springer, Cham, pp 257–274

Perovic I, Chang EP, Lui M, Rao A, Cölfen H, Evans JS (2014) A nacre protein, n16.3, self-assembles to form protein oligomers that dimensionally limit and organize mineral deposits. Biochemistry 53:2739–2748

Politi Y, Levi-Kalisman Y, Raz S, Wilt F, Addadi L, Weiner S, Sagi I (2006) Structural characterization of the transient amorphous calcium carbonate precursor phase in sea urchin embryos. Adv Funct Mater 16:1289–1298

Politi Y, Metzler RA, Abrecht M, Gilbert B, Wilt FH, Sagi I, Addadi L, Weiner S, Gilbert P (2008) Transformation mechanism of amorphous calcium carbonate into calcite in the sea urchin larval spicule. Proc Natl Acad Sci 105:17362–17366

Politi Y, Batchelor DR, Zaslansky P, Chmelka BF, Weaver JC, Sagi I, Weiner S, Addadi L (2010) Role of magnesium ion in the stabilization of biogenic amorphous calcium carbonate: a structure-function investigation. Chem Mater 22:161–166

Pouget EM, Bomans PH, Goos JA, Frederik PM, Sommerdijk NA (2009) The initial stages of template-controlled $CaCO_3$ formation revealed by cryo-TEM. Science 323:1455–1458

Radha A, Forbes T, Killian C, Gilbert P, Navrotsky A (2010) Transformation and crystallization energetics of synthetic and biogenic amorphous calcium carbonate. Proc Natl Acad Sci. 107:16438–16443

Raiteri P, Gale JD (2010) Water is the key to nonclassical nucleation of amorphous calcium carbonate. J Am Chem Soc 132:17623–17634

Rao A, Seto J, Berg JK, Kreft SG, Scheffner M, Cölfen H (2013) Roles of larval sea urchin spicule SM50 domains in organic matrix self-assembly and calcium carbonate mineralization. J Struct Biol 183:205–215

Rao A, Berg JK, Kellermeier M, Gebauer D (2014) Sweet on biomineralization: effects of carbohydrates on the early stages of calcium carbonate crystallization. Eur J Mineral 26: 537–552

Rao A, Vásquez-Quitral P, Fernández MS, Berg JK, Sánchez M, Drechsler M, Neira-Carrillo A, Arias J, Gebauer D, Cölfen H (2016) pH-dependent schemes of calcium carbonate formation in the presence of alginates. Crystal Growth & Design 16:1349–1359

Raz S, Testeniere O, Hecker A, Weiner S, Luquet G (2002) Stable amorphous calcium carbonate is the main component of the calcium storage structures of the crustacean Orchestia cavimana. Biol Bull 203:269–274

Raz S, Hamilton P, Wilt F, Weiner S, Addadi L (2003) The transient phase of amorphous calcium carbonate in sea urchin larval spicules: the involvement of proteins and magnesium ions in its formation and stabilization. Adv Funct Mater 13:480–486

Reichel V, Faivre D (2017) Magnetite nucleation and growth. In: Van Driessche AES, Kellermeier M, Benning LG, Gebauer D (eds) New perspectives on mineral nucleation and growth. Springer, Cham, pp 275–292

Rodriguez-Blanco J, Shaw S, Benning L (2008) How to make 'stable' ACC: protocol and preliminary structural characterization. Mineral Mag 72:283–286

Rodriguez-Blanco JD, Sand KK, Benning LG (2017) ACC and vaterite as intermediates in the solution based crystallization of $CaCO_3$. In: Van Driessche AES, Kellermeier M, Benning LG, Gebauer D (eds) New perspectives on mineral nucleation and growth. Springer, Cham, pp 93–112

Rodriguez-Navarro C, Kudlacz K, Cizer Ö, Ruiz-Agudo E (2015) Formation of amorphous calcium carbonate and its transformation into mesostructured calcite. CrystEngComm 17: 58–72

Schwahn D, Ma Y, Cölfen H (2007) Mesocrystal to single crystal transformation of d, l-alanine evidenced by small angle neutron scattering. J Phys Chem C 111:3224–3227

Sethmann I, Hinrichs R, Wörheide G, Putnis A (2006) Nano-cluster composite structure of calcitic sponge spicules—a case study of basic characteristics of biominerals. J Inorg Biochem 100: 88–96

Seto J, Ma Y, Davis SA, Meldrum F, Gourrier A, Kim Y-Y, Schilde U, Sztucki M, Burghammer M, Maltsev S (2012) Structure-property relationships of a biological mesocrystal in the adult sea urchin spine. Proc Natl Acad Sci 109:3699–3704

Seto J, Rao A, Cölfen H (2014) Hierarchically nanostructured biological materials. Adv Hierarchical Nanostruct Mater 35–70

Smith BL, Schäffer TE, Viani M, Thompson JB, Frederick NA, Kindt J, Belcher A, Stucky GD, Morse DE, Hansma PK (1999) Molecular mechanistic origin of the toughness of natural adhesives, fibres and composites. Nature 399:761–763

Song R-Q, Cölfen H (2010) Mesocrystals: ordered nanoparticle superstructures. Bibliothek der Universität Konstanz, Konstanz

Su X, Kamat S, Heuer A (2000) The structure of sea urchin spines, large biogenic single crystals of calcite. J Mater Sci 35:5545–5551

Sugawara A, Nishimura T, Yamamoto Y, Inoue H, Nagasawa H, Kato T (2006) Self-organization of oriented calcium carbonate/polymer composites: effects of a matrix peptide isolated from the exoskeleton of a crayfish. Angew Chem Int Ed 45:2876–2879

Sun J, Bhushan B (2012) Hierarchical structure and mechanical properties of nacre: a review. RSC Adv 2:7617–7632

Thompson DW (1942) On growth and form. Cambridge University Press, Cambridge

Tompa P, Fuxreiter M (2008) Fuzzy complexes: polymorphism and structural disorder in protein–protein interactions. Trends Biochem Sci 33:2–8

Tritschler U, Zlotnikov I, Zaslansky P, Aichmayer B, Fratzl P, Schlaad H, Cölfen H (2013) Hierarchical structuring of liquid crystal polymer–laponite hybrid materials. Langmuir 29:11093–11101

Tseng Y-H, LIN H-Y, LIU M-H, Chen Y-F, Mou C-Y (2009) Biomimetic synthesis of nacrelike faceted mesocrystals of ZnO − gelatin composite. J Phys Chem C 113:18053–18061

Van De Locht R, Slater TJ, Verch A, Young JR, Haigh SJ, KröGER R (2014) Ultrastructure and crystallography of nanoscale calcite building blocks in rhabdosphaera clavigera coccolith spines. Cryst Growth Des 14:1710–1718

Van Driessche AES, Stawski TM, Benning LG, Kellermeier M (2017) Calcium sulfate precipitation throughout its phase diagram. In: Van Driessche AES, Kellermeier M, Benning LG, Gebauer D (eds) New perspectives on mineral nucleation and growth, Springer. Cham, pp 227–256

Wang T, Cölfen H, Antonietti M (2005) Nonclassical crystallization: mesocrystals and morphology change of $CaCO_3$ crystals in the presence of a polyelectrolyte additive. J Am Chem Soc 127:3246–3247

Wang G, Selbach SM, Yu Y, Zhang X, Grande T, Einarsrud M-A (2009) Hydrothermal synthesis and characterization of KNbO 3 nanorods. CrystEngComm 11:1958–1963

Wang Q, Nemoto M, Li D, Weaver JC, Weden B, Stegemeier J, Bozhilov KN, Wood LR, Milliron GW, Kim CS, Dimasi E, Kisailus D (2013a) Phase transformations and structural developments in the radular teeth of cryptochiton stelleri. Adv Funct Mater 23:2908–2917

Wang S-S, Picker A, Cölfen H, Xu A-W (2013b) Heterostructured calcium carbonate microspheres with calcite equatorial loops and vaterite spherical cores. Angew Chem Int Ed 52:6317–6321

Weiner S, Sagi I, Addadi L (2005) Choosing the crystallization path less traveled. Science 309:1027–1028

Weiss IM, Tuross N, Addadi L, Weiner S (2002) Mollusc larval shell formation: amorphous calcium carbonate is a precursor phase for aragonite. J Exp Zool 293:478–491

Wilt FH (2002) Biomineralization of the spicules of sea urchin embryos. Zool Sci 19:253–261
Wilt FH (2005) Developmental biology meets materials science: morphogenesis of biomineralized structures. Dev Biol 280:15–25
Wise ER, Maltsev S, Davies ME, Duer MJ, Jaeger C, Loveridge N, Murray RC, Reid DG (2007) The organic-mineral interface in bone is predominantly polysaccharide. Chem Mater 19:5055–5057
Wolf SE, Gower LB (2017) Challenges and perspectives of the polymer-induced liquid-precursor process: the pathway from liquid-condensed mineral precursors to mesocrystalline products. In: Van Driessche AES, Kellermeier M, Benning LG, Gebauer D (eds) New perspectives on mineral nucleation and growth. Springer, Cham, pp 43–76
Xu A-W, Ma Y, Cölfen H (2007) Biomimetic mineralization. J Mater Chem 17:415–449
Yamamoto Y, Nishimura T, Sugawara A, Inoue H, Nagasawa H, Kato T (2008) Effects of peptides on $CaCO_3$ crystallization: mineralization properties of an acidic peptide isolated from exoskeleton of crayfish and its derivatives. Cryst Growth Des 8:4062–4065
Yang L, Killian CE, Kunz M, Tamura N, Gilbert PUPA (2011) Biomineral nanoparticles are space-filling. Nanoscale 3:603–609
Yokoo N, Suzuki M, Saruwatari K, Aoki H, Watanabe K, Nagasawa H, Kogure T (2011) Microstructures of the larval shell of a pearl oyster, Pinctada fucata, investigated by FIB-TEM technique. Am Mineral 96:1020–1027
Young JR, Henriksen K (2003) Biomineralization within vesicles: the calcite of coccoliths. Rev Mineral Geochem 54:189–215
Yuwono VM, Burrows ND, Soltis JA, Penn RL (2010) Oriented aggregation: formation and transformation of mesocrystal intermediates revealed. J Am Chem Soc 132:2163–2165
Zhang G, Xu J (2013) From colloidal nanoparticles to a single crystal: new insights into the formation of nacre's aragonite tablets. J Struct Biol 182:36–43
Zhou L, O'brien P (2012) Mesocrystals properties and applications. J Phys Chem Lett 3:620–628

Chapter 9
Nucleation and Growth from a Biomineralization Perspective

Giuseppe Falini and Simona Fermani

9.1 Introduction

Organisms, after hundred million years of evolution, have acquired the ability to grow diverse families of different minerals (Lowenstam and Weiner 1989). The synthesis of these biogenic minerals (biominerals) is governed in space and time by cells that produce specific organic molecules, referred as organic matrix, able to control each step of the mineralization process. This generates as final products biominerals on which the organic matrix perfectly architectures the assembly of the growth units. In Fig. 9.1 (Cuif et al. 2012), an example is given.

The mechanism of crystal nucleation and growth has only been studied for a few biominerals. In this chapter, an overview of the strategies adopted by organisms in the control of nucleation and growth of their mineralized regions is reported, focusing on a few examples for which enough information is available. Moreover, recent studies on growth adaptation strategies of calcifying organisms to ocean acidification are presented.

9.1.1 Biomineralization Control at Nucleation Level

The nucleation process has been well described by classical theories (Liu 2004, e.g., Lutsko 2017, Chap. 2). However, last decade observations strongly suggest that this view could not be appropriate in describing nucleation in biomineralization. The discovery of stable pre-nucleation clusters (PNCs) has represented the basis

G. Falini (✉) • S. Fermani
Dipartimento di Chimica "Giacomo Ciamician", Alma Mater Studiorum – Università di Bologna, via Selmi 2, 40126 Bologna, Italy
e-mail: giuseppe.falini@unibo.it

A.E.S. Van Driessche et al. (eds.), *New Perspectives on Mineral Nucleation and Growth*,
DOI 10.1007/978-3-319-45669-0_9

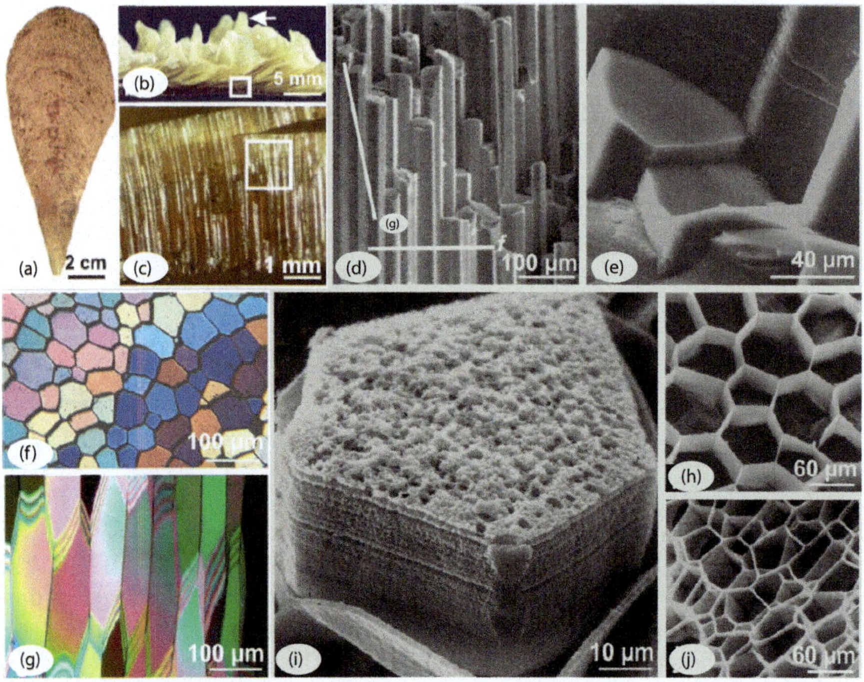

Fig. 9.1 Layered growth in the prisms of the mollusk pelecypod Pinna: the historical reference for biocrystals. The building units of the outer layer of the Pinna shell (**a**, **b**) are closely assembled linear polygonal prisms (**c**, **d**) with massive aspect of fractures (**e**). Thin sections observed under the microscope between cross-nicols (**f**, **g**). Owing to the high refractive index of calcite, such primary colors are obtained far below the standard thickness: here about 5–6 mm. Note the homogeneity of color for each prism on the transverse section (**f**). Difference in color from prism to prism results from variation of orientation of the crystallographic c axis (the elongation axis of the prism). When prisms are sectioned in length (**g**), color is more variable because obliquity of the sections with respect to elongation axis creates variations in the thickness with the possibility of superposed contributions of two prisms. Strong acidic etching, the polygonal network of prism envelopes has been made free (**h**). Layered growth visible without any preparative process of the lateral side of the mineralized unit (**i**). Note also the granular aspect of the growth surface. After a one week decay in water the polygonal network corresponding to a single growth layer has been freed (**j**) (and slightly displaced) from underlying envelopes (Cuif et al. 2012)

of the nonclassical crystallization theory (e.g., Gebauer et al. 2008; Cölfen and Antonietti 2008). Accordingly, during the nucleation process, the relevant species are not ions but PNCs (Gebauer et al. 2008; Demichelis et al. 2011). Despite this alternative view, the crystallization phase diagram can be considered as analogous to that of atomic and molecular systems and can be described under the laws of the classical nucleation theory (Habraken et al. 2013). Accordingly, heterogeneous substrates lower the interfacial energy favoring the nucleation process. In addition, in the presence of heterogeneous substrates, the growth kinetics can be controlled by varying the height of the energy barrier, and non-equilibrium final or intermediate states can be selected (Mann et al. 1993; De Yoreo and Vekilov 2003). For example, in calcium carbonate precipitation pathway, the assembly of PNCs can form amorphous calcium carbonate (ACC) phases, which evolve in distinct crystalline structures, already encoded in PNCs before the precipitation takes place (Cartwright et al. 2012).

In biomineralization, heterogeneous nucleation is ubiquitous. A main issue is how organized organic matrix surfaces (heterogeneous substrates) control the nucleation of inorganic materials by geometric, electrostatic, and stereochemical complementarity between nuclei and functionalized substrates (Mann et al. 1993). The role that organic matrix molecules – proteins, polysaccharides, and lipids – have in the nucleation stage during a biomineralization process is beginning to be understood, and it is likely that over time mechanistic insights will emerge (Zhang et al. 2012). Exploring the early stages that bring to calcium carbonate nucleation, it was shown that organic matrix proteins play an important role manipulating the PNCs structure (Demichelis et al. 2011). In addition, it was shown that in the nucleation pathway to the final crystalline state, ACCs of increasing stability formed (Politi et al. 2010; Habraken et al. 2013).

There are several examples of the mineral nucleation process in organisms through an amorphous precursor, but, at our knowledge, not direct observations of PNCs. Thus, the following case of studies will make an excursus on the most relevant biominerals for which amorphous precursors of crystalline nuclei has been detected. The tooth of chiton (Fig. 9.2) represents the first biomineral in which an amorphous precursor was observed. The outer layer of the tooth contains magnetite, an iron magnetic mineral. It forms from a disordered ferrihydrite precursor phase (Towe and Lowenstam 1967). The inner layer of the tooth contains carbonated

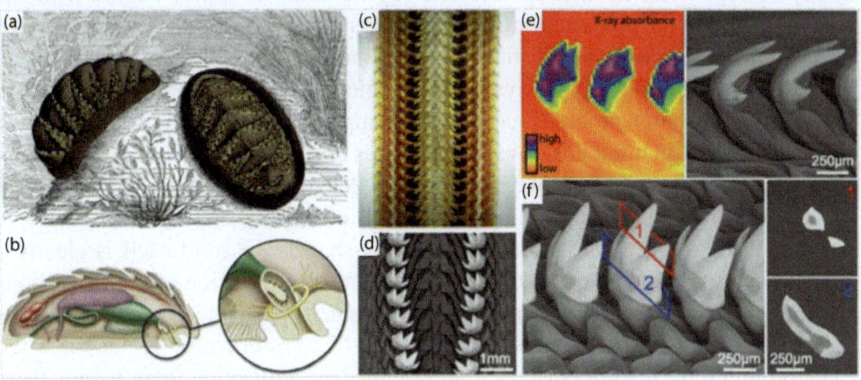

Fig. 9.2 Morphological features of the chiton radula. External (**a**) and internal (**b**) anatomy of a representative chiton showing the location of the radula, a rasping, toothed conveyor belt-like structure used for feeding. Details of the anterior region of the radula from *C. stelleri* (**c**–**f**). Optical (**c**) and backscattered SEM (**d**) imaging and X-ray transmission studies (**e**) reveal the nature of the electron density distribution of the tricuspid tooth caps. Cross-sectional studies through the mature teeth from *C. stelleri* (f1, 2) reveal a concentric biphasic structure (Weaver et al. 2010)

apatite that forms by an amorphous calcium phosphate precursor phase (Lowenstam and Weiner 1985). The three-dimensional chemical maps of organic fibers, which have a diameter of 5–10 nm and surround nanocrystalline magnetite in the tooth, have been unveiled by the use of pulsed-laser atom-probe tomography. Most fibers contain with either sodium or magnesium ions, and the clustering of these cations in the fiber indicates a structural level of hierarchy in which individual organic fibers have probably different functional roles in controlling fiber formation and matrix–mineral interactions (Gordon and Joester 2011). Recently, the involvement of precursors and organic matrix molecules in the formation of magnetite has been reexamined. It has been shown by cryogenic transmission electron microscopy that the nucleation and growth of magnetite proceeds through the rapid agglomeration of nanometric primary particles and no intermediate amorphous precursor phases are involved (Baumgartner et al. 2013).

The spines of echinoderms represent another example of biomineralization control over nucleation that has been extensively investigated. These biominerals, diffracting as single crystals (Dubois and Chen 1989), form within a syncytium, a membrane envelope produced by cells. These cells provide the required raw materials, ions, and macromolecules, for constructing the growing crystals. Insights on the nucleation of these crystals have been gained from studies on calcitic spicule formation in sea urchin larvae (Wilt 2002). Larval spicules grow on a single calcite crystal seed by transformation of a transient ACC phase. ACC is apparently fed into the syncytium by cells in the form of ACC-containing vesicles (Beniash et al. 1999). Accordingly, particles of ACC are delivered to the crystal deposition site and then transform in a controlled manner into calcite single crystals. Moreover, there is no discernible aqueous phase around the growing spicule (Beniash et al. 1999). Most information on spine formation was gained taking advantage of the sea urchins ability to regenerate broken spines. The regeneration of spines begins with the reconstruction of the epidermis around the broken spine; a new syncytium is formed by sklerocytes in contact with the stump of the old spine. The regenerated and old spines together diffract X-rays as one single crystal (Dubois and Chen 1989). Thus, the regeneration process is considered comparable to the original spine growth process. Keeping in mind this observation, it was observed that calcium carbonate is first precipitates as hydrated ACC, which then dehydrates prior to, or concomitant with, crystallization. Since ACC is introduced into the syncytium as an isotropic noncrystalline solid, it can be molded into any shape. The solid itself is clearly a very concentrated source of ions. The subsequent transformation of the amorphous phase into a composite crystalline solid with much better mechanical properties leads to a functional skeleton. The mechanism of transformation is complex: the transforming spine does not display a well-defined crystallization front. In fact, there are three distinct mineral phases: A short-lived, presumably hydrated ACC phase, an intermediate transient form of ACC, and the biogenic crystalline calcite phase. The amorphous and crystalline phases are placed side by side, often appearing in adjacent sites. Thus, the amorphous-crystal transformation may propagate following a winding path through preexisting amorphous units (Politi et al. 2008). The co-orientation of the nanoparticles in the polycrystalline matrix occurs via solid-state

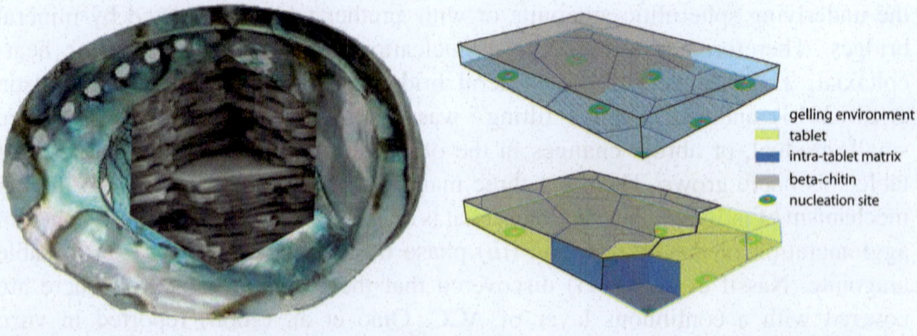

Fig. 9.3 Picture of the nacre (*left*) and schematic representation of the model for nacre formation before (*right top*) and after mineralization (*right bottom*). The assembled organic matrix prior to mineral deposition is formed by two layers of β-chitin, with a gel comprising silklike protein filling the space in between. Acidic protein inducing nucleation of aragonite (from colloidal particles) are adsorbed on the beta-chitin sheet. As the mineral grows, water and silk are displaced (Addadi et al. 2006)

secondary nucleation, propagating out from the previously formed fibers and plates into the amorphous precursor nanoparticles (Killian et al. 2009). During this process, a residual surface layer of ACC and/or macromolecules remains around the crystalline nanoparticle units and contributes to the conchoidal fracture behavior (Seto et al. 2012).

The nacre represents one of the most studied mineralized tissues (more than 600 published papers in the last 10 years, WoK source). It consists of a brick-and-mortar-like structure in which hard aragonite tablets are glued together with soft organic materials to form tiles (Fig. 9.3). Different models have been proposed to describe the mechanisms of nacre nucleation. Weiner and Lowenstam (1986) suggested a structural relationship between organic macromolecules and the tablet-like aragonite crystals, where the organic sheets template aragonite tablet orientation by heteroepitaxy. This model is supported by in vitro experiments that show how organic molecules extracted from the nacre induce aragonite deposition rather than the more stable calcite (Falini et al. 1996). Nudelman et al. (2006) hypothesized by histochemical assays that each tablet crystal nucleates independently by a single, well-defined macromolecule arrangement in the organic matrix sheet, resulting in a peculiar pattern of functional groups highly conserved across species. Schäffer et al. (1997) observed the presence of "mineral bridges" through pores in the organic sheets. This is in agreement with the connected tablet model by which no new nucleation events occurs at each tablet. However, the presence of pores in the organic matrix has been questioned as an effect due to sample preparation (Levi Kalisman et al. 2001). Olson et al. (2013) proposed a mechanism in which there is one mineral bridge at the center of each organic tablet site of nucleation, which looks like as a donut with a hole at its center. Moreover, stacks of co-oriented aragonite tablets arranged into vertical columns, or staggered diagonally, were found. Overgrowing nacre tablet crystals were most frequently co-oriented with the underlying spherulitic aragonite or with another tablet, connected by mineral bridges. Therefore, aragonite crystal nucleation in nacre is epitaxial or near-epitaxial. The presence of one mineral bridge per tablet was proposed (Olson et al. 2013) and that "bridge tilting" was a possible mechanism to introduce small, gradual, or abrupt changes in the orientation of crystals within a stack of tablets as nacre grows. There are three main hypotheses to explain the formation mechanism of individual nacre tablets, that is (*i*) single crystal growth, (*ii*) a coherent agglomeration of nanograins, and (*iii*) phase transformation from ACC to stable aragonite. Nassif et al. (2005) discovered that the aragonite tablets in nacre are covered with a continuous layer of ACC. Qiao et al. (2008) reported in vitro crystal growth on a fresh nacre surface, by which the growth process of layers of nacre-like tablets was monitored in real time. The formation of nacre-like tablets was a complex and multistep process, from an ACC layer, iso-oriented nanostacks, to hexagonal tablets. Zhang and Xu (2013) observed that the early immature tablet consists of closely-packed colloidal nanoparticles, which contain nanocrystals surrounded by ACC. The nanocrystals were generally different in shape, size, and orientation. In this work, it is shown the immature tablet grew via oriented attachment instead of a transformation of the ACC phase; and with growth, the colloidal nanoparticles gradually increased in crystallinity and size until fully crystallized and fused together, leading to a mature tablet that is a monolithic single crystal of aragonite.

Bone possesses a particular hierarchically structured architecture, unique mechanical properties, and remodeling capabilities. Weiner and Wagner (1998) have described diverse levels of hierarchy in bone structure, from the nanoscale to the macroscopic scale (Fig. 9.4). At the most elementary level, bone composition includes molecular and crystalline components, being the main collagen, apatite, and water. The second level is at the nanoscale and concerns mineralized collagen fibrils. Patterns of fibril arrays and cylindrical structures called osteons constitute the

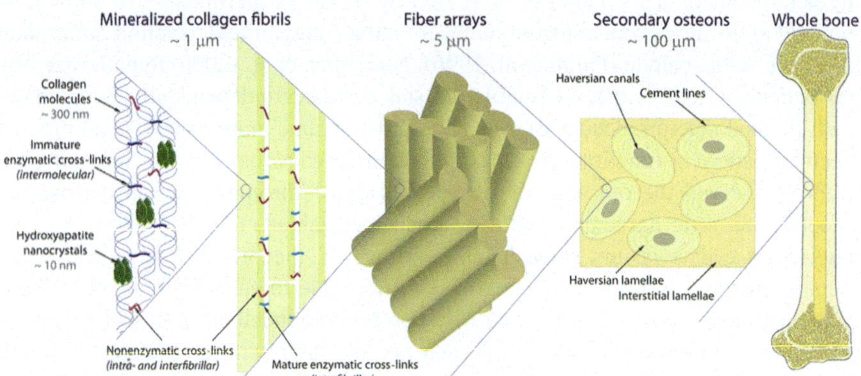

Fig. 9.4 Schematic illustration of the hierarchical levels of bone organization that span from the nano- to the macroscale

microscale level. The upper level is composed of spongy (trabecular or cancellous) or compact (cortical) bone tissues, while the ultimate one is the whole bone. The nucleation, growth, and orientation of the biological crystals of apatite take place at the nanoscale level of the bone hierarchy (e.g., Delgado-Lopez and Guagliardi 2017, Chap. 11). Type I collagen molecules assemble their tropocollagen units giving rise to holes and overlapping areas that form a periodic banding pattern, the so-called D-band pattern. In the gap zone, the first crystalline units of apatite nucleate. A long-standing controversy on whether it is collagen or non-collageneous proteins (NCPs) (Gericke et al. 2005) that initiate mineral nucleation exists. It was proposed that NCPs, while bonded to collagen fibers, acted as promoters of intrafibrillar nucleation. Such a hypothesis was supported by reports showing that NCPs appear during the formation of new bone in a very specific spatiotemporal timeline and also that type-I collagen alone, dissociated from NCPs, is unable to produce bone mineralization. On the other hand, some NCPs, when unlinked from collagen, are inhibitors of calcium phosphate nucleation in solution. Price et al. (2009) proposed a mechanism for fibril mineralization based on inhibitor exclusion, in which macromolecular inhibitors of apatite growth favor fibril mineralization by selectively inhibiting apatite crystal growth in the solution outside the fibril. In 2010, Sommerdijk and coworkers (see Nudelman et al. 2010) used cryoTEM and cryogenic electron tomography to show that collagen functions in synergy with inhibitors of apatite (polyaspartic acid or fetuin instead of NCPs) nucleation to actively control mineralization by a mechanism of electrostatic interactions.

9.1.2 Biomineralization Control at Growth Level

One of the most striking features of many biologically formed crystals is the remarkable morphology. Organisms can produce single crystals with complex shapes and curved surfaces. They have developed mechanisms that override the basic growth form of a crystal. They have crystals whose overall morphologies often bear no relationship with the symmetry of the crystal lattice. Furthermore, it has been observed that mature crystalline biogenic minerals often have a texture composed of nanosphere particles, as was first observed in cephalopod shells. Subsequent high-resolution AFM and SEM studies showed that not only mature mollusk shells but also coral skeletons, echinoderm skeletons, sponge spicules, brachiopod shells, and crustacean cuticles have this most unusual nanoscale morphology of a mature crystal, namely, that they are composed of nanosphere particles. The reasons for these unique characteristics must be related to the differences between biological crystallization pathways and solution-mediated crystal growth. Gal et al. (2015) proposed that many crystal growth mechanisms in biology can involve solution-mediated ion-by-ion growth, or particle accretion, or a combination of both processes. When both processes occur together, the necessary conditions are the presence of small amounts of liquid water, possibly partly being derived from the initial hydrated amorphous phase, and stabilization of the particles in this microenvironment.

The biological control over the growth is mainly exerted by (*i*) interaction of growing crystals with soluble additives, which can be occluded within biominerals, (*ii*) physical constrains of the compartment where mineralization takes place; and (*iii*) changes in the activity or positioning of ion pumps and channels during mineralization that may lead to crystal growth in preferred directions.

The morphological influence of additives was firstly described in the general crystal growth theory, and Weiner and Addadi (1997) were the first to suggest its potential applicability to biomineralization. Since then, it has been widely demonstrated that many additives can alter the morphology of calcite crystals and other biologically relevant minerals. This was indicated to occur by adsorption onto specific crystallographic faces. Atomic force microscopy (AFM) studies suggest that this specificity in binding may actually be occurring at the growing step edges and not the flat crystal faces (De Yoreo et al. 2007). The "sculpted" non-equilibrium morphologies observed in various biominerals was supposed to be as a result of the stereospecific interactions with the soluble organic matrix. The evidence to support this mechanism was partly based on in vitro crystal growth experiments in the presence of additives, including proteins extracted from the biomineral. These experiments show that proteins influence crystal growth and often produce the expression of new crystal faces, which, although rough, have been correlated to the crystallographic planes expressed on the corresponding biomineral. This argument was also supported by the analysis of biomineral crystallographic textures, which arose from anisotropic intercalation of proteins after they had adsorbed to specific crystallographic faces. In addition to biological macromolecules, moderately sized organic molecules and small inorganic molecules behaving as growth modifiers may influence mineralization kinetics and energetic and the shapes of crystals (De Yoreo and Dove 2004).

Crystals with unusual morphologies and curved surfaces are typically found within vesicles (e.g., calcite produced by coccoliths and sea urchin larval spicules). Vesicles have well-defined shapes, and crystals grow until they impinge upon the vesicle, which effectively acts as a mould. The size and shape of the vesicle may be altered during the crystal growth process. Soft organic membranes also impose a form on a crystal. The growth of biogenic single crystals of calcite may also occur inside the cell and the final morphology is determined by the cell membrane. The coccolith (from the algae *Emiliania huxleyi*) comprises about 30–40 units organized in a ring to give a double-rimmed structure. The formation of coccoliths begins with the assembly of vesicles along the rim of an organic base plate scale. These crystal units become interlinked by selective growth along the inside rim, and further radial growth from the base and top of the element produces the proximal and distal shield elements (Paasche 2001).

In contrast to this, there are additives, inorganic or organic, which can modify the crystallization process by transforming the conventional crystal growth into an amorphous precursor process. Notably, the organic–inorganic interactions that lead to shape regulation in this case occur prior to the formation of any crystal structure and, therefore, do not require interactions specific to crystal lattice arrangements. These process-directing agents can have a pronounced effect on crystal morphology, as well as other crystal properties, and thus provide an alternative explanation for the morphogenesis of biominerals (Pouget et al. 2009).

9.1.3 Biomineralization at Ocean Acidification Sites

Anthropogenic CO_2 emissions represent a massive reorganization of global carbon. As this carbon invades the ocean, it changes the acid–base chemistry of seawater impacting a wide range of biological and chemical processes. Many marine organisms reduce their calcification rate in response to this ocean acidification (OA). However, the detailed mechanism responsible for this sensitivity is poorly understood. During biomineralization, organisms orchestrate the flow of ions, manipulate aqueous speciation, and modulate surface chemistry through organic–mineral interactions. Due to differences in the intricate balance between these processes, marine calcifiers exhibit divergent sensitivities to ocean acidification. Furthermore, this response is often modulated by secondary environmental and biological factors, making the impact of ocean acidification on skeletal growth difficult to predict.

It was shown that biological control over mineralization relates to species abundance along a natural pH gradient (Fig. 9.5). As pCO_2 increased, the mineralogy of a scleractinian coral (*Balanophyllia europaea*) and a mollusk (*Vermetus triqueter*)

Fig. 9.5 The natural lab for the study of the OA at Panarea island (Italy). In situ camera picture of the CO_2 vent. (**a**) In proximity of the CO_2 crater, the mollusk *Vermetus triqueter* (**b**), the coral *Balanophyllia europaea* (**c**), and the brown algae *Padina pavonica* (**d**) and *Acetabularia acetabulum* (**e**) were found

did not change. In contrast, two calcifying algae (*Padina pavonica* and *Acetabularia acetabulum*) reduced and changed mineralization with increasing pCO_2, from aragonite to the less soluble, under acidic conditions, calcium sulfate hydrated salts, and whewellite, respectively. As pCO_2 increased, the coral and mollusk abundance was severely reduced, with both species disappearing at $pH < 7.8$. Conversely, the two calcifying and a noncalcifying algae (*Lobophora variegata*) showed less severe, or no, reductions with increasing pCO_2 and were all found at the lowest pH site. The mineralization response to decreasing pH suggests a link with the degree of control over the biomineralization process by the organism, as only species with lower control managed to thrive in the lowest pH (Goffredo et al. 2014).

The response of common blue mussel, *Mytilus edulis*, to OA was investigated by culturing them at projected levels of pCO_2 and increased temperatures. OA caused mussels to produce shell calcite that is stiffer (higher modulus of elasticity) and harder than shells grown in control conditions. The outer shell (calcite) is more brittle in OA conditions, while the inner shell (aragonite) is softer and less stiff in shells grown under OA conditions. The combination of increasing ocean pCO_2 and temperatures as projected for future global ocean appears to reduce the impact of increasing pCO_2 on the material properties of the mussel shell (Fitzer et al. 2015).

The sources of dissolved inorganic carbon (DIC) used to produce scleractinian coral skeletons are essential for understanding coral biomineralization and assessing the potential impacts of OA on coral reefs. The use skeletal boron geochemistry allowed reconstructing the DIC chemistry of the fluid used for coral calcification. Corals concentrated DIC at the calcification site substantially above seawater values. They actively increased the pH of the calcification fluid, decreasing the proportion of DIC present as CO_2 and creating a diffusion gradient favoring the transport of molecular CO_2 from the overlying coral tissue into the calcification site. This chemistry induces high aragonite saturation states, favorable to the precipitation of the skeleton (Allison et al. 2014).

Shells of the bivalve *Arctica islandica* were used to reconstruct paleoenvironmental conditions via biogenic components that were closely related to environmental parameters at the time of shell formation. Growth experiments with *A. islandica* kept under different pCO_2 levels indicated no affect of elevated pCO_2 on shell growth or crystal microstructure, indicating that *A. islandica* shows an adaptation to a wider range of pCO_2 levels than reported for other species (Stemmer et al. 2013).

OA generated striking patterns of genome-wide selection in purple sea urchins (*Strongylocentrotus purpuratus*) cultured under different pCO_2 levels. The larval development and morphology showed little response to elevated pCO_2, but substantial allelic change in functional classes of proteins involving hundreds of loci were found. Pronounced genetic changes, including excess amino acid replacements, were detected in all populations and occurred in genes for biomineralization, lipid metabolism, and ion homeostasis – gene classes that build skeletons and interact in pH regulation. These results demonstrate the capacity for rapid evolution in the face of OA and show that standing genetic variation could be a reservoir of resilience to climate change (Pespeni et al. 2013).

9.1.4 Conclusions and Future Perspectives

Organisms are terrific crystallizers. Their capability to control nucleation and growth processes has no equal in the laboratory syntheses. Species-specific control of crystal shape, morphology, orientation, and composition is achieved with high fidelity. Even more impressive is the control over the polymorphism that organisms exert depositing in adjacent locations diverse phases or hydrates. Although the molecular mechanisms involved in this control are still unknown, it is well established that this occurs through nonclassical pathways of nucleation and growth. Doing this, organisms are able to carry out crystallization in favorable energetic conditions in aqueous media under ambient conditions. In their crystallization processes, organisms also show a great capability of adaptation to diverse chemical–physical conditions, as shown by their acclimation to OA.

All this represents a big source of inspiration for new crystallization strategies. This is particular relevant in an era wise from the environmental point of view, tacking in account that organisms are the best crystallizers in aqueous environment, pH values close to the neutrality and ambient temperatures. However, to date, the possibility to produce in laboratory a biominerals with features close to those of the biogenic ones is still missing. One of the main reasons for this is the ignorance at atomistic resolution of the biochemical processes carried out by organic matrix macromolecules in controlling biomineral formation. To achieve this goal is of primary importance to know first which are the key steps for the biomineral formation, which macromolecules are involved in each step and then how these macromolecules control each step. The research is now clarifying the crucial steps of biomineral formation, and there is some info on few macromolecules involved in these steps. It is desirable that in the near future, thanks to the use of microscopic, spectroscopic, and diffraction techniques increasingly sophisticated both in spatial and temporal resolution, the macromolecule involved in biomineralization could be observed with atomic resolution shading light on how they are made and work. This will produce a terrific effect in the development of new synthetic materials that unlike the biogenic ones may also use elements of the periodic table, toxic to the organisms, but of considerable technological importance.

Acknowledgments This chapter has been carried out under the funding of the European Research Council under the EU FP7/2007-2013Programme (ERC grant agreement no. 249930 – CoralWarm: Corals and global warming: the Mediterranean versus the Red Sea). The CIRCMSB is also thanked for the support.

References

Addadi L, Joester D, Nudelman F et al (2006) Mollusk shell formation: a source of new concepts for understanding biomineralization processes. Chem Eur J 12:980–987

Allison N, Cohen I, Finch AA et al (2014) Corals concentrate dissolved inorganic carbon to facilitate calcification. Nat Commun 5:5741–5748

Baumgartner J, Dey A, Bomans PHH et al (2013) Nucleation and growth of magnetite from solution. Nat Mater 12:310–314

Beniash E, Addadi L, Weiner S (1999) Cellular control over spicule formation in sea urchin embryos: a structural approach. J Struct Biol 125:50–62

Cartwright JHE, Checa AG, Gale JD, Gebauer D, Sainz-Díaz CI (2012) Calcium carbonate polyamorphism and its role in biomineralization: how many ACCs are there? Angew Chem Int Ed Engl 51:11960–11970

Cölfen H, Antonietti M (2008) Mesocrystals and nonclassical crystallization. Wiley, Chichester

Cuif J-P, Dauphin Y, Nehrke G et al (2012) Layered growth and crystallization in calcareous biominerals: impact of structural and chemical evidence on two major concepts in invertebrate biomineralization studies. Minerals 2:11–39

De Yoreo JJ, Dove PM (2004) Shaping crystals with biomolecules. Science 306:1301–1302

De Yoreo JJ, Vekilov PG (2003) Biomineralization. In: Dove PM, De Yoreo JJ, Weiner S (eds) Reviews in mineralogy and geochemistry, vol 54. The Mineralogical Society of America, Washington, DC, pp 57–94

De Yoreo JJ, Wierzbickib A, Dove PM (2007) New insights into mechanisms of biomolecular control on growth of inorganic crystals. CrystEngComm 9:1144–1152

Delgado-López JM, Guagliardi A (2017) Control over nanocrystalline apatite formation: what can the X-ray total scattering approach tell us. In: Van Driessche AES, Kellermeier M, Benning LG, Gebauer D (eds) New perspectives on mineral nucleation and growth, Springer, Cham, pp 211–226

Demichelis R, Raiteri P, Gale JD et al (2011) Stable prenucleation mineral clusters are liquid-like ionic polymers. Nat Commun 590:1–8

Dubois PH, Chen C (1989) Calcification in echinoderms. In: Jangoux MJ, Lawrence JM (eds) Echinoderm studies, vol 3. Balkema, Rotterdam, pp 109–178

Falini G, Albeck S, Weiner S et al (1996) Control of aragonite or calcite polymorphism by mollusk shell macromolecules. Science 271:67–69

Fitzer SC, Zhu W, Tanner KE et al (2015) Ocean acidification alters the material properties of *Mytilus edulis* shells. J R Soc Interface 12:20141227

Gal A, Weiner S, Addadi L (2015) A perspective on underlying crystal growth mechanisms in biomineralization: solution mediated growth versus nanosphere particle accretion. CrystEngComm 17:2606–2615

Gebauer D, Völkel A, Cölfen H (2008) Stable prenucleation calcium carbonate clusters. Science 322:1819–1822

Gericke A, Qin C, Spevak L et al (2005) Importance of phosphorylation for osteopontin regulation of biomineralization. Calcif Tissue Int 77:45–54

Goffredo S, Prada F, Caroselli E et al (2014) Biomineralization control related to population density under ocean acidification. Nat Clim Change 4:593–597

Gordon L, Joester D (2011) Nanoscale chemical tomography of buried organic–inorganic interfaces in the chiton tooth. Nature 469:194–197

Habraken WJEM, Tao J, Brylka LJ et al (2013) Ion association complexes unite classical and non- classical theories for the biomimetic nucleation of calcium phosphate. Nat Commun 4:1507–1510

Killian CE, Metzler RA, Gong YUT et al (2009) Mechanism of calcite co-orientation in the sea urchin tooth. J Am Chem Soc 131:18404–18409

Levi Kalisman Y, Falini G, Addadi L et al (2001) Structure of the nacreous organic matrix of a bivalve mollusk shell examined in the hydrated state using cryo-TEM. J Struct Biol 135:8–17

Liu XY (2004) From solid-fluid interface to nanostructure engineering. In: Liu XY, De Yoreo JJ (eds) Nanoscale structure and assembly at solid–fluid interfaces, vol 1. Springer, London, pp 109–175

Lowenstam HA, Weiner S (1985) Transformation of amorphous calcium phosphate to crystalline dahllite in the radular teeth of chitons. Science 227:51–53

Lowenstam HA, Weiner S (1989) On biomineralization. Oxford University Press, New York

Lutsko JF (2017) Novel paradigms in nonclassical nucleation theory. In: Van Driessche AES, Kellermeier M, Benning LG, Gebauer D (eds) New perspectives on mineral nucleation and growth, Springer, Cham, pp 25–42

Mann S, Archibald DD, Didymus JM et al (1993) Crystallization at inorganic-organic interfaces: biominerals and biomimetic synthesis. Science 261:1286–1292

Nassif N, Pinna N, Gehrke N et al (2005) Amorphous layer around aragonite platelets in nacre. Proc Natl Acad Sci U S A 102:12653–12655

Nudelman F, Gotliv AB, Addadi L et al (2006) Mollusk shell formation: mapping the distribution of organic matrix components underlying a single aragonitic tablet in nacre. J Struct Biol 153:176–187

Nudelman F, Pieterse K, George A et al (2010) The role of collagen in bone apatite formation in the presence of hydroxyapatite nucleation inhibitors. Nat Mater 9:1004–1009

Olson IC, Blonsky AZ, Tamura N et al (2013) Crystal nucleation and near-epitaxial growth in nacre. J Struct Biol 184:454–463

Paasche E (2001) A review of the coccolithophorid Emiliania huxleyi (Prymnesiophyceae), with particular reference to growth, coccolith formation, and calcification-photosynthesis interactions. Phycologia 40:503–529

Pespeni MH, Sanford E, Gaylord B et al (2013) Evolutionary change during experimental ocean acidification. Proc Natl Acad Sci U S A 110:6937–6942

Politi Y, Metzler RA, Abrecht M et al (2008) Transformation mechanism of amorphous calcium carbonate into calcite in the sea urchin larval spicule. Proc Natl Acad Sci U S A 105:17362–17366

Politi Y, Batchelor DR, Zaslansky P et al (2010) Role of magnesium ion in the stabilization of biogenic amorphous calcium carbonate: a structure-function investigation. Chem Mater 22:161–166

Pouget EM, Bomans PHH, Goos JACM et al (2009) The initial stages of template-controlled $CaCO_3$ formation revealed by Cryo-TEM. Science 323:1455–1458

Price PA, Toroian D, Lim JE (2009) Mineralization by inhibitor exclusion: the calcification of collagen with fetuin. J Biol Chem 284:17092–17101

Qiao L, Feng Q, Lu S (2008) In vitro growth of nacre-like tablet forming: from amorphous calcium carbonate, nanostacks to hexagonal tablets. Cryst Growth Des 8:1509–1514

Schäffer TE, Ionescu-Zanetti C, Proksch R et al (1997) Does abalone nacre form by heteroepitaxial nucleation or by growth through mineral bridges? Chem Mater 9:1731–1740

Seto J, Ma YR, Davis SA et al (2012) Structure-property relationships of a biological mesocrystal in the adult sea urchin spine. Proc Natl Acad Sci U S A 109:3699–3704

Stemmer K, Nehrke G, Brey T (2013) Elevated CO_2 levels do not affect the shell structure of the bivalve Arctica islandica from the Western Baltic. PLoS One 8:e70106

Towe KM, Lowenstam HA (1967) Ultrastructure and development of iron mineralization in the radular teeth of Cryptochiton stelleri (Mollusca). J Ultrastruct Res 17:1–13

Weaver JC, Wang Q, Miserez A et al (2010) Analysis of an ultra hard magnetic biomineral in chiton radular teeth. Mater Today 13:42–52

Weiner S, Addadi L (1997) Design strategies in mineralized biological materials. J Mater Chem 7:689–702

Weiner S, Lowenstam H (1986) Organization of extracellularly mineralized tissues: a comparative study of biological crystal growth. Crit Rev Biochem Mol Biol 20:365–408

Weiner S, Wagner HD (1998) The material bone: structure-mechanical function relations. Annu Rev Mater Sci 28:271–298

Wilt FH (2002) Biomineralization of the spicules of sea urchin embryos. Zool Sci 19:253–261

Zhang G, Xu J (2013) From colloidal nanoparticles to a single crystal: new insights into the formation of nacre's aragonite tablets. J Struct Biol 182:36–43

Zhang G, Fang X, Guo X et al (2012) The oyster genome reveals stress adaptation and complexity of shell formation. Nature 490:49–54

Chapter 10
Phase Transformations in Calcium Phosphate Crystallization

Henrik Birkedal

10.1 Introduction

The crystallization of calcium phosphates has fascinated researchers for many decades, especially because of the role of apatites in the biomineralization of bone and teeth but also due to its relevance for geochemistry (Hughes and Rakovan 2015). There is an enormous body of work on calcium phosphates in the literature (Wang and Nancollas 2008; Dorozhkin 2009; Elliott 1994). Several calcium phosphates can be obtained from precipitation experiments, including hydroxyapatite ($Ca_{10}(PO_4)_6(OH)_2$, HAP), octacalcium phosphate ($Ca_8H_2(PO_4)_6 \cdot 5(H_2O)$, OCP), brushite ($CaHPO_4 \cdot 2(H_2O)$), and monetite ($CaHPO_4$). Herein, we will mostly cover aspects related to the mechanisms of crystallization and almost exclusively focus on pathways leading to apatite formation.

Crystallization research has undergone a renaissance over the last decade or two. While attempts have been made to establish a unifying picture of inorganic crystallization processes based on the numerous recent findings (De Yoreo et al. 2015; De Yoreo et al. 2017, Chap. 1), many important fundamental questions remain unanswered, particularly in systems like the calcium phosphates. These questions pertain both to idealized laboratory conditions (Delgado-Lopez and Guagliardi 2017, Chap. 11) and to the formation of biogenic calcium phosphate minerals in vivo (Falini and Fermani 2017, Chap. 9). In the latter context, several new perspectives have emerged recently, including the idea that citrate ions constitute an integral part of nanoparticle aggregates in bone, forming organic bridges between minute inorganic crystals (Davies et al. 2014).

H. Birkedal (✉)
Department of Chemistry and iNANO, Aarhus University, Gustav Wieds Vej 14, 8000 Aarhus, Denmark
e-mail: hbirkedal@chem.au.dk

A.E.S. Van Driessche et al. (eds.), *New Perspectives on Mineral Nucleation and Growth*,
DOI 10.1007/978-3-319-45669-0_10

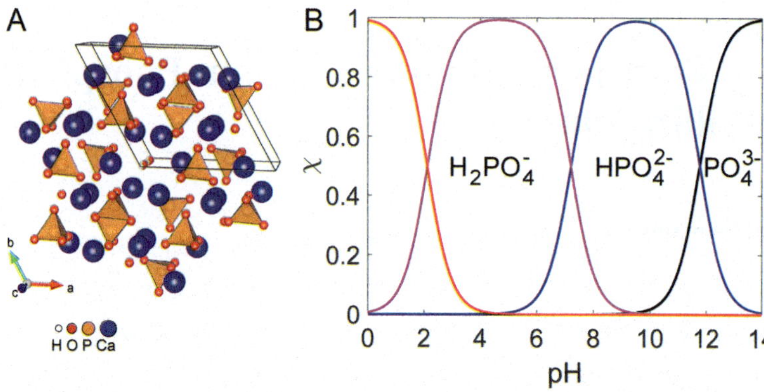

Fig. 10.1 (**a**) Hexagonal apatite crystal structure with disordered hydroxyl ion orientations. (**b**) pH dependence of the speciation of phosphate in aqueous solution

Hydroxyapatite (HAP) has an ideal composition of $Ca_{10}(PO_4)_6(OH)_2$. However, there is a wide range of possible substitutions; for example, carbonate ions can substitute on either the phosphate or hydroxide position (Fig. 10.1a), with the former occurring almost exclusively at low temperature and in biogenic apatites. HAP is typically hexagonal with disordered hydroxide ion orientations, as shown in Fig. 10.1a. The ordered monoclinic form is usually not observed in apatite obtained from syntheses giving nanocrystals.

In aqueous solution, phosphate speciation is strongly dependent on pH as shown in Fig. 10.1b. This means that the free phosphate concentration at pH values below about 9 is negligible. It should be recalled, however, that coordination to metal ions such as calcium changes the effective acidity constant and can strongly influence acid/base equilibria.

In the following, various aspects of calcium phosphate formation will be discussed, starting with a brief account of amorphous calcium phosphates (ACPs) and progressing via prenucleation phenomena to a presentation of recent results from our laboratory on the crystallization of HAP from ACP, to end with a recapitulation of current works on the importance of confinement for calcium phosphate crystallization.

10.2 Amorphous Calcium Phosphates

Amorphous calcium phosphate forms easily from aqueous solution, where it subsequently transforms into, e.g., HAP. There are several excellent reviews on this subject (Combes and Rey 2010; Dorozhkin 2010) and the present discussion will therefore be kept brief. Today it is clear that there are many ways to stabilize ACPs, for example, by the addition of organic additives or foreign ions such as carbonate or pyrophosphate. However, the "true" structure of ACPs is still a matter of debate.

Posner and Betts used X-ray diffraction to derive a structural model for the basic building block of ACP in the form typically obtained under basic conditions (Posner and Betts 1975). Through pair-distribution function (PDF) analysis, they proposed that ACP is built from $Ca_9(PO_4)_6$ clusters that are interspersed with water to form the larger spheroids observed experimentally in electron microscopy. These so-called Posner clusters have been central to most discussions about ACP and apatite crystallization ever since. Meanwhile it is known that there are several different structural forms of ACP (Combes and Rey 2010; Dorozhkin 2010), depending on the way in which they are prepared – much like in the case of calcium carbonate (Fernandez-Martinez et al. 2017, Chap. 4, this volume). It is also clear that while Posner clusters certainly explain part of the phenomena observed, they are not able to account for all features observed for ACPs. Thus there is an urgent need for improved models of the structure and composition of ACPs, especially because their importance as precursor species in HAP and biogenic apatite formation increases, as discussed below and elsewhere in this book (Delgado-Lopez and Guagliardi 2017, Chap. 11, this volume). Recent improvements in PDF analysis methods, including the ability to perform in situ experiments, promise deeper insights into the structure of ACPs (Tyrsted et al. 2014), especially when combined with advanced NMR (Davies et al. 2014) and TEM (Dey et al. 2010; Nudelman et al. 2010; Nielsen and De Yoreo 2017, Chap. 18) techniques.

10.3 Prenucleation Phenomena

At concentrations below saturation, ion pairs have been known to form for a long time (Berry et al. 2000). In 2008, however, it was suggested that somewhat larger clusters, termed prenucleation clusters, occur in the $CaCO_3$ system (Gebauer et al. 2008; Gebauer et al. 2014). They have also been reported for silica (Tobler et al. 2017, Chap. 15) and iron oxyhydroxide (Reichel and Faivre 2017, Chap. 12 and Penn et al. 2017, Chap. 13) and some other systems (Gebauer et al. 2014). In solutions of zirconium acetate, from which zirconia forms upon heating, metastable polymeric states were observed (Bremholm et al. 2015) and have been shown to be directly involved in crystallization through a series of structural rearrangements (Tyrsted et al. 2014).

For calcium phosphates, there were early suggestions by Posner and Betts that such clusters should exist in solution, even though this was based exclusively on indirect evidence from their work on solid ACP (Posner and Betts 1975). Using light scattering and other techniques, Onuma and Ito (1998) detected calcium phosphate clusters in simulated body fluid (SBF) at pH 7.4, but also in acidified SBF solutions at pH 5.3, which are only supersaturated with respect to HAP but undersaturated with respect to octacalcium phosphate and ACP. These clusters were 0.7–1.1 nm in diameter, consistent with the size of Posner clusters. Moreover, the authors suggested a cluster aggregation model for crystal growth of HAP. More recently, Dey et al. (2010) used cryo-TEM to study the early stages of the formation of apatite

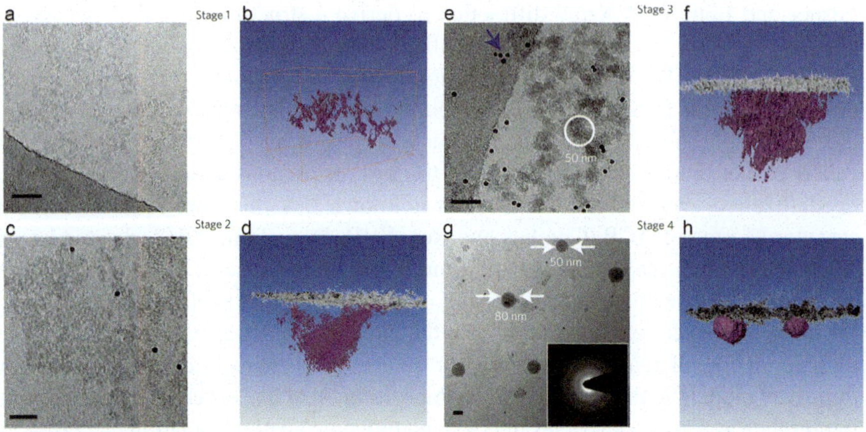

Fig. 10.2 Cryo-TEM studies of prenucleation clusters and the early stages of ACP formation from SBF solution in contact with a monolayer of arachidic acid. Panels *a*, *c*, *e*, and *g* are 2D projected TEM images with scale bars of 50 nm. Panels *b*, *d*, *f*, and h are corresponding 3D computer renderings of cryo-TEM tomograms with the monolayer shown in *gray* and the calcium phosphate in *purple*. The inset in panel *g* is a selected-area electron diffraction image showing that the particles are amorphous. Adapted and reprinted with permission from Macmillan Publishers Ltd: Nature Materials (Dey et al. 2010), copyright 2010

from SBF at 37 °C at the surface of a Langmuir monolayer of arachidic acid in a "frozen-snapshot" manner, where selected stages of the growth process could be resolved by stabilizing them through vitrification. Clusters were observed in solution also in this study, as shown in Fig. 10.2a, b. Their sizes were consistent with those of Posner clusters, although their chemical composition again remained undetermined. Upon contact with the arachidic acid monolayer, the clusters accumulated at the interface forming fractal aggregates (Fig 10.2c, d), which condensed further over time (Fig. 10.2e, f) to form dense ACP globules (Fig. 10.2g, h), shown to be amorphous by electron diffraction (inset of Fig. 10.2g), that ultimately crystallized to yield HAP. It should be stressed that the identification of the observed nanometric species as Posner clusters is based mostly on size. Secondly, in both works cited above, the investigated solutions were supersaturated with respect to HAP. Thus these clusters are not necessarily prenucleation clusters per se (Gebauer et al. 2014), but rather they should be considered as strong indications that prenucleation phenomena are also at play in calcium phosphate systems.

In another recent study, Habraken et al. (2013) conducted experiments where TRIS buffer solutions of calcium and phosphate salts were mixed at an initial pH of 7.4 and a Ca:P molar ratio of 1.43. Subsequently, the authors followed the progress of the reaction by ion-selective electrodes and pH measurements. Based on the data, a model was proposed in which the initially formed clusters are assumed to be highly charged with the composition $[Ca(HPO_4)_3]^{4-}$, while 57 % of these species are protonated ($[Ca(HPO_4)_2(H_2PO_4)]^{3-}$) under the given conditions. With time, these charged complexes were suggested to take up a calcium ion, and the resulting clusters (i.e., $[Ca_2(HPO_4)_3]^{2-}$) would then form the basis of ACP by building up a fractal network. This speciation model has been criticized in particular due to the fact that it is based on ion association constants from the literature, which were obtained without accounting for cluster formation, and by the question of how such highly charged ion complexes would aggregate without an additional charge-compensating step (Gebauer et al. 2014). Again the work of Habraken et al. took place under supersaturated conditions with respect to all crystalline phases.

Thus there are strong – almost "smoking gun" – indications that prenucleation clusters may occur in the calcium phosphate system, but still complete and conclusive evidence in particular for undersaturated conditions is missing. In addition, most studies were performed at pH levels very close to 7.4, even though HAP is known to form much more readily at high pH. This choice has presumably been based on the use of SBF as a biologically relevant fluid. However, in actual body fluid, crystallization is suppressed by a range of mechanisms to avoid mineralization, and hence one is left to wonder how particularly biologically relevant these conditions are (see also below).

10.4 Crystallization of HAP

The formation of HAP crystals demands either a transformation from preformed ACP or the agglomeration/association of prenucleation clusters and/or ions. Which particular pathway is followed must depend on several factors, including concentration and pH (Delgado-Lopez and Guagliardi 2017, Chap. 11).

To understand the crystallization process, it is essential to be able to study it as it happens, i.e., using in situ techniques. The amorphous phase is supposed to have spherical morphology, whereas apatite crystals are strongly anisotropic in shape. From the hexagonal symmetry, a needlelike morphology can be expected, which is indeed observed in synthetic and geological apatites (Hughes and Rakovan 2015). In teeth, the crystals are needle-shaped, while in bone, they are predominantly nanoplates (Mann 2001; Weiner and Wagner 1998). However, it has been an open question at what point during crystal growth the anisotropic morphology is attained, i.e., when one transitions from a sphere to an anisotropic shape. This question is related to the crystal surface energy, and thus it is tightly connected to phase stability and its dependence on nanocrystal size (Jensen et al. 2010). One suitable technique for studying crystallization kinetics is to use synchrotron in situ X-ray diffraction. This allows quantifying how much crystalline material is formed over time, but it can also yield the lattice constants of the forming crystals and, by analysis of the diffraction peak widths, the nanocrystal shape. We have reported several such studies based on a custom-designed stopped-flow apparatus (Ibsen and Birkedal 2010; Ibsen and Birkedal 2012; Jensen et al. 2015; Ibsen et al. 2016a; Ibsen et al. 2016b). We mixed calcium- and phosphate-containing solutions to induce the formation of ACP under the desired temperature and pH conditions. The ACP phase was formed directly in the X-ray diffraction cell, thus allowing its crystallization to be monitored in situ with excellent time resolution. We precipitated ACP both with (Ibsen and Birkedal 2010; Ibsen and Birkedal 2012) and without additives (Ibsen et al. 2016a; Ibsen et al. 2016b); herein we shall only discuss the latter case. Apatite was obtained by the following reaction sequence:

$$10CaCl_2 + 6M_3PO_4 + xMOH \rightarrow ACP \rightarrow HAP \tag{10.1}$$

where M^+ is either Na^+ or K^+ and the initial calcium concentration was 0.2 M, with x equaling either 2 or 0 to correspond to stoichiometric HAP ($x = 2$) or $HPO_4{}^{2-}$-dominated conditions.

Due to the extensive substitution chemistry of apatites, it is essential to carefully consider which counterions to use. With sodium as phosphate counterion, carbonate (from dissolved atmospheric CO_2) can substitute into the apatite lattice, while with potassium this does not happen as evidenced by FTIR spectroscopy (Ibsen et al. 2016b). For $x = 2$ in reaction (1), the initial solution is dominated by phosphate, while with $x = 0$, $HPO_4{}^{2-}$ is the main phosphate species (cf. Fig. 10.1b). In the phosphate-dominated case, the crystals had a small aspect ratio at early stages, suggesting that the initial nuclei were close to spherical in shape. They then rapidly grew along the *c*-axis and only much slower in the *(a,b)* plane, resulting in anisotropic particles. The nanocrystal size and formation kinetics depended on the choice of the counterion, i.e., sodium or potassium (Ibsen et al. 2016b), with potassium yielding more slender and longer needles than sodium.

Under $HPO_4{}^{2-}$-dominated conditions and with potassium as counterion, the observed crystallization behavior was very different than in the phosphate-dominated case discussed above. Figure 10.3 displays some of the kinetic data obtained at 60 °C, which we will discuss in some more detail here to illustrate the large degree of information that can be obtained by high-quality in situ diffraction methods (Ibsen et al. 2016a). The raw diffraction data show strong scattering signal of the aqueous solvent (Fig. 10.3a), but it was nevertheless possible to extract high-quality diffraction data by subtracting the solvent scattering (Fig. 10.3b). Initially, only the scattering signal of ACP was observed as two broad diffuse maxima at $2\theta \approx 12°$ and 18°. After a while, diffraction peaks from HAP appeared, while no other crystalline phases could be observed. The data were treated by Rietveld refinement. The Rietveld scale factor is a measure of the amount of crystalline material present. As seen in Fig. 10.3c, it increased rapidly to a plateau after a short induction time and remained constant thereafter, showing that the amount of crystalline material did not change after this nucleation burst. The nanocrystal size and composition, however, evolved over longer periods of time. In stark contrast to the phosphate-dominated situation, the initially detected nanocrystals were highly anisotropic, only ~3.5 nm wide, and almost ten times longer, as shown in Fig. 10.3e, 10.3f. The shape aspect ratio decreased in two stages, a rapid initial drop followed by a slower decay (Fig. 10.3e). During the early stage of growth, the chemistry of the crystals changes drastically. Initially, they were found to be highly Ca deficient as shown in Figure 10.3d, where the Ca occupation in the HAP unit cell is plotted as a

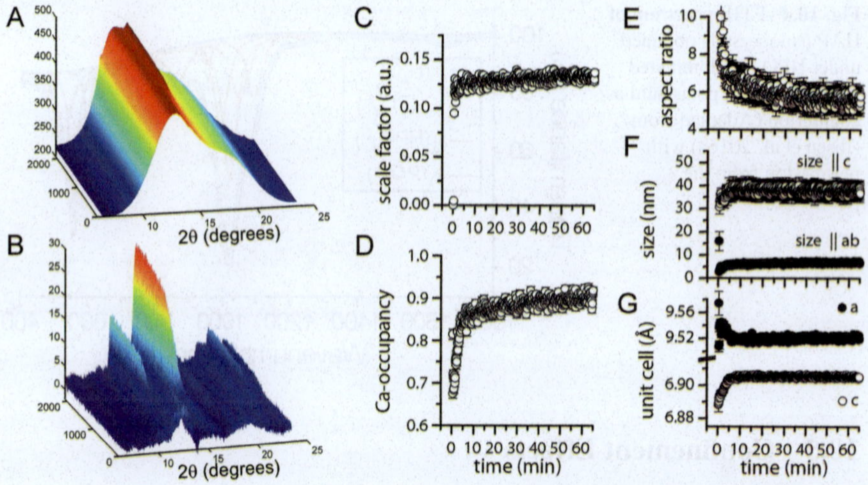

Fig. 10.3 In situ X-ray diffraction data on apatite nanocrystal formation under HPO_4^{2-}-dominated conditions (for explanations see text). Adapted from (Ibsen et al. 2016a) with permission from the publisher

function of time. There is first a rapid increase in the average calcium content of the nanocrystals followed by a stage of slower Ca uptake.

This behavior is also reflected in the time evolution of the unit cell parameters shown in Fig. 10.3g and was interpreted as follows: the initial Ca deficiency is most likely charge-compensated by hydrogen phosphate resulting in compositions of the general formula $Ca_{10-y}(HPO_4)_{2y}(PO_4)_{6-2y}(OH)_2$. The initial highly anisotropic nanocrystals had a composition close to $y = 3$, i.e., with all phosphates being hydrogenated. At later stages of growth, y approached 1. In a simple model, we assumed that stoichiometric HAP was added to a fully hydrogenated initial nanocrystal; this allowed predicting the Ca occupancy with the nanocrystal volumes as obtained from their measured sizes (Fig. 10.3f). A nearly perfect match to the experimentally determined time evolution of the Ca occupancy was found (Ibsen et al. 2016a). The presence of HPO_4^{2-} was confirmed by FTIR, an example of which is shown in Fig. 10.4. As expected with potassium as counterion, no carbonate peaks are observed, but a clear hydrogen phosphate resonance occurs at ca. 871 cm^{-1}.

OCP has long been suggested to be a precursor phase in biological apatite formation (Brown and Chow 1976), and more recent Raman spectroscopic studies of in vitro mineralization experiments point in the same direction (Crane et al. 2006). However, other evidence indicates that ACP is the relevant precursor material in biogenic crystallization, as first shown in chiton teeth (Lowenstam and Weiner 1985) and more recently in the youngest tissue of continuously forming special bones of zebrafish fin rays (Mahamid et al. 2010) as well as in more common bone (Mahamid et al. 2011). There has thus been a controversy about whether the initial phase is indeed ACP/OCP or ACP/nonstoichiometric apatite crystals (Weiner 2006; Grynpas 2007).

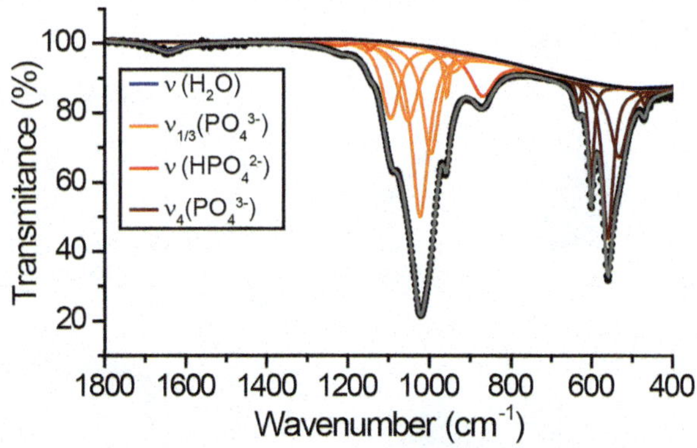

Fig. 10.4 FTIR spectrum of HAP nanocrystals obtained under $HPO_4{}^{2-}$-dominated conditions with potassium as counterion (Adapted from (Ibsen et al. 2016a) with permission from the publisher)

10.5 Confinement Effects

It has become clear that the physical environment plays a key role in deciding how crystallization proceeds, as documented in particular for the case of biogenic apatite formation (Weiner and Wagner 1998; Mann 2001; Olszta et al. 2007; Gower 2008). Rajasekharan and Andersson, for example, have recently reported the formation of crystalline calcium phosphates from amorphous calcium phosphate within the pores of polymerized liquid crystals and found that the pore size of aqueous domains influenced the outcome of crystallization (Rajasekharan and Andersson 2015). Polymerized liquid crystals were obtained by using functionalized Pluronic F127 ($EO_{100}PO_{70}EO_{100}$, where EO is ethylene oxide and PO is propylene oxide), which was crystallized into a hexagonal mesoporous material and then polymerized to yield covalently cross-linked mesoporous networks with excellent mechanical stability (He et al. 2015). These structures were infiltrated with acidic calcium and phosphate solutions at high concentrations. Subsequent exposure to ammonia gas triggered crystallization through a pH increase, leading to ACP formation in the pores of the mesoporous scaffold. With pore sizes below ~10 nm, nanocrystalline HAP was obtained, while a mixture of acidic polymorphs (monetite and brushite) were formed at larger pore sizes (Rajasekharan and Andersson 2015). This method gave purely artificial materials reminiscent of bone (He et al. 2015). Interestingly, confinement can also stabilize ACP and influence the alignment and morphology of the resulting crystals, as shown by Cantaert et al. (2013) using anodized alumina membranes. Likewise, growth of calcium phosphates on patterned, carboxylate-functionalized substrates with gelatin as growth modifier results in ultrahigh aspect ratio nanocrystals oriented perpendicular to the substrate (Tseng et al. 2013). It has also been proven possible to form apatites as chemical gardens using metal-loaded gelatin gels as metal reservoirs (Ibsen et al. 2014). All these findings illustrate how both physical and chemical factors can influence calcium phosphate mineralization.

10.6 Outlook

While great strides have been made in the understanding of calcium phosphate mineralization in general and ACP and HAP formation in particular, a range of open questions remain that need to be addressed in the coming years. As has been argued throughout this chapter, the strong pH dependence of phosphate speciation suggests that there is not one universal mechanism of calcium phosphate mineralization.

The nature of ACP remains a mystery and there is a strong need for improved structural models of ACPs formed under various conditions. With recent developments in, e.g., pair-distribution function analysis and NMR, it seems likely that this goal will be attained within a few years.

The prenucleation state and the steps leading to ACP formation remain contested, and much more work is needed before this part of the calcium phosphate picture is painted. This problem is challenging for several reasons. First and foremost, there is a technical challenge: the low solubility of HAP means that solutions undersaturated with respect to HAP have very low concentrations of the relevant species, which renders the range of applicable experimental techniques limited. Secondly, there is a need for establishing widely accepted methodologies for detecting these phenomena. In the view of the present author, the titration approach employed by Gebauer et al. (2008) is promising, even though it also suffers from drawbacks including a varying ionic strength. The next challenge is to identify which species are present in any detected clusters. As argued above, phosphate speciation changes drastically with pH. Even when taking into account the possible change in effective pK_a due to coordination by, e.g., calcium, there is likely to be several types of cluster compositions that will change with pH. The situation is further complicated by the ability of phosphate to undergo self-oligomerization through condensation, leading to pyrophosphate and higher polyphosphates that are both potent inhibitors of crystallization (Omelon and Grynpas 2008; Omelon et al. 2013) and species that must be considered for a full description of solution composition. These challenges call for novel and above all systematic approaches employing several complementary techniques, both experimental and theoretical.

The crystallization process itself also deserves more attention. An exceedingly large number of additives that affect calcium phosphate crystallization have been identified. Many of these are of relevance to biomineralization, even though it remains unclear which species are the main actors in which type of biomineralization. As shown in this chapter, the crystallization kinetics are very diverse and studying them with in situ techniques provides extremely valuable information, which ultimately can afford very detailed insights into crystallization pathways and (with additional model input) shed light on the relative importance of terms contributing to surface energies (Jensen et al. 2010). Above all there is a need for systematic studies including variations of the Ca:P ratio, pH, temperature, and solution composition. This will pave the way for understanding the factors determining, e.g., nanocrystal morphology evolution.

Thus, while much has been learned about calcium phosphate crystallization in the past years, we have only scratched the surface to reveal that much more is yet to be understood.

References

Anderson HC, Garimella R, Tague SE (2005) The role of matrix vesicles in growth plate development and biomineralization. Front Biosci 10:822–837

Berry RS, Rice SA, Ross J (2000) Physical chemistry. Oxford University Press, Oxford

Boonrungsiman S, Gentleman E, Carzaniga R, Evans ND, Mccomb DW, Porter AE, Stevens MD (2012) The role of intracellular calcium phosphate in osteoblast-mediated apatite formation. Proc Natl Acad Sci U S A 109:14170–14175

Bremholm M, Birkedal H, Iversen BB, Pedersen JS (2015) Structural evolution of aqueous zirconium acetate by time-resolved small-angle X-ray scattering and rheology. J Phys Chem C 119:12660–12667

Brown WE, Chow LC (1976) Chemical properties of bone-mineral. Annu Rev Mater Sci 6:213–236

Cantaert B, Beniash E, Meldrum FC (2013) Nanoscale confinement controls the crystallization of calcium phosphate: relevance to bone formation. Chem Eur J 19:14918–14924

Combes C, Rey C (2010) Amorphous calcium phosphates: synthesis, properties and uses in biomaterials. Acta Biomater 6:3362–3378

Crane NJ, Popescu V, Morris MD, Steenhuis P, Ignelzi MA (2006) Raman spectroscopic evidence for octacalcium phosphate and other transient mineral species deposited during intramembranous mineralization. Bone 39:434–442

Davies E, Müller KH, Wong WC, Pickard CJ, Reid DG, Skepper JN, Duer MJ (2014) Citrate bridges between mineral platelets in bone. Proc Natl Acad Sci U S A 111:E1354–E1363

De Yoreo JJ, Gilbert PUPA, Sommerdijk NAJM, Penn RL, Whitelam S, Joester D, Zhang H, Rimer JD, Navrotsky A, Banfield JF, Wallace AF, Michel FM, Meldrum FC, Cölfen H, Dove PM (2015) Crystallization by particle attachment in synthetic, biogenic, and geologic environments. Science 349(498):aaa6760-1–aaa6760-9

De Yoreo JJ, Sommerdijk NAJM, Dove PM (2017) Nucleation pathways in electrolyte solutions. In: Van Driessche AES, Kellermeier M, Benning LG, Gebauer D (eds) New perspectives on mineral nucleation and growth, Springer, Cham, pp 1–24

Delgado-López JM, Guagliardi A (2017) Control over nanocrystalline apatite formation: what can the X-ray total scattering approach tell us. In: Van Driessche AES, Kellermeier M, Benning LG, Gebauer D (eds) New perspectives on mineral nucleation and growth, Springer, Cham, pp 211–226

Dey A, Bomans PHH, Müller FA, Will J, Frederik PM, De With G, Sommerdijk NAJM (2010) The role of prenucleation clusters in surface-induced calcium phosphate crystallization. Nat Mater 9:1010–1014

Dorozhkin SV (2009) Calcium orthophosphates in nature. Biol Med Mater 2:399–498

Dorozhkin SV (2010) Amorphous calcium (ortho)phosphates. Acta Biomater 6:4457–4475

Elliott JC (1994) Structure and chemistry of the apatite and other calcium orthophosphates. Elsevier, Amsterdam

Falini G, Fermani S (2017) Nucleation and growth from a biomineralization perspective. In: Van Driessche AES, Kellermeier M, Benning LG, Gebauer D (eds) New perspectives on mineral nucleation and growth, Springer, Cham, pp 185–198

Fernandez-Martinez A, Lopez-Martinez H, Wang D (2017) Structural characteristics and the occurrence of polyamorphism in amorphous calcium carbonate. In: Van Driessche AES, Kellermeier M, Benning LG, Gebauer D (eds) New perspectives on mineral nucleation and growth, Springer, Cham, pp 77–92

Gebauer D, Völkel A, Cölfen H (2008) Stable prenucleation calcium carbonate clusters. Science 322:1819–1822

Gebauer D, Kellermeier M, Gale JD, Bergström L, Cölfen H (2014) Pre-nucleation clusters as solute precursors in crystallisation. Chem Soc Rev 43:2348–2371

Gower LB (2008) Biomimetic model systems for investigating the amorphous precursor pathway and its role in biomineralization. Chem Rev 108:4551–4627

Grynpas MD (2007) Transient precursor strategy or very small biological apatite crystals? Bone 41:162–164

Habraken WJEM, Tao J, Brylka LJ, Friedrich H, Bertinetti L, Schenk AS, Verch A, Dmitrovic V, Bomans PHH, Frederik PM, Laven J, Van der Schoot P, Aichmayer B, De With G, Deyoreo JJ, Sommerdijk NAJM (2013) Ion-association complexes unite classical and non-classical theories for the biomimetic nucleation of calcium phosphate. Nat Commun 4:1507

He W-X, Rajasekharan AN, Tehrani-Bagha AR, Andersson M (2015) Mesoscopically ordered bone-mimetic nanocomposites. Adv Mater 27:2260–2264

Hu Y-Y, Rawal A, Schmidt-Rohr K (2010) Strongly bound citrate stabilizes the apatite nanocrystals in bone. Proc Natl Acad Sci U S A 107:22425–22429

Hughes JM, Rakovan JF (2015) Structurally robust, chemically diverse: apatite and apatite supergroup minerals. Elements 11:165–170

Ibsen CJS, Birkedal H (2010) Modification of bone-like apatite nanoparticle size and growth kinetics by alizarin red S. Nanoscale 2:2478–2486

Ibsen CJS, Birkedal H (2012) Influence of poly(acrylic acid) on apatite formation studied by in situ X-ray diffraction using an X-ray scattering reaction cell with high-precision temperature control. J Appl Crystallogr 45:976–981

Ibsen CJS, Mikladal BF, Jensen UB, Birkedal H (2014) Hierarchical tubular structures grown from the gel/liquid interface. Chem Eur J 20:16112–16120

Ibsen CJS, Chernyshov D, Birkedal H (2016a) Apatite formation from amorphous calcium phosphate and mixed amorphous calcium phosphate/amorphous calcium carbonate. Chem Eur J 22:12347–12357

Ibsen CJS, Leemreize H, Mikladal BF, Skovgaard J, Eltzholtz JR, Bremholm M, Iversen BB, Birkedal H (2016b) Crystallization kinetics of bone-like apatite nanocrystals formed from amorphous calcium phosphate in water by in situ synchrotron powder diffraction: counter ions matter (Submitted)

Jensen GV, Bremholm M, Lock N, Deen GR, Jensen TR, Iversen BB, Niederberger M, Pedersen JS, Birkedal H (2010) Anisotropic crystal growth kinetics of anatase TiO2 nanoparticles synthesized in a nonaqueous medium. Chem Mater 22:6044–6055

Jensen ACS, Hinge M, Birkedal H (2015) Calcite nucleation on the surface of PNIPAM-PAAc micelles studied by time resolved *in situ* PXRD. CrystEngComm 17:6940–6946

Lowenstam HA, Weiner S (1985) Transformation of amorphous calcium-phosphate to crystalline dahllite in the radular teeth of chitons. Science 227:51–53

Mahamid J, Aichmayer B, Shimoni E, Ziblat R, Li C, Siegel S, Paris O, Fratzl P, Weiner S, Addadi L (2010) Mapping amorphous calcium phosphate transformation into crystalline mineral from the cell to the bone in zebrafish fin rays. Proc Natl Acad Sci U S A 107:6316–21

Mahamid J, Sharir A, Gur D, Zelzer E, Addadi L, Weiner S (2011) Bone mineralization proceeds through intracellular calcium phosphate loaded vesicles: a cryo-electron microscopy study. J Struct Biol 174:527–535

Mann S (2001) Biomineralization: principles and concepts in bioinorganic materials chemistry. Oxford University Press, Oxford

Nielsen MH, De Yoreo JJ (2017) Liquid phase TEM investigations of crystal nucleation, growth, and transformation. In: Van Driessche AES, Kellermeier M, Benning LG, Gebauer D (eds) New perspectives on mineral nucleation and growth, Springer, Cham, pp 353–371

Nudelman F, Pieterse K, George A, Bomans PHH, Friedrich H, Brylka LJ, Hilbers PAJ, De With G, Sommerdijk NAJM (2010) The role of collagen in bone apatite formation in the presence of hydroxyapatite nucleation inhibitors. Nat Mater 9:1004–1009

Olszta MJ, Cheng X, Jee SS, Kumar R, Kim Y-Y, Kaufman MJ, Douglas EP, Gower LB (2007) Bone structure and formation: a new perspective. Mater Sci Eng R 58:77–116

Omelon S, Grynpas MD (2008) Relationships between polyphosphate chemistry, biochemistry and apatite biomineralization. Chem Rev 108:4694–4715

Omelon S, Ariganello M, Bonnucci E, Grynpas M, Nanci A (2013) A review of phosphate mineral nucleation in biology and geobiology. Calcif Tissue Int 93:382–396

Onuma K, Ito A (1998) Cluster growth model for hydroxyapatite. Chem Mater 10:3346–3351

Posner AS, Betts F (1975) Synthetic amorphous calcium phosphate and its relation to bone mineral structure. Acc Chem Res 8:273–281

Penn RL, Li D, Soltis JA (2017) A perspective on the particle-based crystal growth of ferric oxides, oxyhydroxides, and hydrous oxides. In: Van Driessche AES, Kellermeier M, Benning LG, Gebauer D (eds) New perspectives on mineral nucleation and growth, Springer, Cham, pp 257–274

Rajasekharan AN, Andersson M (2015) Role of nanoscale confinement on calcium phosphate formation at high supersaturation. Cryst Growth Des 15:2775–2780

Reichel V, Faivre D (2017) Magnetite nucleation and growth. In: Van Driessche AES, Kellermeier M, Benning LG, Gebauer D (eds) New perspectives on mineral nucleation and growth, Springer, Cham, pp 275–292

Tobler DJ, Stawski TM, and Benning LG (2017) Silica and alumina nanophases: natural processes and industrial applications. In: Van Driessche AES, Kellermeier M, Benning LG, Gebauer D (eds) New perspectives on mineral nucleation and growth, Springer, Cham, pp 293–316

Tseng Y-H, Birkbak ME, Birkedal H (2013) Spatial organization of hydroxyapatite nanorods on a substrate via a biomimetic approach. Cryst Growth Des 13:4213–4219

Tyrsted C, Lock N, Jensen KMØ, Christensen M, Bøjesen ED, Emerich H, Vaughan G, Billinge SJL, Iversen BB (2014) Evolution of atomic structure during nanoparticle formation. IUCrJ 1:165–171

Wang L, Nancollas GH (2008) Calcium orthophosphates: crystallization and dissolution. Chem Rev 108:4628–4669

Weiner S (2006) Transient precursor strategy in mineral formation of bone. Bone 39:431–433

Weiner S, Wagner HD (1998) The material bone: structure-mechanical function relations. Annu Rev Mater Sci 28:271–298

Chapter 11
Control Over Nanocrystalline Apatite Formation: What Can the X-Ray Total Scattering Approach Tell Us

José Manuel Delgado-López and Antonella Guagliardi

11.1 Introduction

Biomineralization is the process by which living organisms form composites of organic molecules and inorganic minerals, often with surprising properties (e.g. Falini and Fermani 2017, Chap. 9). Examples of mineralized tissues include exoskeleton for protection (shell), endoskeleton for mechanical integrity (bone), cutting tool (teeth, claw) and mechanosensor (statolith or otolith) (Mann 2001; Lowenstam and Weiner 1989). The supramolecular arrangement of inorganic crystals within an organic matrix leads to the exceptional stiffness and elasticity of bone, while enhanced mechanical properties are found in mollusc shells, compared to the rather brittle calcium carbonate mineral (Mann 2001; Lowenstam and Weiner 1989). Understanding the in vivo mechanisms enabling the control over crystal morphology, size, polymorphism and material properties is of paramount relevance on both the fundamental knowledge side and as source of inspiration for designing advanced materials. Indeed, there is an increasing interest in modern chemistry and materials science to synthetically mimic biomineralized structures and create new coatings, novel materials or devices for medical applications (Mann 2001; Lowenstam and Weiner 1989). Bone is one of the most studied mineralized tissues.

J.M. Delgado-López (✉)
Instituto Andaluz de Ciencias de la Tierra, IACT (CSIC-UGR), Av. las palmeras, 4. Armilla, Granada, Spain

Università degli Studi dell'Insubria and To.Sca.Lab, Via Valleggio, 11., Como, Italy
e-mail: jmdl@iact.ugr-csic.es

A. Guagliardi
Istituto di Cristallografia (IC-CNR) and To.Sca.Lab, Via Valleggio, 11., Como, Italy
e-mail: antonella.guagliardi@ic.cnr.it

A.E.S. Van Driessche et al. (eds.), *New Perspectives on Mineral Nucleation and Growth*,
DOI 10.1007/978-3-319-45669-0_11

However, many important aspects of its complex mineralization process need further investigations. A detailed understanding of this complex system requires the use of advanced techniques that allow the design of time-resolved experiments, as will be further described.

X-ray and neutron diffraction techniques have traditionally been used to investigate the apatite crystal structure (Kay et al. 1964; Fratzl 2005). Very recently, innovative wide-angle X-ray total scattering (WAXTS) techniques have emerged as a powerful tool to quantitatively characterize different nanocrystalline systems and nanocomposites (Cernuto et al. 2011; Frison et al. 2013) at the atomic and nanometre length scale, including nanoapatite (Delgado-López et al. 2014). The application of these techniques to citrate-covered apatite nanoparticles at different precipitation times provided a remarkable picture in terms of crystal structure and defects, stoichiometry (in terms of Ca/P ratio), size and morphology of the nanocrystalline component(s) while also describing the evolution of the amorphous component. In this chapter, we briefly present the method of analysis and how it has been applied to model these complex systems. We also show how the results have been used, in combination with atomic force microscopy (AFM), to infer a plausible mechanism able to induce the formation of apatite platelets, in which citrate played a multiple role. The next important step forward is to extend this type of studies to the complex collagen-apatite system and elucidate the early stages of mineral formation within the collagen matrix. However, such challenging research demands the use of techniques with both high temporal and spatial resolutions. The design of time-resolved experiments that monitor in real time the evolution of morphology and crystallinity of the mineral phase, as well as the structural dynamics of the organic phase, is a must (Dey et al. 2010). Therefore, a combination of complementary techniques providing structural, morphological and compositional information while covering a broad range of length and time scales is required. Future perspectives in this direction are also briefly outlined in this chapter, focusing specifically on the combination of wide- and small-angle X-ray scattering. Further combinations with microscopic (AFM) and spectroscopic (Raman) techniques might really open new scenarios in the physicochemical characterization of these materials and strongly impact the field of biomineralization and biomimetic materials.

The chapter is structured in five sections. After this introduction, Sect. 11.2 describes the nucleation and growth mechanisms explaining the formation of apatite platelets in bone and highlighting the role of citrate in controlling the mineral formation. Section 11.3 reveals the strengths of the X-ray total scattering approach for characterizing biomimetic apatites. Section 11.4 is devoted to describing the biological importance of nanocrystalline apatites, and Sect. 11.5 intends to highlight the future perspectives in the characterization of bone biomineralization through the use of time-resolved experiments and combination of complementary techniques.

11.2 The Formation of Apatite Platelets in Bone: A Still Unclear Mechanism

Bone has a hierarchic structure with many levels of organization (Fig. 11.1) (Mann 2001; Lowenstam and Weiner 1989; Rho et al. 1998; Weiner and Wagner 1998; Glimcher and Muir 1984; Olszta et al. 2007; Wegst et al. 2015). Different supra-fibrillar arrangements coexist and form dense structural hierarchies from the nanoscopic to the macroscopic length scales (Mann 2001; Lowenstam and Weiner 1989). The building block of this complex architecture at the atomic-nanometre length scales is the mineralized collagen fibril (Fig. 11.2). Collagen triple helices are secreted from the cells and self-assemble at the molecular level into a periodic staggered array forming fibrils. These fibrils exhibit a characteristic banding pattern of 67 nm, where a densely packed 27 nm-long region (the so-called *overlap zone*) alternates with the less-dense 40 nm-long *gap zone* (Fig. 11.2). This organic matrix acts as a template where mineral deposition occurs both inside the gap zones (*intrafibrillar mineralization)* and between type I collagen (micro)fibrils and fibres, *interfibrillar mineralization*, through apparently distinct events (Fig. 11.2) (Landis and Jacquet 2013). Collagen mineralization has been suggested to form from a transient calcium phosphate amorphous (ACP) precursor that infiltrates into the gap regions, driven by electrostatic interactions, and further transforms into nanocrystalline platelets of apatite (Nudelman et al. 2010; Wang et al. 2012; Olszta et al. 2007).

Elucidating the mechanisms of biologically controlled mineral formation at the molecular level is a key issue to unravel common bone diseases and to design new therapeutic engineered scaffolds for repairing/regenerating hard tissues affected by pathologies, including bone tumours (Fratzl and Weinkamer 2007;

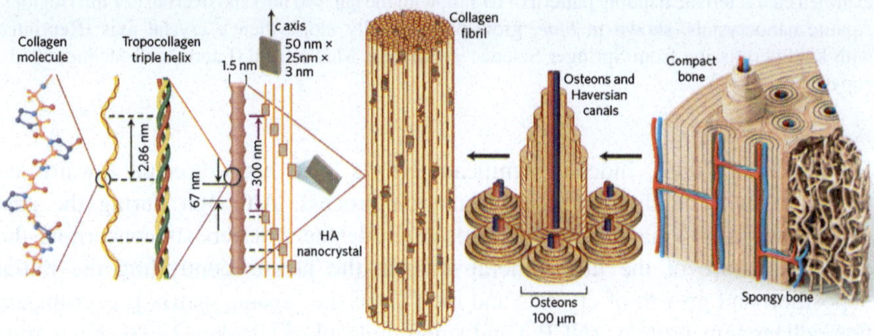

Fig. 11.1 Structural organization in bone with different levels of hierarchy, from the atomic (collagen molecule) to the macroscopic (bone tissue) level. The fibres comprise several mineralized collagen fibrils, composed of collagen molecules (tropocollagen) formed from three chains of amino acids and nanocrystals of hydroxyapatite (HA), and linked by an organic phase to form fibril arrays (Reprinted by permission from Macmillan Publishers Ltd: (Wegst et al. 2015), copyright (2014))

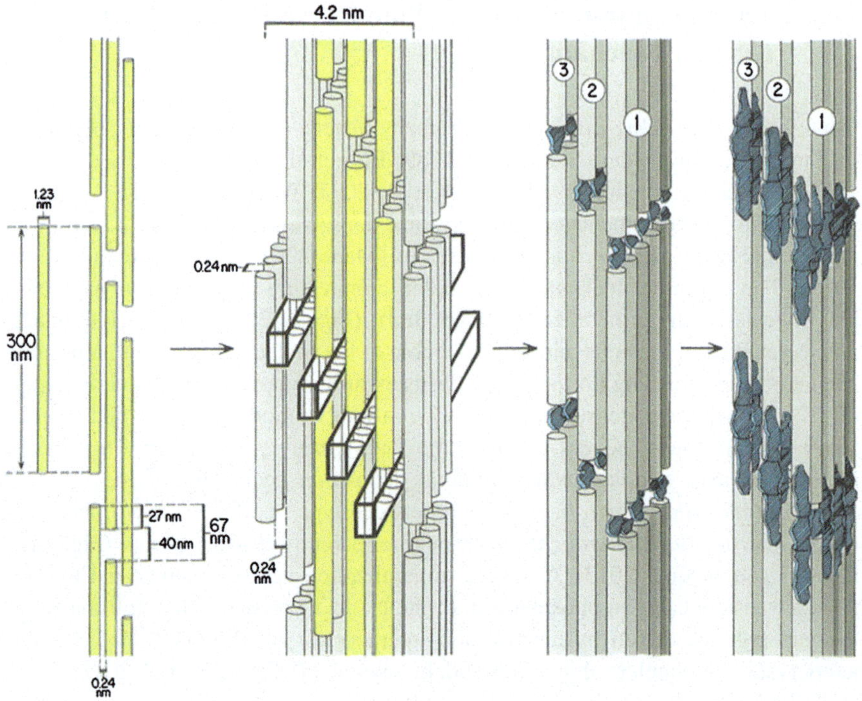

Fig. 11.2 Schematic view of the 2D and 3D collagen organization and intrafibrillar mineralization (from *left* to *right*). This collagen hierarchical organization is based on the Hodge and Petruska (Hodge and Petruska 1963) model that suggest the formation of channels or gaps within collagen arrays. Such channels are considered to be the principal sites in which nucleation of mineral crystals occurs within collagen. A single collagen molecule (in *yellow*) is shown at the *left* of the diagram. They are assembled into a periodic staggered array forming larger fibrils. These fibrils exhibit a characteristic banding pattern of 67 nm, with the gap (40 nm) and overlap (27 nm) regions. Apatite nanocrystals, shown in *blue*, grow preferentially along their c crystal axis (Reprinted with kind permission from Springer Science + Business Media from (Landis and Jacquet 2013), copyright (2013))

Palmer et al. 2008). Indeed, significant efforts have been directed towards an improved understanding of bone formation process. Although during the past years new insights have been reported, much debate still persists concerning the chemical nature of the first mineral formed, the factors controlling the initial deposition and growth of crystals and the role of the organic matrix [i.e. collagen, non-collagenous proteins (NCPs) and small molecules] (Boskey 1998). Numerous studies suggest that both collagen and NCPs may be involved in different steps of bone mineralization, such as the formation of the ACP precursor phase, its further transformation and organization of the final apatite crystals (e.g. Landis and Jacquet 2013; Palmer et al. 2008; George and Veis 2008).

However, many of these issues remain unclear and are still under discussion. For instance, spherical ACP particles are reported at the early stages of the precipitation process, whereas apatite platelets are observed as the final crystalline product (Nudelman et al. 2010; Wang et al. 2012; Mahamid et al. 2010). The platy shape of the mineral plays an important role in determining the unique functional and mechanical properties of bone (Eppell et al. 2001; Fratzl et al. 2004). Nevertheless, the mechanism underlying the ACP-to-apatite transformation and clarifying the origin of platy apatite crystals has never been directly detected neither in vivo nor in vitro experiments. Apatite platelets do not develop normally to the crystallographic *c*-axis of the hexagonal apatite crystal structure (space group $P6_3/m$) (Kay et al. 1964; Elliott et al. 1973; Leventouri 2006), therefore they break the hexagonal crystal symmetry (Fig. 11.3) through a mechanism that, so far, has remained elusive. Interestingly, the role of small organic molecules, such as citrate, has traditionally been overlooked. Nonetheless, it is well documented that citrate is an important component (Hartles 1964) accounting for 2 % weight of bone (Davies et al. 2014), a concentration 5–25 times higher than that occurring in most of other tissues (Hartles 1964). This suggests that citrate might have a broader role to play than it has been thought to date and even directly intervene in bone mineralization. However, how, and to what extent, citrate ions influence nucleation and crystal growth in the absence of other organic constituents is difficult to disentangle in biological systems and only rarely explored in synthetic models (Hu et al. 2011; Delgado-López et al. 2012, 2014; López-Macipe et al. 1998). Recently, solid-state NMR studies on biological samples have confirmed the relatively large amount of citrate in bone and pointed out that it is strongly bound to the surface of apatite nanocrystals, controlling their shape and size (Hu et al. 2010). According to Schmidt-Rohr and co-workers (Hu et al. 2010), the distance matching between carboxylate groups in citrate and pairs of neighbouring Ca^{2+} ions at the (10–10) facet of the apatite crystal causes citrate to be strongly bound to those facets and to block growth along this direction, thus limiting the platelet thickness (Fig. 11.3). In support of this hypothesis, Xie and Nancollas (Xie and Nancollas 2010) outlined the differences between bone and

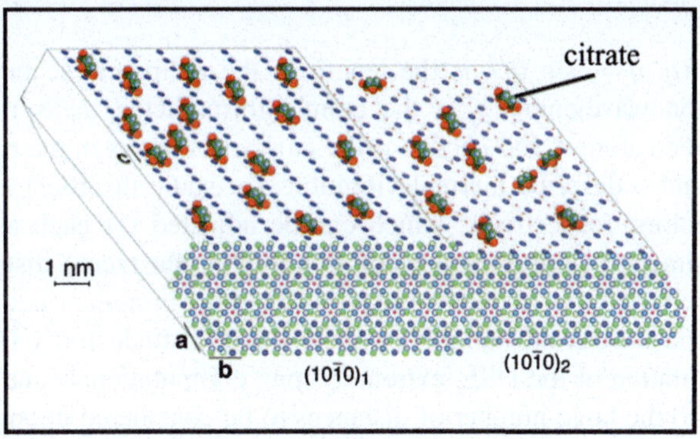

Fig. 11.3 Schematic of apatite-bound citrate interacting with Ca^{2+} on two surfaces of high morphological importance of an idealized bone apatite nanocrystal, at a realistic citrate surface density of ca. $1/(2\ nm)^2$. Calcium ions, blue-filled circles; P, *green* (omitted on the top surfaces); OH^-, *pink dots*; phosphate oxygen, omitted for clarity (Adapted from (Hu et al. 2010), copyright (2010), with permission from the National Academy of Sciences)

tooth enamel. In the latter case, a rod-like crystal morphology is found owing to the much lower content of citrate in saliva and the mismatch between carboxylate groups and the Ca distances at the crystal surface of fluorapatite. However, citrate ions are not expected to drive, alone, the crystal morphology to platelets, as the reported distance-matching criterion would apply in the same manner to all six, symmetry-equivalent, hexagonal facets of the apatite structure, not only to the (10–10) facet, as suggested by Schmidt-Rohr and co-workers (Hu et al. 2010).

11.3 Analysis of Citrate-Controlled Apatite Nanocrystals Through WAXTS Techniques

Wide-angle total scattering techniques offer the unique advantage, compared to conventional diffraction, of treating Bragg and diffuse scattering (originating from long-range order and short-range effects, respectively) on an equal footing (e.g. Cernuto et al. 2011; Frison et al. 2013; Delgado-López et al. 2014). The total scattering method is based on the Debye scattering equation (DSE) (Debye 1915), through its new and fast implementation available in the Debussy suite of programs (Cervellino et al. 2010; Cervellino et al. 2015). Debye's formula is known since 1915 and has the advantage of simultaneously modelling the coherent Bragg and diffuse scattering of a sample, regardless of their ordered or disordered atomic arrangement. The formula can be easily derived by applying the fundamentals of the radiation-matter interaction (Cervellino et al. 2012), eventually providing the average differential cross-section of a collection of randomly oriented nanoparticles (like an ideal powder) as a function of the distribution of interatomic distances, as follows:

$$I(Q) = \sum_{j=1}^{N} f_j(Q)^2 o_j^2 + 2 \sum_{j>i=1}^{N} f_j(Q) f_i(Q) T_j(Q) T_i(Q) o_j o_i \frac{\sin\left(Qd_{ij}\right)}{\left(Qd_{ij}\right)} \tag{11.1}$$

where $Q = 2\pi q$, $q = 2\sin\,\theta/\lambda$ is the length of the reciprocal scattering vector, λ is the radiation wavelength, f_j is the atomic form factor, d_{ij} is the interatomic distance between atoms i and j and N is the number of atoms in the nanoparticle. In Eq. 11.1, T and o describe thermal vibrations (or static disorder) and partial site occupancy factors, respectively, which can be adjusted for each atomic species. The first summation addresses the contribution of the (zero) distances of each atom from itself, and the second summation (the most important interference term) accounts for the non-zero distances between pairs of distinct atoms. However, in the original formulation of the DSE, extremely long computational time is an intrinsic problem due to the huge number of distances to be considered upon increasing the nanoparticle linear size D (from D^2 to D^6 for ordered and fully disordered materials, respectively). It is crucial to make such calculations fast enough to deal with a population of many nanocrystals and to use iterative algorithms for the optimization of model parameters. Cervellino et al. (Cervellino et al. 2006) developed an original approach, which makes use of sampled interatomic distances instead of the original ones, reducing thus the number of terms in the Debye equation by orders of magnitude, without losing accuracy in the pattern calculation.

Using the DSE approach and synchrotron data, Guagliardi et al. characterized bio-inspired apatite nanoparticles, synthesized in the presence of citrate ions as described in Sect. 11.4 and precipitated at increasing maturation times (5 min, 4 h and 96 h), either in the absence (Ap) or presence (cAp) of carbonate (Delgado-López et al. 2014). Briefly, the most important steps of the DSE modelling are here summarized: (1) Two populations of atomistic nanocrystal models were generated using the crystal unit cell of a Ca-deficient hydroxyapatite as the building block (Leventouri et al. 2003). Hexagonal prisms and platy crystals (insets 1 and 2 in Fig. 11.4a) of increasing size were grown through a layer-by-layer construction along two independent directions, one parallel to the sixfold *c*-axis and the second

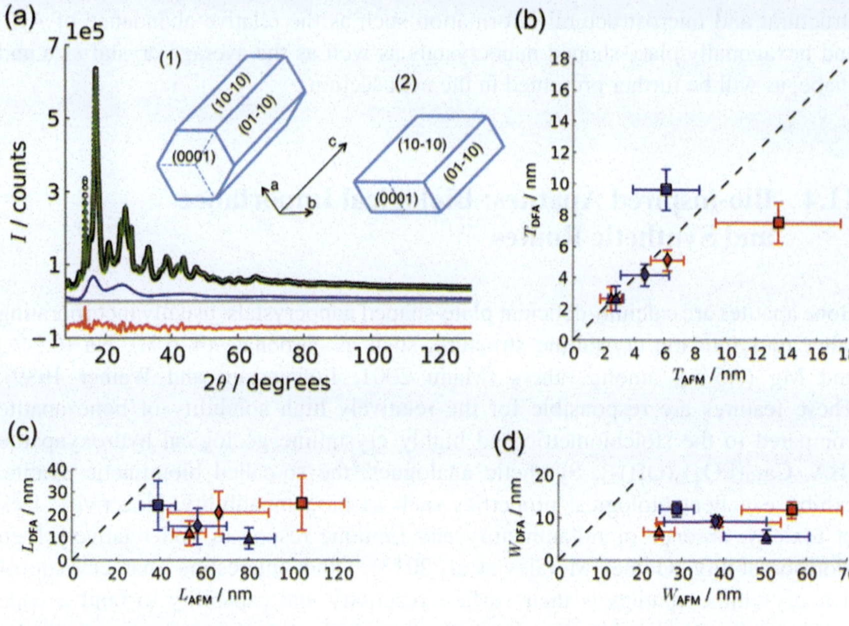

Fig. 11.4 (**a**) Best fit of the cAp 5 min sample provided by the DFA method. The ACP trace is also represented (agreement indices: GoF = 5.34, Rwp = 3.80). The inset shows the hexagonal (1) and platy morphologies (2) of Ca-deficient hydroxyapatite crystals used in the DFA modelling. (h, k, -h-k, l) indices of the most relevant crystal facets are also provided. (**b**) Average thickness of crystal domains provided by the DFA method (T_{DFA}) vs. the average thickness of nanoparticles provided by AFM (T_{AFM}), (**c**) average length of crystal domains (L_{DFA}) vs. the average length of nanoparticles (L_{AFM}) and (**d**) average width of crystal domains (W_{DFA}) vs. the average width of NPs (W_{AFM}). The vertical and horizontal bars depict the DFA- and AFM-derived distributions about each average. The broken line is a guide to the eye to easily identify samples having very similar DFA- and AFM-derived dimensions (Adapted from (Delgado-López et al. 2014), copyright (2014), with permission from John Wiley & Sons)

one in the orthogonal *ab*-plane. In order to avoid the introduction of a too complex trivariate size distribution function, platelets were dealt with through the construction of nanocrystals having anisotropic parallelogram-shaped bases with fixed 1:2 *a:b* ratio, which were used *together* with the hexagonal base prisms and constrained to the same nanocrystal size distribution and stoichiometry values, in a *biphasic* (crystalline) pattern model. This approximation allowed the quantification of the shape anisotropy of the *ab*-base (in addition to the anisotropy due to the crystal elongation in *c* direction) without introducing many unnecessary parameters. (2) The sampled interatomic distances of each nanocrystal were computed and stored in suitable databases. (3) An additional amorphous phase was modelled by suitably scaling the experimental ACP diffraction curve (blue line in Fig. 11.4a). (4) To best match the experimental traces, parameters modelling the nanocrystal size distribution and the site occupancy factors of Ca ions were also refined.

The best pattern model of the cAp 5 min sample is shown in Fig. 11.4a. For each data set, the above-mentioned Debye function analysis (DFA) provided relevant structural and microstructural information such as the relative abundance of ACP and hexagonally/platy-shaped nanocrystals as well as the average crystal size and shape, as will be further presented in the next section.

11.4 Bio-inspired Apatites: Biological Importance and Synthetic Routes

Bone apatites are calcium-deficient plate-shaped nanocrystals, usually incorporating other ions into the crystalline structure, such as carbonate (4–6 %), Na (0.9 %) and Mg (0.5 %), among others (Mann 2001; Lowenstam and Weiner 1989). These features are responsible for the relatively high solubility of bone apatite compared to the stoichiometric and highly crystalline geological hydroxyapatite [HA, $Ca_{10}(PO_4)_6(OH)_2$]. Synthetic analogues, the so-called biomimetic apatite, exhibit excellent biological properties such as biocompatibility, bioactivity, lack of toxicity, absence of inflammatory and immune responses and relatively high bioresorbability (Gómez-Morales et al. 2013). Other interesting characteristic of nanocrystalline apatites is their surface reactivity and capability to bind a wide variety of (bio)molecules. This is due to their high surface-to-volume ratio and a peculiar surface state involving a non-apatitic hydrated ionic layer with labile ions (Rey et al. 2007). All these features make nanocrystalline apatites ideal materials for medical applications spanning from regenerative medicine and bone tissue engineering to drug delivery. The possibility of using nanocrystalline apatites as carriers for local and controlled delivery of drugs is also very attractive, because of their easily tunable physical-chemical features (i.e. size, morphology and surface composition) and pH-dependent solubility (Wang and Nancollas 2008). This latter property is particularly interesting since the pH in the extracellular environments of solid tumours, and in particular within the endosome-lysosome cell compartment, is lower (around 5) than the physiological pH, thus enabling a preferential active drug release from the apatite surface in the tumour microenvironment (Delgado-López et al. 2013; Ren et al. 2011). This controlled drug release can be enhanced by coupling the nanocrystals with specific moieties for cell targeting (Iafisco et al. 2013).

Thus, the design of novel routes, inspired by nature, for the synthesis of nanocrystalline apatites with tailored features can open numerous possibilities of preparing optimal devices in medical sciences. Furthermore, synthetic apatites can be used as model systems for gaining new insights on the biomineralization mechanisms of living organisms, as previously pointed out, still far from being fully understood. In the last years, many strategies have been proposed to prepare synthetic nanosized apatite crystals (e.g. Gómez-Morales et al. 2013; Dorozhkin 2010). However, the preparation of biomimetic apatites is still considered a scientific and technological challenge (Sanchez et al. 2005). The use of organic molecules for the generation of inorganic materials with nanosized dimensions is an interesting strategy which has been receiving increased attention over the last decade. In this view, the role of citrate in controlling and stabilizing the size of bone apatite nanocrystals, as described in the previous section, is a biologically inspired lesson learned from nature, which can be developed into an advanced strategy to control nanomaterial fabrication (Xie and Nancollas 2010). Indeed, the synthesis in the presence of citrate provides bio-inspired citrate-covered nanocrystalline apatite with features very similar to that of biological apatite (Delgado-López et al. 2012, 2014). Citrate anions, which act as calcium-complexing agent, avoid the instantaneous calcium carbonate or calcium phosphate precipitation at room temperature. Temperature increase causes the destabilization of the Ca-citrate complex allowing a gradual release of Ca^{2+} ions into the solution which, in the presence of phosphate groups, form the hydroxyapatite growth units $[CaPO_4]^-$ and $[CaHPO_4]^0$ (Lazic 1995). On this basis, a protocol was proposed to obtain citrate-functionalized nanocrystalline apatite with controlled size, crystallinity and composition (Delgado-López et al. 2012). This protocol has been used to prepare the samples characterized by WAXTS techniques, as described in Sect. 11.3.

The DFA results showed that ACP is a minor component slowly decreasing upon maturation; nanocrystals of platy shape represent nearly the total fraction of the crystalline components after 5 min of maturation and tend to decrease with time; the hexagonally shaped nanocrystals become the major component after 96 h. Upon maturation, more stoichiometric crystals of progressively larger sizes along the two (or three, for platelets) growth directions are found; all samples show a very narrow size dispersion in the *ab*-plane, evidencing the action of citrate in controlling the growth along this direction; in cAp, crystals are systematically smaller and less anisotropic than in Ap.

Average sizes and shape distributions of crystal domains extracted from WAXTS were compared to multidomain nanoparticles measured by AFM. The sub-nanometre vertical resolution (better than 0.1 nm) of the AFM images allowed to obtain reliable values of the nanoparticle thickness (T_{AFM}). Figure 11.4

shows the comparison of the average thickness (T_{DFA}), width (W_{DFA}) and length (L_{DFA}) of *crystal domains* provided by DFA and those determined by AFM on *single nanoparticles* (T_{AFM}, W_{AFM}, L_{AFM}), regardless of their amorphous or crystalline nature. Importantly, crystal thickness systematically fits that of nanoparticles ($T_{DFA} \approx T_{AFM}$) until 4 h maturation (Fig. 11.4a), while width and length were systematically smaller in crystals than in nanoparticles ($W_{DFA} < W_{AFM}$ and $L_{DFA} < L_{AFM}$, Fig. 11.4b, c). This finding indicates that at low and medium maturation time, the nanoparticles *must be* single crystal domains along the thickness and multiple crystal domains along the width and length. The coincidence of T_{DFA} and T_{AFM}, moreover, suggests that crystals are oriented with their (01–10) plane parallel to the nanoparticle surface imaged by AFM. The formation of multidomain nanoparticles along the *c*-axis following a nonclassical oriented attachment mechanism was further observed by high-resolution transmission electron microscopy (HRTEM) (Iafisco et al. 2015).

On the basis of the above-described results, a plausible mechanism was proposed for the formation of apatite platelets with citrate playing multiple roles. Sodium citrate crystals precipitated at the earliest stages (López-Macipe et al. 1998) are thought to act as a template triggering the heterogeneous nucleation of sub-nanometre '*clusters*' which aggregate laterally and form ACP particles with unusual platy morphology (Delgado-López et al. 2014). The partial dissolution of this template, induced by the drop of pH and temperature, favours the release of $Hcit^{2-}/cit^{3-}$ ions, which partially bind to the ACP platelets inhibiting further plate thickening and nanoparticle aggregation. Apatite crystallization starts *within* each ACP platelet. Here, crystals *nuclei* grow along both the *c*-axis and, isotropically, along the *a*- and *b*-axes until the ACP platelet surface is reached. Citrate ions strongly bound to surface Ca^{2+} on the free {10–10} crystal facet stop any further growth in this direction (thickness), while amorphous-to-crystalline transformation can continue in the other two directions (length and width) where untransformed ACP remains. Platy crystals of the same thickness as ACP platelets are thus formed at short maturation times. Upon maturation, dissolution and recrystallization phenomena are likely to take place, and more stoichiometric rod-like hexagonally shaped crystals eventually become the dominant morphology.

Supplementary HRTEM studies on the same samples have also shown the occurrence of oriented attachment phenomena with formation of particles elongated along the *c*-axis and interacting through the {0001} facets (not covered by citrate). The reduction of the overall crystal surface free energy is supposed to act as the driving force of such head-tail attachment (Tao et al. 2007; Iafisco et al. 2015).

However, as far as the origin of the platy morphology of apatite nanocrystals is concerned, other routes in vivo or different biomimetic-like syntheses cannot be excluded. Worth of note is the transient formation of an octacalcium phosphate (OCP) precursor (Drouet 2013; Davies et al. 2014), which also exhibits a platelike morphology and might hydrolyze to apatite in a topotactic process, thus without major morphological changes.

11.5 Outlook: Time-Resolved Experiments and Combination of WAXTS with Complementary Techniques

For an exhaustive characterization of complex systems like bone, a combination of complementary techniques, which cover a broad range of length scales, is needed. Additionally, the greatest challenge of studying the early stages of mineral formation demands the use of techniques with high temporal resolution and the design of in situ experiments.

The limited access to the low-angle region by WAXS techniques can be complemented by small-angle X-ray scattering (SAXS) measurements. This combination is particularly interesting for hybrid materials with multiple length scale such as bone. Indeed, SAXS is sensitive to both the nanoscale apatite size (5–50 nm) as well as the collagen D-band pattern (67 nm) and collagen fibrils packing (Fig. 11.5). SAXS has traditionally been used to investigate the size and morphology of the nanosized mineral in bone tissue (Weiner and Wagner 1998; Fratzl et al. 1996; Fratzl 2005). Pilot ex situ WAXS experiments on synthetic collagen fibrils mineralized in the presence of citrate have shown, in the very low-angle region, the peak originating from the near-neighbour equatorial collagen distance simultaneously to the appearance of the ACP precursor, clearly visible at wider angular regions (Fig. 11.5) (Delgado-López et al. 2016). Therefore, using combined SAXS/WAXS in situ experiments, the structural changes of the mineral component can be investigated while mapping the evolution of collagen fibrils during their mineralization.

Confocal Raman spectroscopy is also a powerful tool for studying the mineral formation in the presence of collagen (Fig. 11.5c). Raman signals can be collected from both the collagen matrix and the mineral components. Additionally, since water is a weak scatterer of Raman signal, this spectroscopy is especially suitable for in situ mineralization experiments in aqueous solutions. The variations in the collagen fibrils caused by the intrafibrillar mineralization can be attained by shifts in the position of the following collagenous peaks: proline, amides, CH_2 deformation (δCH_2) and CH/NH stretching (Buckley et al. 2012) (Fig. 11.5c). On the other hand, the formation of the amorphous precursors can be witnessed by the appearance of the phosphate peak at ca. 952 cm^{-1} (ν_1PO4, symmetric stretching) (Ramírez-Rodríguez et al. 2013). The further amorphous-to-crystalline transformation causes its blueshift from 952 cm^{-1} (ACP) up to 961 cm^{-1} (apatite) (Ramírez-Rodríguez et al. 2013).

Size and morphology data of nanocrystalline apatite can be also extracted from AFM. The sub-nanometre vertical resolution (better than 0.1 nm) is of particular relevance for obtaining reliable values of the nanoparticles thickness (Delgado-López et al. 2014), which are almost impossible to be accurately determined by other techniques such as electron microscopies. Time-lapse AFM is also a very promising approach for observing the biomineralization of collagen in real time with high spatial resolution (in the sub-nanometre range). For instance, using this technique, Cisneros et al. have directly observed the self-assembly of collagen type I molecules into fibrils (Cisneros et al. 2006). The high resolution of their AFM

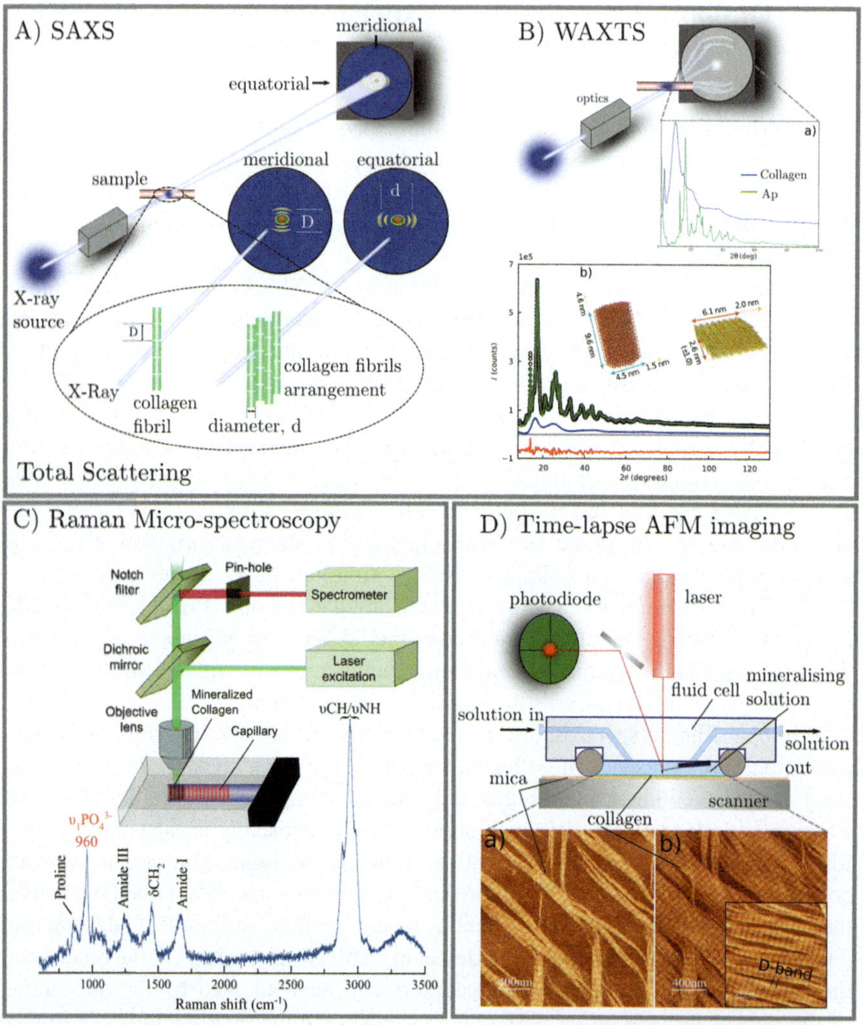

Fig. 11.5 Advanced techniques for characterizing the complex dynamics of bone biomineralization. (**A**) SAXS measurements sensitive to nanoscale mineral size, collagen D-band architecture (D) and later collagen fibrils packing (d). (**B**) WAXTS experiments providing information at the atomic and nanometre scales of both collagen and nanocrystalline apatite. Figure (a) shows the WAXTS patterns of collagen fibrils and mineral component; figure (b) (see also (Delgado-López et al. 2014)) shows the diffraction pattern of nanocrystalline apatite precipitated in the presence of citrate and its best fit as provided by DFA method: experimental powder diffraction pattern (*empty circles*), model pattern with its amorphous precursor (*solid lines*) and residual between experimental and model patterns (*bottom*). The *inset* shows the average size and shape of the nanocrystals as extracted from DFA. (**C**) Raman microspectroscopy set-up to monitor in situ the mineralization of collagen. Raman spectrum of a mineralized collagen fibril showing the typical collagen peaks and the main phosphate band of apatite at 960 cm-1. (**D**) AFM set-up to visualize in real time the collagen self-assembly (*over mica*) and mineralization using a fluid cell. AFM images (**a**, **b**) show the lateral assembly of collagen in contact with a buffer solution (pH 9.2, 50 mM glycine, 200 mM KCl) (unpublished results). The D-band periodicity (67 nm) of single fibrils is clearly observed (*inset*, image b)

topographs revealed substructural details of the D-band architecture of the fibrils forming the collagen matrix in contact with a buffer. However, the self-assembly and mineralization processes have never been simultaneously observed by AFM. Hence, WAXTS, SAXS and high-resolution AFM imaging could be synergistically combined to enhance the fundamental knowledge on the biomineralization of collagen at atomic and nanometre length scales.

References

Boskey AL (1998) Biomineralization: conflicts, challenges, and opportunities. J Cell Biochem Suppl 30–31:83–91

Buckley K, Matousek P, Parker AW, Goodship AE (2012) Raman spectroscopy reveals differences in collagen secondary structure which relate to the levels of mineralisation in bones that have evolved for different functions. J Raman Spectrosc 43:1237–1243

Cernuto G, Masciocchi N, Cervellino A, Colonna GM, Guagliardi A (2011) Size and shape dependence of the photocatalytic activity of TiO_2 nanocrystals: a total scattering debye function study. J Am Chem Soc 133:3114–3119

Cervellino A, Giannini C, Guagliardi A (2006) On the efficient evaluation of Fourier patterns for nanoparticles and clusters. J Comput Chem 27:995–1008

Cervellino A, Giannini C, Guagliardi A (2010) Debussy: a Debye user system for nanocrystalline materials. J Appl Crystallogr 43:1543–1547

Cervellino A, Frison R, Cernuto G, Guagliardi A (2012) Debye function analysis: theoretical and experimental aspects. In: Guagliardi A, Masciocchi N (eds) Crystallography for health and biosciences. IUP, Varese

Cervellino A, Frison R, Bertolotti F, Guagliardi A (2015) Debussy 2.0 – the new release of a Debye user system for nanocrystalline and/or disordered materials. J Appl Crystallogr 48:2026–2032

Cisneros DA, Hung C, Franz CM, Muller DJ (2006) Observing growth steps of collagen self-assembly by time-lapse high-resolution atomic force microscopy. J Struct Biol 154:232–245

Davies E, Müller KH, Wong WC, Pickard CJ, Reid DG, Skepper JN, Duer MJ (2014) Citrate bridges between mineral platelets in bone. Proc Natl Acad Sci 111:E1354–E1363

Debye P (1915) Zerstreuung von Röntgenstrahlen. Ann Phys 351:809–823

Delgado-López JM, Iafisco M, Rodríguez I, Tampieri A, Prat M, Gómez-Morales J (2012) Crystallization of bioinspired citrate-functionalized nanoapatite with tailored carbonate content. Acta Biomater 8:3491–3499

Delgado-López JM, Rodríguez-Ruiz I, Durán-Olivencia MA, Iafisco M, Tampieri A, Colangelo D, Prat M, Gómez-Morales J (2013) pH-responsive delivery of doxorubicin from citrate–apatite nanocrystals with tailored carbonate content. Langmuir 29:8213–8221

Delgado-López JM, Frison R, Cervellino A, Gómez-Morales J, Guagliardi A, Masciocchi N (2014) Crystal size, morphology, and growth mechanism in bio-inspired apatite nanocrystals. Adv Funct Mater 24:1090–1099

Delgado-López JM, Bertolotti F, Lyngsø J, Pedersen JS, Cervellino A, Masciocchi N, Guagliardi A (2016) The synergic role of collagen and citrate in stabilizing amorphous calcium phosphate precursors with platy morphology. Acta Biomater in press. doi:10.1016/j.actbio.2016.11.041

Dey A, de With G, Sommerdijk NAJM (2010) In situ techniques in biomimetic mineralization studies of calcium carbonate. Chem Soc Rev 39:397–409

Dorozhkin SV (2010) Nanosized and nanocrystalline calcium orthophosphates. Acta Biomater 6:715–734

Drouet C (2013) Apatite formation: why it may not work as planned, and how to conclusively identify apatite compounds. Biomed Res Int 2013:1–12

Elliott JC, Mackie PE, Young RA (1973) Monoclinic hydroxyapatite. Science 180:1055–1057

Eppell SJ, Tong W, Katz JL, Kuhn L, Glimcher MJ (2001) Shape and size of isolated bone mineralites measured using atomic force microscopy. J Orthop Res 19:1027–1034
Falini G, Fermani S (2017) Nucleation and growth from a biomineralization perspective. In: Van Driessche AES, Kellermeier M, Benning LG, Gebauer D (eds) New perspectives on mineral nucleation and growth, Springer, Cham, pp 185–198
Fratzl P (2005) Hierarchical structure and mechanical adaptation of biological materials. In: Reis RL, Weiner S (eds) Learning from nature how to design new implantable biomaterials: from biomineralization fundamentals to biomimetic materials and processing routes. Springer, Dordrecht
Fratzl P, Weinkamer R (2007) Nature's hierarchical materials. Prog Mater Sci 52:1263–1334
Fratzl P, Schreiber S, Klaushofer K (1996) Bone mineralization as studied by small-angle X-Ray scattering. Connect Tissue Res 34:247–254
Fratzl P, Gupta HS, Paschalis EP, Roschger P (2004) Structure and mechanical quality of the collagen-mineral nano-composite in bone. J Mater Chem 14:2115–2123
Frison R, Cernuto G, Cervellino A, Zaharko O, Colonna GM, Guagliardi A, Masciocchi N (2013) Magnetite–maghemite nanoparticles in the 5–15 nm range: correlating the core–shell composition and the surface structure to the magnetic properties. A total scattering study. Chem Mater 25:4820–4827
George A, Veis A (2008) Phosphorylated proteins and control over apatite nucleation, crystal growth, and inhibition. Chem Rev 108:4670–4693
Glimcher MJ, Muir H (1984) Recent studies of the mineral phase in bone and its possible linkage to the organic matrix by protein-bound phosphate bonds [and discussion]. Philos Trans R Soc Lond Ser B Biol Sci 304:479–508
Gómez-Morales J, Iafisco M, Delgado-López JM, Sarda S, Drouet C (2013) Progress on the preparation of nanocrystalline apatites and surface characterization: overview of fundamental and applied aspects. Prog Cryst Growth Charact Mater 59:1–46
Hartles RL (1964) Citrate in mineralized tissues. Adv Oral Biol 1:225–253
Hodge AJ, Petruska JA (1963) Recent studies with the electron microscope on ordered aggregates of the tropocollagen macromolecule. In: Ramachandran GN (ed) Aspects of protein structure. Academic Press, New York
Hu YY, Rawal A, Schmidt-Rohr K (2010) Strongly bound citrate stabilizes the apatite nanocrystals in bone. Proc Natl Acad Sci U S A 107:22425–22429
Hu YY, Liu XP, Ma X, Rawal A, Prozorov T, Akinc M, Mallapragada SK, Schmidt-Rohr K (2011) Biomimetic self-assembling copolymer-hydroxyapatite nanocomposites with the nanocrystal size controlled by citrate. Chem Mater 23:2481–2490
Iafisco M, Delgado-López JM, Varoni EM, Tampieri A, Rimondini L, Gomez-Morales J, Prat M (2013) Cell surface receptor targeted biomimetic apatite nanocrystals for cancer therapy. Small 9:3834–3844
Iafisco M, Ramirez-Rodriguez GB, Sakhno Y, Tampieri A, Martra G, Gomez-Morales J, Delgado-Lopez JM (2015) The growth mechanism of apatite nanocrystals assisted by citrate: relevance to bone biomineralization. CrystEngComm 17:507–511
Kay MI, Young RA, Posner AS (1964) Crystal structure of hydroxyapatite. Nature 204:1050–1052
Landis W, Jacquet R (2013) Association of calcium and phosphate ions with collagen in the mineralization of vertebrate tissues. Calcif Tissue Int 93:329–337
Lazic S (1995) Microcrystalline hydroxyapatite formation from alkaline solutions. J Cryst Growth 147:147–154
Leventouri T (2006) Synthetic and biological hydroxyapatites: crystal structure questions. Biomaterials 27:3339–3342
Leventouri T, Bunaciu CE, Perdikatsis V (2003) Neutron powder diffraction studies of silicon-substituted hydroxyapatite. Biomaterials 24:4205–4211
López-Macipe A, Gómez-Morales J, Rodríguez-Clemente R (1998) Nanosized hydroxyapatite precipitation from homogeneous calcium/citrate/phosphate solutions using microwave and conventional heating. Adv Mater 10:49
Lowenstam HA, Weiner S (1989) On biomineralization. Oxford University Press, New York

Mahamid J, Aichmayer B, Shimoni E, Ziblat R, Li C, Siegel S, Paris O, Fratzl P, Weiner S, Addadi L (2010) Mapping amorphous calcium phosphate transformation into crystalline mineral from the cell to the bone in zebrafish fin rays. Proc Natl Acad Sci 107:6316–6321

Mann S (2001) Biomineralization: principles and concepts in bioinorganic materials chemistry. Oxford University Press, New York

Nudelman F, Pieterse K, George A, Bomans PHH, Friedrich H, Brylka LJ, Hilbers PAJ, de With G, Sommerdijk N (2010) The role of collagen in bone apatite formation in the presence of hydroxyapatite nucleation inhibitors. Nat Mater 9:1004–1009

Olszta MJ, Cheng X, Jee SS, Kumar R, Kim Y-Y, Kaufman MJ, Douglas EP, Gower LB (2007) Bone structure and formation: a new perspective. Mater Sci Eng R Rep 58:77–116

Palmer LC, Newcomb CJ, Kaltz SR, Spoerke ED, Stupp SI (2008) Biomimetic systems for hydroxyapatite mineralization inspired by bone and enamel. Chem Rev 108:4754–4783

Ramírez-Rodríguez GB, Delgado-López JM, Gómez-Morales J (2013) Evolution of calcium phosphate precipitation in hanging drop vapor diffusion by in situ Raman microspectroscopy. CrystEngComm 15:2206–2212

Ren D, Kratz F, Wang S-W (2011) Protein nanocapsules containing doxorubicin as a pH-responsive delivery system. Small 7:1051–1060

Rey C, Combes C, Drouet C, Sfihi H, Barroug A (2007) Physico-chemical properties of nanocrystalline apatites: implications for biominerals and biomaterials. Mater Sci Eng C Biomimetic Supramol Syst 27:198–205

Rho JY, Kuhn-Spearing L, Zioupos P (1998) Mechanical properties and the hierarchical structure of bone. Med Eng Phys 20:92–102

Sanchez C, Arribart H, Giraud Guille MM (2005) Biomimetism and bioinspiration as tools for the design of innovative materials and systems. Nat Mater 4:277–288

Tao J, Pan H, Zeng Y, Xu X, Tang R (2007) Roles of amorphous calcium phosphate and biological additives in the assembly of hydroxyapatite nanoparticles. J Phys Chem B 111:13410–13418

Wang L, Nancollas GH (2008) Calcium orthophosphates: crystallization and dissolution. Chem Rev 108:4628–4669

Wang Y, Azaïs T, Robin M, Vallée A, Catania C, Legriel P, Pehau-Arnaudet G, Babonneau F, Giraud-Guille M-M, Nassif N (2012) The predominant role of collagen in the nucleation, growth, structure and orientation of bone apatite. Nat Mater 11:724–733

Wegst UGK, Bai H, Saiz E, Tomsia AP, Ritchie RO (2015) Bioinspired structural materials. Nat Mater 14:23–36

Weiner S, Wagner HD (1998) The material bone: structure-mechanical function relations. Annu Rev Mater Sci 28:271–298

Xie B, Nancollas GH (2010) How to control the size and morphology of apatite nanocrystals in bone. Proc Natl Acad Sci 107:22369–22370

Chapter 12
Calcium Sulfate Precipitation Throughout Its Phase Diagram

Alexander E.S. Van Driessche, Tomasz M. Stawski, Liane G. Benning, and Matthias Kellermeier

12.1 Introduction

Over the past decade, significant progress has been made in our understanding of crystallization phenomena. In particular, a number of precursor and intermediate species, both solutes and solids, and either stable or metastable (or even unstable), have been identified. Their existence extends the simplified picture of classical nucleation and growth theories toward much more complex pathways. These newly found species include various solute clusters, liquid-like phases, amorphous particles, and metastable (often nanosized) crystalline polymorphs, all of which may exist for different periods and may convert into one another depending on the chosen conditions (see, for instance, De Yoreo et al. 2017, Chap. 1; Wolf and Gower 2017, Chap. 3; Rodriguez-Blanco et al. 2017, Chap. 5; Birkedal 2017, Chap. 10; Reichel and Faivre 2017, Chap. 14, and references therein). Quite naturally, most

A.E.S. Van Driessche Driessche (✉)
University Grenoble Alpes, CNRS, ISTerre, F-38000 Grenoble, France
e-mail: alexander.van-driessche@univ-grenoble-alpes.fr

T.M. Stawski
German Research Centre for Geosciences, GFZ, D-14473 Potsdam, Germany

School of Earth and Environment, University of Leeds, LS2 9JT Leeds, UK

L.G. Benning
German Research Center for Geosciences, GFZ, Interface Geochemistry Section, 14473 Potsdam, Germany

Department of Earth Sciences, Free University of Berlin, 12249 Berlin, Germany

School of Earth and Environment, University of Leeds, Leeds LS2 9JT, UK

M. Kellermeier (✉)
Material Physics, BASF SE, Carl-Bosch-Str. 38, D-67056 Ludwigshafen, Germany
e-mail: matthias.kellermeier@basf.com

A.E.S. Van Driessche et al. (eds.), *New Perspectives on Mineral Nucleation and Growth*,
DOI 10.1007/978-3-319-45669-0_12

of the key observations supporting these alternative crystallization mechanisms were derived for (bio)minerals such as calcium carbonate or calcium phosphate, due to their great relevance for global carbon cycling and ocean chemistry as well as biomimetic materials chemistry, as was documented in many of the previous chapters in this book. In the quest to better understand such alternative pathways, other likewise important minerals have been neglected and only slowly these move into the focus of more detailed investigation. One such example is calcium sulfate, a compound of considerable interest for both industrial applications and geological environments. Recent studies have indicated that the formation of this mineral can also occur via a multistage pathway, again confirming that the crystallization of sparsely soluble salts is far more complex than pictured by standard models found in general mineralogy textbooks.

The foremost aim of this chapter is to review the current state of knowledge on the process of calcium sulfate precipitation from solution. Following a brief introduction to the mineralogy and relevance of calcium sulfates, we will focus on two main aspects: (1) the phase diagram of the $CaSO_4$–H_2O system (including solubilities and relative stabilities of the different mineral phases as a function of temperature, pressure, and salinity) and (2) the mechanisms of $CaSO_4$ nucleation and growth, both from a classical perspective and in view of recently gained insights, again considering the influence of different solution conditions. Eventually, an attempt will be made to unify the various reported observations toward an integrated model for the formation of calcium sulfates, not forgetting however to also highlight some central questions still open for further research.

12.1.1 $CaSO_4$ Mineralogy and Structure

Calcium sulfate can exist in various structural forms. In the presence of water, three distinct crystalline phases occur, differing in their degree of hydration: gypsum, the dihydrate ($CaSO_4 \cdot 2H_2O$), bassanite, the hemihydrate ($CaSO_4 \cdot 0.5H_2O$), and anhydrite, the anhydrous form ($CaSO_4$). The crystal structures of these three main phases are shown in Fig. 12.1. In addition to the water content, there are further structural and/or morphological differences between crystals belonging to the respective hydrate family (Table 12.1). Anhydrite, for example, comprises three different polymorphs: (1) the AIII phase, or γ-anhydrite, with hexagonal symmetry (obtained when water is quantitatively removed from bassanite by drying, but generally metastable); (2) the AII phase, or β-anhydrite, with orthorhombic symmetry (the thermodynamically stable phase at temperatures <1200 °C); and (3) the AI phase, or α-anhydrite, with trigonal symmetry (formed when β-anhydrite is heated above 1200 °C and replacing β-anhydrite as stable phase at higher temperatures) (Chang et al. 1996).

Since orthorhombic anhydrite is the only phase that precipitates spontaneously from aqueous solutions, the other polymorphs will not be considered further in this chapter.

In the case of bassanite, there is a distinction into α- and β-hemihydrate based on the way of production (Lewry and Williamson 1994; Singh and Middendorf 2008).

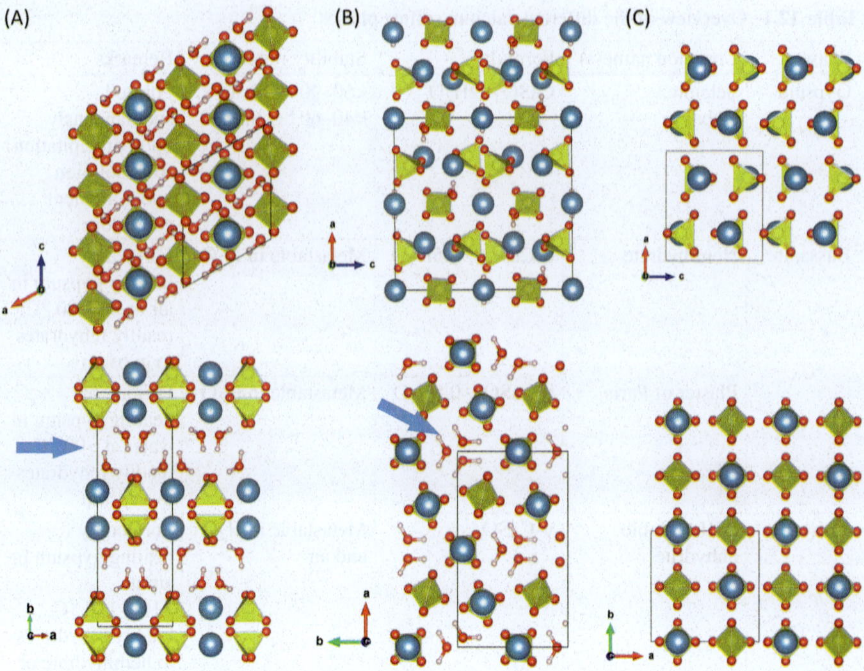

Fig. 12.1 Structural motifs of the three main calcium sulfate phases, gypsum, bassanite, and anhydrite (AII form). Tetrahedra are sulfate ions, large spheres are Ca ions, and arrows indicate water sheets and channels in the gypsum and bassanite structures, respectively (Images were drawn with VESTA (Momma and Izumi 2011))

The two types of bassanite differ primarily with respect to the shape of the crystals (α being more prismatic than β), their reactivity in contact with water (Lewry and Williamson 1994), and the mechanical properties of the final product (Satava 1970). Whether or not there are true structural differences between the two forms is still under debate (Singh and Middendorf 2008, and references therein). In addition, it has been reported that the actual water content in the bassanite structure can vary over a quite broad range ($CaSO_4 \cdot xH_2O$, where $0 \leq x \leq 0.8$, in particular 0.5–0.8), likely due to the fact that variable amounts of water can be accommodated in the channels of the lattice (which at low values of x resembles that of γ-anhydrite) (Abriel 1983; Lager et al. 1984). However, based on single-crystal refinement and structure optimization by DFT calculations, a recent study concluded that hemihydrate should contain exactly 0.5 water molecules per formula unit and that subhydrates with a crystal water content of more than 0.5 per $CaSO_4$ are unlikely (Weiss and Bräu 2009).

12.1.2 Calcium Sulfate in Natural Environments

Gypsum and anhydrite are the most abundant sulfate minerals in the Earth's crust and frequently occur in evaporitic environments (e.g., Warren 2006), such as the

Table 12.1 Overview of the different calcium sulfate phases

Phase	Common name(s)	Formula	Stability range	Remarks
Gypsum	Selenite, alabaster	$CaSO_4 \cdot 2H_2O$	<60–90 °C in air <40–60 °C in H_2O	Formed spontaneously upon precipitation from solution under ambient conditions
Bassanite	Hemihydrate	α-$CaSO_4 \cdot 0.5H_2O$	Metastable in H_2O	Formed by heating gypsum in air at >60–90 °C; readily rehydrates to gypsum
	Plaster of Paris	β-$CaSO_4 \cdot 0.5H_2O$	Metastable in H_2O	Formed by heating gypsum in water at >90 °C; readily rehydrates to gypsum
Anhydrite	AIII, Soluble anhydrite	γ-$CaSO_4$	Metastable in H_2O and air	Formed by heating gypsum in air at >100–130 °C; readily rehydrates to hemihydrate or gypsum
	AII, Insoluble anhydrite	β-$CaSO_4$	Stable <~1200 °C	Natural anhydrite formed at the Earth's surface
	AI	α-$CaSO_4$	Stable >~1200 °C	Converts immediately into β-$CaSO_4$ on cooling below ~1200 °C

Modified after Osinski and Spray, 2003

large deposits formed during the Messinian salinity crisis event (~5.3 million years ago), during which the majority of the Mediterranean evaporated and produced huge amounts of gypsum (Ryan 2009). Both gypsum and anhydrite are also present in low-temperature hydrothermal zones (Blount and Dickson 1969), with one spectacular example being the giant gypsum crystals found along with considerable amounts of anhydrite inside the Naica Mine in Chihuahua, Mexico (Garcia-Ruiz et al. 2007; Van Driessche et al. 2011). Recently, considerable quantities of gypsum have also been discovered on Mars (Langevin et al. 2005). As opposed to gypsum and anhydrite, bassanite is a very rare mineral on Earth (Allen and Kramer 1953; Apokodje 1984; Peckmann et al. 2003), but significant amounts have been detected on Mars (Wray et al. 2010) as well as in Martian meteorites (Ling and Wang 2015).

In addition to the geological occurrences mentioned above, calcium sulfates can also be found as structural components associated with living organisms. For example, two classes of medusae (Scyphozoa and Cubozoa) use bassanite for gravitational sensing (Tienmann et al. 2002; Becker et al. 2005; Boßelmann et al. 2007), and there is evidence for the presence of bassanite in the so-called "toothbrush" tree in Africa (Dongan et al. 2005). Also noteworthy is the frequent occurrence of gypsum-rich microbialites (e.g., stromatolites), both in past- and present-day evaporitic environments (Rouchy and Monty 2000). However, the role that microorganisms played in the formation of these structures is still under debate. In any case, compared to carbonates (see Falini and Fermani 2017, Chap. 9), phosphates (see Birkedal 2017, Chap. 10; Delgado-Lopez and Guagliardi 2017, Chap. 11), silicates (see Tobler et al. 2017, Chap. 15), or iron oxides (see Reichel and Faivre 2017, Chap. 14; Penn et al. 2017, Chap. 13), the number of living organisms that decided on sulfates as their biomineral of choice is minimal.

12.1.3 Relevance of Calcium Sulfate for Industrial Applications

Calcium sulfate is an important industrial material, with more than 100 million tons annually consumed worldwide (Sharpe and Cork 2006). While gypsum and anhydrite are mainly extracted from their abundant natural resources (or gained as by-products of other processes, e.g., in the industrial synthesis of several acids or during flue-gas desulfurization in coal-fired power plants), bassanite is usually obtained by partial dehydration of gypsum through heating at temperatures between 80 and 180 °C. Despite this cost- and energy-intensive process, hemihydrate remains one of the most extensively produced inorganic materials due to its relevance for the construction industry, where it is used on large scales as a binder in cements, mortars, or stucco (bassanite is often referred to as *plaster of Paris*). Anhydrite is likewise a common component of cementitious products like binders or adhesives. Finally, gypsum finds broad application in the agricultural (e.g., as soil conditioner), food (e.g., as flocculant), and pharmaceutical (e.g., as inert filler material) sectors (Sharpe and Cork 2006). Given this wide scope of possible uses, it is evident that there is an industrial demand for the design of simple and efficient protocols providing control over the phase composition, size, and morphology of $CaSO_4$ crystals, which may be achieved via solution precipitation if the underlying mechanisms are properly understood.

On the other hand, solid calcium sulfates also pose a severe problem to the industrial sector, as they represent recurrent scalants in several important processes, such as water purification in desalination plants, oil recovery, and mining activities (Ahmed et al. 2004; Mi and Elimelech 2010, Lu et al. 2012). Here, unwanted precipitation in pipes and on heat exchanger surfaces leads to incrustation, which causes a loss of efficiency or even costly downtimes for cleaning (with anhydrite being the major scale-forming phase at temperatures >100 °C and gypsum prevailing at lower temperatures). Therefore, considerable efforts are made to develop advanced strategies to prevent or retard crystallization under process conditions, which again demands (or at least would benefit from) detailed insights into the mechanisms at work. Another challenge related to calcium sulfates is the role that gypsum crystallization plays in the deterioration of building materials (e.g., concrete, mortar, marble, etc.) during sulfate (Neville 2004) or acid (Charola et al. 2007) attack. The latter is of increasing importance due to the continuing acidification of rainwater.

12.2 The $CaSO_4$–H_2O Phase Diagram

12.2.1 Solubility

In order to understand, and eventually control, the precipitation of calcium sulfates from solution, it is essential to know the phase diagram of the system, as this provides thermodynamic information on relative phase stabilities. Together with kinetic factors, these will determine the outcome of any crystallization process. The intersections of the solubility curves of different possible phases indicate the transition temperatures and thus define their corresponding thermodynamic stability field during precipitation from aqueous solution. Precise knowledge of these transition temperatures is of great relevance for understanding under which conditions evaporites (gypsum/anhydrite) have formed in the geological record, but equally also for optimizing the production/application of calcium sulfate-based materials (gypsum/bassanite and anhydrite/bassanite) in industrial processes. Not surprisingly, the solubility of solid $CaSO_4$ phases has been investigated extensively, with the first systematic works on the $CaSO_4$–H_2O system dating back to the late nineteenth century (e.g., Marignac 1874; Raupenstrauch 1885), followed by a plethora of studies in the twentieth century. Considerable effort was also devoted to the solubility in multicomponent systems, i.e., in the presence of other salts (e.g., Posnjak 1938; Bock 1961), and the effect of pressure (e.g., Dickson et al. 1963; Monnin 1990). Figure 12.2 shows a comprehensive overview of the solubility of the three relevant phases as a function of temperature and salinity, including both experimental and calculated data (using the PHREEQC speciation software (Parkhurst and Appelo 1999)).

Although most experimental measurements are in relatively good agreement with each other, still a considerably broad range of solubility data exists for all three phases (shaded areas in Fig. 12.2a). Quite remarkably, the same conclusion has already been drawn as early as 1902 by Hulett and Allen, who stated: "Although the solubility of the substance $CaSO_4$ has been the subject of investigation by many careful workers, the results vary widely, while the experimental errors are comparatively small." The marked spread among the reported experimental solubility values, as well as those derived from thermodynamic predictions, leads to a distinct uncertainty in the transition temperatures of gypsum to anhydrite and hemihydrate to anhydrite. Consequently, the stability regions of the different phases in the $CaSO_4$–H_2O system are still rather ill-defined.

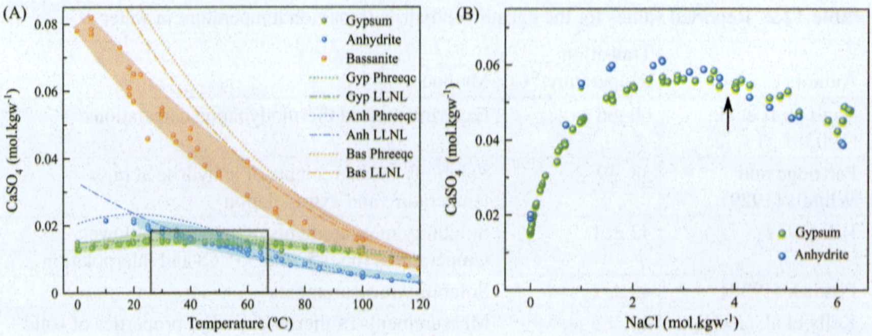

Fig. 12.2 Solubility curves for (**a**) gypsum, bassanite, and anhydrite (AII form) in pure water at different temperatures and (**b**) gypsum and anhydrite (AII form) as a function of salinity at ambient temperature. Both plots contain experimental data extracted from the literature (*dots*). The lines in (**a**) represent solubility curves calculated using the PHREEQC (*dotted lines*) and LLNL (*dot-stripe lines*) databases. The *shaded areas* in (**a**) are a visual aid highlighting the range of the reported experimental solubility data. The *black box* in (**a**) highlights the uncertainty range of the gypsum/anhydrite transition temperature. The *black arrow* in (**b**) marks the expected transition temperature at 25 °C and ~4 M NaCl. Experimental data were taken from: Poggiale (1843), Marignac (1874), Droeze (1877), Tilden and Shenstone (1984), Raupenstrauch (1885), Boyer-Guillon (1900), Cameron (1901), Hulet and Allen (1902), Melcher (1910), Hall et al. (1926), Hill (1937), Hill and Wills (1938), Partridge and White (1929), Roller (1931), D'Ans (1933, 1968), Booth and Bidwell (1950), Madgin and Swayles (1956), Bock (1961), Dickson et al. (1963), Marshall et al. (1964), Zen (1965), Marshall and Slusher (1966), Power et al. (1964, 1966), Block and Waters (1968), Blount and Dickson (1969, 1973), Culberson et al. (1978), Innorta et al. (1980), Kontrec et al. (2002), and Azimi and Papangelakis (2010)

12.2.2 Phase Transition Temperatures

Gypsum/Anhydrite The temperatures for the gypsum/anhydrite transition have been extensively studied due to its relevance for understanding the evolution of evaporitic deposits. Van't Hoff and coworkers were the first to address this issue in detail using experimental measurements and thermodynamic calculations and proposed a range of 60–66 °C for the transition temperature (van't Hoff et al. 1903; see also black box in Fig. 12.2a). Since then, a large number of related experimental and theoretical studies have been performed. The resulting transition temperatures are summarized in Table 12.2. They can be roughly divided into two groups, centered around 42 and 58 °C, respectively. The main challenge, standing in the way of a more precise determination of the transition temperature, is the very slow dissolution/growth kinetics of anhydrite at temperatures lower than ~80 °C, i.e., anhydrite remains metastable in either under- or supersaturated solution and true thermodynamic equilibrium is difficult (or even impossible) to reach in a reasonable period of time. One way to circumvent this problem is to precisely measure anhydrite solubility at higher temperatures (faster kinetics) and then extrapolate to lower temperatures. However, this method also suffers from obvious drawbacks (such as the extrapolation of data) and depends on precise knowledge

Table 12.2 Reported values for the gypsum/anhydrite transition temperature in water

Authors	Transition temperature/°C	Method
Van't Hoff et al. (1903)	60–66	Experiments and thermodynamic calculations
Partridge and White's (1929)	38–39	Solubility measurements of anhydrite at high temperature and extrapolation
Hill (1937)	42 ± 1	Solubility measurements of anhydrite at lower temperatures (65, 45, and 35 °C) and interpolation
Posnjak (1938)	42 ± 2	Solubility measurements
Kelly et al. (1941)	40	Measurements of thermochemical properties of solid gypsum and anhydrite and calculation of the transition temperature
Bock (1961)	42	Solubility measurements of anhydrite and gypsum
Marshall et al. (1964)	42	Solubility measurements
Zen (1965)	46 ± 25	Calculations based on revised thermodynamic data of Kelly et al.
Power et al. (1966)	41 ± 1	Solubility measurements of gypsum and anhydrite
Hardie (1967)	58 ± 2	Water activity measurements
Blount and Dickson (1973)	56 ± 3	Solubility measurements of anhydrite at high temperature (>70 °C) and extrapolation
Knacke and Gans (1977)	55.5 ± 1.5	Equilibration experiments in solutions containing both gypsum and anhydrite
Innorta et al. (1980)	49.5 ± 2.5	Solubility measurements
Corti and Fernandez-Prini (1983)	42.6 ± 0.4	Thermodynamic calculations
Hamad (1985)	46	Solubility measurements at high pressure and extrapolations to 1 atm
Raju and Atkinson (1990)	59.9	Thermodynamic calculations
Kontrec et al. (2002)	40	Solubility measurements and phase transition experiments
Azimi et al. (2007)	40 ± 2	Modeling of $CaSO_4$ solubility

of the solubility of gypsum, which itself is subject to uncertainties (cf. Fig. 12.2a). Using thermodynamic parameters to calculate the transition temperature is also not a more reliable approach, because these parameters themselves rely on solubility data. In summary, there is still considerable ambiguity with respect to the correct temperature for the gypsum-to-anhydrite transition, and as discussed by Freyer and Voigt (2003), there are no obvious reasons to prefer one group of values over the other.

Gypsum/Bassanite Considerably less attention has been paid to the gypsum/bassanite transition temperature due to the fact that at Earth's surface conditions, bassanite is metastable, i.e., at the conditions where gypsum transforms to bassanite, anhydrite is actually the stable phase (cf. Fig. 12.2a). However, anhydrite does not readily form due to much slower nucleation and growth kinetics (Ossorio et al. 2014). Moreover, the exact value of the gypsum/bassanite transition temperature seems to be of little relevance to natural phenomena. However, it is very well important for the production process of plaster of Paris and other industrial hemihydrate materials. Within the range of reported solubility data, the possible gypsum/bassanite transition temperature may vary from less than 80 to nearly 110 °C (cf. Fig. 12.2a). From dilatometric and tensiometric measurements, van't Hoff and coworkers (1903) derived a transition temperature of 106 °C, while Posnjak (1938) obtained 97 ± 1 °C using solubility data of bassanite and gypsum, which is close to the 98–100 °C reported by Partridge and White (1929). By modeling the solubility of $CaSO_4$ phases, Azimi et al. (2007) determined the transition temperature to be 99 ± 2 °C. These studies suggest that there is generally much better agreement about the transition temperature of gypsum to bassanite compared to that of gypsum to anhydrite.

Anhydrite/Bassanite In the case of the transition between bassanite and anhydrite, it is assumed that bassanite remains metastable (i.e., more soluble than anhydrite) over the entire relevant temperature range (~50–1200°C). Thus, no distinct crossover temperature has been reported so far (Kontrec et al. 2002). Although solubility data of hemihydrate at higher temperatures (>100 °C) are very close to those of anhydrite, stability experiments show that bassanite transforms into anhydrite in contact with aqueous solution when given enough time (e.g., 1 week at 99 °C and 0.8 M NaCl) (Ossorio et al. 2014).

12.2.3 *Influence of Salt and Pressure on Calcium Sulfate Solubility*

So far, we have only considered the $CaSO_4$–H_2O system at atmospheric pressure (except for the solubility data of bassanite and anhydrite above 100 °C) and in relatively pure systems without any additional salts or other solutes. Indeed, there is a vast array of studies that assessed the solubility of gypsum in ternary and quaternary systems (Azimi et al. 2007, Azimi and Papangelakis 2010, and references therein). The best-documented case is the solubility of $CaSO_4$ in the presence of sodium chloride (Fig. 12.2b). As NaCl is added to solutions of gypsum, a sharp increase in solubility is observed from 0 to 1 M; this can be ascribed to the concurrent decrease in the activity coefficients of Ca^{2+} and SO_4^{2-}, which increases the dissolved concentration at a given constant solubility product (as indicated by calculations with PHREEQC). At higher NaCl concentrations, first a maximum is reached between 2 and 3 M, before a slight decrease occurs at still higher salt contents. The solubility of gypsum in the presence of a large variety of other salts has also been measured and modeled, but a detailed discussion of these effects is beyond the scope of this chapter.

The dependency of solubility on ionic strength is slightly different for anhydrite, especially at low temperatures, leading to a crossover at high salt concentrations (cf. Fig. 12.2b) and suggesting that anhydrite should be the stable phase at room temperature and high salinity. Hence, with increasing NaCl concentration the gypsum/anhydrite transition temperature is shifted progressively to lower temperatures (Bock 1961). Interestingly, so far no experiments have been reported to confirm the spontaneous formation of anhydrite under these conditions; quite the contrary, Cruft and Chao (1970) and Ossorio et al. (2014) found that up to ~70 °C gypsum or bassanite are always the primary phases obtained due to kinetic inhibition of anhydrite formation.

Finally, it is well established that the solubilities of both gypsum and anhydrite increase with pressure. For example, at 50 °C the solubility of gypsum is about 15 mM at 1 bar and ca. 35 mM at 1000 bar, while at 80 °C anhydrite has a solubility of ca. 9 and 25 mM at 1 and 1000 bar, respectively (Blount and Dickson 1973). In the case of anhydrite, the presence of other salts strongly influences this dependency, making the pressure-solubility relation more complex (Blount and Dickson 1969).

12.3 $CaSO_4$ Formation from Solution

Similar to the solubility of calcium sulfate, its precipitation from solution has been the subject of extensive studies, again due to the importance of corresponding processes in both natural and industrial settings. In this section, we will discuss the state of the art of $CaSO_4$ mineralization from different perspectives. The first, and most comprehensive part, deals with precipitation from aqueous environments at low temperatures (<60 °C), where gypsum is the thermodynamically stable phase. The second part will focus on precipitation at higher temperatures, where bassanite and anhydrite are the dominant phases. In the third part, we discuss the influence of additives on $CaSO_4$ mineralization, before finally we attempt to reconcile the various observations and integrate them into a general model for the crystallization of calcium sulfate.

12.3.1 Nucleation and Growth of Gypsum

12.3.1.1 The "Classical" Picture

The first systematic studies on the precipitation of $CaSO_4$ from aqueous solutions were performed during the late nineteenth and early twentieth century, mostly focusing on solubility measurements (as described above). Later on, driven by the industrial and geological interests, a large number of papers were published on the nucleation and growth of gypsum, the kinetics of the transformation of bassanite to gypsum (i.e., the hydration process of corresponding binders in construction materials), and the relative phase stabilities in the $CaSO_4$–H_2O system. These early studies broadly agreed that gypsum precipitation from a supersaturated solution proceeded via a two-stage process (Conley and Boundy 1958; Schierholtz 1958; Smith and Sweett 1971): (I) formation of incipient (more or less) crystalline nuclei and (II) growth of these nuclei into gypsum crystals – i.e., the classical view of crystallization (Kashchiev 2000).

Induction time measurements performed by Liu and Nancollas (1973) suggested that the nucleation of gypsum is associated with an appreciable activation barrier, involving a critical nucleus composed of approximately six ions, as predicted based on a "nonclassical nucleation model" introduced by Christiansen and Nielsen (1952). On the other hand, Packter (1971) measured induction times as well as numbers and final sizes of crystals to conclude that gypsum formed through a heterogeneous process, in which nuclei are generated rapidly (almost immediately after mixing the reagents) on dust particles suspended in the supersaturated solution. Growth was then suggested to occur initially through a slow "mononuclear" process, yielding small crystallites that persisted over prolonged induction periods (which is equivalent to the large activation barrier mentioned above). These primary particles were finally thought to grow into large gypsum crystals by a more rapid "polynuclear" mechanism. Overall, this proposed scheme appears to be inspired by the precipitation model for sparingly soluble metal salts in solutions of low supersaturation presented by Nielsen (1964).

More recently, a large number of studies were performed with the aim to precisely measure induction times of gypsum precipitation as a function of parameters like supersaturation or temperature and/or in the presence of additives at different concentrations (e.g., Klepetsanis and Koutsoukos 1989; Hamdona et al. 1993; He et al. 1994; Klepetsanis et al. 1999; Lancia et al. 1999; Prisciandaro et al. 2001a, b; Alimi et al. 2003; Prisciandaro et al. 2003; Rashad et al. 2004; Fan et al. 2010). In essence, these studies concluded that the formation of gypsum appears to be reasonably well described by the classical nucleation theory (CNT), with nucleation kinetics being strongly dependent on temperature and solution speciation (like for many other insoluble salts). In most cases, the experimental data were fitted by the CNT equation and separated the into two regimes, i.e., homogeneous nucleation at high levels of supersaturation and heterogeneous nucleation at lower supersaturation. From these fits, the effective surface free energies of the forming nuclei was obtained, yielding rather realistic values of $\sim$40 mJ/m^2 and $\sim$14 mJ/m^2 for homogeneous (He et al. 1994; Lancia et al. 1999; Alimi et al. 2003; Fan et al. 2010) and heterogeneous (Alimi et al. 2003) nucleation, respectively.

This highlights that even though the particular mechanisms underlying nucleation may involve species that are fundamentally distinct from those envisaged in the classical picture of crystallization (see below), CNT can still provide a good description of the kinetics of the process – i.e., "nonclassical" nucleation does not necessarily mean that CNT is not valid (note that CNT was actually developed for the case of liquid condensation from vapors (e.g., Becker and Döring 1935) and hence it should be equally well, or even more, applicable to liquid/amorphous nuclei than to crystalline nuclei). This notion is of course not unique to calcium sulfate, but may generally be considered for a large variety of mineral systems (see e.g. De Yoreo et al. 2017, Chap. 1).

Indeed, none of the aforementioned investigations provided direct insight into the actual pathway(s)/mechanism(s) of the nucleation process. Thus, despite the large amount of data on gypsum precipitation collected over the years, the nature of the relevant species and their spatiotemporal evolution were still largely unknown at the beginning of this century. Instead, it was inherently assumed that $CaSO_4$ nucleation is a one-step process, which according to the classical paradigm directly produces nuclei that have the same characteristics (e.g., order, density, composition, etc.) as the fully grown (macroscopic) crystals (Kashchiev 2000).

12.3.1.2 Recent Observations: "Non-Classical" Processes

Sparked by the significant advances made for prominent mineral systems like calcium carbonate (De Yoreo et al. 2017, Chap. 1; Wolf and Gower 2017, Chap. 3; Fernandez-Martinez et al. 2017, Chap. 4; Rodriguez-Blanco et al. 2017, Chap. 5; Demichelis et al. 2017, Chap. 6; Andreassen and Lewis 2017, Chap. 7; Rao and Cölfen 2017, Chap. 8, and references therein) or calcium phosphate (Birkedal 2017, Chap. 10; Delgado-Lopez and Guagliardi 2017, Chap. 11, and references therein), various groups started to explore if a more complex crystallization pathway may underlie the precipitation of calcium sulfate. And indeed, the past few years have seen a growing body of evidence supporting the notion of multistage processes governing the formation of gypsum from aqueous solutions. In the following, we provide an overview of these new insights and discuss their implications for the emerging new picture of calcium sulfate mineralization at ambient conditions.

In 2012, Wang et al. (2012) reported on the existence of precursor phases during gypsum crystallization. They precipitated $CaSO_4$ at room temperature by mixing equimolar aqueous solutions of $CaCl_2$ and Na_2SO_4, and the formed solid particles were isolated at regular time intervals from the reacting solutions and characterized in the dry state by TEM and a number of complementary methods. Under the conditions tested, evidence was found for a short-lived amorphous phase appearing almost immediately after mixing solutions containing 25 mM $CaSO_4$. This conclusion was based on the fact that the isolated solids did not produce distinct crystalline spots when analyzed by selected-area electron diffraction (SAED) and that they readily transformed into gypsum upon prolonged exposure to the electron beam. Furthermore, samples isolated from a 50 mM solution (supersaturated with respect to gypsum but undersaturated with respect to bassanite) suggested that bassanite can form under conditions where it should not and remained stable for at least 1 h after mixing. Therefore, it was proposed that initially an amorphous calcium sulfate (ACS) phase formed that then converted into gypsum *via* bassanite as an intermediate crystalline stage. Moreover, the data suggested that both the amorphous precursor and the bassanite intermediate became progressively less stable (shorter lifetimes) with increasing supersaturation, and thus it was hypothesized that at high supersaturation, gypsum may form directly from solution.

At the same time, Van Driessche et al. (2012) used time-resolved turbidity measurements combined with ex situ characterization by high-resolution TEM and nano-diffraction to probe the early stages of calcium sulfate precipitation from mixtures of Na_2SO_4 and $CaCl_2$ at different levels of supersaturation. Based on the turbidity results, the precipitation process was arrested at different times by quenching the reaction through flash-freezing of solution aliquots or by addition of ethanol to wet particles obtained by micro-filtration. From detailed HR-TEM analyses, it was inferred that gypsum was formed via a three-stage process: (1) homogeneous nucleation of ca. 5 nm primary nanocrystalline bassanite particles and growth of these particles to about 100 nm; (2) self-assembly of the as-formed bassanite nanoparticles into elongated aggregates, co-oriented along their *c*-axis (i.e., oriented attachment); and (3) collective transformation of these aggregates into gypsum by merging of the bassanite units into a common crystallographic register along high-energy faces (Zhang and Banfield 2012). In the study of Van Driessche et al. (2012), no evidence of an amorphous $CaSO_4$ precursor phase was found. Moreover, the exact mechanism by which the oriented aggregates transformed into gypsum remained unresolved. However, it was suggested that there might be a crossover in the thermodynamic stability of calcium sulfate phases at small particle sizes, with bassanite becoming more stable than gypsum at the nanoscale under ambient conditions. This notion is similar to what has been proposed for the $CaCO_3$ system in the pioneering work of Navrotsky (2004).

Further observations supporting the idea of a multistage precipitation pathway for gypsum were reported by Saha et al. (2012). They used time-resolved cryogenic transmission electron microscopy (cryo-TEM) to probe the early stages of particle formation in a solution supersaturated with respect to gypsum, obtained by dissolution of bassanite. They observed small nanoclusters (2–4 nm in size) that evolved to seemingly amorphous particles (ca. 100 nm in diameter), which later reorganized to yield crystalline gypsum. On the whole, this multistep process took place on timescales on the order of tens of seconds.

Finally, at the end of 2012, Jones reported about an interesting study where attenuated total reflection (ATR) Fourier transform infrared (IR) spectroscopy was used to follow in situ the crystallization behavior of $CaSO_4$ in solution upon evaporation (Jones 2012). It was found that the intensity of IR bands corresponding to structural water increased over time, indicating that water molecules initially occupied disordered positions (in an assumed amorphous $CaSO_4$ phase). In later stages, the water molecules adopted more defined configurations characteristic of the forming of crystalline solids (inferred to be bassanite and/or gypsum). It was also noted that the presence of water might influence the stability of the initial disordered phase. Thus, Jones (2012) concluded that the crystallization process started with a disordered ("amorphous") phase that slowly transformed into crystalline gypsum, possibly via a bassanite intermediate. During this transformation, water seems to play a key role, shifting from disordered to ordered positions and thus inducing the emergence of a defined crystal lattice.

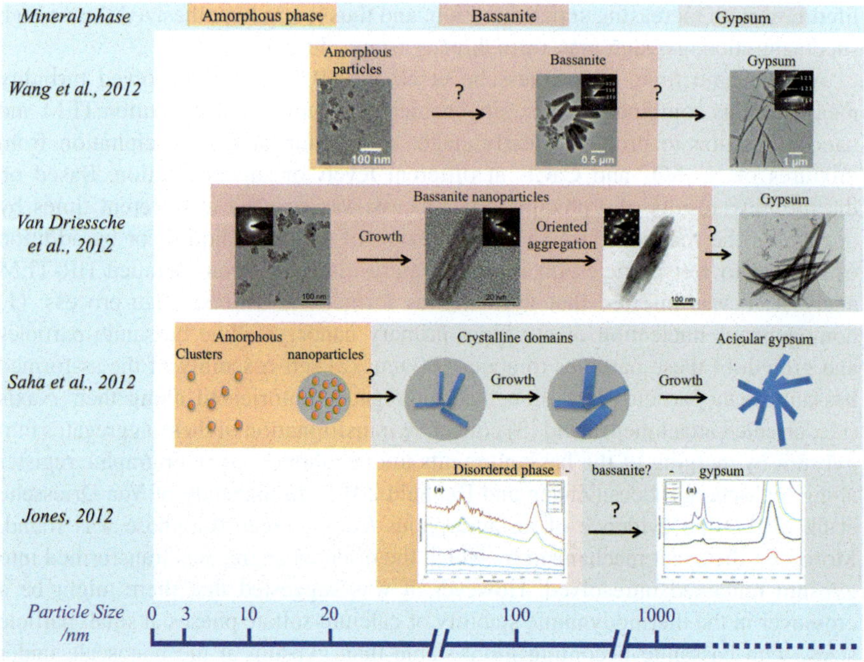

Fig. 12.3 Schematic overview of the various precursor phases and pathways leading to gypsum formation from aqueous solutions at room temperature, as recently proposed by Wang et al. (2012), Van Driessche et al. (2012), Saha et al. (2012) and Jones (2012). Note that the in the online version different phases are color-coded and that the respective phase sequences are arranged according to changes in size (bottom axis), not time

Although all these studies addressed the same problem, the results still show discrepancies, likely because of inherent differences in the used experimental and analytical approaches. Nevertheless, it seems clear that there are alternative pathways to crystalline gypsum from supersaturated aqueous solutions other than direct nucleation and growth as envisaged in the classical picture. The general conclusion is that the phase(s) formed initially upon precipitation from solution can be quite distinct from the final crystalline state (gypsum), or in other words that gypsum formation indeed involves one or more precursor/intermediate phases. Figure 12.3 provides an overview of the reaction sequences proposed in the recent studies discussed above.

It is important to note that the reported data are ambiguous with respect to the question whether amorphous precursors or bassanite nanoparticles, or both, play a key role in the formation of gypsum. The main reason for these ambiguities lies in the fact that it is extremely difficult to capture the elusive early stages of precipitation with sufficient temporal and spatial resolution, while at the same time not affecting the state of the reacting solids and solutions. Although some of the described studies used in situ methods (i.e., IR spectroscopy or turbidity measurements), these lack specific resolution and provide limited information about size, shape, or crystallinity of the formed solids. In turn, the applied ex situ techniques provide sufficient resolution (e.g., (cryo)-TEM) but suffer from the obvious drawback of possible sample preparation artifacts, along with other problems like beam damage. These difficulties may lead to a situation where the detected particles are actually not present in solution but rather formed/transformed during isolation and/or analysis. Indeed, the strategies used for sample preparation in the abovementioned studies (including micro-filtration, cryo-quenching, or direct analysis of cryo-vitrified samples) are designed to minimize undesirable effects, but still they do not offer true in situ and time-resolved characterization of the occurring processes, as would be needed to fully address these rapidly evolving reactions in the $CaSO_4$ system at the proper length scale.

This limitation was just recently circumvented by Stawski et al. (2016), who combined synchrotron-based small- and wide-angle X-ray scattering (SAXS/WAXS) to investigate gypsum formation from supersaturated solutions. These in situ experiments were carried out over a temperature range of 12–40 °C, by direct mixing of equimolar Na_2SO_4 and $CaCl_2$ solutions and circulating the resulting samples through a capillary centered in the X-ray beam. In-depth evaluation of the obtained data led to the conclusion that crystalline gypsum forms via a multistep process, involving four well-defined stages (see Fig. 12.4):

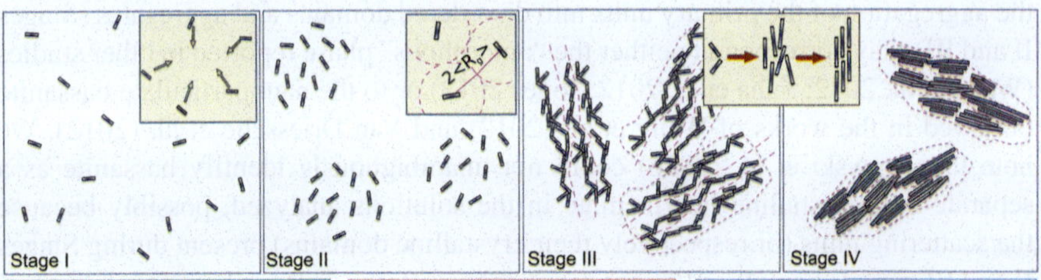

Fig. 12.4 Schematic representation of the four stages of $CaSO_4$ precipitation from solution as proposed by Stawski et al. (2016) based on the results of in situ SAXS/WAXS experiments: Stage I, formation of sub-3 nm primary species. Stage II, assembly of the primary species into loose domains; the inset shows that individual scatterers are separated by a distance larger than two times the radius of gyration. Stage III, further aggregation into large surface-fractal morphologies. Stage IV, growth and coalescence of the primary species within the aggregates and transformation to gypsum. The inset between stages III and IV shows the successive evolution at the nano- and mesoscale, where the primary units first grow/coalesce in length and subsequently in all dimensions, yielding structural domains of increasing size and crystallinity. The three panels on the bottom show tentative molecular structures for the anhydrous rodlike primary species based on the gypsum, bassanite, and anhydrite lattice

(Stage I) Formation of elongated primary species, <3 nm in length and ~0.5 nm in diameter and composed of "anhydrous" Ca–SO_4 cores (as indicated by considerations of SAXS volume fractions and electron densities)
(Stage II) Assembly of these primary species via density fluctuations into loose domains, where the interparticle distance remains still fairly large, i.e., >2 times the radius of gyration of the individual primary particles
(Stage III) Further condensation of these domains into larger and denser aggregates, still consisting of the same primary units
(Stage IV) Growth of the primary units within the aggregates, mutual alignment, and eventual collective reorganization/transformation into gypsum, as confirmed through the simultaneously collected WAXS data

These results suggest that $CaSO_4$ precipitation from solution relies on the formation and aggregation of nanosized primary units with a rather well-defined size and composition – which may or may not be analogous to the so-called "pre-nucleation clusters" reported for the calcium carbonate and phosphate systems (Gebauer et al. 2008; Dey et al. 2010; Kellermeier et al. 2014). So far no data are available on how stable or metastable these primary $CaSO_4$ species are. In any case, the structure of the primary units cannot be uniquely assigned to any of the crystalline $CaSO_4$ phases (cf. bottom panels in Fig. 12.4). In other words, the primary species are equally similar to all three possible phases, and hence they may be considered as a universal precursor phase to any possible subsequent crystallization process. Under the conditions investigated by Stawski et al. (2016), the aggregation of the primary units into disordered domains and aggregates (Stages II and III) may correspond to either the "amorphous" phase reported in other studies (Wang et al. 2012; Saha et al. 2012; Jones 2012) or to the nanoparticulate bassanite observed in the works of Wang et al. (2012) and Van Driessche et al. (2012). We note that Stawski et al. (2016) could not unambiguously identify bassanite as a separate nanocrystalline intermediate in the solutions analyzed, possibly because the scattering units (or respectively their crystalline domains) present during Stages II and III were too small and/or not dense enough (large distance between individual units) to be detected by WAXS – or simply because bassanite is not an intermediate on the way to gypsum under the tested conditions.

A possible clue to the answer to this question has recently been reported by Tritschler et al. (2015a, b), who investigated the influence of organic (co)solvents on the precipitation behavior of calcium sulfate. By mixing supersaturated (but optically clear, i.e., metastable) aqueous $CaSO_4$ solutions with ethanol, nanoparticles of phase-pure bassanite were obtained (Fig. 12.5a) (Tritschler et al. 2015a). Similar observations were already made in an earlier study by Yang et al. (2011). These findings challenge the results of experiments where ethanol was used to supposedly quench the precipitation process in aqueous media (Wang et al. 2012; Van Driessche et al. 2012), as addition of the organic solvent may in fact induce the formation of bassanite rather than preserving structures that were thought to be present in solution before. Systematic variations of the water/ethanol ratio further showed that there is a critical water content (between 40 and 50 wt%), below which bassanite becomes

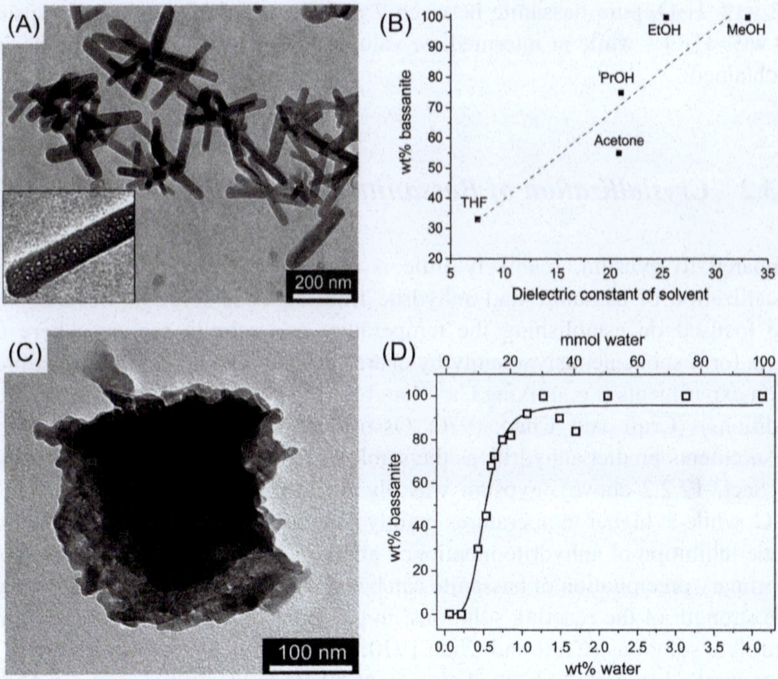

Fig. 12.5 Precipitation of calcium sulfate from mixtures of water and organic solvents. (**a**) Bassanite nanorods obtained by quenching 25 mM $CaSO_4$ solutions in an excess of ethanol. Note the morphological and structural similarity of the particles to those shown in Fig. 12.3. (**b**) Fraction of bassanite in mixtures with gypsum, as obtained by mixing aqueous $CaSO_4$ solutions with solvents of different polarity. (**c**) Phase-pure anhydrite produced by addition of concentrated sulfuric acid to solutions of $CaCl_2$ in methanol at an effective water content of 0.07 wt%. (**d**) Plot of the fraction of bassanite in mixtures with anhydrite as a function of the amount of water in the reaction medium (Figures are reproduced from (Tritschler et al. 2015a, b) with permission by the Royal Society of Chemistry and Wiley, respectively)

progressively favored over gypsum (Tritschler et al. 2015a). Another major factor for phase selection was found to be the type of organic solvent used for quenching, with more polar solvents yielding higher fractions of bassanite (Fig. 12.5b). This strongly suggests that the availability of water and the degree of hydration of any precursors (which is likely to depend on the polarity of the medium) are key factors for the outcome of the $CaSO_4$ precipitation process (as discussed in more detail below).

In a subsequent study, Tritschler et al. (2015b) extended this concept to significantly lower water contents and demonstrated that pure anhydrite (Fig. 12.5c) forms spontaneously at room temperature if less than 0.2 wt% water is present in mixtures with alcohol (Fig. 12.5d). Thus, it is possible to fine-tune the phase composition of precipitated $CaSO_4$ powders in a one-step process at ambient conditions simply by adjusting the water content of the solvent medium – yielding pure anhydrite at

<0.2 wt% H_2O, pure bassanite between 2 and 30 wt% H_2O, and pure gypsum at >50 wt% H_2O – while at intermediate values mixtures with defined compositions are obtained.

12.3.2 *Crystallization of Bassanite and Anhydrite*

Compared to gypsum, relatively little is known about the early stages of the crystallization of bassanite and anhydrite from aqueous solution. Several studies have focused on establishing the temperature and salinity regions where these phases form spontaneously, mainly by characterizing the outcome of direct precipitation experiments (i.e., mixing Ca^{2+}- and SO_4^{2-}-containing solutions at different conditions) (Cruft and Chao 1970; Ossorio et al. 2014). Although solubility measurements predict anhydrite as the stable phase above 40–60 °C (cf. Fig. 12.2a and Sect. 12.2.2 above), gypsum was always the major product obtained up to 90 °C, while at higher temperatures mainly bassanite was formed (likely due to the kinetic inhibition of anhydrite mentioned above). Interestingly, the temperature for the primary precipitation of bassanite can be significantly lowered by increasing the ionic strength of the reacting solutions, in particular by using highly concentrated electrolyte solutions (Cruft and Chao 1970; Jiang et al. 2013; Ossorio et al. 2014). For example, bassanite is formed already at 80 °C in the presence of 4.5 M NaCl (Ossorio et al. 2014), while a temperature of around 50 °C is sufficient to produce pure hemihydrate from solutions containing >6 M $CaCl_2$ and small amounts of Na_2SO_4 (Cruft and Chao 1970). This again hints at the importance of hydration effects during the crystallization process, which are likely to change substantially as the activity of water is modulated at high salt contents. Another observation pointing in the same direction is the fact that bassanite forms spontaneously during the evaporation of droplets of $CaSO_4$ solutions at room temperature (Qian et al. 2012; Shahidzadeh et al., 2015). In those cases, the relative humidity ($RH < \sim 80\,\%$) and the time-dependent availability of water – potentially along with confinement effects during the evaporation process – seem to be the controlling factors for phase selection. Although the conditions needed to induce bassanite formation in purely aqueous systems appear to be fairly well known, so far distinct insight into the details of the nucleation mechanisms and pathways is still missing.

To the best of our knowledge, there is only one published study that reports on the kinetics of direct anhydrite precipitation in aqueous media at temperatures between 100 and 200 °C and at a constant salt content of 1 M NaCl (Fan et al. 2010). Although no precipitation was observed at 118 and 148 °C, induction times could be measured as a function of supersaturation at 178 °C. Based on these results, the authors estimated the effective surface free energy of anhydrite to be 92.9 mJ/m^2. This value is roughly twice as high as that for gypsum and explains why anhydrite does not readily form in its own thermodynamic stability region (Ossorio et al. 2014). Since the nucleation rate depends strongly (to the power of 3 according to CNT) on the surface free energy of the emerging mineral phase, it is reasonable that gypsum (~40 mJ/m^2) and bassanite (~9 mJ/m^2) (Guan et al. 2010) form much more readily than anhydrite across large parts of the phase diagram (Ossorio et al. 2014). Similar to the case of bassanite, the presence of high concentrations of salt in the reacting solutions can facilitate the formation of anhydrite. This was recently shown by Dixon et al. (2015), who studied the dissolution of jarosite, $(K,Na,H_3O)Fe_3(SO_4)_2(OH)_6$, in saturated $CaCl_2$ brines under ambient conditions. In their experiments, the release of sulfate and the high concentration of calcium in the medium inevitably led to precipitation of calcium sulfates. Interestingly, the resulting phase composition varied depending on the particular procedure applied, with experiments conducted in a batch reactor (i.e., a closed system) yielding a mixture of gypsum and bassanite, whereas in a flow-through setup the main product was anhydrite. This is the first experimental evidence that anhydrite can spontaneously nucleate in aqueous environments at room temperature (i.e., without organic (co)solvents), confirming the previously predicted stability region of anhydrite in high-salinity brines (Bock 1961; Hardie 1967). Initial nucleation of gypsum and subsequent transformation into anhydrite represent an unlikely scenario under these conditions, due to the slow dehydration rates of gypsum near the predicted anhydrite-gypsum transition temperature (~18 °C at the given salinity) (Hardie 1967). Again, one could argue that water activity and corresponding changes in the hydration of solute precursors are crucial for phase selection in the $CaSO_4$ system – with very high salt contents leading to anhydrite, much like extremely low water contents caused anhydrite formation in the work of Tritschler et al. (2015b).

12.3.3 Influence of Additives on Nucleation and Phase Transformation

Based on numerous studies for a variety of mineral systems, it is a well-established fact that both organic and inorganic additives can have a substantial influence on the outcome of crystallization processes, in some cases already at very low concentrations (e.g., Gebauer et al. 2009). Although this subject has not been frequently addressed for the calcium sulfate system in the past, there are a few interesting observations, which shall be summarized in the following.

Probably the most compelling example is found in biomineralization, where two classes of medusae use bassanite for their gravitational sense (Tienmann et al. 2002; Becker et al. 2005; Boßelmann et al. 2007). The *tour de force* achieved by these organisms is twofold: (I) they induce the nucleation, growth, and stabilization of well-defined macroscopic bassanite crystals at room temperature – something that in the lab can only be realized above 50 °C and at extreme high salinity (see Sect. 12.3.2) (Cruft and Chao 1970; Ossorio et al. 2014); (II) they manage to stabilize this highly hygroscopic material in a water-rich environment – usually bassanite transforms immediately into gypsum when put in contact with moisture (explaining its scarcity on the Earth's surface). At present any conclusive mechanistic insights into how this is achieved are lacking, but it is reasonable to assume that molecules of biological origin (such as proteins, peptides, or polysaccharides) and/or confinement effects may be responsible for this impressive level of control. Support for this hypothesis has been reported by Tartaj et al. (2015), who performed $CaSO_4$ mineralization in carboxy- and amino-functionalized reverse micelles and observed that strong binding between calcium and carboxylate groups can lead to long-term (up to 5 months) stabilization of nano-bassanite.

Already in 1980, Cody and Hull studied the influence of organic crystallization inhibitors on the precipitation pathway of calcium sulfate close to the gypsum/anhydrite transition temperature (i.e., 60 °C) and at moderate salinity (Cody and Hull 1980). First, they tested the effect of a range of different additives in diffusion-controlled crystallization experiments, where in the absence of any additives only gypsum was formed (over a remarkable period of up to 4 years). The presence of polycarboxylates and phosphate esters changed the picture completely, as anhydrite was the only product detected after 60 days. In a second set of controlled mixing experiments, anhydrite was formed both by direct precipitation from solution and via transformation of initially nucleated gypsum and/or bassanite, again under the influence of the mentioned organic species. The authors proposed that the primary role of the polymeric additives was to inhibit gypsum and bassanite crystallization, rather than accelerating anhydrite nucleation. However, the particular mechanism and the question whether (and under which conditions) the formation of anhydrite involves a precursor remained unresolved. Nevertheless, it seems clear that certain polymers can promote anhydrite crystallization at relatively low temperatures, which is not possible in pure $CaSO_4$ solutions.

Along with the more recent finding that amorphous calcium sulfate and/or bassanite may precede the formation of gypsum in aqueous solutions at ambient conditions (Wang et al. 2012; Van Driessche et al. 2012; Saha et al. 2012; Jones 2012), the influence of additives on the precipitation pathway and the (meta)stability of the different precursors received further attention. In the cryo-TEM work of Saha et al. (2012), added citrate ions were found to stabilize the initial disordered phase and thus delayed its transformation into gypsum. Wang and Meldrum (2012) performed another study in which $CaSO_4$ precipitation was carried out in the presence of polyacrylic acid (PAA), polystyrene sulfonate (PSS), sodium triphosphate, and Mg^{2+} ions. In essence, the results suggested that all these additives, except PSS, can retard the transformation of both amorphous calcium sulfate and bassanite to a greater or lesser extent. For instance, 200 μg/mL PAA were found to increase the lifetime of ACS and bassanite particles to a few minutes and more than 3 days, respectively. This effect was also observed by Rabizadeh et al. (2014), who showed that even at lower concentrations (5–20 μg/mL) carboxylic acids can change the induction times and stabilize bassanite prior to its transformation to gypsum. In the work of Wang and Meldrum (2012), sodium triphosphate showed similar behavior at 20 μg/mL, while the presence of magnesium (at a Mg^{2+}/Ca^{2+} ratio of 2:1) stabilized mainly bassanite (up to 2 days). Although this appears to be a promising concept to control $CaSO_4$ mineralization, we note that the procedure used in these studies for isolation and analysis of the precursor phases from solution involved quenching in ethanol – a step that has proven to be critical and may affect phase composition or even induce the formation of structures that were actually not present before (Tritschler et al. 2015a, b). A different approach to the stabilization of ACS and bassanite was reported by Nissinen et al. (2014), who dissolved calcium sulfate in *N*-methylmorpholine *N*-oxide (NMMO) along with cellulose at 80 °C and prepared thin films by spin-coating and subsequent hydration. The resulting $CaSO_4$ phase was determined by the rate of hydration – once more highlighting the key role of water availability in the precipitation process.

12.3.4 Toward a General Model for $CaSO_4$ Precipitation from Solution

In the previous sections, we have discussed a number of early and more recent studies on the crystallization of calcium sulfate from solution. While at first sight there seems to be quite some disagreement between the various reported observations, closer examination and consideration of the details intrinsic to each work suggest that all the evidence is converging into a unified general picture of $CaSO_4$ mineralization. In Fig. 12.6 we have sketched out our opinion about the key stages along the pathway from homogeneous solutions to crystalline phases.

In undersaturated solutions, calcium and sulfate occur both as individual ions and ion pairs (Garrels and Christ 1965). Whether they also form larger species, akin to the pre-nucleation clusters suggested for other salt systems, is still unclear and the respective equilibrium constants of ion association remain to be determined. The evidence so far suggests that when supersaturation is established, primary nanosized units are formed by the assembly of multiple ions into rather well-defined elongated entities, the structure of which can be related to any of the possible crystalline phases. These primary units thus represent sort of a universal $CaSO_4$ proto-structure (see Fig. 12.4). With time and through fluctuations, these primary units begin to aggregate and condense, yielding domains that initially are mostly disordered (possibly corresponding to amorphous calcium sulfate, although this is still debated). These domains consist of the primary units separated by appreciable amounts of solvent. In order to transform into well-ordered sheets of $CaSO_4$ cores with more (gypsum), less (bassanite), or no (anhydrite) interspersed H_2O layers, the disordered precursors must reorganize by progressive coalescence of the primary species into larger units. During this internal rearrangement, there will be local domains that adopt configurations that are not (yet) fully hydrated and ordered, especially during the early stages of the transformation process. Therefore, the local availability of water and the particular degree of hydration of the precursor units

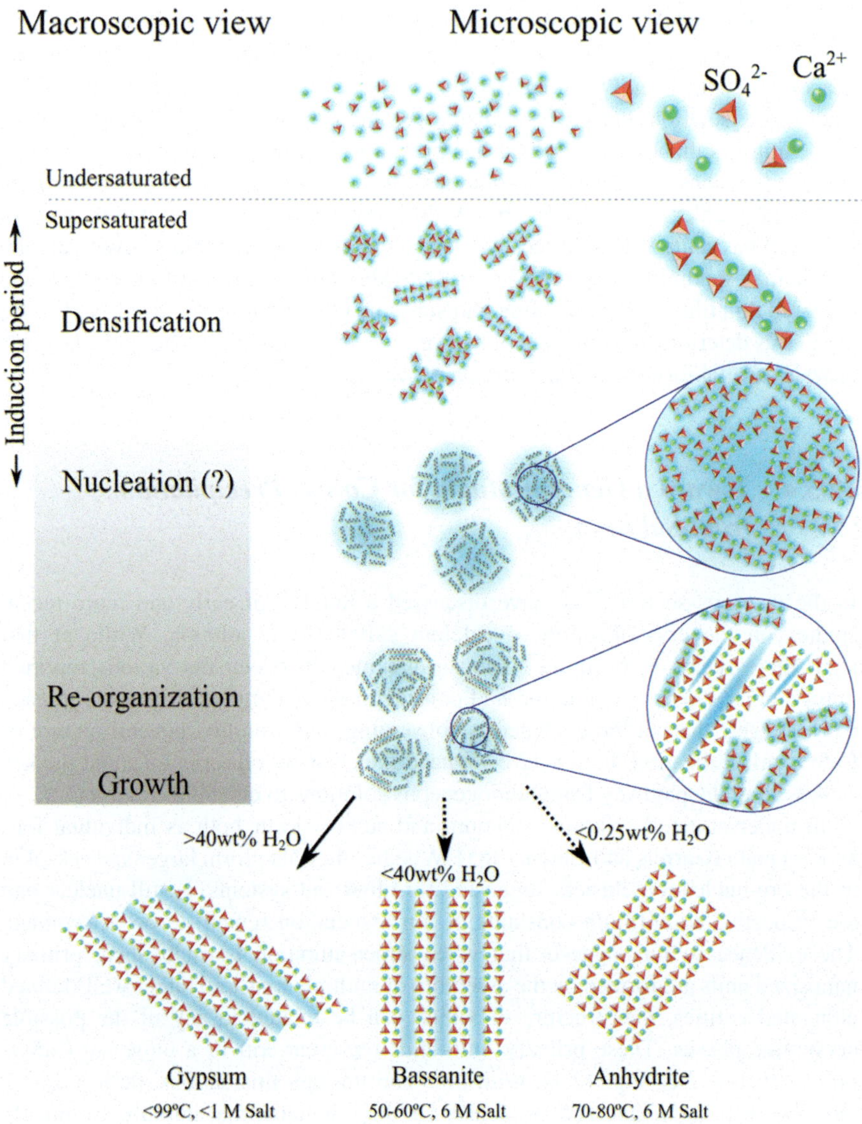

Fig. 12.6 Proposed mechanism for the nucleation and early growth of calcium sulfate from solution. Note that the nature of the formed crystalline phase essentially depends on the availability of water in the medium (respective water contents are quoted) and the hydration of the precursor units, which is further influenced by temperature and salinity (tentative conditions are indicated)

control which crystalline phase is favored under the given conditions. As both of these factors are inherently time-dependent, kinetics are likely to have a strong impact on the final outcome. Once structural rearrangement has afforded an ordered lattice (i.e., the crystalline state has nucleated), growth takes over and rapidly yields larger particles due to the high concentration of (still disordered) $CaSO_4$ units in the local microenvironment. Nevertheless, the actual mechanism of growth remains to be clarified.

Based on this precipitation scheme, most of the recent (and older) experimental evidence can be rationalized. Beyond that, this pathway reveals an intriguing strategy to guide the precipitation process in the calcium sulfate system, namely, by controlling the available amount of water or, respectively, by modulating the hydration of the precursor units. Using defined mixtures of organic solvents and water is a viable approach to achieve this, as demonstrated in the work of Tritschler et al. (2015a, b). The importance of kinetics in the process has been exemplified by the experiments of Nissinen et al. (2014), where the differences in the hydration rate in pure water and water/ethanol mixtures proved to determine phase selection. Apart from the use of organic solvents, the hydration of the precursor units can also be influenced in purely aqueous media, namely, by changing salinity and/or temperature. Increasing any of the two parameters is expected to decrease the degree of hydration, as higher temperatures should favor the return of hydration water into the bulk (for entropic reasons) (Paula et al. 1995; Damasceno et al. 2012; Kellermeier et al. 2016), while microscopic osmotic forces should also remove water from the precursors at high salt content (Di Tommaso et al. 2014). This hypothesis is supported by the well-established trend that bassanite replaces gypsum as the primary phase formed at high enough temperatures and/or salinities (cf. Sect. 12.3.2) (Cruft and Chao 1970; Jiang et al. 2013; Ossorio et al. 2014). Moreover, it may explain the kinetic inhibition of anhydrite crystallization observed under various conditions: regardless of any thermodynamic stability fields, it is very difficult to remove enough water from the precursors in an aqueous environment for anhydrite to be able to compete with bassanite and gypsum. Drastically decreasing the availability of water – for example, by working in organic solvents that contain only trace amounts of H_2O (Tritschler et al. 2015b) or by increasing salinity to extreme values (Cruft and Chao 1970) – can change the situation and induce direct anhydrite formation, although the particular precipitation procedure seems to play an important role here as well (Dixon et al. 2015). Finally, it is feasible that additives such as organic polymers (Cody and Hull 1980; Wang and Meldrum 2012) or Mg^{2+} ions (Wang and Meldrum 2012) also act upon the initial process of phase selection and/or the later kinetic stability of nucleated phases by modulating local hydration phenomena. All these hypotheses await confirmation by further studies.

12.4 Outlook

Although significant advances have been made in our understanding of the precipitation of solid phases in the $CaSO_4$–H_2O system, there is still a considerable dearth of basic information that is crucial for a complete and comprehensive model of nucleation, growth, and transformation of the different $CaSO_4$ phases. In our opinion, the following key issues should be addressed in future work:

- Unambiguous determination of both bulk and nanoparticle solubilities (and thus relative stabilities) for the different calcium sulfate phases
- Actual measurements of the degree of hydration at different precursor stages under various conditions to verify the role of water in the process of phase selection, along with the influence of additives and other factors like confinement
- Characterization of ion association phenomena in undersaturated solutions with respective equilibrium constants, existence (or not) of $CaSO_4$ pre-nucleation clusters (i.e., larger species beyond ion pairs), relation of any such clusters to the elongated primary species occurring in supersaturated solutions
- Determination of the actual structure and (meta)stability of the primary species and the driving force triggering their initial aggregation and later crystallization to one or the other $CaSO_4$ phase
- Identification of the actual nucleation step in the $CaSO_4$–H_2O system (i.e., the stage where an interface emerges), distinction of the observed structures into solution species (pre-nucleation) and solid particles (post-nucleation)
- More elaborate study of the nucleation pathways of anhydrite and bassanite to confirm, or refute, the hypotheses made above
- Confirming the existence of a truly disordered (ACS) phase and characterizing its properties with respect to parameters like local order, solubility, density, etc.
- Clarifying the role of the observed primary species in the growth of macroscopic calcium sulfate crystals
- Verification of the relevance of the mechanisms described above for $CaSO_4$ formation under real application conditions (e.g., for the case of construction materials or scaling from hard water)
- Comparison of the precipitation pathways found for calcium sulfate with those of other important sulfate minerals, such as $BaSO_4$ and $SrSO_4$

While obviously much is left to be unveiled, the progress made over the past few years promises deeper insights in the near future that will hopefully help to improve existing calcium sulfate-based materials and possibly become integrated into a still larger unified picture of crystallization mechanisms from solution in general.

Acknowledgments The authors thank Dr. Alejandro Fernandez-Martinez (ISTerre, France) and Dr. Luc Nicoleau (BASF) for valuable discussions. TMS and LGB were funded by a Helmholtz Recruiting Initiative Grant to LGB for this work.

References

Abriel W (1983) Calcium sulfate subhydrate, $CaSO_4.0,8H_2O$. Acta Crystallogr Sect C 39:956–958

Ahmed SB, Tlili M, Amor MB, Bacha HB, Eullech B (2004) Calcium sulphate scale prevention in a desalination unit using the SMCEC technique. Desalination 167:311–318

Alimi F, Elfil H, Gadri A (2003) Kinetics of the precipitation of calcium sulfate dihydrate in a desalination unit. Desalination 157:9–16

Allen RD, Kramer H (1953) Occurrence of bassanite in two desert basins in southeastern California. Am Mineral 38:1266–1268

Andreassen JP, Lewis AE (2017) Classical and nonclassical theories of crystal growth. In: Van Driessche AES, Kellermeier M, Benning LG, Gebauer D (eds) New perspectives on mineral nucleation and growth, Springer, Cham, pp 137–154

Apokodje EG (1984) The occurrence of bassanite in some Australian arid-zone soils. Chem Geol 47:361–364

Azimi G, Papangelakis VG (2010) The solubility of gypsum and anhydrite in simulated laterite pressure acid leach solutions up to 250 °C. Hydrometallurgy 102:1–13

Azimi G, Papangelakis VG, Dutrizac JE (2007) Modelling of calcium sulphate solubility in concentrated multi-component sulphate solutions. Fluid Phase Equilib 260:300–315

Becker R, Doring W (1935) Kinetic treatment of grain-formation in super-saturated vapours. Ann Phys 24:719–752

Becker A, Sötje I, Paulmann C, Beckmann F, Donath T, Boese R, Prymak O, Tiemann H, Epple M (2005) Calcium sulphate hemihydrate is the inorganic mineral in statoliths of scyphozoan medusae (Cnidaria). Dalton Trans 8:1545–1550

Birkedal H (2017) Phase transformations in calcium phosphate crystallization. In: Van Driessche AES, Kellermeier M, Benning LG, Gebauer D (eds) New perspectives on mineral nucleation and growth, Springer, Cham, pp 199–210

Block J, Waters OB (1968) The CaSO4-Na2SO4-NaCl –H2O system at 25 to 100 °C. J Chem Eng Data 13:336–344

Blount CW, Dickson FW (1969) The solubility of anhydrite ($CaSO_4$) in NaCl–H_2O from 100 to 450°C and 1 to 1000 bars. Geochim Cosmochim Acta 33:227–245

Blount CW, Dickson FW (1973) Gypsum-anhydrite equilibria in systems $CaSO_4$ –H_2O and $CaSO_4$ –NaCl–H_2O. Am Mineral 58:323–331

Boßelmann F, Epple M, Sötje I, Tiemann H (2007) Statoliths of calcium sulfate hemihydrate are used for gravity sensing in rhopaliophoran medusae (Cnidaria). In: Baeuerlein E (ed) Biomineralisation: biological aspects and atructure formation. Wiley-VCH, Weinheim, pp 261–272

Bock E (1961) On the solubility of anhydrous calcium sulphate and of gypsum in concentrated solutions of sodium chloride at 25 °C. Can J Chem 39:1746–1751

Booth HS, Bidwell RM (1950) Solubilities of salts in water at high temperatures. J Am Chem Soc 72:2567–2575

Boyer-Guillon M (1900) Etude sur la solubilite du sulfate de chaux. Extrait des Annales du Conservatoire des Arts et Metiers, 3e serie, tome II

Cameron FK (1901) Solubility of gypsum in aqueous solutions of sodium chloride. J Phys Chem 5:556–576

Chang LLY, Howie RA, Zussman J (1996) Rock-forming minerals, Vol. 5B: Non-silicates, longman scienti¢c and technical. Harlow, 383 pp

Charola AE, Pühringer J, Steiger M (2007) Gypsum: a review of its role in the deterioration of building materials. Environ Geol 52:339–352

Christiansen JA, Nielsen AE (1952) The interplay between nucleus formation and crystal growth. Z Elektrochem 56:465

Cody RD, Hull AB (1980) Experimental growth of primary anhydrite at low temperatures and water salinities. Geology 8:505–509

Conley RF, Bundy WM (1958) Mechanism of gypsification. Geochim Cosmochim Acta 15:57–72

Corti HR, Fernandez-Prini R (1983) Thermodynamics of solution of gypsum and anhydrite in water over a wide temperature range. Can J Chem 62:484–488

Cruft EF, Chao PC (1970) Nucleation kinetics of the gypsum-anhydrite system. In: 3rd symposium on salt, northern Ohio geological society proceedings, vol 1, pp 109–118

Culberson CH, Lathman G, Bates RG (1978) Solubilities and activity coefficients of calcium and strontium sulfate in synthetic seawater at 0.5 and 25 °C. J Phys Chem 82:2693–2699

D'Ans J (1933) Die Losungsgleichgewichte der Systeme der Salze ozeanischer Salzablagerungen. Verlagsgesellschaft fur Ackerbau, Berlin, p 5

D'Ans J (1968) Der Ubergangspunkt Gips-Anhydrit. Kali Steinsalz 5:109–111

Damasceno PF, Engel M, Glotzer SC (2012) Crystalline assemblies and densest packings of a family of truncated tetrahedra and the role of directional entropic forces. ACS Nano 6:609–614
De Yoreo JJ, Sommerdijk NAJM, Dove PM (2017) Nucleation pathways in electrolyte solutions. In: Van Driessche AES, Kellermeier M, Benning LG, Gebauer D (eds) New perspectives on mineral nucleation and growth, Springer, Cham, pp 1–24
Delgado-López JM, Guagliardi A (2017) Control over nanocrystalline apatite formation: what can the X-ray total scattering approach tell us. In: Van Driessche AES, Kellermeier M, Benning LG, Gebauer D (eds) New perspectives on mineral nucleation and growth, Springer, Cham, pp 211–226
Demichelis R, Raiteri P, Gale JD (2017) Ab Initio modelling of the structure and properties of crystalline calcium carbonate. In: Van Driessche AES, Kellermeier M, Benning LG, Gebauer D (eds) New perspectives on mineral nucleation and growth, Springer, Cham, pp 113–136
Dey A, Bomans PHH, Müller FA, Will J, Frederik PM, de With G, Sommerdijk NAJM (2010) The role of prenucleation clusters in surface-induced calcium phosphate crystallization. Nat Mater 9:1010–1014
Dickson FW, Blount CW, Tunell G (1963) Use of hydrothermal solution equipment to determine the solubility of anhydrite in water from 100 °C to 275 °C and from 1 bar to 1000 bars pressure. Am J Sci 261:61–78
Di Tommaso D, Ruiz-Agudo E, de Leeuw NH, Putnis A, Putnis CV (2014) Modelling the effects of salt solutions on the hydration of calcium ions. Phys Chem Chem Phys 16:7772–7785
Dixon EM, Elwood Madden AS, Hausrath E, Elwood Madden ME (2015) Assessing hydrodynamic effects on jarosite dissolution rates, reaction products, and preservation on Mars. J Geophys Res 120:625–642
Dongan AU, Dogan M, Chan DCN, Wurster DE (2005) Bassanite from *Salvadora persica*: a new evaporitic biomineral. Carbonates Evaporites 20:2–7
Droeze H (1877) Solubility of gypsum in water and in saline solutions. Bericht d. deutsch, chem. Ges. in Berlin. 10:330–343
Fan C, Kan AT, Fu G, Tomson MB (2010) Quantitative evaluation of calcium sulfate precipitation kinetics in the presence and absence of scale inhibitors. SPE J 15:977–988
Falini G, Fermani S (2017) Nucleation and growth from a biomineralization perspective. In: Van Driessche AES, Kellermeier M, Benning LG, Gebauer D (eds) New perspectives on mineral nucleation and growth, Springer, Cham, pp 185–198
Fernandez-Martinez A, Lopez-Martinez H, Wang D (2017) Structural characteristics and the occurrence of polyamorphism in amorphous calcium carbonate. In: Van Driessche AES, Kellermeier M, Benning LG, Gebauer D (eds) New perspectives on mineral nucleation and growth, Springer, Cham, pp 77–92
Freyer D, Voigt W (2003) Crystallization and phase stability of $CaSO_4$ and $CaSO_4$-based salts. Monatsh Chem 134:693–719
Garcia-Ruiz JM, Villasuso R, Ayora C, Canals A, Otalora F (2007) Formation of natural gypsum megacrystals in Naica, Mexico. Geology 35:327–330
Garrels RM, Christ CL (1965) Solutions, minerals and equilibria. Harper and Row, New York, 450 p
Gebauer D, Völkel A, Cölfen H (2008) Stable prenucleation calcium carbonate clusters. Science 322:1819
Gebauer D, Cölfen H, Verch A, Antonietti M (2009) The multiple roles of additives in $CaCO_3$ crystallization: a quantitative case study. Adv Mater 21:435
Guan B, Yang L, Wu Z (2010) Effect of Mg^{2+} ions on the nucleation kinetics of calcium sulfate in concentrated calcium chloride solutions. Ind Eng Chem Res 49:5569–5574
Hall RE, Robb JA, Coleman CE (1926) The solubility of calcium sulfate at boiler-water temperatures. J Am Chem Soc 48:927–938
Hamad S el D (1985) The transition temperature of gypsum and anhydrite. Sudan J Sci 1:48
Hamdona SK, Nessim RB, Hamza SM (1993) Spontaneous precipitation of calcium sulfate dihydrate in the presence of some metal ions. Desalination 94:69–80

Hardie LA (1967) The gypsum-anhydrite equilibrium at one atmosphere pressure. Am Mineral 52:171–200

He S, Oddo JE, Tomson MB (1994) The nucleation kinetics of calcium sulfate dihydrate in NaCl solutions up to 6 m and 90°C. J Collloid Interface Sci 162:297–303

Hill AE (1937) The transition temperature of gypsum to anhydrite. J Am Chem Soc 59:2242–2244

Hill AE, Wills JH (1938) Ternary systems. XXIV. Calcium sulfate, sodium sulfate and water. J Am Chem Soc 60:1647–1655

Hulett GA, Allen LE (1902) The solubility of gypsum. J Am Chem Soc 24:667–679

Innorta G, Rabbi E, Tomadin L (1980) The gypsum-anhydrite equilibrium by solubility measurements. Geochim Cosmochim Acta 44:1931–1936

Jiang G, Mao J, Fu H, Zhou X, Guan B (2013) Insight into metastable lifetime of a-calcium sulfate hemihydrate in $CaCl_2$ solution. J Am Ceram Soc 96:3265–3271

Jones F (2012) Infrared investigation of barite and gypsum crystallization: evidence for an amorphous to crystalline transition. CrystEngComm 14:8374–8381

Kashchiev D (2000) Nucleation: basic theory with applications. Butterworth-Heinemann, Oxford

Kellermeier M, Picker A, Kempter A, Cölfen H, Gebauer D (2014) A straightforward treatment of activity in aqueous $CaCO_3$ solutions and the consequences for nucleation theory. Adv Mater 26:752

Kellermeier M, Raiteri P, Berg JK, Kempter A, Gale JD, Gebauer D (2016) Entropy drives calcium carbonate ion association. Chem Phys Chem. doi:10.1002/cphc.201600653

Kelley KK, Southard JC, Anderson CT (1941) Thermodynamic properties of gypsum and its dehydration products, vol 625, US bureau mines technical paper. U.S. G.P.O, Washington, DC

Klepetsanis PG, Koutsoukos PG (1989) Precipitation of calcium sulfate dihydrate at constant calcium activity. J Cryst Growth 98:480

Klepetsanis PG, Dalas E, Koutsoukos PG (1999) Role of temperature in the spontaneous precipitation of calcium sulfate dihydrate. Langmuir 15:1534–1540

Knacke O, Gans W (1977) The thermodynamics of the system $CaSO_4$–H_2O. Z Phys Chem NF 104:41–48

Kontrec J, Kralj D, Breèeviæ L (2002) Transformation of anhydrous calcium sulphate into calcium sulphate dihydrate in aqueous solutions. J Cryst Growth 240:203–211

Lager GA, Armbruster T, Rotella FJ, Jorgensen JD, Hinks DG (1984) A crystallographic study of the low-temperature dehydration products of gypsum, $CaSO_4.2H_2O$: hemihydrate $CaSO_4.0.5H_2O$, and γ-$CaSO_4$. Am Mineral 69:910–918

Lancia A, Musmarra D, Prisciandaro M (1999) Measuring induction period for calcium sulfate dihydrate precipitation. AIChE J 45:390–397

Langevin Y, Poulet F, Bibring JP, Gondet B (2005) Sulfates in the north polar region of Mars detected by OMEGA/Mars express. Science 307:1584–1586

Lewry AJ, Williamson J (1994) The hydration of calcium sulphate hemihydrate. J Mat Sci 29:5279–5284

Ling Z, Wang A (2015) Spatial distributions of secondary minerals in the Martian meteorite MIL 03346,168 determined by Raman spectroscopic imaging. J Geophys Res. doi:10.1002/2015JE004805

Liu ST, Nancollas GH (1973) Linear crystallization and induction-period studies of the growth of calcium sulphate dihydrate crystals. J Colloid Interface Sci 44:422

Lu H, Kan A, Zhang P, Yu J, Fan C, Work S, Tomson MB (2012) Phase stability and inhibition of calcium sulfate in the system NaCl/Monoethylene Glycol/H2O. SPE J 17:187–198

Madgin WM, Swayles DA (1956) Solubilities in the system $CaSO_4$-NaCl-H_2O at 25° and 35° C. J Appl Chem 6:482–487

Marignac C (1874) Ueber die Löslichkeit des schwefelsauren kalkes in wasser. Z Anal Cam 13: 57–59

Marshall WL, Slusher R (1966) Thermodynamics of Calcium Sulfate Dihydrate in Aqueous sodium chloride solutions, 0– 110°. J Phys Chem 70:4015–4027

Marshall WL, Slusher R, Jones EV (1964) Aqueous systems at high temperature. XIV. Solubility and thermodynamic relationships for CaSO4 in NaCl-H2O solutions from 40 °C to 200 °C, 0 to 4 molal NaCl. J Chem Eng Data 9:187
Melcher AC (1910) The solubility of silver chloride, barium sulphate, and calcium sulphate at high temperatures. J Am Chem Soc 32:50–66
Mi B, Elimelech M (2010) Gypsum scaling and cleaning in forward osmosis: measurements and mechanisms. Environ Sci Technol 44:2022–2028
Momma K, Izumi F (2011) VESTA 3 for three-dimensional visualization of crystal, volumetric and morphology data. J Appl Crystallogr 44:1272–1276
Monnin C (1990) The influence of pressure on the activity-coefficients of the solutes and on the solubility of minerals in the system Na-Ca- Cl-SO_4-H_2O to 200°C and 1 kbar, and to high NaCl concentration. Geochim Cosmochim Acta 54:3265–3282
Navrotsky A (2004) Energetic clues to pathways to biomineralization: precursors, clusters, and nanoparticles. Proc Natl Acad Sci U S A 101:12096–12101
Neville A (2004) The confused world of sulphate attack on concrete. Cem Concr Res 34:1275–1296
Nielsen AE (1964) Kinetics of precipitation. Pergamon Press, Oxford
Nissinen T, Li M, Davis SA, Mann S (2014) In situ precipitation of amorphous and crystalline calcium sulphates in cellulose thin films. CrystEngComm 16:3843–3847
Osinski GR, Spray JG (2003) Impact-generated carbonate melts: evidence from the Haughton structure, Canada. Earth Planet Sci Lett 194:17–29
Ossorio M, Van Driessche AES, Pérez P, García-Ruiz JM (2014) The gypsum-anhydrite paradox revisited. Chem Geol 386:16–21
Packter A (1971) The precipitation of calcium sulphate dihydrate from aqueous solution induction period, crystal numbers and final size. J Cryst Growth 21:191
Parkhurst DL, Appelo CAJ (1999) User's guide to PHREEQC (version 2). A computer program for speciation, batch-reaction, one-dimensional transport, and inverse geochemical calculations. Water Resour Invest Rep US Geol Surv 99(4259):4312
Partridge E, White AH (1929) The solubility of calcium sulfate from 0 to 200°C. J Am Chem Soc 51:360–370
Paula S, Süs W, Tuchtenhagen J, Blume A (1995) Thermodynamics of micelle formation as a function of temperature: a high sensitivity titration calorimetry study. J Phys Chem 99:11742–11751
Peckmann J, Goedert JL, Heinrichs T, Hoefs J, Reitner J (2003) The late eocene 'Whiskey Creek' methane-seep deposit (western Washington State)—part II: petrology, stable isotopes, and biogeochemistry. Facies 48:241–254
Penn RL, Li D, Soltis JA (2017) A perspective on the particle-based crystal growth of ferric oxides, oxyhydroxides, and hydrous oxides. In: Van Driessche AES, Kellermeier M, Benning LG, Gebauer D (eds) New perspectives on mineral nucleation and growth, Springer, Cham, pp 257–274
Poggiale M (1843) Memoire sur la solubilite des sels dans l'eau. Ann Chim Phys 3:463–478
Posnjak E (1938) The system $CaSO_4$–H_2O. Am J Sci 35:247–272
Power WH, Fabuss BM, Satterfield CN (1964) Transient solubilities in the calcium sulfate-water system. J Chem Eng Data 9:437
Power WH, Fabuss BM, Satterfield CN (1966) Transient solute concentrations and phase changes of calcium sulfate in aqueous sodium chloride. J Chem Eng Data 11:149–154
Prisciandaro M, Lancia A, Musmarra D (2001a) Calcium sulphate dihydrate nucleation in the presence of calcium and sodium chloride salts. Ind Eng Chem Res 40:2335–2339
Prisciandaro M, Lancia A, Musmarra D (2001b) Gypsum nucleation into sodium chloride solutions. AIChE J 47:929–934
Prisciandaro M, Lancia A, Musmarra D (2003) The retarding effect of citric acid on calcium sulfate nucleation kinetics. Ind Eng Chem Res 42:6647–6652
Qian Z, Wang F, Zheng Y, Yu J, Zhang Y (2012) Crystallization kinetics of sea-salt aerosols studied by high-speed photography. Chin Sci Bull 57:591–594

Rabizadeh T, Peacock CL, Benning LG (2014) Carboxylic acids: effective inhibitors for calcium sulfate precipitation. Mineral Mag 78:1465–1472
Raju KUG, Atkinson G (1990) The thermodynamics of "scale" mineral solubilities. 3. Calcium sulfate in aqueous NaCl. J Chem Eng Data 35:361
Rao A, Cölfen H (2017) Mineralization schemes in the living world: mesocrystals. In: Van Driessche AES, Kellermeier M, Benning LG, Gebauer D (eds) New perspectives on mineral nucleation and growth, Springer, Cham, pp 155–184
Rashad MM, Mahmoud MHH, Ibrahim IA, Abdel-Aal EA (2004) Crystallization of calcium sulfate dihydrate under simulated conditions of phosphoric acid production in the presence of aluminum and magnesium ions. J Cryst Growth 267:372–379
Raupenstrauch GA (1885) Uber die Bestimmungder loslichkeit einiger salze in wasser bei verschiedenen temperaturen. Monatsh Chem 6:563–591
Reichel V, Faivre D (2017) Magnetite nucleation and growth. In: Van Driessche AES, Kellermeier M, Benning LG, Gebauer D (eds) New perspectives on mineral nucleation and growth, Springer, Cham, pp 275–292
Roller PS (1931) Chemical activity and particle size. The rate of solution of anhydrite below 70 microns. J Phys Chem 35:1132–1142
Rodriguez-Blanco JG, Sand KK, Benning LG (2017) ACC and vaterite as intermediates in the solution-based crystallization of $CaCO_3$. In: Van Driessche AES, Kellermeier M, Benning LG, Gebauer D (eds) New perspectives on mineral nucleation and growth, Springer, Cham, pp 93–112
Rouchy JM, Monty C (2000) Gypsum microbial sediments: neogene and modern examples. In: Riding RE, Awramik SM (eds) Microbial sediments. Springer, Berlin, pp 209–216
Ryan WBF (2009) Decoding the Mediterranean salinity crisis. Sedimentology 56:95–136
Saha A, Lee J, Pancera SM, Bräeu MF, Kempter A, Tripathi A, Bose A (2012) New insights into the transformation of calcium sulphate hemihydrate to gypsum using time-resolved cryogenic transmission electron microscopy. Langmuir 28:11182–11187
Satava V (1970) Ist α- oder β-gips better. Sprechsaal Keramik, Glas, Email 103:792–798
Schierholtz OJ (1958) The crystallization of calcium sulphate dihydrate. Can J Chem 36:1057–1063
Shahidzadeh N, Schut MFL, Desarnaud J, Prat M, Bonn D (2015) Salt stains from evaporating droplets. Sci Rep 5:10335
Sharpe R, Cork G (2006) Gypsum and anhydrite. In: Kogel JE, Kogel JE et al (eds) Industrial minerals & rocks. Society for Mining, Metallurgy, and Exploration, Inc, Littletown, pp 519–540
Singh NB, Middendorf B (2008) Calcium sulphate hemihydrate hydration leading to gypsum crystallization. Prog Cryst Growth Charact Mater 53:57–77
Smith BR, Sweett F (1971) The crystallization of calcium sulfate dihydrate. J Colloid Interface Sci 37:612–618
Stawski T, Van Driessche AES, Ossorio M, Rodriguez-Blanco JD, Besselink R, Benning LG (2016) Formation of calcium sulfate through the aggregation of sub-3 nanometre primary species. Nat Comm 7:10177
Tartaj P, Morales J, Fernandez-Diaz L (2015) $CaSO_4$ mineralization in carboxy- and amino-functionalized reverse micelles unravels shape-dependent transformations and long-term stabilization pathways for poorly hydrated nanophases (bassanite). Cryst Growth Des 15:2809–2816
Tiemann H, Sötje I, Jarms G, Paulmann C, Epple M, Hasse B (2002) Calcium sulphate hemihydrate in statoliths of deep-sea medusae. J Chem Soc Dalton Trans 7:1266–1268
Tilden WA, Shenstone WA (1984) On the solubility of salts in water at high temperatures. Philos Trans R Soc 175A:31
Tobler DJ, Stawski TM, and Benning LG (2017) Silica and alumina nanophases: natural processes and industrial applications. In: Van Driessche AES, Kellermeier M, Benning LG, Gebauer D (eds) New perspectives on mineral nucleation and growth, Springer, Cham, pp 293–316

Tritschler U, Van Driessche AE, Kempter A, Kellermeier M, Cölfen H (2015a) Controlling the selective formation of calcium sulfate polymorphs at room temperature. Angew Chem Int Ed 54:4083–4086
Tritschler U, Kellermeier M, Debus C, Kempter A, Coelfen H (2015b) A simple strategy for the synthesis of well-defined bassanite nanorods. CrystEngComm 17:3772–3776
Van Driessche AES, Garcia-Ruiz JM, Tsukamoto K, Patino-Lopez LD, Satoh H (2011) Ultraslow growth rates of giant gypsum crystals. Proc Natl Acad Sci U S A 108:15721–15726
Van Driessche AES, Benning LG, Rodriguez-Blanco JD, Ossorio M, Bots P, Gárcia-Ruiz JM (2012) The role and implications of bassanite as a stable precursor phase to gypsum precipitation. Science 336:69–72
Van't Hoff JH, Armstrong EF, Hinrichsen W, Weigert F, Just G (1903) Gips und anhydrit. Z Phys Chem 45:257
Wang YW, Meldrum FC (2012) Additives stabilize calcium sulfate hemihydrate (bassanite) in solution. J Mater Chem 22:22055–22062
Wang YW, Kim YY, Christenson HK, Meldrum FC (2012) A new precipitation pathway for calcium sulfate dihydrate (gypsum) via amorphous and hemihydrate intermediates. Chem Commun 48:504–506
Warren JK (2006) Evaporites: sediments, resources and hydrocarbons. Springer, Berlin
Weiss H, Bräu MF (2009) How much water does calcined gypsum contain? Angew Chem Int Ed 48:3520–3524
Wolf SE, Gower LB (2017) Challenges and perspectives of the polymer-induced liquid-precursor process: the pathway from liquid-condensed mineral precursors to mesocrystalline products. In: Van Driessche AES, Kellermeier M, Benning LG, Gebauer D (eds) New perspectives on mineral nucleation and growth, Springer, Cham, pp 43–76
Wray JJ, Squyres SW, Roach LH, Bishop JL, Mustard JF, Dobrea NEZ (2010) Identification of the Ca-sulfate bassanite in Mawrth Vallis, Mars. Icarus 209:416–421
Yang L-X, Meng Y-F, Yin P, Yang Y-X, Tang Y-Y, Qin L-F (2011) Shape control synthesis of low-dimensional calcium sulfate. Bull Mater Sci 34:233–237
Zen E (1965) Solubility measurements in the system CaSO4– NaCl–H2O at 35° 50° and 70°C and one atmosphere pressure. J Petrol 6:124–164
Zhang H, Banfield JF (2012) Energy calculations predict nanoparticle attachment orientations and asymmetric crystal formation. J Phys Chem Lett 3:2882–2886

Chapter 13
A Perspective on the Particle-Based Crystal Growth of Ferric Oxides, Oxyhydroxides, and Hydrous Oxides

R. Lee Penn, Dongsheng Li, and Jennifer A. Soltis

13.1 Introduction

The diversity of iron oxide, oxyhydroxide, and hydroxide materials in natural settings is remarkable. Hereafter referred to simply as the iron oxides, these materials exhibit myriad textures and morphologies, and such features provide evidence that classical growth cannot adequately explain the formation and growth of iron oxides.

The iron oxides often form as a result of iron leaching from iron-containing minerals (e.g., biotite) through both abiotic and biotic weathering processes (Barker et al. 1998). Furthermore, iron oxides can form when natural water containing Fe(II) encounters oxidizing conditions (e.g., Waychunas et al. 2005). With its abundance in near-surface materials and its redox reactivity, iron plays important roles in biogeochemical cycling of a wide range of species, including metals and molecular species. The capacity of the iron oxides for sorption of metals and polyatomic anions makes these materials important players in the fate and transport of a wide range of contaminants (Waychunas et al. 2005). Elucidating how these minerals form, transform, aggregate, and grow is critical to understanding their geochemical reactivity.

Particle-based crystallization has been featured prominently in the recent crystal growth literature (De Yoreo et al. 2015 and references therein). Of the diverse crystal growth mechanisms known, both classical crystal growth and particle-based crystal growth are particularly important in the iron oxides. Classical crystal growth

R.L. Penn (✉) • J.A. Soltis
Department of Chemistry, University of Minnesota, 207 Pleasant St SE, Minneapolis, MN 55455, USA
e-mail: rleepenn@umn.edu

D. Li
Pacific Northwest National Laboratory, 902 Battelle Boulevard, Richland, WA 99352, USA

A.E.S. Van Driessche et al. (eds.), *New Perspectives on Mineral Nucleation and Growth*,
DOI 10.1007/978-3-319-45669-0_13

can be simply described as the monomer-by-monomer addition of molecular-scale species to a growing crystal. Oriented attachment is a special case of particle-based crystal growth and has been recognized since at least the late nineteenth century (Ivanov et al. 2014). In oriented attachment, primary particles associate to reversibly form complexes that are analogous to the outer sphere complexes described in inorganic chemistry. The primary particles composing these complexes lack direct contact, with solvent molecules and other molecular-scale species residing in the spaces separating them. The primary particles can rearrange and reorient through Brownian motion within this intermediate structure. If the primary particles achieve a common crystallographic orientation, the intermediate structures, which are sometimes referred to as mesocrystals (Cölfen and Mann 2003; Yuwono et al. 2010; Rao and Cölfen 2017, Chap. 8), can either dissociate or irreversibly bond together to form new secondary crystals. These new crystals can have symmetry-defying morphologies and contain defects like dislocations, stacking faults, and twin boundaries (Penn 2004).

Numerous reviews describing crystal growth by oriented attachment have appeared in the relatively recent literature (De Yoreo et al. 2015; Ivanov et al. 2014; Xiong and Tang 2012; Dalmaschio et al. 2010; Zhang et al. 2010; Zhang et al. 2009; Niederberger and Cölfen 2006; Penn 2004), among others. These reviews provide concise descriptions of the fundamental mechanism as well as observations of oriented attachment, reporting numerous examples of oriented attachment in synthetic materials, such as titanium dioxide, iron oxides, metal selenides and sulfides, and more. In addition, evidence for oriented attachment (OA) has been observed in natural environment (Banfield et al. 2000; Hochella et al. 2008; Penn et al. 2001b).

Currently, there is no universal description of how the iron oxides grow. In fact, the iron oxides literature is rife with contradictions, even when observed morphologies, textures, and microstructures are similar. An excellent example is the case of pseudocubic hematite crystals, which have been prepared by several research groups. Kandori et al. (1991) prepared synthetic hematite by aging an acidic solution of ferric chloride at 100 °C for 20 days, and they described their product pseudocubic crystals as polycrystals composed of smaller and oriented subcrystals. They concluded that the pseudocubic hematite formed by aggregation of hydrous ferric oxide crystallites is followed by recrystallization and dehydration. Whether the hydrous ferric oxides precursor particles were oriented prior to recrystallization and dehydration was not addressed (Kandori et al. 1991). Similarly, Sugimoto et al. (1993) described their synthetic pseudocubic hematite particles, which were prepared by aging a partially neutralized solution of ferric chloride at 100 °C for 8 days, as consisting of oriented subcrystals and even stated that their product appeared similar to that of Kandori et al. (1991). However, they stated that single crystals could not be produced by an aggregative mechanism and concluded that their pseudocubic hematite must have grown by addition of monomeric species (e.g., dissolved ferric complexes) from the solution phase. They concluded that the observed textures arose because individual two-dimensional surface nuclei could not progress to form a continuous surface layer due to surface sites blocked by adsorbed species like protons, chloride anions, and chloro ferric complexes (Sugimoto et al. 1993).

Texture, morphology, microstructure, and the absence or presence of defects can serve as clues regarding the crystal growth mechanisms at play. Relics of a particle-mediated crystal growth mechanism can include dimpled surfaces, symmetry-defying morphologies, dislocations, twin boundaries, stacking faults, and internal porosity, and many of these features are often observed in reports describing iron oxide crystal growth and phase transformations. Indeed, features apparent in images of iron oxide particles shown in the classic Cornell and Schwertmann (Cornell et al. 1989; Cornell and Schwertmann 2003) books are often consistent with particle-mediated crystal growth. Recent work has demonstrated the importance of aggregation, including oriented attachment, not only in iron oxide crystal growth (e.g., Burleson and Penn 2006; Burrows et al. 2012; Burrows et al. 2013; Penn et al. 2006) but also during iron oxide phase inter-transformations (e.g., Davidson et al. 2008; Frandsen et al. 2014). Aggregation, including the special case of oriented attachment, can lead to final crystals with rough surfaces as well as the incorporation of pore spaces, defects, stacking faults, all of which can dramatically impact the chemical and physical properties of the iron oxides.

Nucleation of particles with a structure distinct from the thermodynamically most stable phase given the conditions (e.g., pH, temperature) is also common. Phase transformation subsequently occurs to produce the more stable phase at the expense of the initially nucleated phase. An excellent example is goethite produced from six-line ferrihydrite nanoparticles in moderately acidic aqueous suspensions (e.g., Burleson and Penn 2006; Burrows et al. 2012; Burrows et al. 2013; Penn et al. 2006). Forced hydrolysis of dissolved ferric nitrate results in formation of ferrihydrite nanoparticles that are a few nanometers in diameter. With time, the ferrihydrite nanoparticles loosely aggregate into fractal aggregates, within which the primary particles rearrange and reorient to form linear strings of nanoparticles. These linear strings range from one to a few primary particles wide and several to tens of particles long. In this intermediate state, the primary particles lack direct contact with one another, with water and possibly other dissolved species residing in the spaces between them. Further, the fraction of particles residing in the linear strings increases with time (Yuwono et al. 2010), as does the fraction of goethite. At some point, the phase transformation from ferrihydrite to goethite occurs. Based on high-resolution imaging and X-ray diffraction results (Burleson and Penn 2006) in combination with high-resolution imaging of the loose aggregates in vitrified water (Yuwono et al. 2010), the current hypothesis is that phase transformation to goethite precedes rearrangement from random appearing fractal aggregates to linear strings of nanocrystals (Fig. 13.1 upper scheme). Indeed, linear strings with overall morphologies consistent with goethite twins were frequently observed (e.g., Fig. 13.2), with the angle between the crystalline arms residing on either side of the twin boundary matching that observed in twinned goethite crystals (Yuwono et al. 2010).

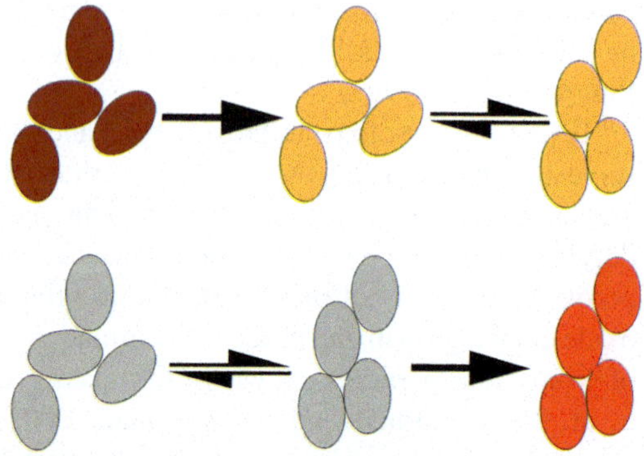

Fig. 13.1 Simple scheme illustrating the two end member possibilities for phase transformation after (*top*) or before (*bottom*) primary particles are crystallographically aligned with respect to one another

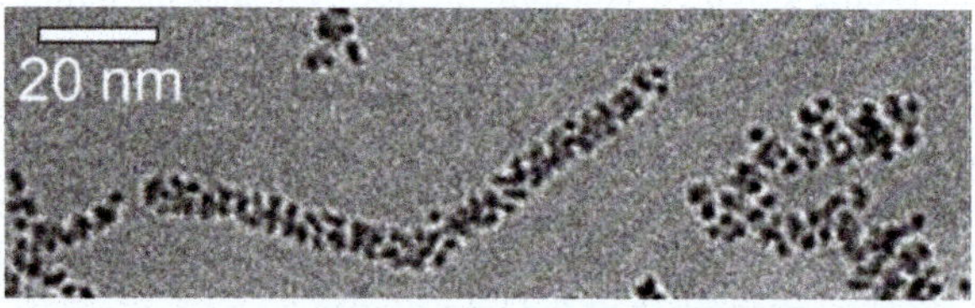

Fig. 13.2 Cryo-TEM image of aggregates of iron oxide nanoparticles. The aggregate with the shape of "v" has a morphology and size that matches those of goethite twins observed in the product material. High-resolution images of similar objects demonstrate lattice fringes that span the entire object and with spacings consistent with the goethite crystal structure (Yuwono et al. 2010)

Alternatively, the phase transformation could occur after a threshold size is achieved (Fig. 13.1 lower scheme). Indeed, size-dependent thermodynamic relationships between initial and product phases can mean that the new phase is not thermodynamically favored until the initially nucleated crystals reach a threshold size (Navrotsky et al. 2008). This results from the interplay between surface and bulk energies, with the material with lower surface energy favored at small size and the material with lower bulk energy favored at larger size. A beautiful example of phase transformation occurring after oriented attachment is hematite produced from oriented aggregates of akaganeite under hydrothermal conditions. Cryo-TEM, in combination with XRD results, leads to the proposal that akaganeite nanorods grew by oriented attachment and that the initial formation of hematite was driven by a change in phase stability due to the increasing size of the akaganeite crystallites. That is to say, once an akaganeite crystallite passed a threshold size, phase transformation was driven by the comparative stability of hematite versus akaganeite (Frandsen et al. 2014). Such size-dependent phase stability has been demonstrated in a number of systems, such as the titanium dioxides (Navrotsky 2011), calcium sulfate (Van Driessche et al. 2012), as well as the iron oxides system (Navrotsky et al. 2008).

In fact, considering the size-dependent phase stability of the iron oxides means reevaluation of the results presented above regarding the growth of goethite from six-line ferrihydrite. Examining the thermodynamic relationships presented by Navrotsky et al. (2008) does not, unfortunately, yield a clear answer regarding the relative phase stability of ferrihydrite compared to goethite at extremely small particle size. Ferrihydrite presents unique challenges because its crystal structure, its hydrated state, and the probability that this material's composition may vary with particle size are not yet well understood. A challenge of the high-resolution transmission electron microscopy (HRTEM) images presented by Burleson and Penn (2006) is that the particles were dried before inserting into the high vacuum of the transmission electron microscope (TEM). Thus, conclusions drawn about the structure of individual particles as observed via HRTEM may not match the structure of individual particles while still suspended in aqueous solutions. Nevertheless, the relative stability for akaganeite and hematite as a function of particle size (Navrotsky et al. 2008) seems consistent with the conclusions described by Frandsen et al. (2014).

13.2 The Critical Role of Materials Characterization

Generally speaking, the final morphology, microstructure, and texture of a crystalline object are path dependent. That is to say, the detailed crystal growth mechanism leads to the presence and/or absence of specific features in the growing crystal. However, the dominant crystal growth mechanism(s) can change over time. Later crystal growth can effectively erase evidence for early crystal growth mechanisms (De Yoreo et al. 2015). The above disparate descriptions of how hematite crystals composed of smaller subcrystals form highlight the challenges of connecting the properties of the final product to its crystal growth history. Interestingly, Sugimoto et al. (1993) stated that drawing conclusions from ex situ TEM images is problematic because the dispersed state of the particles will change upon drying on the TEM grid. Indeed, drawing conclusions based on data obtained once the final crystal has been produced is risky, although sometimes these are the only data available (e.g., crystals formed long ago in natural environments). Thus, there is a pressing need for reliable characterization methods that can enable holistic determination of the path to a final crystal. The most robust approaches will combine methods, both in situ and ex situ, with particular focus on techniques that enable characterization as a function of time (Penn and Soltis 2014).

13.2.1 Imaging Using Transmission Electron Microscopy (TEM)

Direct imaging of nanoparticles using TEM and high-resolution TEM has long been the primary technique for characterizing the often nanoscale features typically associated with particle-based crystal growth mechanisms. Features such as crystal

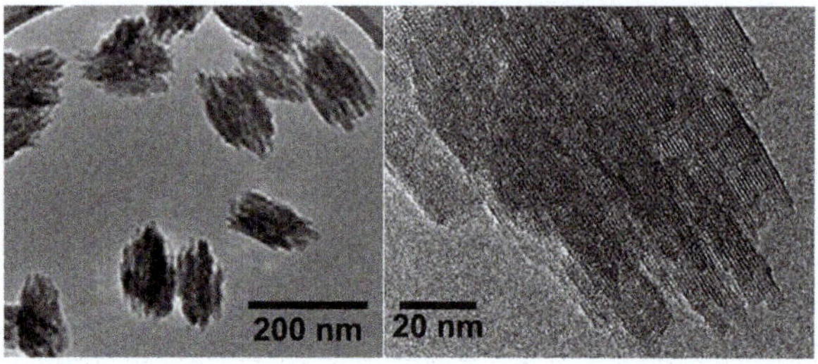

Fig. 13.3 TEM images of silicalite-1 prepared using bis-1,6-(tripropyalammonium)hexamethylene dihydroxide as the structure-directing agent. The sample was prepared after repeated washing and centrifugation to remove dissolved species and amorphous material from the aqueous suspension. The high-resolution image highlights the internal porosity and very complicated microstructure arising from aggregation of structural diverse precursor nanoparticles followed by recrystallization (Kumar et al. 2011)

twinning, dislocations, and dimpled boundaries can often be readily detected and even quantified in TEM images (Penn and Soltis 2014). However, detecting such features does not represent sufficient evidence for concluding a crystal has formed by oriented attachment or another particle-based growth mechanism. Indeed, features like dimples and twins could form by random particle attachment followed by recrystallization (Kumar et al. 2008; Kumar et al. 2011) (Fig. 13.3); alternatively, this can proceed by monomer-by-monomer growth onto surface nuclei that cannot extend beyond some physical or chemical barrier residing on the crystal surface, as hypothesized by Sugimoto et al. (1993). Likewise, the absence of such features is not a conclusive indicator of classical crystal growth, since rough surfaces and defects can essentially be erased when particle-based crystal growth is accompanied or followed by recrystallization or monomer-by-monomer crystal growth. It is therefore critical to employ correlative methods, time-resolved methods, and in situ methods like X-ray scattering and cryogenic TEM (Penn and Soltis 2014).

13.2.2 Cryogenic and Fluid Cell TEM

Cryogenic (cryo) and fluid cell TEM enable direct imaging of particles as they exist suspended in solvents (De Yoreo et al. 2017, Chap. 1; Nielsen and De Yoreo 2017, Chap. 18). This removes the potential for artifacts that can arise due to drying a sample on a TEM grid. The methods differ in that cryo-TEM provides a snapshot of the sample at the instant of cryogenic vitrification, while fluid cell TEM enables capture of the movement of particles in their suspended state. To prepare a sample for cryo-TEM, a thin film of suspension is applied to a TEM grid and then plunged into liquid cryogen (typically liquid ethane for aqueous suspensions) (Burrows and Penn 2013). The rapid cooling (ca. 100,000 K per second) achieved by plunging

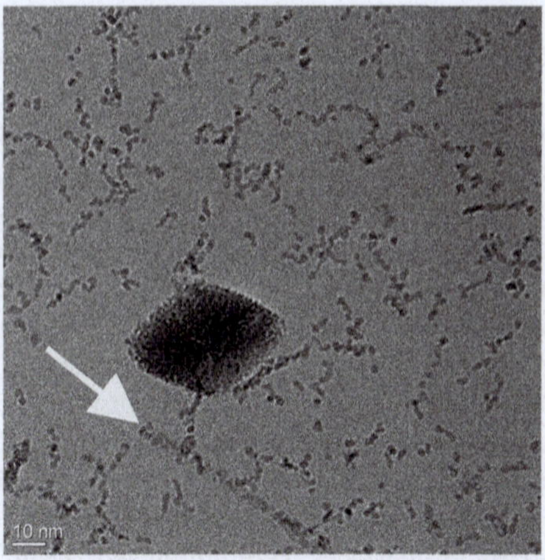

Fig. 13.4 Cryo-TEM image of an aqueous suspension of three types of iron oxide particles: (1) hematite (large *diamond-shaped* particle), (2) linear arrangements of primary particles with size and shape similar to goethite crystals produced by OA, and primary particles that are aggregated (3) or isolated (4). In the lower center of the field of view, the linear arrangement of particles is consistent with the size and shape of final goethite crystals (*white arrow*), and the working hypothesis is that the primary particles within that linear arrangement have the goethite crystal structure

into liquid cryogen ideally serves to preserve the arrangement of objects in the solvent and prevent crystallization of solvent molecules. By maintaining cryogenic conditions during imaging and minimizing the electron dose, one can obtain high-resolution images of crystals as they existed in the liquid state at the instant of vitrification. Figure 13.4 shows a cryo-TEM image of an aqueous suspension aged at 80 °C. Several different nanostructures can be discerned in this image: (1) a large crystal of hematite, with texture consistent with particle-mediated crystal growth; (2) an aggregate of particles with morphology, aspect ratio, and size consistent with product goethite crystals (arrow); (3) fractal aggregates with random orientation of primary particles, which are hypothesized to be ferrihydrite; and (4) isolated primary particles, which are also hypothesized to be ferrihydrite.

To prepare a sample for fluid cell TEM characterization, one need only ensure that the mass loading of particles is suitable for characterization. The two basic options are a static cell and a flow-through cell. An aliquot of suspension is injected into a static cell and the cell sealed, which means only a single aliquot of suspension can be examined with each cell. In the flow-through cell, the cell has both inlet and outlet, and fresh suspension can be flowed into the cell. Both cells are bounded by two electron-transparent windows (Grogan et al. 2012; Penn and Soltis 2014; Nielesen and de Yoreo 2017, Chap. 18). Either way, imaging by fluid cell TEM can yield unprecedented insights into the dynamics of particle-particle interactions, enabling quantitative characterization of particle motion, rearrangement of particles in loosely bound aggregates, crystal growth, and crystal dissolution. Generally speaking, the thickness of the liquid layer is thinner for the static than for the flow-through cell, which means higher resolution and better contrast are obtained with the static cell.

De Yoreo and coworkers successfully captured an instance of oriented attachment between two ferrihydrite crystals in an aqueous solution (Li et al. 2012). They directly imaged two ferrihydrite crystallites moving in suspension. The ferrihydrite crystallites achieved close approach, and the authors were able to directly observe the particles rearranging with respect to one another. A particle-particle attachment occurred once crystallographic alignment was achieved, and continued crystal growth by coarsening immediately after the oriented attachment event was observed (Li et al. 2012). This resulted in smoothening out of the high radius of curvature dimples that formed upon oriented attachment (Li et al. 2012). In addition to quantifying the translational and rotational velocity of the smaller particle, they also quantified the rate of coarsening by measuring the rate at which the curved surfaces at particle-particle interfaces were eliminated. Their experimental rate of coarsening was consistent with the dependence of chemical potential on interface curvature (Cao 2004; Li et al. 2012).

The advantages of obtaining direct images of particles as they exist in liquid suspension by cryo and fluid cell TEM come with some serious challenges. As with all TEM methods, beam damage is a potential and serious complication. Both radiolysis and knock-on damage can result in substantive changes to samples during imaging. Unique to cryo-TEM, exposure to the beam can induce crystallization of the vitrified solvent, which can result in the introduction of diffraction contrast that can obscure the objects of interest. One main advantage of vitrification is that the amorphous solvent contributes a uniform background, which means that nanoparticles of sufficient size (with respect to the total thickness of the solvent film) can be readily discerned. In addition, the amorphous solvent can liquefy or sublime, which can cause the particles to move (Burrows and Penn 2013; Burrows et al. 2014). In fluid cell TEM, additional considerations are numerous (Ross 2015) and include the possibility that particles could interact with the cell windows, with irreversible adhesion to the cell window or modified particle-particle interactions both possibilities (Liu et al. 2013; Yuk et al. 2012; Zheng et al. 2009). Finally, exposure to a sufficiently intense beam can yield substantial reduction of Fe^{3+} to Fe^{2+} (Pan et al. 2006), which could result in the reductive dissolution of ferric oxide during imaging. Figure 13.5 shows a series of images of goethite crystals in water captured from a video collected while using a fluid cell TEM. Upon exposure to the beam (unquantified intensity), the ca. one to two hundred nanometer goethite crystals dissolved within approximately 70 s. Moving to a nearby location revealed that crystals not exposed to the beam remained intact and appeared unaltered, although onset of dissolution was observed soon after selecting a new field of view. Furthermore, under similar conditions in the laboratory, no detectable dissolution occurs. Thus, we conclude that the goethite reductively dissolved due to exposure to the electron beam (Fig. 13.5).

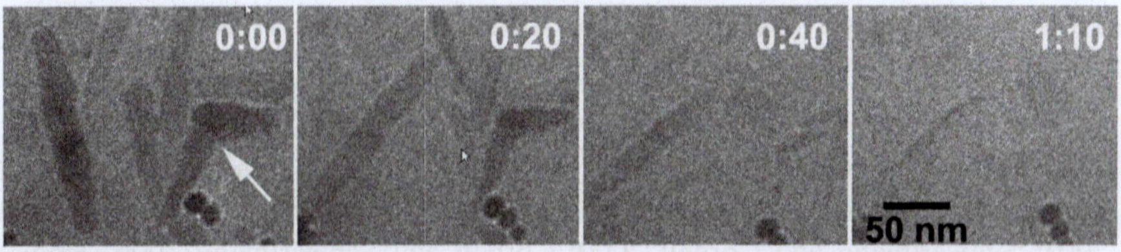

Fig. 13.5 Fluid cell TEM images obtained as snapshots from a video recording initiated immediately after the *left-hand*, acicular particles were in the field of view. The dark-appearing round objects are unknown contamination. The ca. 100–200 nm long by 20–50 nm wide particles are goethite crystals, and one twinned particle can be seen on the *right-hand* size of the time zero image (*large white arrow*). The ca. 200 nm long goethite crystal on the *left-hand* side of the field of view rotated as it dissolved

13.2.3 Correlative Methods

One major problem with the above TEM methods is vanishingly small amounts of material are imaged. In Fig. 13.5, the amount of material imaged in the left-most panel is a fraction of a femtogram. High-resolution micrographs typically involve just attograms of material. Thus, techniques that can produce data for comparatively bulk samples are essential to ensuring the data are suitably representative of the overall sample. X-ray diffraction (XRD) (Chiche et al. 2008; Hapiuk et al. 2013; Huang et al. 2003) and other diffraction techniques can be used to identify minerals and quantify crystallite size, size distribution, and morphology. Scattering techniques, such as small angle X-ray scattering (SAXS) (Davis et al. 2006; Drews and Tsapatsis 2007; Kumar et al. 2008; Stawski and Benning 2013), small angle neutron scattering (SANS) (Schwahn et al. 2007), and dynamic light scattering (DLS) (Davis et al. 2006; Mintova et al. 2002; Schwahn et al. 2007) can be used to quantify particle size and size distribution of particles in the suspended state. A more detailed description of some of these methods for characterizing growth via oriented attachment can be found in a review by Penn and Soltis (2014).

Kinetic models serve as an important method for detecting crystal growth by particle-based mechanisms, especially when more than one crystal growth mechanism operates. Several models employed for quantifying crystal growth that involves aggregative processes, including oriented attachment, have been reviewed elsewhere (Burrows et al. 2010; Xue et al. 2014), and describing those models is beyond the scope of this chapter. However, the addition of semiquantitative and quantitative models improves drastically our ability to detect crystal growth by particle-based mechanisms, even when features that could serve as indicators of such growth have been erased by subsequent processes.

13.3 Highlights of Particle-Mediated Crystal Growth in the Iron Oxides Literature

The work of Cornell and Schwertmann dramatically changed our understanding of the occurrence and formation of iron oxides. Their books (Cornell and Schwertmann 2003; Schwertmann and Cornell 2000; Cornell and Schwertmann 2006) feature descriptions of iron oxide minerals and reliable recipes for synthesizing them, complete with results from careful materials characterization. Below are some highlights of reports describing iron oxide crystal growth by nonclassical, particle-based mechanisms.

13.3.1 Goethite

Goethite (α-FeOOH) has been described as the dominant reactive iron oxyhydroxide in lake and marine sediments (van der Zee et al. 2003). Physical properties like microstructure, particle size, and aggregation state are expected to strongly impact the behavior of this ubiquitous and redox reactive material.

Evidence for goethite crystal growth by a particle-based mechanism was described in the 1970s. Murphy et al. (1976a) prepared synthetic goethite from partially neutralized solutions of ferric nitrate, and in their paper they describe goethite growth by spheres linking together to form rods. They describe the spheres as discrete, with sizes in the 1.5–3 nm size range and refer to them as polycations that initially form in the partially neutralized ferric nitrate solutions. In related work, they (Murphy et al. 1976b) describe goethite growth from ferric perchlorate solutions that also starts with spheres linking together to form rods. They again describe the spheres as polycations, which initially form in the ferric perchlorate solutions as discrete spheres in the 1.5–3 nm size range, and they further discuss link between ionic strength and rod formation. Whether their growth mechanism involves a phase transformation to goethite that precedes or follows the aggregation step is unclear.

Burleson and Penn (2006) proposed a two-step growth mechanism for goethite crystal growth under aqueous conditions. They suggested that primary particles first transformed from ferrihydrite to goethite followed by secondary assembly via oriented attachment (Burleson and Penn 2006). This hypothesis was supported by the cryo-TEM work published by Yuwono et al. (2010), in which images of loose aggregates composed of oriented goethite primary particles were presented. The kinetics of goethite growth by oriented attachment as well as the final goethite crystal size are sensitive to temperature and pH (Burleson and Penn 2006; Burrows et al. 2013), ionic strength (Burrows et al. 2012), precursor particle size (Penn et al. 2006; Penn et al. 2007), and the presence of chemical additives (Yuwono et al. 2012).

The aforementioned studies on goethite crystal growth were mainly performed at mildly acidic conditions. In contrast, Schwertmann and Murad (1983) performed experiments at higher pH, and they concluded that goethite had been formed by a dissolution and reprecipitation mechanism. No evidence for goethite growth by oriented attachment has been reported, and their general conclusion seems to be dissolution and precipitation dominates at high pH due to increased Fe(III) solubility (Davidson et al. 2008; Shaw et al. 2005).

13.3.2 Hematite

The case of the pseudocubic hematite is but one example of disparate interpretations based on very similar observed crystal textures. Similarly, spindle-shaped hematite particles consisting of oriented subcrystals have been described as growing by two different mechanisms. Morales et al. (1992) concluded that the spindle-shaped hematite was formed by an aggregation process involving hematite crystallites. In contrast, Sugimoto et al. (1993) stated that single crystals could not be constructed by aggregation of primary crystallites and concluded that these structures were formed by direct monomer-by-monomer addition of ferric complexes from the solution phase.

Bailey et al. (1993) used time-resolved TEM and cryo-TEM to examine the evolution of iron oxide particles produced from aqueous solutions of ferric chloride. They concluded that their 1.5 μm hematite nanocubes formed from raft-like aggregates of rod-shaped akaganeite nanoparticles aligned along <001>. They proposed that hematite nucleated within the structure of the raft-shaped oriented aggregate. Ocaña et al. (1995) described the "ordered aggregation" of small ellipsoidal hematite nanoparticles to form larger hematite crystals. Finally, Fischer and Schwertmann (1975) described the coalescence of "amorphous particles" into larger aggregates that eventually transform into single crystals of hematite (Fischer and Schwertmann 1975). Their results may be consistent with crystal growth by oriented attachment preceded by or followed by phase transformation from ferrihydrite to hematite.

Hematite produced by aging an acidic ferric chloride solution at elevated temperature exhibited a texture consistent with crystal growth by oriented attachment, resulting in the formation of lobed single crystals containing significant defects, such as edge dislocations (Penn et al. 2001a). In addition, the recent work of Frandsen et al. (2014) demonstrated hematite growth that indirectly involves crystal growth by oriented attachment. In their work, akaganeite crystals grew by oriented attachment, and the larger secondary crystals eventually phase transformed to hematite. Other examples of particle-based hematite growth undoubtedly exist.

13.3.3 Akaganeite

Akaganeite is a less common iron-bearing mineral that is not precisely an iron oxide mineral since it contains structural chloride. This mineral often occurs in the nanoscale size range and is often a precursor phase to the more stable hematite (e.g., Frandsen et al. (2014)). It has been described as occurring in the vicinity of biological organisms, with biological polymers serving to template the crystallization of akaganeite (Chan et al. 2004). It is less frequently studied (Barron and Torrent 2013), but the material frequently exhibits features that may indicate particle-mediated crystal growth.

An early report describing akaganeite crystal growth featured the conclusion that akaganeite grew from ferric chloride solutions by addition of subcrystals to produce the final larger structure. Watson et al. (1962) prepared ultrathin sections of akaganeite collected from aged solutions of ferric chloride (Fig. 13.6). Images of cross sections of their akaganeite crystals demonstrate that they are composed of oriented arrays of tetragonal prisms of crystalline akaganeite.

Similar to their work with goethite, Murphy et al. (1976c) described akaganeite growth from partially neutralized solutions of ferric chloride, starting with spheres that link together to form rods. They described the initial spheres as polycations that are in the 1.5–3 nm size range. Whether their growth mechanism involves a phase transformation to goethite that precedes or follows the aggregation step is once again (and understandably) unclear.

Recent work by Frandsen et al. (2014), which is described above, demonstrates akaganeite crystal growth by oriented attachment, using cryo-TEM tomography to produce exquisite three-dimensional images of the crystals.

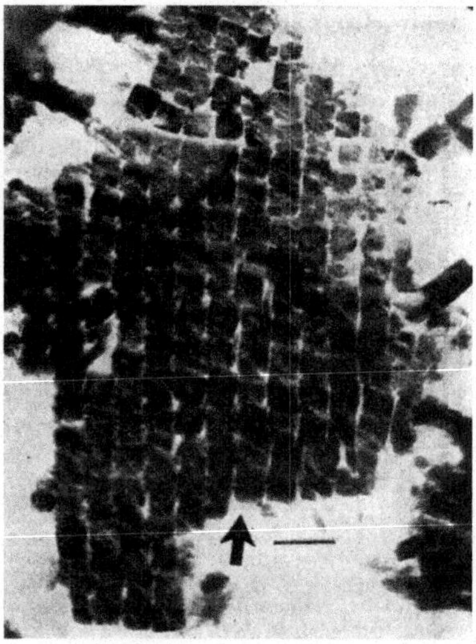

Fig. 13.6 TEM image of a cross-sectional ultra thin section of an akaganeite crystal impregnated with polymethracrylate. The scale bar represents 100 nm. High-resolution images of the individual tablets confirms their crystalline nature as well as the parallel crystallographic alignment (Watson et al. 1962)

13.3.4 Feroxyhite

Feroxyhite (δ-FeOOH) nanocrystals, which were produced upon oxidation of an Fe(II)-bearing precipitate prepared by neutralizing a solution of ferrous chloride, were described as high aspect ratio particles, with substantial dimpling, porosity, and incorporation of defects like edge dislocations. The authors concluded that the feroxyhite crystals had formed by oriented attachment (Penn et al. 2001a).

13.3.5 Ferrihydrite

Ferrihydrite is in a class of its own. There is a lack of consensus regarding its crystal structure, homogeneity, and even composition (Drits et al. 1993; Gilbert et al. 2013; Janney et al. 2000, 2001; Michel et al. 2007). This iron oxyhydroxide occurs only as nanoparticles, with sizes typically ranging from one to several nanometers, and its structure continues to be poorly understood. This material is often the first solid material to form upon partial neutralization of solutions containing ferric ions, and it is often prepared as a precursor to the more thermodynamically stable iron oxides like goethite and hematite (e.g., Cornell et al. 1989; Cornell and Schwertmann 2003, 2006). In fact, aqueous suspensions containing ferrihydrite nanoparticles typically are not stable, with the more stable iron oxide minerals, such as goethite and hematite, forming even after quite short aging times. The new phases often exhibit textures and morphologies consistent with particle-mediated crystal growth. Mechanisms for conversion of ferrihydrite to the more stable phases range from dissolution of ferrihydrite followed by precipitation of the more stable phase as new nuclei or onto existing nuclei to aggregation-induced phase transformations. Elucidating these mechanisms is complicated by the limits of current characterization methods as well as our poor understanding of the nature of this ubiquitous material.

13.4 Outlook

Iron oxides are important players in the geochemical cycling of the elements as well as synthetic and naturally occurring chemical species. They can participate in redox reactions and can be exploited in remediation of contaminants in a wide range of settings. Particle size, morphology, composition, and microstructure on chemical behavior can all dramatically affect the chemical behavior of iron oxides. Furthermore, these materials are used in a wide range of industrial settings, magnetic recording media, water treatment facilities, pigments, and more. Thus, the importance of elucidating how the iron oxides form, transform, grow, and even dissolve is far reaching. Detailed examination of direct images of final crystals has yielded dramatic insights into the mechanisms by which iron oxide crystals form, but reliance on such images is risky since features produced during the early stages of crystal growth can effectively be erased by later processes. Thus, the best approach combines a suite of time-resolved and in situ methods in order to elucidate the formation and growth history of the crystalline iron oxides. Elucidating the mechanisms by which iron oxides grow and transform in complex environments will require advancements in in situ materials characterization methods.

Acknowledgments We acknowledge the University of Minnesota, the National Science Foundation (No. NSF-0957696), and the Nanostructural Materials and Processes Program at the University of Minnesota for the financial support. We also thank Characterization Facility at the University of Minnesota, a member of the NSF-funded Materials Research Facilities Network (www.mrfn.org) via the MRSEC program (Figs. 13.2, 13.3, and 13.4). In addition, the TEM images shown in Fig. 13.5 were obtained using a Tecnai TF20 FEI microscope located at the Pacific Northwest National Laboratory, which is operated by Battelle Memorial Institute for the US Department of Energy under Contract DE-AC05-76RL01830.

References

Bailey JK, Brinker CJ, Mecartney ML (1993) Growth mechanisms of iron oxide particles of differing morphologies from the forced hydrolysis of ferric chloride solutions. J Colloid Interface Sci 157:1–13

Banfield JF, Welch SA, Zhang H, Thomsen Ebert T, Lee Penn R (2000) Aggregation-based crystal growth and microstructure development in natural iron oxyhydroxide biomineralization products. Science 289:751–754

Barker WW, Welch SA, Chu S, Banfield JF (1998) Experimental observations of the effects of bacteria on aluminosilicate weathering. Am Mineral 83:1551–1563

Barron V, Torrent J (2013) Iron, manganese and aluminium oxides and oxyhydroxides. Miner Nanoscale 14:297–336

Burleson DJ, Penn RL (2006) Two-step growth of goethite from ferrihydrite. Langmuir 22:402–409

Burrows ND, Penn RL (2013) Cryogenic transmission electron microscopy: aqueous suspensions of nanoscale objects. Microsc Microanal 19:1542–1553

Burrows ND, Yuwono VM, Penn RL (2010) Quantifying the kinetics of crystal growth by oriented aggregation. Mater Res Sci Bull 35:133–137

Burrows ND, Hale CRH, Penn RL (2012) Effect of ionic strength on the kinetics of crystal growth by oriented aggregation. Cryst Growth Des 12:4787–4797

Burrows ND, Hale CRH, Penn RL (2013) Effect of pH on the kinetics of crystal growth by oriented aggregation. Cryst Growth Des 13:3396–3403

Burrows ND, Kesselman E, Sabyrov K, Stemig A, Talmon Y, Penn RL (2014) Crystalline nanoparticle aggregation in non-aqueous solvents. CrystEngComm 16:1472–1481

Cao G (2004) Nanostructures and nanomaterials: synthesis, properties and applications. Imperial College Press, London

Chan CS, De Stasio G, Welch SA, Girasole M, Frazer BH, Nesterova MV, Fakra S, Banfield JF (2004) Microbial polysaccharides template assembly of nanocrystal fibers. Science 303:1656–1658

Chiche D, Digne M, Revel R, Chaneac C, Jolivet JP (2008) Accurate determination of oxide nanoparticle size and shape based on X-ray powder pattern simulation: application to boehmite AlOOH. J Phys Chem C 112:8524–8533

Cölfen H, Mann S (2003) Higher-order organization by mesoscale self-assembly and transformation of hybrid nanostructures. Angew Chem Int Ed 42:2350–2365

Cornell RM, Schwertmann U (2003) The iron oxides: structures, properties, reactions, occurrences and uses. Wiley-VCH, Weinheim

Cornell RM, Schwertmann U (2006) The iron oxides: structure, properties, reactions, occurrences and uses, 2nd, completely revised and extended edition. Wiley-VCH, Weinheim

Cornell RM, Schneider W, Giovanoli R (1989) Phase transformations in the ferrihydrite/cysteine system. Polyhedron 8:2829–2836

Dalmaschio CJ, Ribeiro C, Leite ER (2010) Impact of the colloidal state on the oriented attachment growth mechanism. Nanoscale 2:2336–2345

Davidson LE, Shaw S, Benning LG (2008) The kinetics and mechanisms of schwertmannite transformation to goethite and hematite under alkaline conditions. Am Mineral 93:1326–1337

Davis TM, Drews TO, Ramanan H, He C, Dong JS, Schnablegger H, Katsoulakis M, Kokkoli E, Mccormick A, Penn RL, Tsapatsis M (2006) Mechanistic principles of nanoparticle evolution to zeolite crystals. Nat Mater 5:400–408

De Yoreo JJ, Gilbert PUPA, Sommerdijk NAJM, Penn RL, Whitelam S, Joester D, Zhang HZ, Rimer JD, Navrotsky A, Banfield JF, Wallace AF, Michel FM, Meldrum FC, Cölfen H, Dove PM (2015) Toward a comprehensive picture of crystallization by particle attachment. doi: 10.1126/science.aaa6760

De Yoreo JJ, Sommerdijk NAJM, Dove PM (2017) Nucleation pathways in electrolyte solutions. In: Van Driessche AES, Kellermeier M, Benning LG, Gebauer D (eds) New perspectives on mineral nucleation and growth, Springer, Cham, pp 1–24

Drews TO, Tsapatsis M (2007) Model of the evolution of nanoparticles to crystals via an aggregative growth mechanism. Micropor Meospor Mater 101:97–107

Drits VA, Sakharov BA, Salyn AL, Manceau A (1993) Structural model for ferrihydrite. Clay Miner 28:185–207

Fischer WR, Schwertmann U (1975) The formation of hematite from amorphous iron(III) hydroxide. Clays Clay Miner 23:33–37

Frandsen C, Legg BA, Comolli LR, Zhang H, Gilbert B, Johnson E, Banfield JF (2014) Aggregation-induced growth and transformation of b-FeOOH nanorods to micron-sized a-Fe_2O_3 spindles. CrystEngComm 16:1451–1458

Gilbert B, Erbs JJ, Penn RL, Petkov V, Spagnoli D, Waychunas GA (2013) A disordered nanoparticle model for 6-line ferrihydrite. Am Mineral 98:1465–1476

Grogan JM, Schneider NM, Ross FM, Bau HH (2012) The nanoaquarium: a new paradigm in electron microscopy. J Indian Inst Sci 92:295–308

Hapiuk D, Masenelli B, Masenelli-Varlot K, Tainoff D, Boisron O, Albin C, Mélinon P (2013) Oriented attachment of ZnO nanocrystals. J Phys Chem C 117:10220–10227

Hochella MF Jr, Lower SK, Maurice PA, Penn RL, Sahai N, Sparks DL, Twining BS (2008) Nanominerals, mineral nanoparticles, and earth systems. Geochim Cosmochim Acta 72:A382

Huang F, Zhang HZ, Banfield J (2003) Two-stage crystal-growth kinetics observed during hydrothermal coarsening of nanocrystalline ZnS. Nano Lett 3:373–378

Ivanov VK, Fedorov PP, Baranchikov AY, Osiko VV (2014) Oriented attachment of particles: 100 years of investigations of non-classical crystal growth. Russ Chem Rev 83:1204–1222

Janney DE, Cowley JM, Buseck PR (2000) Structure of synthetic 2-line ferrihydrite by electron nanodiffraction. Am Mineral 85:1180–1187

Janney DE, Cowley JM, Buseck PR (2001) Structure of synthetic 6-line ferrihydrite by electron nanodiffraction. Am Mineral 86:327–335

Kandori K, Kawashima Y, Ishikawa T (1991) Characterization of monodispersed hematite particles by gas-adsorption and Fourier-transform infrared-spectroscopy. J Chem Soc Faraday Trans 87:2241–2246

Kumar S, Wang Z, Penn RL, Tsapatsis M (2008) A structural resolution cryo-TEM study of the early stages of MFI growth. J Am Chem Soc 130:17284–17286

Kumar S, Penn RL, Tsapatsis M (2011) On the nucleation and crystallization of silicalite-1 from a dilute clear sol. Micropor Mesopor Mater 144:74–81

Li D, Nielsen MH, Lee JR, Frandsen C, Banfield JF, De Yoreo JJ (2012) Direction-specific interactions control crystal growth by oriented attachment. Science 336:1014–1018

Liu Y, Lin XM, Sun Y, Rajh T (2013) In situ visualization of self-assembly of charged gold nanoparticles. J Am Chem Soc 135:3764–3767

Michel FM, Ehm L, Antao SM, Lee PL, Chupas PJ, Liu G, Strongin DR, Schoonen MAA, Phillips BL, Parise JB (2007) The structure of ferrihydrite, a nanocrystalline material. Science 316:1726–1729

Mintova S, Olson NH, Senker J, Bein T (2002) Mechanism of the transformation of silica precursor solutions into Si-MFI zeolite. Angew Chem Int Ed 41:2558–2561

Morales MP, Gonzalezcarreno T, Serna CJ (1992) The formation of alpha-Fe_2O_3 monodispersed particles in solution. J Mater Res 7:2538–2545

Murphy PJ, Posner AM, Quirk JP (1976a) Characterization of partially neutralized ferric nitrate solutions. J Colloid Interface Sci 56:270–283

Murphy PJ, Posner AM, Quirk JP (1976b) Characterization of partially neutralized ferric perchlorate solutions. J Colloid Interface Sci 56:298–311

Murphy PJ, Posner AM, Quirk JP (1976c) Characterization of partially neutralized ferric-chloride solutions. J Colloid Interface Sci 56:284–297

Navrotsky A (2011) Nanoscale effects on thermodynamics and phase equilibria in oxide systems. ChemPhysChem 12:2207–2215

Navrotsky A, Mazeina L, Majzlan J (2008) Size-driven structural and thermodynamic complexity in iron oxides. Science 319:1635–1638

Niederberger M, Cölfen H (2006) Oriented attachment and mesocrystals: non-classical crystallization mechanisms based on nanoparticle assembly. Phys Chem Chem Phys 8:3271–3287

Nielsen MH, De Yoreo JJ (2017) Liquid phase TEM investigations of crystal nucleation, growth, and transformation. In: Van Driessche AES, Kellermeier M, Benning LG, Gebauer D (eds) New perspectives on mineral nucleation and growth, Springer, Cham, pp 353–371

Ocaña M, Morales MP, Serna CJ (1995) The growth mechanism of a-Fe_2O_3 ellipsoidal particles in solution. J Colloid Interface Sci 171:85–91

Pan Y, Brown A, Brydson R, Warley A, Li A, Powell J (2006) Electron beam damage studies of synthetic 6-line ferrihydrite and ferritin molecule cores within a human liver biopsy. Micron 37:403–411

Penn RL (2004) Kinetics of oriented aggregation. J Phys Chem B 108:12707–12712

Penn RL, Soltis JA (2014) Characterizing crystal growth by oriented aggregation. CrystEngComm 16:1409–1418

Penn RL, Oskam G, Strathmann TJ, Searson PC, Stone AT, Veblen DR (2001a) Epitaxial assembly in aged colloids. J Phys Chem B 105:2177–2182

Penn RL, Zhu C, Xu H, Veblen DR (2001b) Iron oxide coatings on sand grains from the Atlantic coastal plain: high-resolution transmission electron microscopy characterization. Geology 29:843–846

Penn RL, Erbs J, Gulliver D (2006) Controlled growth of alpha-FeOOH nanorods by exploiting-oriented aggregation. J Cryst Growth 293:1–4

Penn RL, Tanaka K, Erbs J (2007) Size dependent kinetics of oriented aggregation. J Cryst Growth 309:97–102

Rao A, Cölfen H (2017) Mineralization schemes in the living world: mesocrystals. In: Van Driessche AES, Kellermeier M, Benning LG, Gebauer D (eds) New perspectives on mineral nucleation and growth, Springer, Cham, pp 155–184

Ross FM (2015) Opportunities and challenges in liquid cell electron microscopy. Science 350:aaa9886-1–aaa9886-9

Schwahn D, Ma Y, Cölfen H (2007) Mesocrystal to single crystal transformation of d, l-alanine evidenced by small angle neutron scattering. J Phys Chem C 111:3224–3227

Schwertmann U, Cornell RM (2000) Iron oxides in the laboratory: preparation and characterization. Wiley-VCH, Weinheim

Schwertmann U, Murad E (1983) Effect of pH on the formation of goethite and hematite from ferrihydrite. Clays Clay Miner 31:277–284

Shaw S, Pepper SE, Bryan ND, Livens FR (2005) The kinetics and mechanisms of goethite and hematite crystallization under alkaline conditions, and in the presence of phosphate. Am Mineral 90:1852–1860

Stawski TM, Benning LG (2013) SAXS in inorganic and bioinspired research. In: De Yoreo JJ (ed) Research methods in biomineralization science. Academic, San Diego

Sugimoto T, Muramatsu A, Sakata K, Shindo D (1993) Characterization of hematite particles of different shapes. J Colloid Interface Sci 158:420–428

Van der Zee C, Roberts DR, Rancourt DG, Slomp CP (2003) Nanogoethite is the dominant reactive oxyhydroxide phase in lake and marine sediments. Geology 31:993–996

Van Driessche AES, Benning LG, Rodriguez-Blanco JD, Ossorio M, Bots P, Garcia-Ruiz JM (2012) The role and implications of bassanite as a stable precursor phase to gypsum precipitation. Science 336:69–72

Watson JHL, Cardell RR Jr, Heller W (1962) The internal structure of colloidal crystals of beta-FeOOH and remarks on their assemblies in schiller layers. J Phys Chem 66:1757–1763

Waychunas GA, Kim CS, Banfield JF (2005) Nanoparticulate iron oxide minerals in soils and sediments: unique properties and contaminant scavenging mechanisms. J Nanopart Res 7:409–433

Xiong Y, Tang Z (2012) Role of self-assembly in construction of inorganic nanostructural materials. Sci China-Chem 55:2272–2282

Xue X, Penn RL, Leite ER, Huang F, Lin Z (2014) Crystal growth by oriented attachment: kinetic models and control factors. CrystEngComm 16:1419–1429

Yuk JM, Park J, Ercius P, Kim K, Hellebusch DJ, Crommie MF, Lee JY, Zettl A, Alivisatos AP (2012) High-resolution EM of colloidal nanocrystal growth using graphene liquid cells. Science 336:61–64

Yuwono V, Burrows ND, Soltis JA, Penn RL (2010) Oriented aggregation: formation and transformation of mesocrystal intermediates revealed. J Am Chem Soc 132:2163–2165

Yuwono VM, Burrows ND, Soltis JA, Do T, Penn RL (2012) Aggregation of ferrihydrite nanoparticles in aqueous systems. Faraday Discuss 159:235–245

Zhang Q, Liu S, Yu S (2009) Recent advances in oriented attachment growth and synthesis of functional materials: concept, evidence, mechanism, and future. J Mater Chem 19:191–207

Zhang J, Huang F, Lin Z (2010) Progress of nanocrystalline growth kinetics based on oriented attachment. Nanoscale 2:18–34

Zheng H, Smith RK, Jun YW, Kisielowski C, Dahmen U, Alivisatos AP (2009) Observation of single colloidal platinum nanocrystal growth trajectories. Science 324:1309–1312

Chapter 14
Magnetite Nucleation and Growth

Victoria Reichel and Damien Faivre

14.1 Introduction

Magnetite (Mgt) is a fascinating material, which plays a role in our daily life and accompanies us in various situations: Imagine, for example, going to the hospital to have an organ screening in a magnetic resonance imaging device or just listening to music from an old cassette. In the aforementioned situations, magnetic materials, especially magnetite, play an important role: In the magnetic resonance imaging device, magnetic nanoparticles are injected as contrast agents to visualize, e.g., tumors in the human body. In cassettes, magnetic tapes are used for data storage. Alternatively, the magnetic properties of iron oxides in general, but of magnetite in particular, are used in paleomagnetic and geomagnetic reconstruction. The minerals are indeed preserving the main vectors of the Earth's magnetization and are therefore used to reconstruct past climates and continent positioning (Acton 2011). In ancient times, sailors were using magnetic materials as well. The earliest known compass, called lodestone, was used for navigation (Fig. 14.1) (Annett 1921). Magnetite ($Fe^{II}\ Fe_2^{III}\ O_4$, Fe_3O_4), which is one of a number of iron oxide phases (Cornell and Schwertmann 2003), can be found in nature and synthesized in the laboratory. In this chapter, we will discuss magnetite and its occurrence, its properties in general and of magnetite nanoparticles in particular, and how magnetite can be formed synthetically. We will present how magnetite is biomineralized in magnetotactic bacteria (MTB). Finally, we will show how biological additives can be combined with magnetite synthesis to influence the properties of magnetite nanoparticles.

V. Reichel • D. Faivre (✉)
Department of Biomaterials, Max Planck Institute of Colloids and Interfaces, Science Park Golm, 14424 Potsdam, Germany
e-mail: damien.faivre@mpikg.mpg.de

A.E.S. Van Driessche et al. (eds.), *New Perspectives on Mineral Nucleation and Growth*,
DOI 10.1007/978-3-319-45669-0_14

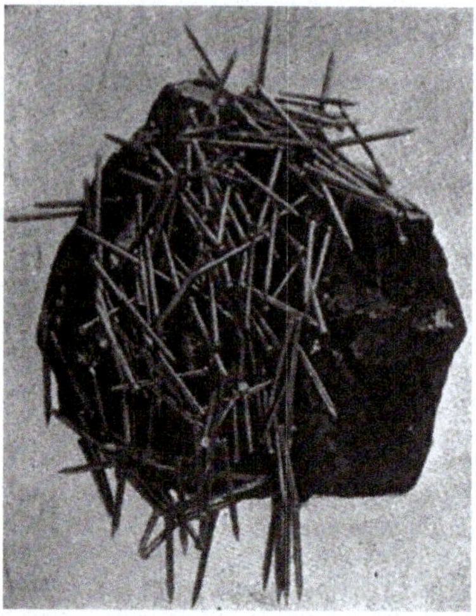

Fig. 14.1 An old photograph of lodestone. Lodestones are naturally occurring magnets, which can attract iron-based materials. They were used as the earliest compass for navigation (Annett 1921)

14.1.1 Occurrence of Magnetite

Geological magnetite can be found all over the world, e.g., in the Austrian Central Alps, in Shark Bay and Lake Clifton in Australia, in parts of Iran, in China, and in the United States (Zhao et al. 2014; Hüsing et al. 2011; Pechersky et al. 2009; Grachev et al. 2008; Stolz et al. 1989). The iron content of magnetite is 70 %wt, and it is therefore a valuable iron source. Magnetite is not only found in geological samples, it also occurs in living organisms, e.g., magnetotactic bacteria, mollusks, termites, bees, birds, and dolphins (Kirschvink and Gould 1981). In the case of magnetotactic bacteria, the magnetite particles enable the cells to passively orient themselves along the Earth's magnetic field lines (Frankel and Bazylinski 1994). In birds and fish, magnetite has been associated with magnetoreceptive cells (Solov'yov and Schulten 2012). Finally, in mollusks (chitons), magnetite is found in the radula where it is used to harden the teeth that scrape rocks (Brooker and Shaw 2012).

14.1.2 Properties of Magnetite

Nanoparticles of magnetite can grow to different sizes and have different morphologies due to environmental or synthetic conditions. In magnetite, the iron ions are distributed into the tetrahedral (Td) and octahedral (Oh) sites of the face-centered cubic structure, stacking oxygen according to ($[Fe^{3+}]_{Td}[Fe^{3+}+Fe^{2+}]_{Oh}O_4$) as shown in Fig. 14.2. Magnetite can easily oxidize to maghemite, which is its natural weathering product (Widdrat et al. 2014). Maghemite is similar to magnetite in its crystallographic structure (Fig. 14.2), in that it consists in the inverse spinel group with a cubic unit cell. The magnetite unit cell contains 32 cubic closely packed O^{2-} ions. One half of the octahedral positions are filled with eight Fe^{3+}

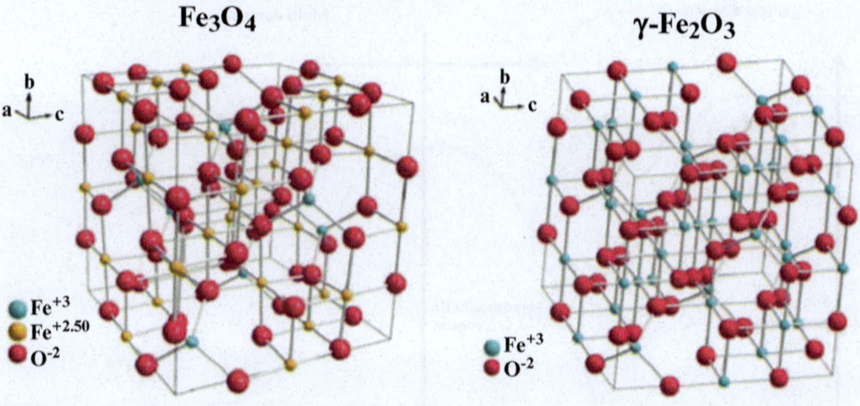

Fig. 14.2 Crystal structure of magnetite (Fe_3O_4) and maghemite (ϒ-Fe_2O_3) (Modified from Zhang et al. (2013) with permission of CrystEngComm)

ions and eight Fe^{2+} ions. Additionally, eight of the tetrahedral positions are filled with Fe^{3+} ions. Two different sublattices are formed regarding to the octahedrally and the tetrahedrally bound iron ions, which is shown in Fig. 14.2. Therefore, magnetite is a ferrimagnetic material (Cornell and Schwertmann 2003). In the case of maghemite, the unit cell consists of 21 1/3 Fe^{3+} ions and 32 O^{2-} ions and forms the same crystalline structure as magnetite with the exception of observed vacancies in the crystal lattice (γ–Fe_2O_3) ($[Fe^{3+}]_{Td}[Fe^{3+}{}_{5/3}X_{1/3}]_{Oh}O_4$ (X stands for a cationic vacancy) (Jolivet et al. 2004). The two antiparallel sublattices of maghemite have a nonzero net magnetic moment, which leads to maghemite also being ferrimagnetic (Cornell and Schwertmann 2003; Widdrat et al. 2014). Due to these similarities, magnetite and maghemite are hard to differentiate by analytical means: the lattice parameter for maghemite is, e.g., 0.83419 nm, slightly reduced in comparison to that of magnetite, which is 0.83969 nm (Widdrat et al. 2014). The reduction is due to the presence of the vacancies that replace the ferrous iron in the lattice. This difference enables the differentiation of these minerals only by high-resolution techniques such as high-resolution X-ray diffraction (Kim et al. 2012).

The magnetic properties of magnetite are influenced by their nanoparticulate dimensions. Below a size of 30 nm, such nanoparticles exhibit a superparamagnetic behavior (SP) (Dunlop and Özdemir 2001). In this case, magnetite nanoparticles have to be placed in an external magnetic field to exhibit measurable magnetic properties. Magnetite nanoparticles between ca. 30 and 80 nm are called stable single domain. A ferrimagnet in which the magnetization does not shift across the magnet has the stable single domain behavior. Particles above ~80 nm are multidomain. In this case, several magnetic domains coexist within a single particle, thereby reducing the net volume magnetization of the sample (Fig. 14.3).

Finally, the hardness of magnetite is also an important property of this mineral, as it is, e.g., used by mollusks to reinforce their teeth. Magnetite, which occurs in the chiton radula of, e.g., *Acanthopleura gaimardi*, shows a Mohs hardness score around 6 (Brooker and Shaw 2012). The Mohs score helps to characterize the visible scratch resistance of minerals by other minerals. As a comparison, diamond is one

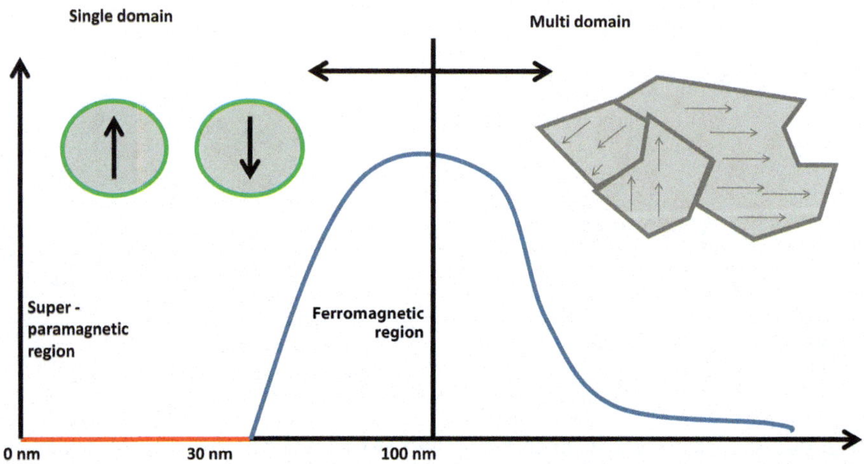

Fig. 14.3 Sketch of the magnetic properties of magnetite nanoparticles as a function of their size

of the hardest materials with a Mohs score of 10(Ricks 2010). With a Mohs hardness around 6, the chiton can easily use magnetite inside its teeth to scratch minerals such as calcite or fluorite without damaging its teeth. These teeth are one of the hardest biomineral structures, and they are known to exhibit three times the hardness of human teeth or of typical mollusk shells (Brooker and Shaw 2012).

14.1.3 Applications of Magnetite Nanoparticles

Magnetite nanoparticles have found applications in domains as wide as coloring pigments, magnetic materials, catalysis, data storage, biological assays, purification processes, MRI contrast agents, agents for hyperthermia treatment, and magnetically guided drug delivery (Guo et al. 2009). An overview of possible applications is shown in Fig. 14.4. A principal criterion when utilizing magnetite nanoparticles for these applications is the colloidal stability and monodispersity they display as those directly influence their behavior. To stabilize magnetite nanoparticles and create new morphologies and structures, different additives are used. These additives can be biological compounds, polymers, and more, and their roles will be detailed below (McBain et al. 2008; Ito et al. 2005; Shimizu et al. 2007).

14.2 Synthetic Formation of Magnetite

14.2.1 The Different Synthetic Routes

Synthetic magnetite nanoparticles can be formed via many different routes (Laurent et al. 2008). Possible approaches include reactions in constrained environments, hydrothermal reactions, flow injection synthesis, polyol methods, sol-gel reactions,

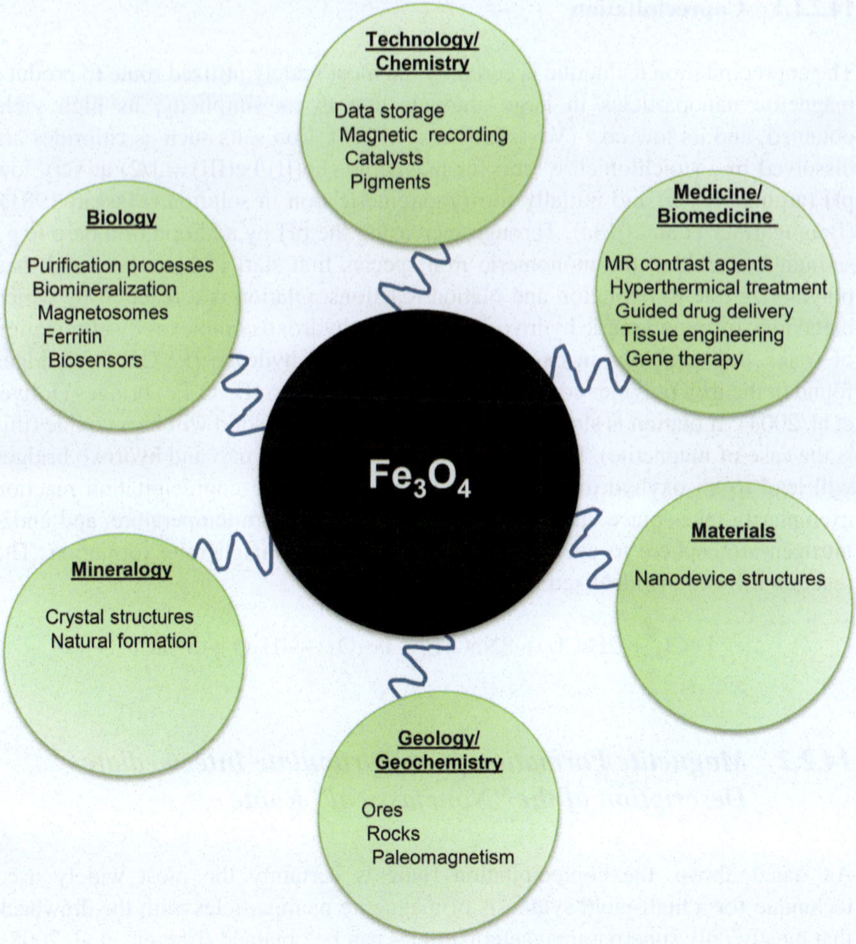

Fig. 14.4 Possible applications and scientific fields that rely on the properties and uses of magnetite nanoparticles

aerosol/vapor methods, electrochemical methods, sonolysis, and the classical coprecipitation to name a few. In one important method, the sol-gel method, the materials are produced from small monomers, which are converted into a colloidal solution, solution that acts as a precursor for an integrated network of particles. Sol-gel syntheses are performed at room temperature, and the product acquires the final crystalline state via aging (Vinogradov and Vinogradov 2014). High-temperature and hydrothermal reactions are also used to form magnetite nanoparticles (Laurent et al. 2008). In these cases, ultrafine powders are formed at temperatures above 200 °C. The particle size increases with time, which can result in larger magnetite aggregates. In the following, we focus on the coprecipitation technique for reason detailed below.

14.2.1.1 Coprecipitation

The coprecipitation technique is certainly the most widely utilized route to produce magnetite nanoparticles in large amounts due to its simplicity, its high yield obtained, and its low cost (Vayssières et al. 1998). Iron salts such as chlorides are dissolved in a stoichiometric ratio for magnetite (Fe(II)/Fe(III) = 1/2) at very low pH (around 1–1.5) and initially form monomeric iron in solution (Massart 1981); (Baumgartner et al. 2013a). Through increasing the pH by addition of a base (e.g., sodium hydroxide), the monomeric iron species first start to hydrolyze and then polymerize due to oxolation and olation reactions. Olation is a reaction by which metal ions form polymeric hydroxides from aquohydroxo complexes by elimination of water. Subsequently, in oxolation reaction, these hydroxo (Fe-OH-Fe) bridges found in the iron polymer complexes are replaced by oxo (Fe-O-Fe) bridges (Jolivet et al. 2004). If olation is slow compared to oxolation, the solid will be an oxide (this is the case of magnetite). If it is faster, the coexistence of oxo and hydroxo bridges will lead to an oxyhydroxide (such as ferrihydrite). The coprecipitation reaction exemplarily takes place under continuous stirring, at room temperature, and under nitrogen atmosphere to avoid oxidation and therefore maghemite formation. The reaction follows the idealized formula:

$$FeCl_2 + 2FeCl_3 + 8NaOH \rightarrow Fe_3O_4 + 4H_2O + 8NaCl$$

14.2.2 Magnetite Formation from Particulate Intermediate Description of the "Nonclassical" Route

As stated above, the coprecipitation route is certainly the most widely used technique for a high-yield synthesis of magnetite nanoparticles with the drawback that mostly only superparamagnetic particles can be obtained (Laurent et al. 2008). Therefore, efforts were developed in recent years to control the particle size of the magnetite nanoparticles by producing them under physiological conditions (Baumgartner et al. 2013a). The standard coprecipitation technique was modified, and nanoparticles were grown beyond the superparamagnetic/stable single domain threshold value. We describe below the mechanism of formation of such large particles based on our recent work (Baumgartner et al. 2013a, b).

Using a computer-controlled titration setup, iron oxide nanoparticles (Baumgartner et al. 2013a) are formed from iron solutions (again with the stoichiometric ratio Fe(II)/Fe(III) = 1/2) that were added into a reaction vessel, kept at a constant temperature. The pH is controlled and kept stable by addition of a sodium hydroxide solution under continuous stirring. The dynamics of the process was followed by cryogenic transmission electron microscopy (Baumgartner et al. 2013b). Samples were extracted from the reaction vessel and vitrified through plunging in liquid ethane at 100 % to humidity and under a nitrogen atmosphere to prevent the samples

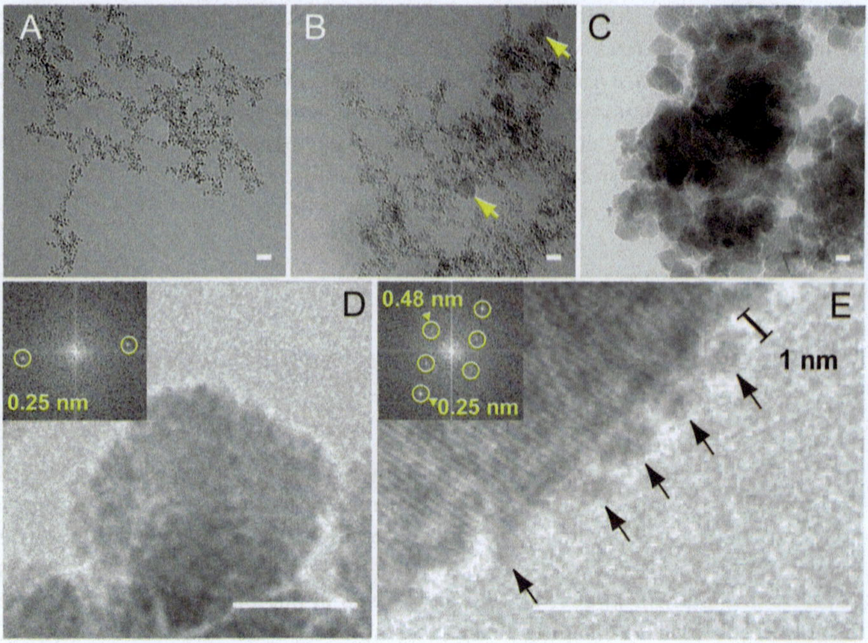

Fig. 14.5 Cryo-transmission electron microscopy images of the nucleation and growth of coprecipitated magnetite. Images show the time-resolved evolution of primary particle and magnetite nanoparticle aggregates after 2 min (**a**), 6 min (**b**), and 82 min (**c**). The early formed crystalline magnetite nanoparticles can be seen in image **b** (*yellow arrows*). Magnetic nanoparticles can be seen in image **d**, and image e shows primary particles which attach to the surface of a magnetite nanoparticle (*black arrows*). The *insets* in images **d** and **e** are fast Fourier transforms, which demonstrated the crystallinity of these nanoparticles. The *scale bar* in all images is 10 nm (Taken from Baumgartner et al. (2013b) with permission of Nature Materials)

from oxidation or drying-induced crystallization. This method is particularly well suited for avoiding artifacts in sample preparation (Nudelman and Sommerdijk 2012).

The following patterns were observed: Shortly after the initial addition of the dilute sodium hydroxide solution to the iron chloride solution, so-called primary particles are observed (Baumgartner et al. 2013b). These primary particles are highly monodisperse with a dimension of about 2 nm in diameter. The primary particles subsequently aggregate into branched networks, which become denser, forming spheroidal nanoparticles of magnetite as shown by the reflections on the high-resolution images (Fig. 14.5b). These first magnetite particles measure about 10 nm in diameter. With increasing reaction time, the magnetite nanoparticles increase in size. Low-resolution images (Fig. 14.5c) depict well-defined cuboctahedral surfaces typical of magnetite crystals. However, a closer look at high-resolution micrographs shows that the crystal surfaces exhibit high surface roughness attributed to the presence of primary particles on the surface of the growing magnetite crystals

(Baumgartner et al. 2013b). In the moment of accretion, the dimension of the primary particles decreases to around 1 nm on the large magnetite surface. Potential explanations of this size decrease include contraction due to water loss concomitant with olation or oxolation bond formation and/or partial redissolution. Therefore, the primary particles are not only involved in the nucleation of synthetic magnetite nanoparticles but are also critical for their growth.

14.3 Biomineralization of Magnetite

14.3.1 Generalities on Magnetotactic Bacteria and Magnetosomes

Magnetotactic bacteria (MTB) (Fig. 14.6) are aquatic and motile prokaryotes found at the oxic–anoxic transition zone, where oxygenated water/sediment meets oxygen-deficient water/sediment (Bazylinski et al. 2014). MTB are gram-negative bacteria and consist of various phyla of different proteobacteria, among others (Lefèvre and Bazylinski 2013). MTB are diverse in phylogeny, physiology, mode of magnetotaxis, and mineral and cellular morphology (Faivre and Schüler 2008). What all MTB have in common is the magnetosome organelle, which consists of an organic and an inorganic phase. The magnetosome comprises ferrimagnetic stable single domain nanoparticles (either of the iron oxide magnetite (Fe_3O_4) or the iron sulfide greigite (Fe_3S_4)), which are surrounded by a phospholipid bilayer called the magnetosome membrane (Lefèvre and Bazylinski 2013; Faivre and Schüler 2008). These magnetosomes are aligned by a bundle of filaments along the long axis of the bacteria to form the magnetosome chain and act as miniature compass needles. The magnetosome chains have the ability to passively align the bacteria along the Earth's magnetic field lines, and the bacterium itself can swim actively through its

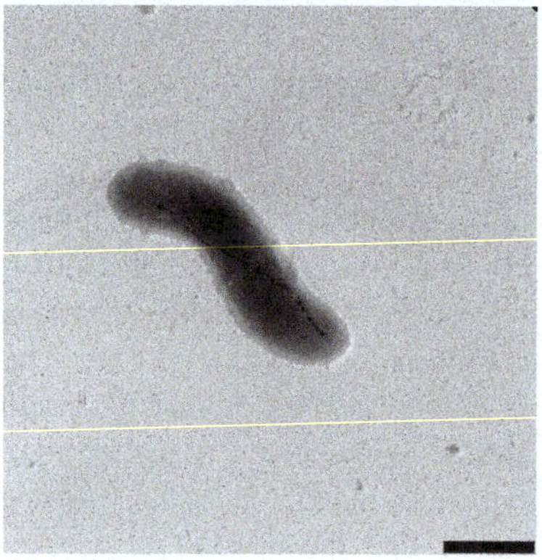

Fig. 14.6 *Magnetospirillum magneticum AMB-1* (scale bar, 1000 nm)

flagellar system. This behavior is called magnetotaxis and combines the alignment of the cells in the magnetic field and the chemotaxis along the oxygen gradient (Lower and Bazylinski 2013).

14.3.2 The Reaction Pathway of Magnetite Biomineralization

Either magnetite or greigite is formed in magnetotactic bacteria. Before these iron oxides are formed within the cell, the elements, which are necessary for mineral formation, have to be taken up from the environment. Iron can be taken up as Fe^{3+} or Fe^{2+} (Faivre et al. 2007), and oxygen originates from water (Mandernack et al. 1999). A strict control of the iron uptake is necessary (Bazylinski et al. 2014; Grünberg et al. 2001). Several scenarios have been proposed: Either the iron can be transported into invaginated magnetosome organelles via crossing the outer membrane to the periplasmic space or the iron has to be transported into the magnetosome vesicle, which is physically separated from the cytoplasmic membrane via the periplasm of the bacteria (Schüler 2008).

14.3.2.1 Chemical Pathway of Biomineralization

Four main steps have been proposed for the mechanism of magnetite formation in *Magnetospirillum magneticum* (Baumgartner et al. 2013c) (Fig. 14.7): after iron uptake from the environment, a highly disordered phosphate-rich iron(III) hydroxide phase is first observed outside the magnetosome organelle. This precursor phase is consistent with prokaryotic ferritins as was shown by transmission electron microscopy and X-ray absorption (Baumgartner et al. 2013c). In the next step, a transient ferrihydrite-like intermediate, which is a purely inorganic material, is found within the magnetosome organelle. Basically, the ferritin-like precursor is transformed by the loss of the phosphate. The third step is the partial reduction of Fe^{3+} or the direct addition of Fe^{2+}. Indeed, ferrihydrite is a Fe(III) species, whereas magnetite is a mixed Fe(II)-Fe(II) phase. Therefore, the formation of magnetite from a ferrihydrite intermediate requires a partial reduction. Magnetite is formed by the aggregation of small particles in a process reminiscent of what was described in the nonclassical synthetic pathway (Sect. 14.2.2). The final step is particle growth toward mature size to reach the magnetic stable single domain dimension (Baumgartner et al. 2013c), since the first observed crystals are about 10 nm in size and therefore superparamagnetic.

14.3.3 Influence of Proteins on Magnetite Formation In Vivo

The magnetosome formation is biologically controlled by *mam* (for *ma*gnetosome *m*embrane) and *mms* (for *m*agnetosome *m*embrane-*s*pecific) genes. The genes

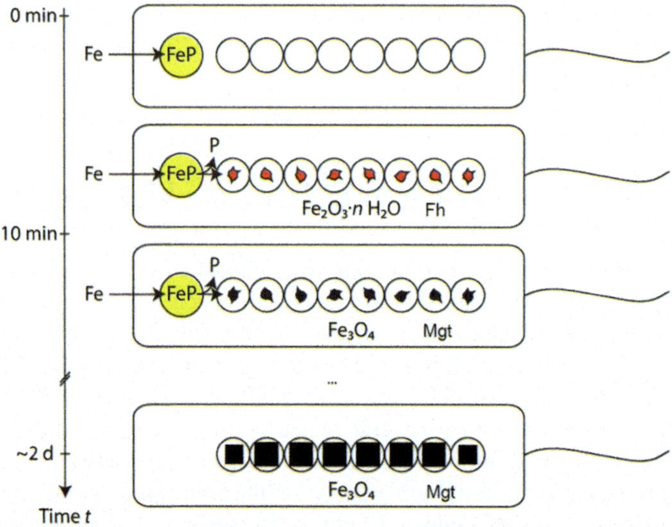

Fig. 14.7 Scheme of the proposed magnetite biomineralization pathway in magnetotactic bacteria. After iron uptake, storage in a phosphate-rich ferric hydroxide phase (FeP) occurs. Phosphorus and iron are separated into the magnetosome compartments in a transfer process. This leads to the formation of ferrihydrite (Fh) ($Fe_2O_3 \cdot n\ H_2O$). The reduction of Fh results in magnetite (Mgt), which undergoes further growth to mature crystals (Modified from Baumgartner et al. (2013c) with permission from PNAS)

responsible for biomineralization and organization of magnetosomes are clustered in the magnetosome genomic island (Schübbe et al. 2003; Bazylinski and Frankel 2004; Ullrich et al. 2005). Recently, further bacterial strain was isolated in pure culture and their genome sequenced and annotated (Jogler and Schüler 2009). Comparative analysis of the data led to the discovery of *mad* genes, which are specific to strains forming elongated magnetosome particles, and therefore the genes were putatively associated with the control of the magnetosome morphology (Lefèvre et al. 2013).

To understand the molecular mechanism of magnetite biomineralization in more detail, two model organisms are particularly studied: *M. gryphiswaldense* (MSR-1) and *M. magneticum* (AMB-1). In these strains, the role of particular proteins has been studied in details and is briefly reviewed below (Table 14.1). The proteins MamG, MamF, MamD, and MamC are the most abundant proteins found in the magnetosome membranes, and they help in the control of the magnetosome size (Scheffel et al. 2008). MamK is a bacterial actin-like protein, which stabilizes the magnetosome chain (Katzmann et al. 2010). MamJ interacts with MamK as it anchors the magnetosomes to the MamK filaments (Lin et al. 2014). MamE and MamO are proteins, which are involved in the arrangement of other proteins into functional complexes within the magnetosome membrane (Lower and Bazylinski 2013). MamP is involved in the control of the redox condition

Table 14.1 Different magnetosome protein groups. Different functions of magnetosome membrane proteins are in *italics*

Name of protein	Origin of the name (abbreviation)	Name of protein	Function of protein
Mam	Magnetosome membrane	*MamB, MamI, MamL, MamQ, MamY*	Membrane formation
Mme	Magnetosome membrane	*MamA, MamM, MamN, MamO, MamP*	Magnetite crystallization
Mms	Magnetosome membrane specific	*MamC, MamD, MamE, MamF, MamG, MamR, MamS, MamT, MmsF, Mms6*	Control of particle size
Mtx	Magnetotaxis	*MamJ, MamK*	Chain assembly
Mad	Magneto-deltaproteobacteria specific		

Lower and Bazylinski (2013)

within the magnetosome organelle and thereby helps in the establishment of the physicochemical condition needed for magnetite precipitation (Siponen et al. 2013). The most important proteins and their roles are summed up in Table 14.1.

14.4 Biomimetic Formation of Magnetite

14.4.1 Magnetite Synthesis in the Presence of Magnetosome-Related Additives

As discussed in the previous section, magnetosome formation is strictly controlled within the cell by different proteins. MTB can produce magnetite nanoparticles at room temperatures and atmospheric pressures and with the help of biological determinants. In this section, we focus on three of these dedicated proteins and discuss their function when using them as biological additives for magnetite nanoparticle synthesis in vitro.

Historically, Mms6 originating from *M. magneticum* AMB-1 was the first protein identified to potentially affect magnetite formation in vitro (Arakaki et al. 2003). The protein was indeed found to be tightly bound to the magnetosome particle in vivo, and when used as additive in coprecipitation experiments, the size was qualitatively observed to be larger than in the control (Arakaki et al. 2003). Later, the same group of authors showed that the particle size could be decreased when the partial oxidation technique was used (Arakaki et al. 2010). The authors concluded that Mms6 controls magnetite size to obtain stable single domain particles both in vivo and in vitro. Mms6 C-terminus is hydrophilic and was postulated to accumulate iron ions, thereby triggering crystal nucleation and interacting with the magnetite nanoparticle surface, while the crystal is growing in vivo. However, the biological function and the influence of the protein on crystal nucleation and growth are not yet confirmed (Murat et al. 2012).

The second protein we focus on is MamJ, which is known as magnetosome connector to the magnetosome filament MamK. Due to the acidic residues of MamJ, binding moieties for cationic Fe^{II}/Fe^{III} are eventually provided. When using MamJ as an additive in magnetite synthesis, a strong effect on crystallinity, particle size, aggregation, morphology, and phase of the precipitate is obtained (Baumgartner et al. 2014). When using low concentrations of MamJ, extremely small magnetite nanoparticles are formed at pH 11, whereas an unidentifiable mixture of minerals is obtained at pH 9. When using higher concentrations of MamJ, either at pH 9 or pH 11, a poorly ordered pattern is observed, potentially correlating with goethite (Baumgartner et al. 2014).

Finally, MamP represents a new class of c-type cytochromes, which are exclusively found in MTB (Siponen et al. 2013). MamP acts as an iron oxidase, which contributes to the formation of ferrihydrite that is eventually needed for magnetite crystal growth in vivo. Ferrihydrite would indeed be partially reduced to finally obtain magnetite as described in Sect. 14.3.2.1 above (chemical pathway of biomineralization) (Siponen et al. 2013). Such pathway can also be compared with the synthetic pathway presented in part 3.2 (Baumgartner et al. 2013b). As a precise Fe^{III}/Fe^{II} ratio is required in magnetite crystal formation, MamP enables the formation of the ferrihydrite precursor, which, when Fe^{II} is present in addition, develops toward magnetite in vitro or in the magnetosome (Siponen et al. 2013).

14.4.2 Magnetite Synthesis in the Presence of Other Additives

Polymers, such as polyethylenimine (Laurent et al. 2008), and fatty acids, e.g., oleic acid (Kovalenko et al. 2007), are widely used additives during the synthesis of magnetite nanoparticles. Furthermore capsids, which are empty virus shell proteins (Qazi et al. 2013), and micelles are used to synthesize magnetite nanoparticles inside nanocontainers (Qazi et al. 2013; Reichhardt et al. 2011). In this section, we focus on proteins not specific to MTB and polypeptides, which influence magnetite particle properties during in vitro synthesis.

Ferritin is a ubiquitous protein used for iron homeostasis in nearly every organism. The proteins can be divided in two components: apoferritin, which is the spherical polypeptide shell, and a 6 nm inorganic core made of the hydrated iron oxide ferrihydrite. The shell can also be used in vitro as a nanoreactor. Under controlled conditions, a 6 nm magnetite nanoparticle has been synthesized within the nanodimensional cavity of a horse spleen ferritin (Meldrum et al. 1992). The construct consisting of magnetically filled apoferritin is called magnetoferritin (Meldrum et al. 1992; Lenders et al. 2014).

Polypeptides such as poly-L-glutamic acid (polyE) and poly-L-arginine (polyR) are also used for in vitro studies. The main difference in these polypeptides is

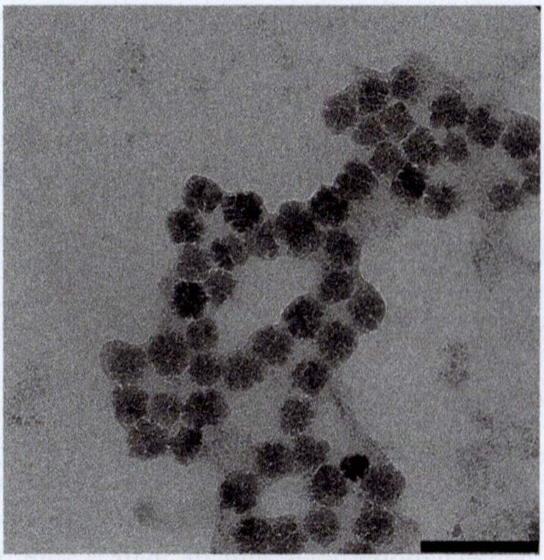

Fig. 14.8 Transmission electron microscopy image of magnetite nanoparticles synthesized in the presence of polypeptides (*scale bar*, 100 nm)

the availability of different charged groups, which can nonspecifically interact with iron or iron (oxyhydr) oxide species in different manners. It has been shown that polycationic polypeptides preferably attach to magnetite crystals (Baumgartner et al. 2014). In the case of poly-L-arginine (polyR), its cationic guanidinium group can electrostatically interact with negatively charged iron (oxyhydr) oxide crystal surfaces in alkaline environment (Lai et al. 2010). PolyR affects the size, morphology, and aggregation behavior of the formed magnetite nanoparticles. Single domain-sized and highly monodispersed stable nanoparticles are formed, which are able to assemble into a chain structure in solution (Fig. 14.8) (Baumgartner et al. 2014). The colloidal stability of these synthetic magnetite nanoparticles is similar to the colloidal stabilization of magnetite crystals by magnetosome membranes in magnetotactic bacteria. In magnetosomes, the compartmentalization is based on a lipid bilayer, where the lipids expose mainly positively charged amines toward the intracellular magnetite crystal (Baumgartner et al. 2014).

In a recent paper, it was shown that nucleation and growth of magnetite crystals using a slow coprecipitation method based on a ferrihydrite precursor can be manipulated by co-polypeptides (Lenders et al. 2015). These polypeptides are used to mimic the function of biomineralization proteins. The ferrihydrite precursor phase was stabilized, and the negatively charged amino acids of the co-polypeptides were proposed to act as inhibitors of magnetite nucleation. In addition, the size and shape of the magnetite crystals could be regulated by varying the amino acid composition of the peptide (Lenders et al. 2015). The dimension of the formed magnetite nanocrystals was affected as demonstrated by magnetic measurements: A ferrimagnetic behavior was measured for particles synthesized with co-polypeptides containing low glutamic acid content and alternatively, a superparamagnetic behavior was observed for particles synthesized in the presence of polypeptides with high glutamic acid content showing that a high glutamic acid content reduces the particle size. When varying the lysine content of the co-polypeptides, the electrostatic stabilization upon protonation of lysine residues increased. This was inferred to protect the particles against uncontrolled agglomeration. A low lysine content leads to stable dispersions at high pH (Lenders et al. 2015). Higher lysine content enables the crystals to self-organize in chain-like assemblies in a similar manner to what was observed with the use of poly-L-arginine (Baumgartner et al. 2014).

14.5 Outlook

Magnetite is a fascinating material, and magnetite nanoparticles were used for many applications in recent years. Especially in the field of nanomedicine, magnetite nanoparticles are of great interest due to their magnetic properties and size ranges.

As we have seen, magnetite occurs in nature where the properties of magnetite nanoparticles and of their assemblies can be controlled. Along this way, understanding the concepts of how biomineralization of magnetite nanoparticles in magnetotactic bacteria is achieved is of particular interest since these simple organisms are able to form particles with a controlled but large library of shape and size but always with monodispersity and colloidal stability. If different additives have already been tested and their roles in magnetite nucleation and growth deciphered, we are still far away from being able to control both size and shape as bacteria do, even further away of doing so with limited conditions that are required for physiological processes.

In addition, the synthetic pathway presented here is already exhibiting high similarities with the biological pathway. Still, and in order to potentially go toward morphology control, it will be useful to determine the structure of the primary particles. Knowing what are these 1–2 nm particles might indeed enable to tune their interactions and orient their aggregation toward anisotropic structures.

Thereby, routes for the synthesis of magnetite particles with controlled properties are emerging, which should pave the road toward an even wider applicability of these fascinating particles.

References

Acton G (2011) Essentials of paleomagnetism. Eos Trans AGU 92:166–166

Annett FA (1921) Electrical machinery; a practical study course on installation, operation and maintenance. McGraw-Hill Book Company, Inc., New York/London

Arakaki A, Webb J, Matsunaga T (2003) A novel protein tightly bound to bacterial magnetic particles in Magnetospirillum magneticum strain AMB-1. J Biol Chem 278:8745–8750

Arakaki A, Masuda F, Amemiya Y, Tanaka T, Matsunaga T (2010) Control of the morphology and size of magnetite particles with peptides mimicking the Mms6 protein from magnetotactic bacteria. J Colloid Interface Sci 343:65–70

Baumgartner J, Bertinetti L, Widdrat M, Hirt AM, Faivre D (2013a) Formation of magnetite nanoparticles at low temperature: from superparamagnetic to stable single domain particles. PLoS One 8:e57070

Baumgartner J, Dey A, Bomans PHH, Le Coadou C, Fratzl P, Sommerdijk N, Faivre D (2013b) Nucleation and growth of magnetite from solution. Nat Mater 12:310–314

Baumgartner J, Morin G, Menguy N, Gonzalez TP, Widdrat M, Cosmidis J, Faivre D (2013c) Magnetotactic bacteria form magnetite from a phosphate-rich ferric hydroxide via nanometric ferric (oxyhydr) oxide intermediates. Proc Natl Acad Sci 110:14883–14888

Baumgartner J, Carillo MA, Eckes KM, Werner P, Faivre D (2014) Biomimetic magnetite formation: from biocombinatorial approaches to mineralization effects. Langmuir 30:2129–2136

Bazylinski DA, Frankel RB (2004) Magnetosome formation in prokaryotes. Nat Rev Microbiol 2:217–230

Bazylinski DA, Lefèvre CT, Lower BH (2014) Magnetotactic bacteria, magnetosomes, and nanotechnology. In: Barton LL, Bazylinski DA, Xu H (eds) Nanomicrobiology. Springer, New York

Brooker LR, Shaw JA (2012) The chiton radula: a unique model for biomineralization studies. INTECH Open Access Publisher, Rijeka

Cornell RM, Schwertmann U (2003) The iron oxides: Structure, properties, reactions, occurrences and uses. Wiley-VCH, Weinheim. ISBN 3-527-30274-3

Dunlop DJ, Özdemir Ö (2001) Rock magnetism: fundamentals and frontiers. Cambridge University Press, Cambridge

Faivre D, Schüler D (2008) Magnetotactic bacteria and magnetosomes. Chem Rev 108:4875–4898

Faivre D, Böttger LH, Matzanke BF, Schüler D (2007) Intracellular magnetite biomineralization in bacteria proceeds by a distinct pathway involving membrane-bound ferritin and an iron(II) species. Angew Chem Int Ed 46:8495–8499

Frankel R, Bazylinski D (1994) Magnetotaxis and magnetic particles in bacteria. Hyperfine Interact 90:135–142

Grachev AF, Pechersky DM, Borisovskii SE, Tselmovich VA (2008) Magnetic minerals in sediments at the Cretaceous/Paleogene boundary (the Gams Section, Eastern Alps). Izv Phys Solid Earth 44:789–803

Grünberg K, Wawer C, Tebo BM, Schüler D (2001) A large gene cluster encoding several magnetosome proteins is conserved in different species of magnetotactic bacteria. Appl Environ Microbiol 67:4573–4582

Guo S, Li D, Zhang L, Li J, Wang E (2009) Monodisperse mesoporous superparamagnetic single-crystal magnetite nanoparticles for drug delivery. Biomaterials 30:1881–1889

Hüsing SK, Deenen MHL, Koopmans JG, Krijgsman W (2011) Magnetostratigraphic dating of the proposed Rhaetian GSSP at Steinbergkogel (Upper Triassic, Austria): implications for the Late Triassic time scale. Earth Planet Sci Lett 302:203–216

Ito A, Shinkai M, Honda H, Kobayashi T (2005) Medical application of functionalized magnetic nanoparticles. J Biosci Bioeng 100:1–11

Jogler C, Schüler D (2009) Genomics, genetics, and cell biology of magnetosome formation. Annu Rev Microbiol 63:501–521

Jolivet J-P, Chaneac C, Tronc E (2004) Iron oxide chemistry. From molecular clusters to extended solid networks. Chem Commun 5:481–483

Katzmann E, Scheffel A, Gruska M, Plitzko JM, Schüler D (2010) Loss of the actin-like protein MamK has pleiotropic effects on magnetosome formation and chain assembly in Magnetospirillum gryphiswaldense. Mol Microbiol 77:208–224

Kim W, Suh C-Y, Cho S-W, Roh K-M, Kwon H, Song K, Shon I-J (2012) A new method for the identification and quantification of magnetite–maghemite mixture using conventional X-ray diffraction technique. Talanta 94:348–352

Kirschvink JL, Gould JL (1981) Biogenic magnetite as a basis for magnetic field detection in animals. Biosystems 13:181–201

Kovalenko MV, Bodnarchuk MI, Lechner RT, Hesser G, Schäffler F, Heiss W (2007) Fatty acid salts as stabilizers in size- and shape-controlled nanocrystal synthesis: the case of inverse spinel iron oxide. J Am Chem Soc 129:6352–6353

Lai Y, Yin W, Liu J, Xi R, Zhan J (2010) One-pot green synthesis and bioapplication of L-arginine-capped superparamagnetic Fe3O4 nanoparticles. Nanoscale Res Lett 5:302–307

Laurent S, Forge D, Port M, Roch A, Robic C, Vander Elst L, Muller RN (2008) Magnetic iron oxide nanoparticles: synthesis, stabilization, vectorization, physicochemical characterizations, and biological applications. Chem Rev 108:2064–2110

Lefèvre CT, Bazylinski DA (2013) Ecology, diversity, and evolution of magnetotactic bacteria. Microbiol Mol Biol Rev 77:497–526

Lefèvre CT, Trubitsyn D, Abreu F, Kolinko S, Jogler C, De Almeida LGP, De Vasconcelos ATR, Kube M, Reinhardt R, Lins U, Pignol D, Schüler D, Bazylinski DA, Ginet N (2013) Comparative genomic analysis of magnetotactic bacteria from the Deltaproteobacteria provides new insights into magnetite and greigite magnetosome genes required for magnetotaxis. Environ Microbiol 15:2712–2735

Lenders JJM, Altan CL, Bomans PHH, Arakaki A, Bucak S, De With G, Sommerdijk NAJM (2014) A bioinspired coprecipitation method for the controlled synthesis of magnetite nanoparticles. Cryst Growth Des 14:5561–5568

Lenders JJM, Zope HR, Yamagishi A, Bomans PHH, Arakaki A, Kros A, De With G, Sommerdijk NAJM (2015) Bioinspired magnetite crystallization directed by random copolypeptides. Adv Funct Mater 25:711–719

Lin W, Benzerara K, Faivre D, Pan Y (2014) Intracellular biomineralization in bacteria. Front Microbiol 5:293

Lower BH, Bazylinski DA (2013) The bacterial magnetosome: a unique prokaryotic organelle. J Mol Microbiol Biotechnol 23:63–80

Mandernack KW, Bazylinski DA, Shanks WC, Bullen TD (1999) Oxygen and iron isotope studies of magnetite produced by magnetotactic bacteria. Science 285:1892–1896

Massart R (1981) Preparation of aqueous magnetic liquids in alkaline and acidic media. IEEE Trans Magn 17

Mcbain SC, Yiu HHP, Dobson J (2008) Magnetic nanoparticles for gene and drug delivery. Int J Nanomedicine 3:169–180

Meldrum F, Heywood B, Mann S (1992) Magnetoferritin: in vitro synthesis of a novel magnetic protein. Science 257:522–523

Murat D, Falahati V, Bertinetti L, Csencsits R, Körnig A, Downing K, Faivre D, Komeili A (2012) The magnetosome membrane protein, MmsF, is a major regulator of magnetite biomineralization in Magnetospirillum magneticum AMB-1. Mol Microbiol 85:684–699

Nudelman F, Sommerdijk NAJM (2012) Biomineralization as an inspiration for materials chemistry. Angew Chem Int Ed 51:6582–6596

Pechersky DM, Nurgaliev DK, Sharonova ZV (2009) Magnetic properties of the boundary layer at the Cretaceous/Tertiary boundary in the Gams section, Eastern Alps, Austria. Izv Phys Solid Earth 45:482–494

QAZI S, Liepold LO, Abedin MJ, Johnson B, Prevelige P, Frank JA, Douglas T (2013) P22 viral capsids as nanocomposite high-relaxivity MRI contrast agents. Mol Pharm 10:11–17

Reichhardt C, Uchida M, O'neil A, Li R, Prevelige PE, Douglas T (2011) Templated assembly of organic–inorganic materials using the core shell structure of the P22 bacteriophage. Chem Commun 47:6326–6328

Ricks R (2010) Removal of a tungsten carbide wedding ring with a diamond tipped dental drill. J Plast Reconstr Aesthet Surg 63:e701–e702

Scheffel A, Gärdes A, Grünberg K, Wanner G, Schüler D (2008) The major magnetosome proteins MamGFDC are not essential for magnetite biomineralization in *Magnetospirillum gryphiswaldense*, but regulate the size of magnetosome crystals. J Bacteriol 190:377–386

Schübbe S, Kube M, Scheffel A, Wawer C, Heyen U, Meyerdierks A, Madkour M, Mayer F, Reinhardt R, Schüler D (2003) Characterization of a spontaneous nonmagnetic mutant of Magnetospirillum gryphiswaldense reveals a large deletion comprising a putative magnetosome island. J Bacteriol 185:5779–5790

Schüler D (2008) Genetics and cell biology of magnetosome formation in magnetotactic bacteria. FEMS Microbiol Rev 32(4):654–672

Shimizu K, Ito A, Yoshida T, Yamada Y, Ueda M, Honda H (2007) Bone tissue engineering with human mesenchymal stem cell sheets constructed using magnetite nanoparticles and magnetic force. J Biomed Mater Res B Appl Biomater 82B:471–480

Siponen MI, Legrand P, Widdrat M, Jones SR, Zhang W-J, Chang MCY, Faivre D, Arnoux P, Pignol D (2013) Structural insight into magnetochrome-mediated magnetite biomineralization. Nature 502:681–684

Solov'yov IA, Schulten K (2012) Reaction kinetics and mechanism of magnetic field effects in cryptochrome. J Phys Chem B 116:1089–1099

Stolz JF, Chang S-BR, Kirschvink JL (1989) Biogenic magnetite in stromatolites. I. Occurrence in modern sedimentary environments. Precambrian Res 43:295–304

Ullrich S, Kube M, Schübbe S, Reinhardt R, Schüler D (2005) A hypervariable 130-kilobase genomic region of *Magnetospirillum gryphiswaldense* comprises a magnetosome island which undergoes frequent rearrangements during stationary growth. J Bacteriol 187:7176–7184

Vayssières L, Chanéac C, Tronc E, Jolivet JP (1998) Size tailoring of magnetite particles formed by aqueous precipitation: an example of thermodynamic stability of nanometric oxide particles. J Colloid Interface Sci 205:205–212

Vinogradov AV, Vinogradov VV (2014) Low-temperature sol–gel synthesis of crystalline materials. RSC Adv 4:45903–45919

Widdrat M, Kumari M, Tompa É, Pósfai M, Hirt AM, Faivre D (2014) Keeping nanoparticles fully functional: long-term storage and alteration of magnetite. ChemPlusChem 79:1225–1233

Zhang X, Niu Y, Meng X, Li Y, Zhao J (2013) Structural evolution and characteristics of the phase transformations between [small alpha]-Fe2O3, Fe3O4 and [gamma]-Fe2O3 nanoparticles under reducing and oxidizing atmospheres. CrystEngComm 15:8166–8172

Zhao Y, Feng C, Li D (2014) The major ore clusters of super-large iron deposits in the world, present situation of iron resources in China, and prospect. Acta Geol Sin Engl Ed 88: 1895–1915

Chapter 15
Silica and Alumina Nanophases: Natural Processes and Industrial Applications

Dominique J. Tobler, Tomasz M. Stawski, and Liane G. Benning

15.1 Introduction

Silicon (Si), aluminium (Al) and oxygen (O) are the three most abundant elements in the Earth's crust. The oxides these elements form constitute the main building block for silicate minerals, and they are most often the end member oxides in rock weathering processes. Due to their ubiquity, the variety of polymorphs they form, each unique in structure, property and reactivity, they are among the most useful natural materials.

Silica, usually denoted as SiO_2, can crystallise in many different crystal systems, forming a variety of polymorphs with stabilities governed by variable pressure and temperature conditions (Heaney et al. 1994). Among these, quartz is the most common and most stable form of SiO_2. When dissolved, the amount of silica in natural water is controlled by physicochemical parameters that are linked to

D.J. Tobler (✉)
Nano-Science Center, Department of Chemistry, University of Copenhagen, Copenhagen, Denmark
e-mail: dominique.tobler@nano.ku.dk

T.M. Stawski
German Research Centre for Geosciences, GFZ, D-14473 Potsdam, Germany

School of Earth and Environment, University of Leeds, LS2 9JT Leeds, UK

L.G. Benning
German Research Center for Geosciences, GFZ, Interface Geochemistry Section, 14473 Potsdam, Germany

Department of Earth Sciences, Free University of Berlin, 12249 Berlin, Germany

School of Earth and Environment, University of Leeds, Leeds LS2 9JT, UK

A.E.S. Van Driessche et al. (eds.), *New Perspectives on Mineral Nucleation and Growth*,
DOI 10.1007/978-3-319-45669-0_15

water–rock interactions (e.g. weathering) or biomineral growth (e.g. diatoms or radiolarians). These can vary significantly between different natural environments, e.g. 100 ppb in seawater to >1000 ppm in near-boiling pressurised geothermal waters. At Earth surface conditions, when SiO_2 levels become supersaturated, silica precipitates as an amorphous phase. Such amorphous silica, often referred to as opal-A, has a local structural coherence smaller than 20 Å, and its solubility, density, hardness and composition (e.g. water content can vary between 1 and 15 %) are highly dependent on precipitation conditions (Perry and Keeling-Tucker 2000). Yet, over time opal-A is unstable and gradually transforms into more crystalline phases and eventually to quartz. The time scale for this transformation process is highly dependent on the physicochemical conditions (i.e. temperature, pressure, pH conditions) (Williams and Crerar 1985) and can take tens of thousands of years (Herdianita et al. 2000).

The alumina system (Al_2O_3–H_2O) is notoriously more complicated, with different polymorphs forming under different temperature and pressure conditions but also at different solution pH. Aqueous Al^{3+} oxy-species (i.e. the aluminates) form as products of aqueous dissolution of aluminosilicates (which make up ~75 % of the Earth's crust) or of simpler $Al(OH)_3$, AlOOH and Al_2O_3 phases, with the highest dissolution rates occurring at acidic and alkaline conditions. In solutions supersaturated with regard to aluminate, typically only $Al(OH)_3$ and AlOOH phases precipitate, whereas Al_2O_3 or complex aluminosilicate phases are kinetically inhibited and usually only form at higher temperatures. In the pure alumina system, about half a dozen of metastable oxide and oxyhydroxide polymorphs are known to exist, and these only transform at >1000 °C to the thermodynamically stable trigonal corundum (α-Al_2O_3). For a full review, see Levin and Brandon 1998 and references therein.

In both the Si and Al systems, the formation of amorphous and (micro)crystalline phases is preceded by hydrolysis and condensation reactions leading to various mono-, di- or polynuclear species (Iler 1979; Brinker and Scherer 1990). In the silica system, solution speciation (Fig. 15.1a) and the type and abundance of polynuclear silica species do not affect polymorph selection, although they are believed to influence the formation kinetics (Iler 1979; Rothbaum and Rohde 1979; Belton et al. 2012). In contrast, the versatility of hydrolysed alumina species (Fig. 15.1b) and the polynuclear complexes it forms, most likely affects alumina polymorph selection (Wefers and Misra 1987; Casey 2006; Sipos 2009). In this chapter, we review the current knowledge on the mechanisms and kinetics of polymerisation and precipitation for these two systems. For simplicity, they will be discussed separately, first focusing on the silica system, considering both processes in natural waters and industrial applications. For the alumina system, we will take into account the amphoteric character of the solid phases, and the precipitation process will be discussed separately for acidic and alkaline solutions.

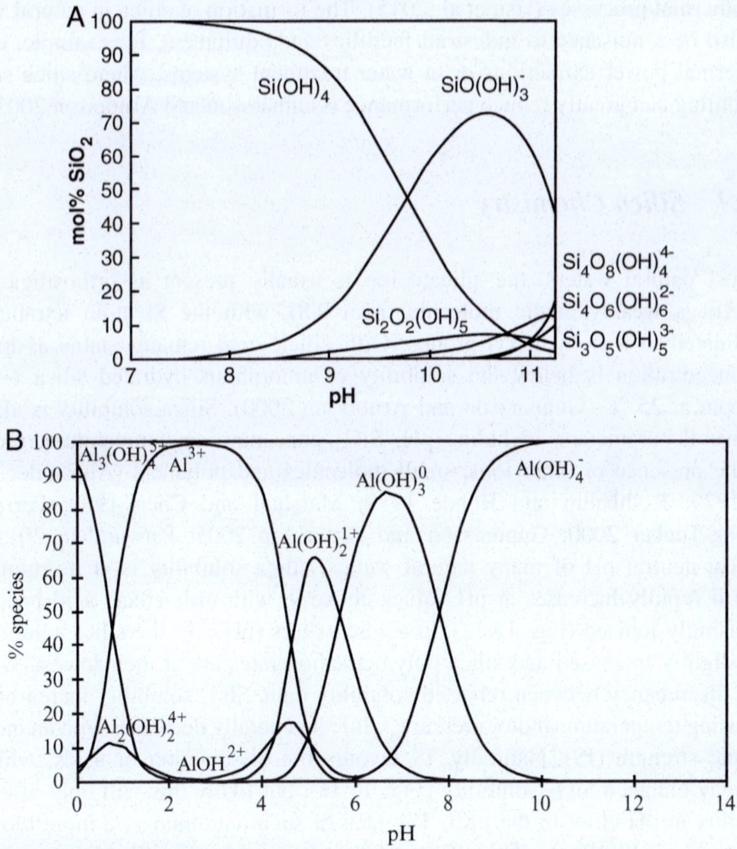

Fig. 15.1 Partial speciation diagrams of dissolved silica (**a**) and alumina (**b**) as a function of pH at 25 °C and 1 atm (Printed with permission from Elsevier from Dietzel (2000) for silica and Panias et al. (2001) for alumina)

15.2 Silica in Natural Waters

Silica polymerisation and the subsequent precipitation of silica occur in many modern terrestrial environments (e.g. hot springs, brines, marine sediments, rivers), and these processes are critical to a variety of processes including sinter formation, silica diagenesis and biosilicification (e.g. diatoms, sponges and plants) (Pancost et al. 2005; Perry and Keeling-Tucker 2000; Tobler et al. 2008), as well as the transport and fate of nutrients and contaminants, with silica nanoparticles acting as adsorbents or mobile carriers (Ryan and Elimelech 1996). Silica also played an important role in ancient geological settings; for example, most Archean fossils were preserved in silica cherts (Westall and Walsh 2000). Furthermore, such processes were likely also crucial in the formation of silica-rich deposits on Mars (Squyres et al. 2008) or on other extraterrestrial bodies with the recent discovery of silica-rich nanoparticles ejected from Enceladus, Saturn’s sixth largest moon, pointing towards subsurface hydrothermal processes (Hsu et al. 2015). The formation of silica in natural waters can also be a nuisance to industrial facilities and equipment, for example, during geothermal power extractions or in water treatment systems, where silica scaling and fouling can greatly reduce performance (Gunnarsson and Arnorsson 2005).

15.2.1 Silica Chemistry

In most natural waters, the silicate ion is usually present as orthosilicic acid, $Si(OH)_4$, a weakly acidic molecule (pKa 9.8), with the Si atom tetrahedrally coordinated to four hydroxyl groups. Orthosilicic acid remains stable as long as its concentration is below the solubility of amorphous hydrated silica (~110–120 ppm at 25 °C; Gunnarsson and Arnorsson 2000). Silica solubility is affected by several parameters, including pH, SiO_2 concentration, temperature, pressure and the presence of other ions, small molecules and polymers (Alexander 1954; Iler 1979; Rothbaum and Rohde 1979; Marshall and Chen 1982; Perry and Keeling-Tucker 2000; Gunnarsson and Arnorsson 2005; Patwardhan 2011). At the near-neutral pH of many natural waters, silica solubility is at its minimum, while it rapidly increases at pH values above 9, with orthosilicic acid becoming increasingly ionised (Fig. 15.1a). In acidic waters ($pH < 3$), the silica solubility is only slightly increased and silica polymerisation rates are at their lowest. Despite some discrepancy between reported solubility data, SiO_2 solubility increases with increasing temperature and/or pressure, while it generally decreases with an increase in ionic strength (IS). Naturally, IS favours the dissociation of acids, which in turn may enhance SiO_2 solubility (Fig. 15.1a), but likely this will only affect the solubility at pH close to the pKa. The lack of such data hinders a more thorough discussion on IS effects. The effects of inorganic and organic additives on silica solubility and polymerisation are highly variable, and the extent and the pathways in which they affect silica formation are still poorly quantified, but good insights can be found in Iler (1979) and Patwardhan (2011).

In solutions supersaturated with respect to amorphous silica, orthosilicic acid undergoes condensation reactions via the coalescence of orthosilicic acid molecules and the concurrent release of water:

$$Si(OH)_4 + Si(OH)_4 \leftrightarrow (HO)_3Si{-}O{-}Si(OH)_3 + H_2O$$

Further oligomerisation leads to the formation of trimers, tetramers, cyclic species and other polynuclear entities, thereby maximising the number of Si–O–Si bonds. Under most experimental conditions, the orthosilicic acid remains the dominant dissolved silica species (Icopini et al. 2005), while the composition and abundance of oligomers are primarily governed by pH and SiO_2 concentration but also temperature, ionic strength and the presence of other cations (Belton et al. 2012). These polynuclear species undergo condensation and dissolution processes up to a point where a critical cluster size is reached, and cluster growth becomes more likely than cluster dissolution. This is also termed the nucleation barrier. Once reached, it leads to the formation of three-dimensional, internally condensed and highly hydrated silica nuclei, a few nanometres in size (Iler 1979). These nuclei can grow through monomer addition and/or through particle aggregation/coalescence (discussed in Sect. 15.2.2) and interact with the biochemical environment in which silica will be deposited. Overall, silica precipitates forming in surface waters are mostly amorphous (i.e. opal-A) and not affected by oligomer type or abundance (contrary to the alumina system, see Sect. 15.4). However, in some studies, silica precipitation rate laws have been inferred as being a consequence of polymerisation of certain oligomers (e.g. a fourth-order reaction corresponding to tetramer condensation) (Rothbaum and Rohde 1979). This is further discussed in the next section where the current understanding of silica formation mechanisms and kinetics is presented.

15.2.2 Silica Formation

The mechanisms of silica polymerisation and silica nanoparticle formation in natural systems have been widely investigated in both laboratory (Alexander 1954; Iler 1979; Carroll et al. 1998, 2007; Tobler et al. 2009; Tobler and Benning 2013; Kley et al. 2014) and field settings (Carroll et al. 1998; Mountain et al. 2003; Tobler et al. 2008; Meier et al. 2014), as well as through computational approaches (Noguera et al. 2015). Silica polymerisation and silica nanoparticle formation are generally described by a three-stage process where (1) silica polymerisation and nucleation of silica nanospheres are followed by (2) particle growth via accretion of silica monomers/oligomers and/or by Ostwald ripening and (3) particle aggregation.

In most natural waters (i.e. $pH > 3$), silica nanoparticles carry a negative charge, which becomes more negative with an increase in solution pH and/or ionic strength (Kobayashi et al. 2005; Barisik et al. 2014). This is explained by the deprotonation of surface silane groups as pH increases and a decrease in surface H^+ concentration with increasing ionic strength. Barisik et al. further observed that for particles with diameters < 100 nm, the negative surface charge increases with a decrease in particle size (Barisik et al. 2014). The higher the negative charge of the silica nanoparticles, the more likely it is that they remain in suspension (because of the electrostatic repulsion) and can get transported with the water flow. However, in most natural waters, silica nanoparticles are not stable within the polymerising solution and they aggregate into dense 3D networks. This is mostly due to IS effects, i.e. the presence of charged species (e.g. coagulating cations such as $Fe^{2+/3+}$ or Al^{3+}) or of "sticky" organic molecules (e.g. polysaccharides) that reduce and even neutralise the negative surface charge (Iler 1979; Perry and Keeling-Tucker 2000; Benning et al. 2005).

Early studies estimated the size of the first silica nanoparticles forming within the polymerising solution to be around 2–3 nm (Iler 1979). This was confirmed and visualised through atomic force microscopy analyses (Conrad et al. 2007). More recently, advances in scattering methods have made it possible to follow

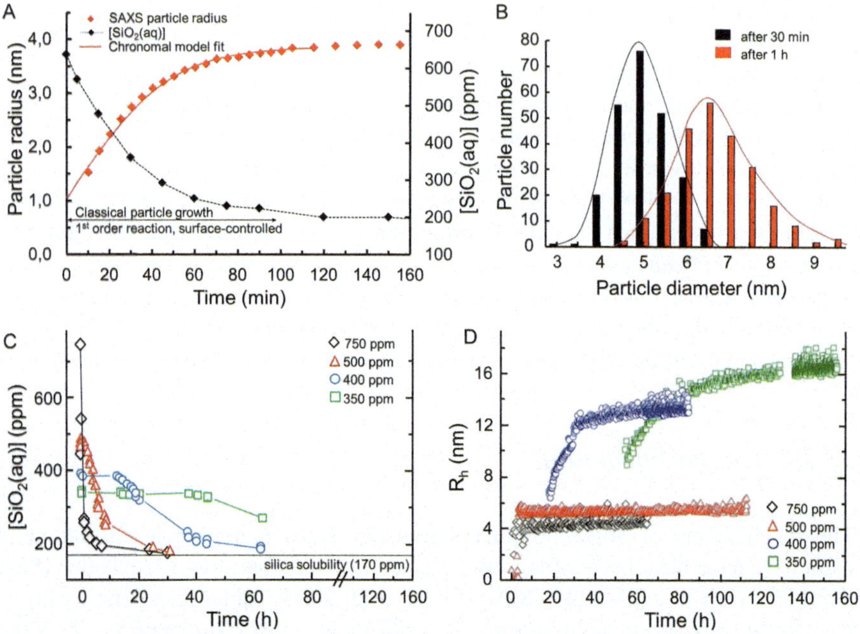

Fig. 15.2 (**a**) Initial steps of silica nanoparticle formation from supersaturated silica solutions monitored by SAXS (640 ppm SiO_2, IS = 0.22, pH = 7). Homogeneous nucleation followed by initial fast particle growth concomitant with the decrease in molybdate-reactive silica, [SiO_2(aq)] (modified after (Tobler et al. 2009). (**b**) Silica nanoparticle size distribution in polymerising solutions for two different ageing times (1600 ppm SiO_2, IS = 0.22, modified after Tobler et al. (2009)). (**c**) Decrease in monomeric silica and (**d**) concomitant growth of silica nanoparticles in solutions with varying initial silica concentrations as indicated (pH = 7; modified after Kley et al. (2014))

in situ and in a time-resolved manner the formation of silica nanoparticles from a polymerising solution. In Tobler et al. (2009), we combined synchrotron radiation small-angle X-ray scattering (SAXS) (Stawski and Benning 2013) and dynamic light scattering (DLS) flow-through experiments with (cryo)electron microscopy and monitored in situ the formation of silica nanoparticles from polymerising solutions where supersaturation was induced by a pH drop from about 12 to 7. These particles grew from 3 nm to approximately 8 nm, which was mirrored by a fast decrease in the orthosilicic acid concentration in solution (Fig. 15.2a, b). The size of the silica nanoparticles after 3h of polymerisation was not greatly affected by the tested silica concentrations (640 and 1600 ppm SiO_2) and ionic strengths (IS = 0.02–0.22), while particle aggregation initiated considerably earlier at higher SiO_2 concentrations and IS. Kley et al. (2014) followed a similar approach, employing both static and dynamic light scattering to monitor silica nanoparticle formation in pH-induced silica polymerising solutions at a wider range of SiO_2 concentrations (350–3000 ppm) and at two pH regimes (7 and 8). They showed that the particle density increased and the final particle size decreased with increasing silica content (Fig. 15.2c, d) and pH, as expected from classical nucleation and growth theory. Similar to the work reported in Tobler et al. (2009), Kley et al. found that particle aggregation quickly set in at high SiO_2 concentrations and that at pH 7 aggregation was more enhanced than at pH 8 (Kley et al. 2014). They explained this as a consequence of an increase in negative surface charge on silica particles with increasing pH and with decreasing size and suggested that this leads to repulsive forces, inhibiting particle aggregation (Barisik et al. 2014; Kley et al. 2014). In an attempt to better mimic natural processes, where silica polymerisation and silica nanoparticle formation are the result of cooling (often extremely fast) of a high-temperature, silica-supersaturated near-neutral fluid to ambient (geothermal) or low temperatures (deep sea), we employed an approach similar to that used in Tobler et al. (2009) but used a high-temperature flow-through experimental system to monitor silica nanoparticle formation in temperature-induced silica polymerising solutions (Tobler and Benning 2013). Our results revealed that the rate at which silica nanoparticles formed was substantially lower (around 50 %) when polymerisation was induced by fast cooling as opposed to pH change (both systems having identical initial solution composition, i.e. $[SiO_2]$). This was evidenced by the occurrence of a lag time in the onset of particle growth, the formation of larger critical nuclei and the absence of particle aggregation in the temperature-induced experiments. This retardation in silica formation is explained by the differences in time to establish supersaturation: the radical change in pH (<30 s) imposed a faster attainment of supersaturated conditions compared to the slower (2–3 min) and more gradual change in T.

Although there are some differences in measured particle sizes and particle growth profiles between these three in situ studies (Fig. 15.2), they do however agree in the mechanistic interpretation of silica nanoparticle formation with silica polymerisation leading to the formation of condensed critical nuclei, which grow by addition of monomers according to classical growth theory up to the point where silica solubility is reached and/or particle aggregation sets in (Fig. 15.2a). Ostwald ripening has been suggested to be an important growth mechanism during the late stages of silica particle growth (Iler 1979; Perry and Keeling-Tucker 2000); however this process could not be identified at the conditions tested in these studies. Ostwald ripening seems particularly unlikely at higher SiO_2 concentrations (>1000 ppm SiO_2), where particle aggregation quickly follows particle growth.

Several studies have observed the occurrence of a lag time (i.e. induction period), where no silica polymerisation and particles are detected (i.e. no significant X-ray/light scattering above background), although at the start of these experiments, the experimental solutions were already at maximum supersaturation (Rothbaum and Rohde 1979; Tobler and Benning 2013; Kley et al. 2014). The length of this induction period is controlled by the same factors that determine silica solubility (i.e. T, pH, IS and $[SiO_2]$), and it decreases with an increase in silica supersaturation (Iler 1979; Rothbaum and Rohde 1979; Gunnarsson and Arnorsson 2005; Conrad et al. 2007). It is generally accepted that induction periods, recently also described as supersaturation plateaus (Noguera et al. 2015), represent the time during which larger polymeric cluster structures and/or particles form and dissolve again (Perry and Keeling-Tucker 2000), in agreement with classical nucleation theory. Such large structures could be similar to the pre-nucleation clusters suggested for other mineral systems (e.g. $CaCO_3$; De Yoreo et al. 2017, Chap. 1). In both the silica and calcium carbonate systems, particle nucleation likely occurs once such cluster structures become dense enough to form the first solid particle.

This shows that despite the extensive research carried out in this field, there are still a few gaps in the molecular-level mechanistic understanding of the nucleation and growth of primary silica particles in natural aqueous solution. This is further illustrated by the variety of kinetic models that have been derived from measurements of the time-dependent decrease in orthosilicic acid concentration or silica nanoparticle growth profiles, with reaction orders ranging between 1 and 6 (Tobler et al. 2009; Tobler and Benning 2013; Kley et al. 2014; Noguera et al. 2015) and with varying dependencies on, for example, pH, [SiO_2], particle surface area or oligomer type. Due to the massive interest of industry in the production of monodisperse silica nanoparticle, the bulk of the research has focused on the mechanistic and kinetic understanding of silica nanoparticles synthesised using organosilanes as a precursor (see Sect. 15.2.3). However, differences in the formation process may prevent the resulting models from being applied to inorganic natural systems.

15.2.3 Silica for Industrial Applications

Silica nanoparticles are highly desirable materials for industrial applications (e.g. electronics, biotechnology, catalysis, water purification and chromatography) because of their specific structural properties (e.g. swelling capacity, strength, durability, thermal stability, dielectric properties) (Ab Rahman and Padavettan 2012; Hench and West 1990; Bagwe et al. 2006; Wang et al. 2008). In natural waters, silica is formed from an inorganic building block, i.e. orthosilicic acid, while most synthetic silica nanoparticles are produced through the hydrolysis and condensation of alkoxysilanes, for example, tetraethoxysilane (TEOS) and tetramethoxysilane (TMOS). Highly monodisperse, spherical silica particles have been obtained through the so-called Stöber method (the base catalysed hydrolysis and condensation of alkoxysilanes in low-molecular-weight alcohols) (Stöber et al. 1968), but other processes such as the diffusion of alkoxysilanes into a stirred aqueous solution of lysine have also been investigated (Yokoi et al. 2006). In aqueous solutions with acidic or basic catalysts, alkoxyl groups are hydrolysed to form silanol groups and alcohol. The condensation between silanol groups (Si–OH) or between silanol and alkoxides (Si–OR) group creates siloxane bridges (Si–O–Si). These processes are affected by the type of alkoxysilane, the nature and composition of solvent and catalyst, temperature, pH, as well as reactant mixing procedures (Ab Rahman and Padavettan 2012; Besselink et al. 2013), which can be tuned to produce specific particle sizes and distributions. Silica nanoparticles prepared with alkoxysilanes have much smoother surfaces compared to silica particles formed by the condensation of orthosilicic acid in water. This is argued to be due to the reversible exchange of alkoxyl (Si–OR) and hydroxyl (Si–OH) groups, thereby creating a relaxation mechanism for further smoothening, while condensation of orthosilicic acid produces fixed Si−O−Si bonds (Carcouet et al. 2014).

The formation conditions for industrial silica nanoparticles greatly differ from those in nature, thus the reaction kinetics and mechanisms, to form highly size-controlled, smooth and monodispersed silica nanoparticles during the hydrolysis and condensation of alkoxysilanes, naturally deviate from those that occur in nature (Sect. 15.2.2). However, there are quite a few similarities discussed below. Scattering and spectroscopy techniques combined with (cryogenic) electron microscopy and molecular modelling have provided evidence that silica polymers with an open fractal structure are the early species, which, once sufficient levels of supersaturation are reached, transform into a dense sphere, with a size of approximately 2 nm (Boukari et al. 2000; Green et al. 2003; Fouilloux et al. 2011; Carcouet et al. 2014). However, there is still debate on the mechanistic steps that lead to the monodispersed mature particles formed during the hydrolysis and condensation of alkoxysilanes, possibly also because of the large number of parameters that impact these processes (i.e. type and concentration of alkoxysilane, solvent and catalyst, mixing procedure, etc.). Similar to silica nanoparticle formation in natural systems, growth by monomer addition (classical growth theory) has been a suggested mechanism (Matsoukas and Gulari 1989). Several studies have argued, however, that the observed particle growth trends could also be explained by particle aggregation (Bogush and Zukoski 1991). In the monomer addition model, nucleation occurs in the initial stages leading to the formation of a set number of nuclei, which then grow by the addition of hydrolysed monomers. On the other hand, in the aggregation model, nucleation occurs continuously throughout the reaction, producing nuclei of a certain size, which first aggregate to form larger particles and then grow by further addition of the primary nuclei. A recent study combining high-resolution cryogenic transmission electron microscopy (TEM) with ^{29}Si magic angle spinning (MAS) nuclear magnetic resonance (NMR) gives convincing evidence of the formation of monodisperse silica nanoparticles through the aggregation of uniform 2.3 nm sized primary particles to larger (>10 nm) units (Fig. 15.3) (Carcouet et al. 2014). These larger units grow by further addition of primary particles, which collapse upon association from 2.3 nm in solution to about 1.3 nm when bound to an aggregate (associated with the expulsion of water). They further showed that free primary particles were present at all stages of particle growth (Fig. 15.3), which demonstrates that nucleation occurs throughout the reaction. Further support for the aggregation model comes from a seeded growth experiment (Bogush and Zukoski 1991), where the decrease in orthosilicic acid concentration was unaffected by the number density of particles, which suggested that particle growth did not occur via direct condensation of monomers onto the particle surface. In the aggregation model, the fast initial decrease in orthosilicic acid concentration (i.e. fast consumption of monomers once the reaction is induced) is explained by the formation of ionised,

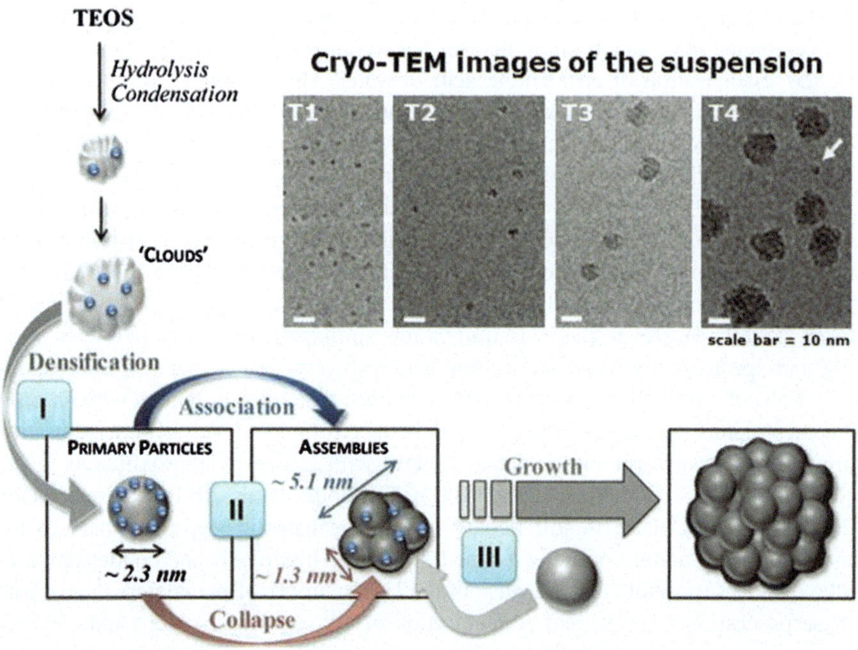

Fig. 15.3 Schematic representation of silica nanoparticle formation mechanism modified after Carcouet et al. (2014). Initial open structures densify to primary particles (Stage I; T1) that collapse upon association into assemblies (Stage II; T2, T3). After association, no further new assemblies are formed, and growth occurs via the addition of primary particles (Stage III; T4, arrows point at a free primary particle)

open structures of low density that collapse to form primary particles. This leads to a steady increase in particle density up to a point where these primary particles start to aggregate into more stable assemblies, leading to a sharp decrease in particle density. Following formation of these assemblies, the particle number remains stable, and the decrease in orthosilicic acid concentration proceeds at a much lower rate (producing new primary particles for growth through aggregation). Similarly, Fouilloux et al. reported an initial fast increase in the particle number density up to a plateau, where particle numbers remained constant, but particle growth still occurred (Fouilloux et al. 2011). They argued that this result supports a mechanism with two consecutive phases: (a) nucleation and growth and then followed by (b) phase densification. However, as pointed out by Carcouet et al., it could also be that primary particles with sizes lower than 3 nm were present but simply not visible by the employed techniques of these previous studies (Carcouet et al. 2014). Also, the usually observed plateau in particle number density occurring during the particle growth process can be easily explained with the aggregation processes described above.

15.3 Alumina in Natural Waters and in Aqueous Industrial Environments

Aluminium is contained in more than 250 different minerals, most noteworthy in the form of common aluminosilicates (e.g. feldspars or clays) and the economically important bauxite ore (mixed Al oxides and hydroxides). Rock weathering and soil formation processes are to a large extent driven by the interaction of fluids with rocks leading to the dissolution of these various phases containing Al. Similar to silica nanoparticles, colloids in the alumina–water system control the cycling of various species in natural waters by introducing a charged solid-liquid interface to which a multitude of inorganic (e.g. phosphate, arsenic, lead, iron, chromium) and organic (e.g. humic acid) compounds can adsorb (Kasprzyk-Hordern 2004). This way alumina colloids are involved in cycling of anthropogenic pollutants in soils and natural waters and have many applications in water treatment (Hu et al. 2006). Al-containing compounds and their processing are also essential for various industries, for example, metallurgy and welding, ceramic production, catalysis, electronics, pharmaceutics, cosmetics, fire prevention, paint production, etc. (Wefers and Misra 1987; Levin and Brandon 1998).

The main difference between the alumina–water and silica–water systems is that in the alumina system, there are a multitude of possible Al-oxyhydroxide phases that can precipitate from solution. Typically, amorphous colloids form at the beginning and then transform directly in solution to crystalline polymorphs (Wefers and Misra 1987). It is generally accepted for the alumina system that the reaction pathways and the products strongly depend on the Al speciation in solution. The various aqueous Al^{3+} oxy-species (i.e. the aluminates) form through dissolution of amphoteric Al phases under acidic and alkaline conditions, as illustrated in Fig. 15.1b. In natural waters, the precipitation of Al phases usually occurs through the neutralisation of Al-rich acidic solutions ($pH > 3$–4), while in industrial settings, the neutralisation is most often from alkaline liquors ($pH < 9$–10, i.e. the Bayer process used for processing of bauxite ores). Unsurprisingly, due to the geochemical and industrial importance of the alumina system, a substantial body of literature on precipitation from solution as well as speciation of aqueous entities exists (e.g. see the reviews by Wefers and Misra (1987), Swaddle et al. (1994), Casey (2006), Sipos (2009) and references therein). Below we present a brief summary of the current knowledge.

15.3.1 Solid Phases in the Alumina–Water System

The water–alumina system is characterised by a plethora of polymorphs, and thus it is inevitable to consider first the possible solid products of precipitation from the aluminate solutions. Excluding pure Al_2O_3 phases (since they typically do not precipitate from solution), these include (Wefers and Misra 1987; Ruan et al. 2001; Kloprogge et al. 2006; Demichelis et al. 2009; Smith et al. 2013):

- Colloidal amorphous and gelatinous hydrated alumina(s) formed through condensation of aluminate complexes (Bale and Schmidt 1959; Petz 1968; Nail et al. 1976a, b, c; Rousseaux et al. 2002) and including all nanometre-sized amorphous oxides, hydroxides and oxyhydroxides.
- Poorly crystalline pseudo-spinel (Bradley and Hanna 1994) and pseudo-boehmite structures formed through ageing of the amorphous phases mentioned above and containing up to 30 wt% water with respect to the Al_2O_3 stoichiometry (Tettenhorst and Hofmann 1980; Wefers and Misra 1987; Brinker and Scherer 1990). Further ageing of pseudo-boehmite leads to either trihydroxide(s) (at $T < \sim 75$ °C) or oxyhydroxide(s) ($T > 75$°C) (Wefers and Misra 1987).
- Crystalline Al-trihydroxides ($Al(OH)_3$), in which the Al^{3+} is octahedrally coordinated by OH^- groups, and each OH^- is bound to two cations with a vacant third octahedron, thus forming neutral sheet structures (Demichelis et al. 2009). The most studied $Al(OH)_3$ phases are:

  - Gibbsite, the most common and thermodynamically stable phase among hydroxides in natural systems (major component in bauxite ore). Structurally it contains OH^- ions in consequent and opposite layers with a sequence arranged as {–AB-BA-AB-BA-…} in the direction perpendicular to the layers – it means that OH^- ions of the adjacent groups are located directly in front of each other (Fig. 15.4a) (Saalfeld and Wedde 1974).
  - Bayerite, which crystallises under highly basic conditions during processing of bauxites (i.e. the Bayer process). Its layers follow the sequence {–AB-AB-…} – OH^- ions from one layer are shifted in respect to the previous one, so that the ions from one layer fall into spaces in-between hydroxyls from the neighbouring ones (Fig. 15.4a) (Rothbauer et al. 1967).
  - Nordstrandite, which crystallises upon ageing of precipitates at mildly basic conditions (pH ~ 7.5–9). Its structure follows the bayerite layer pattern, but with oppositely located hydroxyl groups, hence yielding the sequence {–AB-AB-BA-BA-…}(Fig. 15.4a) (Saalfeld and Jarchow 1968).
  - Doyleite, only discovered recently (Caho et al. 1985), has a similar layer stacking as bayerite (Fig. 15.4a), but the hydroxyls of the two consequent double layers are located in intermediate positions between those of gibbsite and nordstrandite (Demichelis et al. 2009).

- Al-oxyhydroxides (AlOOH) are composed of double layers, in which the primary motif 2 AlOOH forms chains (Wefers and Misra 1987):

  - Boehmite, precipitating upon neutralisation of aluminate solutions at elevated temperatures, in which 2 AlOOH-based chains are arranged in a cubic packing (Fig. 15.4a) (Christoph et al. 1979).
  - Diaspore, in which the 2 AlOOH-based chains are arranged in a hexagonal packing (Fig. 15.4a) (Busing and Levy 1958). This phase is typically found as the product of weathering or high-temperature transformations (Keller 1978).

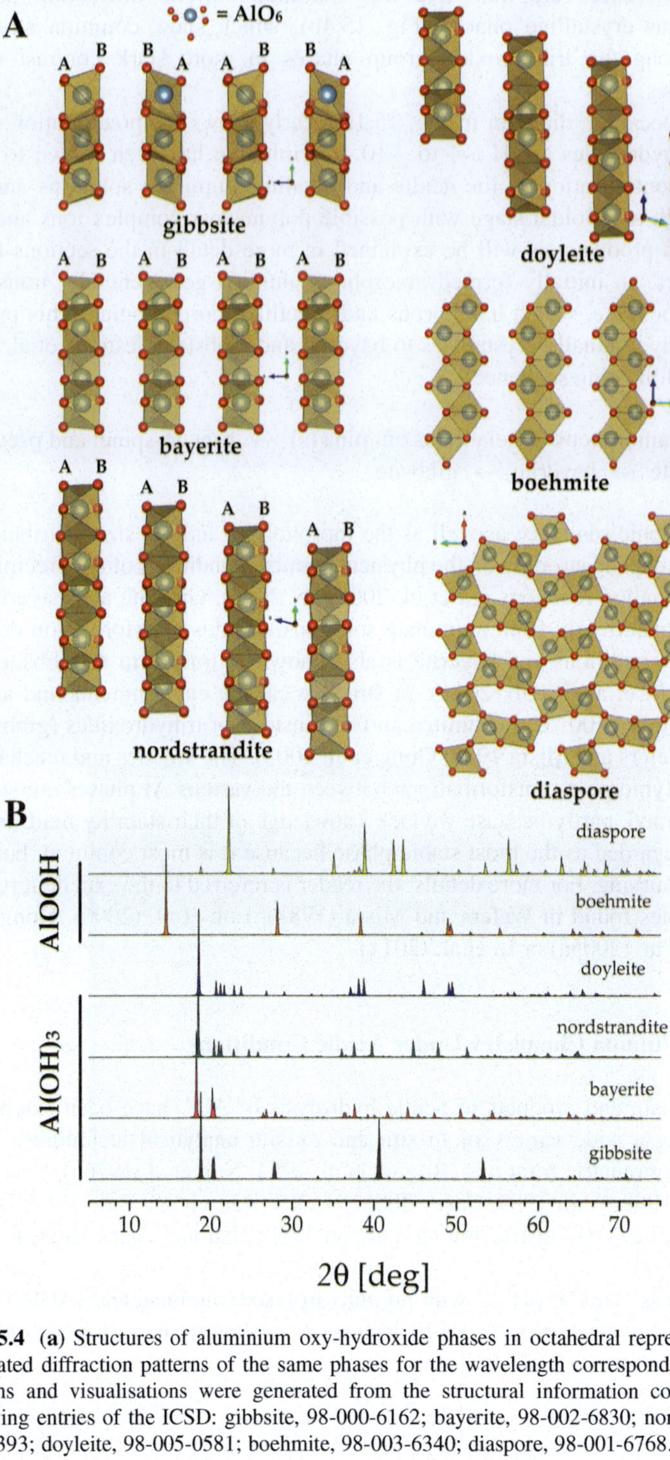

Fig. 15.4 (**a**) Structures of aluminium oxy-hydroxide phases in octahedral representations. (**b**) Simulated diffraction patterns of the same phases for the wavelength corresponding to Cu $K_{\alpha 1}$. Patterns and visualisations were generated from the structural information contained in the following entries of the ICSD: gibbsite, 98-000-6162; bayerite, 98-002-6830; nordstrandite, 98-003-4393; doyleite, 98-005-0581; boehmite, 98-003-6340; diaspore, 98-001-6768. Visualisations and diffraction pattern simulations were prepared in VESTA 3 (Momma and Izumi 2011)

These differences are translated into variations in the diffraction patterns of the various crystalline phases (Fig. 15.4b), which show common structural features among the trihydroxide group phases in more stark contrast to the oxyhydroxides.

The Al speciation diagram in Fig. 15.1b clearly shows the precipitation region for solid trihydroxides at pH ~4 to ~10. Precipitation has been shown to occur during the neutralisation of the acidic and alkaline aluminate solutions and proceeds through a colloidal stage with possible polynuclear complex ions and their condensation products, as will be explained in more detail in the sections below. In both cases, an initially formed amorphous alumina gel eventually transforms to pseudo-boehmite, which has fibrous and sheetlike morphologies. This pseudo-boehmite only gradually crystallises to bayerite and gibbsite (Cesteros et al. 1999) through the following sequence:

Colloidal amorphous or gelatinous alumina (s) → pseudo-spinel and pseudo-boehmite → bayerite → gibbsite

The polymorphic sequence as well as the morphology and the size distribution of the products depend strongly on the physicochemical conditions of the precipitation and transformation reactions (Li et al. 2005a, b, 2011). Gibbsite and bayerite can crystallise concurrently from aluminate solutions, yet this transformation depends on chemical conditions and bayerite is also known to transform to gibbsite (Loh et al. 2000; Li et al. 2005b, 2011). In strongly caustic environments and at temperatures close to 100 °C, boehmite can form instead of trihydroxides (gibbsite or bayerite) (Wefers and Misra 1987; Gong et al. 2003). The kinetics and mechanisms for these polymorphic transformations between the various Al phases are still not fully understood, partly because we lack knowledge of their stability field. Indeed, gibbsite is regarded as the most stable phase because it is most common, but clear evidence is missing. For more details, the reader is referred to the excellent reviews and summaries found in Wefers and Misra (1987), Loh et al. (2000), Gong et al. (2003), Li et al. (2005a) or Li et al. (2011).

15.3.1.1 Alumina Chemistry Under Acidic Conditions

The mechanism and products of acidic hydrolysis of Al^{3+} have been extensively studied using a wide variety of in situ and ex situ analytical techniques. These include potentiometric titrations (Brosset et al. 1954; Nail et al. 1976a), ^{27}Al NMR (Akitt et al. 1988; Akitt 1989; Casey 2006), SAXS (Rausch and Bale 1964; Bottero et al. 1982) and X-ray diffraction (Johansson 1960; Hsu and Bates 1964; Rowsell and Nazar 2000).

The process starts at pH < 3 with an unhydrolysed, mononuclear $[Al(H_2O)_6]^{3+}$ species, which with increasing pH hydrolyses according to the following idealised equations (Brinker and Scherer 1990):

$$[Al(H_2O)_6]^{3+} + xH_2O \rightarrow [Al(OH)_x(H_2O)_{6-x}]^{(3-x)+} + xH_3O^+$$

where x defines the extent of hydrolysis (i.e. the OH/Al ratio ranging from 0 to 3).

The further polycondensation of the so formed $[Al(OH)_x(H_2O)_{6-x}]^{(3-x)+}$ species and how these relate to the final solid Al-polymorphs have been extensively studied, yet the mechanisms are still controversial (Bi et al. 2004 and references therein). Essentially, two condensation mechanisms have been proposed: the "core-links" and the "cagelike" models (Bi et al. 2004).

According to the "core-links" model, the condensation of hexameric $[Al_6(OH)_{12}(H_2O)_{12}]^{6+}$ rings or $[Al_{10}(OH)_{22}(H_2O)_{16}]^{8+}$ double rings leads to larger sheetlike structures. In this case, $[Al_{54}(OH)_{144}]^{18+}$ was proposed to be the last species in the series, before the precipitation of a solid colloidal entity of the $[Al(OH)_3]_n$ type takes place (Bi et al. 2004). These colloids are expected to exhibit a local sheet-structure resembling that of the crystalline $Al(OH)_3$ phases. This model is primarily based on potentiometric titrations (Brosset et al. 1954; Stol et al. 1976) combined with X-ray diffraction analyses (Hsu and Bates 1964). Although the "core-links" model explains well the various titration characteristics, it lacks sufficient structural evidence for the proposed hexameric units and their consequent condensation products.

On the other hand, the "cagelike" model considers the well-characterised tridecamer $[(AlO_4)Al_{12}(OH)_{24}(H_2O)_{12}]^{7+}$ complex ion (Johansson 1960; Johansson et al. 1960) and other similar structures (Casey 2006) to be the actual primary building units in the alumina system. These primary structures are thought to be involved in the nucleation and growth of amorphous colloidal particles and their transformation to various crystalline polymorphs (Bottero et al. 1987; Fu et al. 1991). The tridecamer unit is also known as the Keggin Al_{13} structure, and it has been shown that such units exhibit a very high stability (up to several years) (Casey 2006). The presence of the Keggin ion in aluminate solutions can be easily identified through ^{27}Al NMR, and this method has thus been the primary tool to study the growth mechanisms and structures involving Keggin complex ions (Akitt 1989). Similarly, the structure of another large Al_{30} cation, $[Al_{30}O_8(OH)_{56}(H_2O)_{26}]^{18+}$, composed of two δ-Keggin units linked via four AlO_6 units (Fig. 15.5) has also been revealed by ^{27}Al NMR (Rowsell and Nazar 2000; Allouche et al. 2000; Casey 2006).

It is important to emphasise that depending on reaction conditions, the formation of other polynuclear species has been predicted. However, due to physical limitations of analytical tools (insufficient specificity allowing to distinguish various AlO_6-based species) such as the aforementioned ^{27}Al NMR, many of these species have so far not been characterised or even identified (Casey 2006). Hence, results from the experiments conducted under similar conditions are often interpreted by using either the "core-links" or the "cagelike" model (see the next paragraph). There have been attempts to consolidate the "core-links" and the "cagelike" models by reinterpreting the various experimental data and creating a "continuous" model (Bi et al. 2004). This unified model assumes that the polynuclear species transform

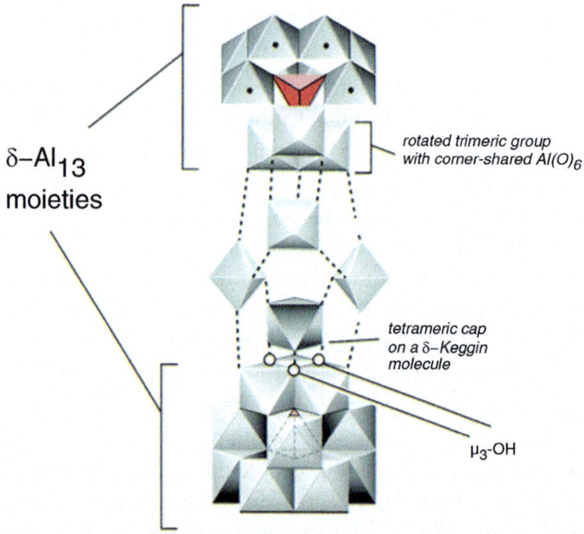

Fig. 15.5 $[Al_{30}O_8(OH)_{56}(H_2O)_{26}]^{18+}$ (aq) (Al_{30}) complex ion shown in a polyhedral exploded view. This structure can be viewed as two Keggin units (Al_{13}) joined by four AlO_6 octahedra. Al_{13} and Al_{30} species are 1 and 2 nm in size, respectively (Printed with permission from Casey (2006). Copyright American Chemical Society 2006)

dynamically through different intermediate stages, where the long-lived Keggin complex ions form upon ageing (i.e. the ion is stable as long as pH or temperature does not drastically change), but the validity of this combined model is still debated.

For instance, Nail et al. studied the early stages of amorphous alumina gel formation from $AlCl_3$ and $Al_2(SO_4)_3$ solutions and postulated a mechanism in which the evolution of polynuclear species to a solid colloidal precipitate was a stepwise process leading to the condensation of hexamers (Nail et al. 1976a). Depending on the original counter anion in the Al^{3+} solution, the colloidal precipitates were suggested to be composed of ~10 six-membered rings for Cl^- and only ~3 for SO_4^{2-}. Leetmaa et al. studied products forming under nearly identical conditions by means of ^{27}Al NMR and infrared/Raman spectroscopy (Leetmaa et al. 2014). Yet, their data also revealed that the products were primarily composed of polynuclear chains of highly distorted aluminium-based octahedra for the Cl^- system and of sulphate-stabilised Keggin Al_{13} clusters for the SO_4^{2-} system. Perry and Shafran evaluated the optimum conditions for the formation of polynuclear Al complexes and found that the neutralisation of $AlCl_3$ and $Al(NO_3)_3$ by KOH and $KHCO_3$ at low ionic strengths favours the formation of Al_{13} clusters and larger polymers (Perry & Shafran 2001). In contrary, in the presence of sulphates, the resulting products were primarily Al_{13} clusters, which was expected due to the stabilisation of the clusters by SO_4^{2-}.

The above-cited studies show that we have been learning more and more about the structures and the synthetic routes for the various polynuclear species for alumina under acidic conditions. Such insight is very important for the systematic analysis of the process, because the mechanistic details of the interplay of the various clusters on the path the crystalline phase remain rather vague at the moment.

15.3.1.2 Alumina Chemistry Under Alkaline Conditions

Processing of bauxite ores (aluminium-rich ores containing primarily gibbsite, boehmite and diaspore) under alkaline conditions is known as the Bayer method, and it is used to form also gibbsite and nordstrandite (Counter et al. 1997). The speciation of Al in basic solutions is rather complex and unsurprisingly a subject of controversy (Sipos 2009). Furthermore, little is known about any polynuclear complex ions that may form under these conditions. In fact, under highly alkaline conditions, the aluminate systems are considered to be actual solutions (see Fig. 15.1b), rather than colloidal systems, as was demonstrated, for instance, by light scattering or infrared and Raman spectroscopies (Li et al. 2003; Sipos 2009). Typically among the aqueous species, only tetrahedral $[Al(OH)_4]^-$ is expected to be present at pH > ~8–9 (Moolenaar et al. 1970; van Straten et al. 1984; Swaddle et al. 1994; Li et al. 2003; Sipos 2009). Some other mononuclear species such as linear $[AlO_2]^-$ and square planar $[Al(OH)_4(H_2O)_2]^-$ complexes may also be present (Gerson et al. 1996), but at insignificant concentrations (Swaddle et al. 1994; Sipos et al. 1998; Sipos 2009). At pH > 13, Al^{3+} can exist as either $[Al(OH)_5]^{2-}$ or $[Al(OH)_6]^{3-}$ (Akitt and Gessner 1984; Barcza and Pálfalvi-Rózsahegyi 1989; Buvári-Barcza et al. 1998), but again these two species are minority species (Sipos 2009) compared to the main $[Al(OH)_4]^-$ species. Using computational approaches, Gale et al. showed that at alkaline conditions dimeric complexes like $[(HO)_3Al(OH)_2Al(OH)_3]^{2-}$ should also be expected (Gale et al. 1998), while the presence (or significance) of higher polynuclear species is still a subject of controversy (an insightful discussion is given in Sipos (2009) and the references therein).

It is known that aluminium in amorphous precipitates is octahedrally coordinated in the same way as it is later in crystalline phases (Tettenhorst and Hofmann 1980; Wefers and Misra 1987). Upon precipitation from acidic solutions, independent of the actual condensation mechanism and possible speciation, one can assume that Al^{3+} also condenses into solids with a sixfold coordination. However, in the Bayer process, the dominant solution species, i.e. the $[Al(OH)_4]^-$ ion, exhibits a fourfold coordination. Hence, one might rationalise the nucleation stage in terms of a transition product, in which the realignment of ligands would occur to establish the octahedral geometry by a swap from the tetrahedral one. The question arises whether abundant $[Al(OH)_4]^-$ species constitute a monomer in crystal growth or whether there are actually other minority species which control the process. Gerson et al. carried out semiempirical quantum modelling of various mononuclear, dimeric, trimeric and tetrameric species in caustic aluminate solutions and suggested a mechanism and possible pathway in which a minority tetrahedral $[Al(OH)_3(H_2O)]$ complex could dimerise and then transform to the tetramer, via a transient trimer moiety (Gerson et al. 1996). In such a tetramer, a central octahedral Al^{3+} ion would be surrounded by three tetrahedral counterparts. However, results from conductivity studies combined with viscosity measurements of caustic aluminate solution were interpreted in terms of the formation of the

$[Al_6(OH)_{24}]^{6-}$ (Barcza and Pálfalvi-Rózsahegyi 1989; Buvári-Barcza et al. 1998) or $[Al_6(OH)_{22}]^{4-}$ moieties (Sipos et al. 1998). In such species, octahedral Al^{3+} geometry dominates, yet actual structural evidence for the proposed mechanisms is still lacking to say the least (Sipos 2009). In more dilute caustic solutions (i.e. [NaOH] / [Al] $\sim$ 1, and [NaOH] $\sim$ [Al] $\sim$ 1.0 mol/L), a polynuclear species in a size range comparable with that of Keggin Al_{13}-like species was proposed based on primarily light scattering evidence (Li et al. 2003, 2005a). Similar colloidal polynuclear aluminate structures were suggested based on optically clear supersaturated aluminate Bayer liquors analysed by cryo-vitrification of solutions at different stages and subsequent imaging by transmission electron microscopy (Counter et al. 1999). In these systems, the Keggin Al_{13}-like species were assumed to be primarily six-coordinated, and their existence in the Bayer process could provide a conceptual link between the mechanisms of $Al(OH)_3$ precipitation from the acidic and alkaline solutions. Finally, the critical nuclei size for gibbsite that forms as an end product in the Bayer process was calculated to have a radius of $\sim$ 1.2 nm (Rossiter et al. 1998), which is in line with the size range of Keggin ions and other small polynuclear oligomers (cf. Fig. 15.5). This might suggest that Keggin ions are sort of a universal "precursor" phase in alumina system regardless of conditions. This is an interesting hypothesis, but yet it has to be confirmed.

To illustrate the influence of physicochemical conditions and speciation onto the nucleation of phases precipitating from aluminate solutions, let us consider the investigations reported by Li et al. concerning the growth of gibbsite and bayerite through the Bayer process (Li et al. 2003, 2005a). The authors studied precipitation using dynamic light scattering to derive particle size distribution and X-ray diffraction for phase identification. Their data showed that dilute solutions ($[Al^{3+}] \leq 0.82$ M and [NaOH] $\leq$ 1.0 M) yielded bayerite and the concentrated ones ($[Al^{3+}] > 2.05$ M and [NaOH] $>$ 2.5 M) gibbsite, whereas a dimorphic product precipitated at intermediate concentrations ($0.82 < [Al^{3+}] \leq 2.05$ M and $1.0 <$ [NaOH] $\leq$ 2.5 M). To explain their results, the authors proposed a growth mechanism that took into account differences in speciation in the original aluminate solutions and was dependent on concentration. For dilute systems (Fig. 15.6a), they suggested that the process started with Keggin-like complex ions (1) that further aggregated and oligomerised to larger clusters (2); these coalesced to amorphous particles (3), which then lead to the rapid growth of nuclei (of pseudoboehmite) and/or pseudo-spinel phases (4), which transformed into pseudoboehmite (5) and then crystallised to a more stable bayerite phase. However, for concentrated solutions (Fig. 15.6b), the process originally involved small aluminate species such as ion pairs, dimers and trimers rather than the large Keggin ions (1), followed by formation of a loose Al^{3+}-containing oligomeric network (2), its initial densification to form a cluster (3), further densification to form a crystalline core (4), densification of the core and nuclei agglomeration (5) and final formation of a gibbsite crystal (6).

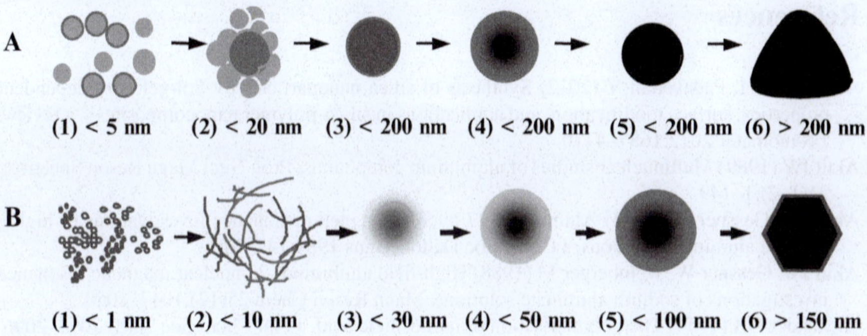

Fig. 15.6 A schematic representation of the formation of crystalline aluminium hydroxide from sodium aluminate solutions: (**a**) dilute solutions and (**b**) concentrated solutions. The size ranges indicated are speculative (Adapted and printed with permission from Li et al. (2005a). Copyright Elsevier 2005)

15.4 Outlook

This chapter highlights the extensive research carried out to unravel the mechanisms and pathways of silica and alumina colloid formation, with regard to both natural processes and industrial applications. Substantial advances in in situ scattering and imaging techniques have allowed to characterise the early stages of colloidal growth and aggregation; however, we are still facing substantial gaps in the molecular-level mechanistic understanding of the condensation reactions leading to the nucleation of the primary particles. This is particularly true for the alumina system where analytical limitations have so far hindered clear identification of all the polynuclear species forming in aluminate solutions, making it even harder to define their role in the formation of alumina colloids. For the silica system, the presence of polynuclear species seems of minor importance and colloid formation is more dependent on the behaviour of the monomeric silica. Whether silica nanoparticle growth occurs via the aggregation of primary particles or the addition of monomeric silica to existing particles is however still debated. It is evident that before we can formulate a fully coherent picture of silica and alumina colloid formation, more studies need to focus on the molecular characterisation of the condensation reactions and the nucleation of the primary particles. This is challenging with the current analytical limitations and the technical difficulties associated with these observations. However, in situ scattering and imaging techniques are constantly improving in resolution, with regard to both time and space. Also, the importance of silica and alumina nanophases for industrial processes further accelerates progress in this field, suggesting that a major breakthrough may come soon.

References

Ab Rahman I, Padavettan V (2012) Synthesis of silica nanoparticles by Sol-gel: size-dependent properties, surface modification, and applications in silica-polymer nanocomposites—a review. J Nanomater 2012:1687–4110

Akitt JW (1989) Multinuclear studies of aluminium compounds. Prog Nucl Magn Reson Spectrosc 21(1–2):1–149

Akitt JW, Gessner W (1984) Aluminium-27 nuclear magnetic resonance investigations of highly alkaline aluminate solutions. J Chem Soc Dalton Trans 1984:147–148

Akitt JW, Gessner W, Weinberger M (1988) High-field aluminium-27 nuclear magnetic resonance investigations of sodium aluminate solutions. Magn Reson Chem 25(12):1047–1050

Alexander GB (1954) The polymerisation of monosilicic acid. J Am Chem Soc 76(8):2094–2096

Allouche L, Gérardin C, Loiseau T, Férey G, Taulelle F (2000) Al30: a giant aluminum polycation. Angew Chem Int Ed 39(3):511–514

Bagwe RP, Hilliard LR, Tan WH (2006) Surface modification of silica nanoparticles to reduce aggregation and nonspecific binding. Langmuir 22:4357–4362

Bale HD, Schmidt PW (1959) Small angle X-ray scattering from aluminium hydroxide gels. II. J Chem Phys 31(6):1612–1618

Barcza L, Pálfalvi-Rózsahegyi M (1989) The aluminate lye as a system of equilibria. Mater Chem Phys 21(4):345–356

Barisik M, Atalay S, Beskok A, Qian S (2014) Size dependent surface charge properties of silica nanoparticles. J Phys Chem C 118(4):1836–1842

Belton DJ, Deschaume O, Perry CC (2012) An overview of the fundamentals of the chemistry of silica with relevance to biosilicification and technological advances. FEBS J 279(10):1710–20

Benning LG, Phoenix V, Mountain BW (2005) Biosilicification: the role of cyanobacteria in silica sinter deposition. In: Gadd GM, Semple KT, Lappin-Scott HM (eds) Micro-organisms and earth systems: advances in geomicrobiology, SGM Symposium. Cambridge University Press, Cambridge, pp 131–150

Besselink R, Stawski TM, Castricum HL, ten Elshof JE (2013) Evolution of microstructure in mixed niobia-hybrid silica thin films from sol-gel precursors. J Colloid Interface Sci 404:24–35

Bi S, Wang C, Cao Q, Zhang C (2004) Studies on the mechanism of hydrolysis and polymerization of aluminum salts in aqueous solution: Correlations between the "Core-links" model and "Cage-like" Keggin-Al 13 model. Coord Chem Rev 248:441–455

Bogush GH, Zukoski CF (1991) Uniform silica particle-precipitation – an aggregation growth model. J Colloid Interface Sci 142(1):19–34

Bottero JY, Tchoubar D, Cases JM, Fiessinger F (1982) Investigation of the hydrolysis of aqueous solutions of aluminum chloride. 2. Nature and structure by small-angle x-ray scattering. J Phys Chem 86(18):3667–3673

Bottero JY, Axelos M, Tchoubar D, Cases JM, Fripiat JJ, Fiessinger F (1987) Mechanism of formation of aluminium trihydroxide from Keggin Al13 polymers. J Colloid Interface Sci 117(1):47–57

Boukari H, Long GG, Harris MT (2000) Polydispersity during the formation and growth of the Stober silica particles from small-angle X-ray scattering measurements. J Colloid Interface Sci 229(1):129–139

Bradley SM, Hanna JV (1994) 27Al and 23Na MAS NMR and powder X-ray diffraction studies of sodium aluminate speciation and the mechanistics of aluminium hydroxide precipitation upon acid hydrolysis. J Am Chem Soc 116:7771–7783

Brinker CJ, Scherer GW (1990) Sol-gel science: the physics and chemistry of sol-gel processing. Academic, London

Brosset C, Biedermann G, Gunnar SL (1954) Studies on the hydrolysis of metal ions. XI. The aluminium ion, Al3+. Acta Chem Scand 8:1917–1926

Busing WR, Levy HA (1958) A single crystal neutron diffraction study of diaspore, AlO(OH). Acta Crystallogr 11:798–803

Buvári-Barcza Á, Rózsahegyi M, Barcza L (1998) Hydrogen bonded associates in the Bayer process (in concentrated aluminate lyes): the mechanism of gibbsite nucleation. J Mater Chem 8:451–455

Caho GY, Baker J, Sabina AP, Roberts AC (1985) Doyleite, a new polymorph of Al(OH)3, and its relationship to bayerite, gibbsite and nordstrandite. Can Mineral 23:21–28

Carcouet CCMC, van de Put MWP, Mezari B, Magusin PCMM, Laven J, Bomans PHH et al (2014) Nucleation and growth of monodisperse silica nanoparticles. Nano Lett 14(3):1433–1438

Carroll S, Mroczek E, Alai M, Ebert M (1998) Amorphous silica precipitation (60 to 120 degrees C): comparison of laboratory and field rates. Geochim Cosmochim Acta 62(8):1379–1396

Casey WH (2006) Large aqueous aluminium hydroxide molecules. Chem Rev 106(1):1–16

Cesteros Y, Salagre P, Medina F, Sueiras JE (1999) Several factors affecting faster rates of gibbsite formation. Chem Mater 11:123–129

Christoph GG, Corbató CE, Hofman DA, Tettenhorst R (1979) The crystal structure of boehmite. Clay Clay Miner 27:81–86

Conrad CF, Icopini GA, Yasuhara H, Bandstra JZ, Brantley SL, Heaney PJ (2007) Modeling the kinetics of silica nanocolloid formation and precipitation in geologically relevant aqueous solutions. Geochim Cosmochim Acta 71(3):531–542

Counter J, Gerson A, Ralston J (1997) Caustic aluminate liquors: preparation and characterisation using static light scattering and in situ X-ray diffraction. Colloids Surf A Physicochem Eng Asp 126(2–3):103–112

Counter JA, Addai-Mensah J, Raiston J (1999) The formation of Al(OH)3 crystals from supersaturated sodium aluminate solutions revealed by cryovitrification-transmission electron microscopy. Colloids Surf A Physicochem Eng Asp 154(3):389–398

De Yoreo JJ, Sommerdijk NAJM, Dove PM (2017) Nucleation pathways in electrolyte solutions. In: Van Driessche AES, Kellermeier M, Benning LG, Gebauer D (eds) New perspectives on mineral nucleation and growth, Springer, Cham, pp 1–24

Demichelis R, Catti M, Dovesi R (2009) Structure and stability of the Al(OH)3 polymorphs doyleite and nordstrandite: a quantum mechanical ab initio study with the CRYSTAL06 code. J Phys Chem C 113:6785–6791

Dietzel M (2000) Dissolution of silicates and the stability of polysilicic acid. Geochim Cosmochim Acta 64(19):3275–3281

Fouilloux S, Tache O, Spalla O, Thill A (2011) Nucleation of silica nanoparticles measured in situ during controlled supersaturation increase: restructuring toward a monodisperse nonspherical shape. Langmuir 27(20):12304–12311

Fu G, Nazar LF, Bain AD (1991) Ageing processes of alumina sol-gels: characterization of new aluminium polyoxycations by 27Al NMR spectroscopy. Chem Mater 3:602–610

Gale JD, Rohl AL, Watling HR, Parkinson GM (1998) Theoretical investigation of the nature of aluminum-containing species present in alkaline solution. J Phys Chem B 102(50):10372–10382

Gerson AR, Raiston J, Smart RSC (1996) An investigation of the mechanism of gibbsite nucleation using molecular modelling. Colloids Surf A Physicochem Eng Asp 110(1):105–117

Gong X, Nie Z, Qian M, Liu J, Pederson LA, Hobbs DT, McDuffle NG (2003) Gibbsite to boehmite transformation in strongly caustic and nitrate environments. Ind Eng Chem Res 42:2163–2170

Green DL, Lin JS, Lam YF, Hu MZC, Schaefer DW, Harris MT (2003) Size, volume fraction, and nucleation of Stober silica nanoparticles. J Colloid Interface Sci 266(2):346–358

Gunnarsson I, Arnorsson S (2000) Amorphous silica solubility and the thermodynamic properties of H4SiO4 degrees in the range of 0 degrees to 350 degrees C at P-sat. Geochim Cosmochim Acta 64(13):2295–2307

Gunnarsson I, Arnorsson S (2005) Impact of silica scaling on the efficiency of heat extraction from high-temperature geothermal fluids. Geothermics 34(3):320–329

Heaney PJ, Prewitt CT, Gibbs GV (1994) Silica: Physical behaviour, geochemistry & materials applications. Reviews in mineralogy Vol 29, Princeton, pp. 606

Hench LL, West JK (1990) The sol-gel process. Chem Rev 90:33–72

Herdianita NR, Browne PRL, Rodgers KA, Campbell KA (2000) Mineralogical and textural changes accompanying ageing of silica sinter. Miner Deposita 35(1):48–62
Hsu PH, Bates TF (1964) Formation of X-ray amorphous and crystalline aluminium hydroxides. Mineral Mag 33:749–768
Hsu H-W, Postberg F, Sekine Y, Shibuya T, Kempf S, Horányi M, Juhász A, Altobelli N, Suzuki K, Masaki Y, Kuwatani T, Tachibana S, Sirono S, Moragas-Klostermeyer G, Srama R (2015) Ongoing hydrothermal activities within Enceladus. Nature 519:207–210
Hu C, Liu H, Qu J, Wang D, Ru J (2006) Coagulation behavior of aluminium salts in eutrophic water: significance of Al13 species and pH control. Environ Sci Technol 40(1):325–331
Icopini GA, Brantley SL, Heaney PJ (2005) Kinetics of silica oligomerization and nanocolloid formation as a function of pH and ionic strength at 25 C. Geochim Cosmochim Acta 69:293–303
Iler RK (1979) The colloid chemistry of silica and silicates. Cornell University Press, Ithaca
Johansson G (1960) On the crystal structure of some basic aluminium salts. Acta Chem Scand 14:771–773
Johansson G, Lundgren G, Sillen LG, Soderquist R (1960) The crystal structure of a basic aluminum sulfate and the corresponding selenite. Acta Chem Scand 14:769–771
Kasprzyk-Hordern B (2004) Chemistry of alumina, reactions in aqueous solution and its application in water treatment. Adv Colloid Interface Sci 110:19–48
Keller WD (1978) Diaspore recrystallized at low temperature. Am Mineral 63:326–329
Kley M, Kempter A, Boyko V, Huber K (2014) Mechanistic studies of silica polymerization from supersaturated aqueous solutions by means of time-resolved light scattering. Langmuir 30(42):12664–12674
Kloprogge JT, Duong LV, Wood BJ, Frost RL (2006) XPS study of the major minerals in bauxite: gibbsite, bayerite and (pseudo-) boehmite. J Colloid Interface Sci 296:572–576
Kobayashi M, Juillerat F, Galletto P, Bowen P, Borkovec M (2005) Aggregation and charging of colloidal silica particles: effect of particle size. Langmuir 21(13):5761–5769
Leetmaa K, Gomez MA, Becze L, Guo F, Demopoulos GP (2014) Comparative molecular characterization of aluminum hydroxy-gels derived from chloride and sulphate salts. J Chem Technol Biotechnol 89(2):206–213
Levin I, Brandon D (1998) Metastable alumina polymorphs: crystal structure and transition sequences. J Am Ceram Soc 81(8):1995–2012
Li H, Addai-Mensah J, Thomas JC, Gerson AR (2003) A study of colloidal Al(III)-containing species in fresh/caustic aluminate solutions. Colloids Surf A Physicochem Eng Asp 223:83–94
Li H, Addai-Mensah J, Thomas JC, Gerson AR (2005a) The crystallization mechanism of Al(OH)3 from sodium aluminate solutions. J Cryst Growth 279:508–520
Li H, Addai-Mensah J, Thomas JC, Gerson AR (2005b) The influence of Al(III) supersaturation and NaOH concentration on the rate of crystallization of Al(OH)3 precursor particles from sodium aluminate solutions. J Colloid Interface Sci 286:511–519
Li Y, Zhang Y, Chen F, Yang C, Zhang Y (2011) Polymorphic transformation of aluminum hydroxide precipitated from reactive NaAl(OH)4-NaHCO3 solution. Cryst Growth Des 11:1208–1214
Loh JSC, Fogg AM, Watling HR, Parkinson GM, O'Hare D (2000) A kinetic investigation of gibbsite precipitation using in situ time resolved energy dispersive X-ray diffraction. Phys Chem Chem Phys 2:3597–3604
Marshall WL, Chen CTA (1982) Amorphous silica solubilities. 5. Predictions of solubility behaviour in aqueous electrolyte solutions of 300°C. Geochim Cosmochim Acta 46(2):289–291
Matsoukas T, Gulari E (1989) Monomer-addition growth with a slow initiation step: a growth model for silica particles from alkoxides. J Colloid Interface Sci 132(1):13–21
Meier D, Gunnlaugsson E, Gunnarsson I, Jamtveit B, Peacock CL, Benning LG (2014) Microstructural and chemical variations in silica-rich precipitates at the Hellisheidi geothermal power plant. Min Mag 78(6):1381–1389
Momma K, Izumi F (2011) VESTA 3 for three-dimensional visualization of crystal, volumetric and morphology data. J Appl Crystallogr 44:1272–1276

Moolenaar RJ, Evans JC, McKeever LD (1970) Structure of the aluminate ion in solutions at high pH. J Phys Chem 74(20):3629–3636

Mountain BW, Benning LG, Boerema JA (2003) Experimental studies on New Zealand hot spring sinters: rates of growth and textural development. Can J Earth Sci 40(11):1643–1667

Nail SL, White JL, Hem SL (1976a) Structure of aluminium hydroxide gel I: initial precipitate. J Pharm Sci 65(8):1189–1191

Nail SL, White JL, Hem SL (1976b) Structure of aluminium hydroxide gel II: aging mechanism. J Pharm Sci 65(8):1192–1195

Nail SL, White JL, Hem SL (1976c) Structure of aluminium hydroxide gel III: mechanism of stabilization by sorbitol. J Pharm Sci 65(8):1195–1198

Noguera C, Fritz B, Clement A (2015) Precipitation mechanism of amorphous silica nanoparticles: a simulation approach. J Colloid Interface Sci 448:553–563

Pancost RD, Pressley S, Coleman JM, Benning LG, Mountain BW (2005) Lipid biomolecules in silica sinters: indicators of microbial biodiversity. Environ Microbiol 7(1):66–77

Panias D, Asimidis P, Paspaliaris I (2001) Solubility of boehmite in concentrated sodium hydroxide solutions: model development and assessment. Hydrometallurgy 59:15–29

Patwardhan SV (2011) Biomimetic and bioinspired silica: recent developments and applications. Chem Commun 47(27):7567–7582

Perry CC, Keeling-Tucker T (2000) Biosilicification: the role of the organic matrix in structure control. J Biol Inorg Chem 5(5):537–550

Perry CC, Shafran KL (2001) The systematic study of aluminium speciation in medium concentrated aqueous solutions. J Inorg Biochem 87:115–124

Petz JII (1968) Structure of aluminium hydroxide gel. J Chem Phys 48(2):909–911

Rausch WV, Bale HD (1964) Small-angle X-ray scattering from hydrolyzed aluminium nitrate solutions. J Chem Phys 40:3391–3394

Rossiter DS, Fawell PD, Ilievski D, Parkinson GM (1998) Investigation of the unseeded nucleation of gibbsite, Al(OH)3, from synthetic bayer liquors. J Cryst Growth 191:525–536

Rothbauer R, Zigan F, O'Daniel H (1967) Verfeinerung der struktur des bayerits, Al(OH)3. Z Kristallogr Krist 125:317–331

Rothbaum HP, Rohde AG (1979) Kinetics of silica polymerization and deposition from dilute-solutions between 5°C and 180°C. J Colloid Interface Sci 71(3):533–559

Rousseaux JM, Weisbecker P, Muhr H, Plasari E (2002) Aging of precipitated amorphous alumina gels. Ind Eng Chem Res 41(24):6059–6069

Rowsell J, Nazar LF (2000) Speciation and thermal transformation in alumina sols: structures of the polyhydroxyoxoaluminum cluster [Al30O8(OH)56(H2O)26]18+ and its δ-Keggin moieté. J Am Chem Soc 122(15):3777–3778

Ruan HD, Frost RL, Kloprogge JT (2001) Comparison of Raman spectra in characterizing gibbsite, bayerite, diaspore and boehmite. J Raman Spectrosc 32:745–750

Ryan JN, Elimelech M (1996) Colloid mobilization and transport in groundwater. Colloids Surf A Physicochem Eng Asp 107:1–56

Saalfeld H, Jarchow O (1968) Die Kristallstruktur von Nordstrandit, Al(OH)3. Neues Jahrb Mineral Abh 109:185–191

Saalfeld H, Wedde M (1974) Refinement of the crystal structure of gibbsite, Al(OH)3. Z Kristallogr Krist 139:129–135

Sipos P (2009) The structure of Al(III) in strongly alkaline solutions – a review. J Mol Liq 146:1–14

Sipos P, Capewell SG, May PM, Hetter G, Laurenczy G, Lukács F, Roulet R (1998) Spectroscopic studies of the chemical speciation in concentrated alkaline solutions. J Chem Soc Dalton Trans 1998:3007–3012

Smith SJ, Amin S, Woodfield BF, Boerio-Goates J, Campbell BJ (2013) Phase progression of γ-Al2O3 nanoparticles synthesized in a solvent deficient environment. Inorg Chem 52:4411–4423

Squyres SW, Arvidson RE, Ruff S, Gellert R, Morris RV, Ming DW et al (2008) Detection of silica-rich deposits on Mars. Science 320(5879):1063–1067

Stawski TM, Benning LG (2013) Chapter 5: SAXS in inorganic and bioinspired research. Methods Enzymol 532:95–127
Stöber W, Fink A, Bohn E (1968) Controlled growth of monodisperse silica spheres in micron size range. J Colloid Interface Sci 26(1):62–69
Stol RJ, van Helden AK, de Bruyn PL (1976) Hydrolysis-precipitation studies of aluminum (III) solutions. 2. A kinetic study and model. J Colloid Interface Sci 57(1):115–131
Swaddle TW, Salerno J, Tregloan PA (1994) Aqueous aluminates, silicates, and aluminosilicates. Chem Soc Rev 23:319–325
Tettenhorst R, Hofmann DA (1980) Crystal chemistry of boehmite. Clay Clay Miner 28(5):373–380
Tobler DJ, Benning LG (2013) In situ and time resolved nucleation and growth of silica nanoparticles forming under simulated geothermal conditions. Geochim Cosmochim Acta 114:156–168
Tobler DJ, Stefansson A, Benning LG (2008) In-situ grown silica sinters in Icelandic geothermal areas. Geobiology 6(5):481–502
Tobler DJ, Shaw S, Benning LG (2009) Quantification of initial steps of nucleation and growth of silica nanoparticles: an in-situ SAXS and DLS study. Geochim Cosmochim Acta 73(18):5377–5393
van Straten HA, Holtkamp BTW, de Bruyn PL (1984) Precipitation from supersaturated aluminate solutions: I. Nucleation and growth of solid phases at room temperature. J Colloid Interface Sci 98(2):342–362
Wang L, Zhao W, Tan W (2008) Bioconjugated silica nanoparticles: development and applications. Nano Res 1:99–115
Wefers K, Misra C (1987) Oxides and hydroxides of aluminium, Alcoa Technical Paper 19. Aluminium Company of America, Pittsburgh
Westall F, Walsh MM (2000) The diversity of fossil microorganisms in Archaen-age rocks. In: Seckbach J (ed) Journey to diverse microbial worlds. Kluwer, Amsterdam
Williams LA, Crerar DA (1985) Silica diagenesis 2: general mechanisms. J Sediment Petrol 55(3):312–321
Yokoi T, Sakamoto Y, Terasaki O, Kubota Y, Okubo T, Tatsumi T (2006) Periodic arrangement of silica nanospheres assisted by amino acids. J Am Chem Soc 128(42):13664–13665

Chapter 16
Crystal Nucleation of Small Organic Molecules

Huaiyu Yang and Joop H. ter Horst

16.1 Introduction

Small molecular organic compounds have a wide range of application areas in food, pharmaceutical, and life science products. These compounds are often applied as dry particulate products or as particles within a formulation; Fig. 16.1 shows an example of a crystalline particulate product. Usually the particulate product has to satisfy quality specifications in terms of crystal form, particle size, particle shape, and purity. These quality specifications are strongly influenced by crystal nucleation (Davey et al. 2013; Kashchiev 2000).

Due to the nucleation and growth of an unknown, more stable form in the production, ritonavir (an antiviral compound for AIDS) had to be redeveloped and reformulated costing over $250 million (Bučar et al. 2015). This shows that it can be catastrophic to produce an unwanted polymorph in pharmaceutical manufacturing. Experimental screening (Aaltonen et al. 2009) and structure predictions (Ismail et al. 2013) can to some extent control the risk of new polymorph appearance in the production of a new drug product.

The final product size is strongly related to the crystal nucleation rate through the number of nuclei formed, which grow due to the amount of crystalline material formed. For product performance such as dissolution behavior and bioavailability, the crystal size has to be well controlled, especially for the particle size of inhalation drugs (below 5 μm) (Patton and Byron 2007).

The industrial suspension crystallization process (Mullin 2001) to make the desired crystal particles therefore should be operated so that the right specifications

H. Yang • J.H. ter Horst (✉)
EPSRC Centre for Innovative Manufacturing in Continuous Manufacturing and Crystallisation (CMAC), Strathclyde Institute of Pharmacy and Biomedical Sciences, Technology and Innovation Centre, University of Strathclyde, 99 George Street, Glasgow G1 1RD, UK
e-mail: Joop.terHorst@strath.ac.uk

A.E.S. Van Driessche et al. (eds.), *New Perspectives on Mineral Nucleation and Growth*,
DOI 10.1007/978-3-319-45669-0_16

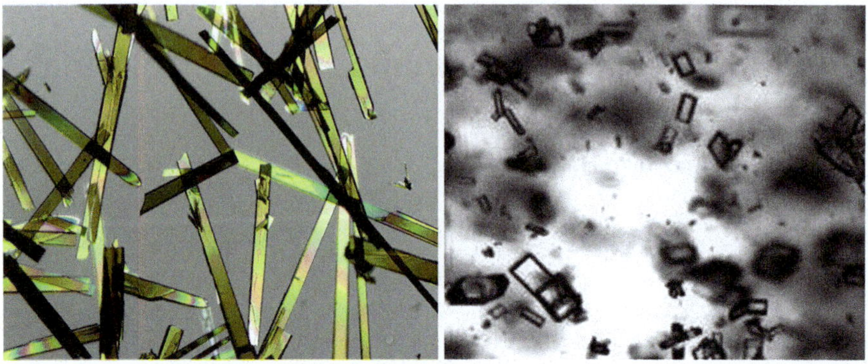

Fig. 16.1 Two crystal forms (polymorphs) of isonicotinamide strongly differing in size and morphology. The quality of a crystalline material can be characterized in terms of crystal form, size, shape/morphology, and purity. These would determine product properties like bioavailability. The quality of a crystalline material is determined by the crystallization process, in which crystal nucleation is an important factor

are achieved. Such crystallizations usually are performed from a solution in which the product solute is dissolved. Prior to beginning the crystallization, due to poorly controlling of the nucleation, usually milled seeds are added to reduce the final particle size.

The future will show a paradigm shift from batch to continuous manufacturing in pharmaceutical and specialty chemicals industry (ter Horst et al. 2015) (Lawton et al. 2009) (Chen et al. 2011). This is initiated in order to cope with increased molecular size and complexity, quality and purity specifications, and sustainability demands of future manufacturing processes in these industries. The crystallization process in which crystalline particles with controlled properties and high purity are produced from complex multicomponent mixtures constitutes a major bottleneck in continuous manufacturing. The drive toward continuous crystallization processes therefore puts a strong demand on understanding, controlling, and predicting crystal nucleation rates under various conditions.

This chapter will highlight recent advances in the area of the crystal nucleation of small organic compounds like pharmaceuticals. It will describe the state of the art of this subfield and a vision on the needed future research.

16.2 Crystallization of Small Organic Molecules

One of the ways to establish crystallization behavior is to measure clear and cloud points for which now commercial equipment is available from several suppliers. These measurements make use of the temperature dependence of the solubility. The clear point is the temperature at which, upon linearly increasing the temperature, a stirred suspension turns into a clear solution. If the heating rate

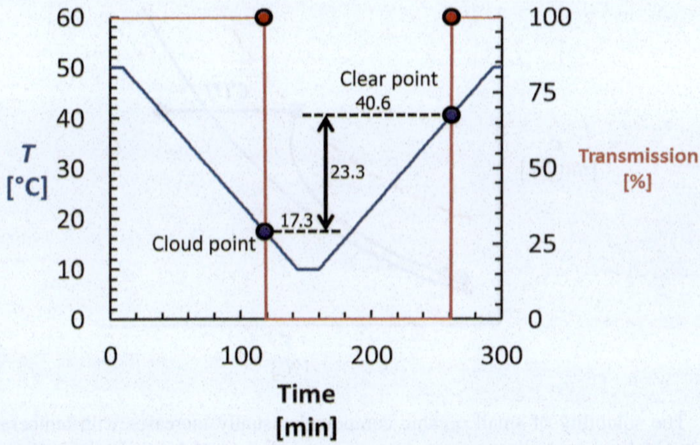

Fig. 16.2 A measurement of the clear and cloud point temperature of a stirred sample containing a certain amount of solid material and solvent. Upon decreasing the temperature (*blue line*) of the clear solution, crystals are detected at a cloud point temperature of 17.3 °C. Upon increasing the temperature of the suspension, all crystals are dissolved at the clear point temperature of 40.6 °C. The metastable zone width then is 23.3 °C. In this case the detection of crystals is done by the transmission of light (*red line*) through the sample. If crystals are present, the transmission of light is less than that through a clear solution (100 %)

is sufficiently slow, the clear point approaches the saturation temperature of the measured composition. The cloud point is the temperature at which, upon linearly decreasing the temperature, crystals are detected in a stirred solution.

Figure 16.2 shows that clear and cloud point of a specific sample composition can differ significantly, showing substantial hysteresis. A solution has to be undercooled considerably to provoke crystallization. The cloud point is a function of crystal nucleation, and thus the difference reflects that crystal nucleation is an activated process associated with an energy barrier that has to be overcome in order for a solution to nucleate.

Industrial crystallization processes for the production of a particulate product of small molecular organic compounds are conventionally performed in large stirred vessels (>1 m^3). Cooling a clear solution in which the compound is dissolved, a crystallization driving force is created. The driving force for crystallization is usually reported as a supersaturation ratio $S = c/c^*$ of concentration c over solubility c^*. To create a supersaturation, several options are available.

In case of cooling crystallization, the driving force is created by reducing the temperature and with it the solubility $c^*(T)$, shown in Fig. 16.3. For evaporative crystallization the solvent is evaporated and the supersaturation is increased by an increase in the concentration c at constant temperature. In antisolvent crystallization the supersaturation is created by adding an antisolvent to a solution, which decreases concentration through dilution but also solubility through the strongly nonideal effect of the mixture of solvents on the solute behavior. For antisolvent

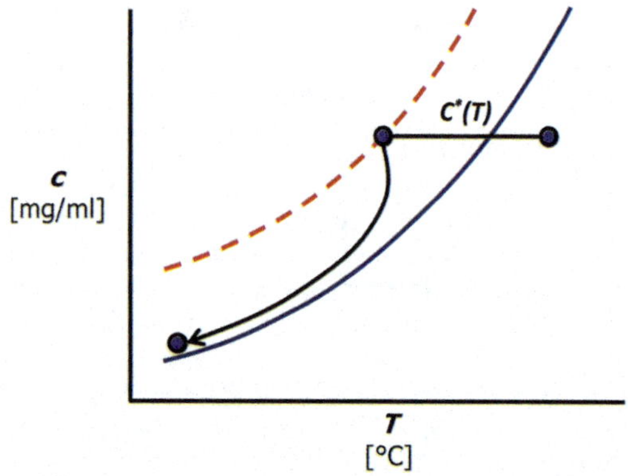

Fig. 16.3 The solubility of small organic compounds usually increases with temperature (*blue solid line*). Cooling crystallization can be employed if the solubility has a sufficiently large dependence on temperature and if the final solubility is sufficiently low. Supersaturation is created by decreasing the solubility below the prevailing concentration in the crystallizer. Between the solubility line (*blue solid line*) and the metastable zone (MSZ) limit (*red dashed line*), no detectable nucleation occurs and the solution will have to be cooled down to the MSZ limit to invoke crystal nucleation. Within the metastable zone width, no nucleation occurs but crystals can grow. By adding seed crystals within the metastable zone and by gently further cooling, seed crystals will consume the supersaturation by growth, crystal nucleation is avoided, and control is obtained over the number of crystals in the process and thus over the final product size

crystallization to occur, the nonideal effect of the antisolvent should strongly outweigh the dilution effect. Sometimes precipitation is used in which a reaction is combined with a crystallization process. An acid can, for instance, be precipitated by going from a dissociated state in water solutions to an associated state by generating a pH shift in the solution through the addition of a dissolved acid. Usually, however, solubility of organic compound is very low in water solutions, and therefore organic solvents are used for industrial crystallization.

During industrial crystallization processes, a suspension of crystalline particles is created. This process involves the subprocesses of crystal nucleation and crystal growth while usually agglomeration is to be avoided. Crystal nucleation is important since it produces the crystals that will be grown to larger sizes. In case of a batch cooling or evaporative crystallization at the initial stage of the batch, crystal primary nucleation sets the initial population of crystals. These crystals then are to be grown out during the remainder of the batch. The number of initially formed crystals strongly affects the final crystal size produced. In addition, these initial crystals should be of the right form. Preferably, the crystal nucleation should take place in the bulk of the solution of the crystallizer and not on its walls (fouling).

In industrial suspension crystallization, secondary nucleation is defined as the formation of new crystals in the presence of larger crystals (Mullin 2001). This can, for instance, be due to collision of larger crystals with the fast-moving stirrer tip. Secondary nucleation falls outside the scope of this chapter.

Crystal nucleation in industrial crystallization processes is hard to control and predict. It therefore is often avoided by adding seed crystals at the initial stages of a batch crystallization process. At slightly supersaturated conditions, a sufficient amount of seeds is added that will grow out in the remainder of the batch. The extent with which seeding is applied in pharmaceutical industry shows that crystal nucleation under industrial conditions is impossible to control and predict. There is therefore a strong drive toward understanding the crystal nucleation process within industrial suspension crystallization processes and with it to come to new predictive nucleation theories that allow industrial control.

16.3 Crystal Nucleation Theory

Crystal nucleation is an activated process in which the crystal building units (atoms, molecules, ions, etc.) attach to and detach from clusters. In the Szilard–Farkas model, nucleation is assumed to be a consecutive series of attachments and detachments of building units to form differently sized clusters of the nucleating phase in the supersaturated phase (Kashchiev 2000). Because nucleation is an activated process, a nucleating particle has to cross an energy barrier located at a specific cluster size n^*, the nucleus size, along its path from a cluster of size $n = 1$ building unit to a macroscopic size.

The nucleation rate J can be defined as the result of a forward net flux (in units of number per unit of volume and time) over the energy barrier of clusters that change size from n^* to $(n^* + 1)$ by the addition of a single building unit:

$$J = f_{n^*} X_{n^*} - g_{n^*+1} X_{n^*+1} \tag{16.1}$$

in which X_n is the concentration of clusters of size n, f_n is the attachment frequency of building units to a cluster of size n, and g_n is the detachment frequency of building units from a cluster of size n.

We can define a hypothetical concentration C_n of clusters in equilibrium with the supersaturated solution for which the net flux of clusters that change size from n to $(n + 1)$ is zero:

$$0 = f_n C_n - g_{n+1} C_{n+1} \tag{16.2}$$

This equation can be used to remove the detachment frequency g_{n^*+1} from Eq. 16.1:

$$J = \left[\frac{X^*}{C^*} - \frac{X_{n^*+1}}{C_{n^*+1}}\right] f^* C^* \tag{16.3}$$

We now can define the nucleation rate as the product of the equilibrium concentration C^* of nuclei, the attachment frequency f^* of building units to the nucleus, and the Zeldovich factor z replacing the part between the brackets in Eq. 16.3:

$$J = zf^*C^* \tag{16.4}$$

The Zeldovich factor accounts for both the use of equilibrium concentration C^* rather than actual concentration X^* and for supernuclei of size $(n^* + 1)$ that decay rather than grow out to macroscopic sizes. This equation is the basis for the nucleation rate equation from the classical nucleation theory (CNT) often used to analyze nucleation rate data.

While CNT is theoretically sound, predictions of nucleation rates can be orders of magnitude off. In recent years CNT has been challenged by observations in experiments and simulations. New pathways for crystal nucleation were defined (Fig. 16.4) that challenge some underlying assumptions of the CNT. However, the new nucleation theories do not yet have predictive power. To come to improved predictive power, we will have to find and overcome the deficits of the CNT. In order to do so, the three constituents of Eq. 16.4 will need to be discussed further. In separate subsections we will reflect on their importance in crystal heterogeneous nucleation of small organic molecular compounds from solutions. The overall expression comes back in Sect. 16.5 where it is discussed in relation to experimental nucleation rate data.

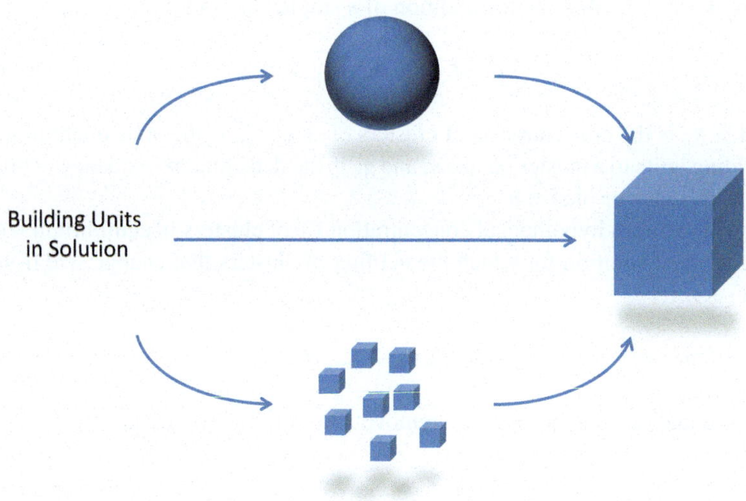

Fig. 16.4 The classical nucleation mechanism (*middle*) has recently been challenged by the two-step nucleation mechanism (*top*) and the mechanism involving the aggregation of clusters: nanosized pre-nucleation clusters form which aggregate into a large single crystal (*bottom*)

16.3.1 Nucleus Equilibrium Concentration

While homogeneous nucleation would occur in the bulk of a clear solution, heterogeneous nucleation is induced by the presence of foreign surfaces such as dust particles, bubbles, or foreign molecules (Kashchiev 2000). In industrial practice usually such surfaces, for instance, from dust particles, are present and unavoidable. Due to the heterogeneous particle onto which the nucleus forms, the energy barrier for nucleation decreases and nucleation can proceed faster. Therefore, in practice it is assumed that crystal nucleation follows the heterogeneous nucleation pathway.

We are therefore concerned with obtaining an expression for C^*, the equilibrium concentration of nuclei which are formed on the active sites of the heterogeneous particles in the system. The equilibrium concentration $C(n)$ can be described through a Boltzmann process in which the work to form the n-sized cluster to form, the heterogeneous nucleation work $W(n)$, is incorporated. If C_0 then is the concentration of active sites onto which the nucleus can form due to the presence of heterogeneous particles, the cluster concentration $C(n)$ can be written:

$$C(n) = C_0 \exp\left(-W(n)/kT\right) \tag{16.5}$$

The fraction of nucleation sites that contain a nucleus is then given by the ratio $C(n)/C_0$. In case of homogeneous nucleation (HON) from solution, the nucleus can appear anywhere in the volume and the concentration of nucleation sites $C_0 = v^{-1}$ with v the molecular volume of the nucleating phase. While C_0 for HON is much higher than that for HEN, the nucleation work for HEN is much smaller, making HEN the dominant nucleating mechanism.

The work $W(n)$ to form an n-sized cluster consists of two parts:

$$W(n)/kT = -n\Delta\mu/kT + \Gamma n^{2/3} \tag{16.6}$$

The first part describes a free energy gain due to the creation of the cluster which scales with the cluster size n and the chemical potential difference $\Delta\mu$. For nucleation in solution, the chemical potential difference $\Delta\mu = kT \ln S$ can be written as a function of the supersaturation ratio S. The second part describes the free energy loss due to the creation of the cluster surface which scales with the cluster surface area and a factor Γ (Ter Horst et al. 2002) containing the interfacial energies γ and shape factors of the various interfaces present between the nucleus and solution and, in case of heterogeneous nucleation, also between the nucleus and the heterogeneous surface as well as between the heterogeneous surface and the solution. The term Γ can be written as a summation over all surfaces i with shape factor c_i and interfacial energy γ_i:

$$\Gamma = v^{2/3} \sum_{i}^{hkl} \frac{c_i \gamma_i}{kT} \tag{16.7}$$

If only a single interfacial energy γ is involved in the work, $\Gamma = cv^{2/3}g/kT$ and the expression for the work $W(n)$ changes which is, e.g., valid for liquid droplet nucleation from a vapor (Kashchiev 2000). The work $W(n)$ for forming a cluster has a maximum value at $n = n^*$, which is the nucleus size:

$$n^* = \frac{8\Gamma}{27\ln^3 S} \tag{16.8}$$

The nucleation work for forming a nucleus then is

$$\frac{W^*}{kT} = \frac{4\Gamma^3}{27\ln^2 S} \tag{16.9}$$

The expression for the nucleus concentration C^* then becomes

$$C^* = C_0 \exp\left(-\tfrac{4}{27}\frac{\Gamma^3}{\ln^2 S}\right) = C_0 \exp\left(-\frac{B}{\ln^2 S}\right) \tag{16.10}$$

where $B = 4\Gamma^3/27$ is used to simplify the equation.

16.3.2 Attachment Frequency

In order to analyze nucleation rate data, we have to find the supersaturation dependence of the attachment frequency f^* (Kashchiev 2000). The attachment frequency f^* is the number of attachments of building units per unit of time to the nucleus of size n^*. It is thus a function of nucleus surface area $a^* = c(vn^*)^{2/3}$, which can be written as a function of the nucleus size using a shape factor c and the molecular volume v. The attachment frequency $f^* = c(vn^*)^{2/3} f_a$ then is a function of the attachment frequency f_a per unit of nucleus surface.

We can understand the attachment frequency f_a per unit of area being a linear function of the concentration of building units. A higher concentration of building units increases the attachment frequency. This concentration of building units can be assumed a linear function of the actual concentration of solute molecules. The attachment frequency f_a then is a linear function of the supersaturation S:

$$f^* = f_{eq} S = f_{a,eq} S c\left(vn^*\right)^{2/3} \tag{16.11}$$

where f_{eq} is the attachment frequency to the nucleus if there would be equilibrium in the solution and $f_{a,\,eq}$ is the attachment frequency per unit of nucleus area if there would be equilibrium in the solution. Using Eq. 16.8 we see that the attachment frequency therefore has a supersaturation dependence of

$$f^* = f_0 \frac{S}{\ln^2 S} \tag{16.12}$$

with f_0 being a supersaturation-independent part of the attachment frequency.

The pathway of a building unit will start in the bulk solution and end incorporated into the cluster. On its way toward incorporation, several energy barriers have to be crossed. First it has to diffuse to the cluster, and then adsorption onto the surface has to take place, after which it undergoes surface diffusion toward the position at which it will be incorporated. The incorporation itself can also be associated with an energy barrier when, for instance, the building unit needs to change its conformation. The building unit may need to lose (part of) its solvent shell on its way. We can thus assume that there is a rate-limiting energy barrier E_a that the building unit has to cross to incorporate in the nucleus. This then means that the supersaturation-independent part of the attachment frequency f^* contains this energy barrier:

$$f_0 = f \exp\left(-\frac{E_a}{kT}\right) \tag{16.13}$$

where f would be the attachment frequency of building units to the nucleus per unit of nucleus surface at equilibrium without an energy barrier.

16.3.3 Zeldovich Factor

The Zeldovich factor z accounts for both the supernuclei that decay rather than grow and the use of equilibrium rather actual nucleus concentration. From a comparison of Eqs. 16.3 and 16.4, it follows that the Zeldovich factor z can be written as

$$z = \frac{X^*}{C^*} - \frac{X_{n^*+1}}{C_{n^*+1}} \tag{16.14}$$

The ratio between actual and equilibrium concentration is well approximated around the nucleus size using (Kashchiev 2000)

$$\frac{X_n}{C_n} = \tfrac{1}{2} - \frac{\beta}{\pi^{1/2}}\left(n - n^*\right) \tag{16.15}$$

It is interesting to note that the decay probability $Q(n)$, the probability that a cluster of size n decays rather than grows out to macroscopic sizes, is equal to the ratio between actual and equilibrium concentration $Q(n) = X_n/C_n$ (ter Horst and Kashchiev 2003). This shows that the Zeldovich factor $z = Q^* - Q\,(n^* + 1)$ is actually just the difference between the decay probability of the nucleus and the supernucleus of size $(n^* + 1)$. Using Eq. 16.15 the decay probability Q^* of the nucleus is exactly ½: at the top of the barrier, a nucleus can lower its energy by both increasing and decreasing sizes. Therefore, only half of the nuclei of size n^* (at the top of the barrier) grow out to become macroscopically large. The other half decay and dissolve. Since the decay probability $Q(n^* + 1)$ cannot be smaller than 0 at which all supernuclei of size $(n^* + 1)$ grow out, the Zeldovich factor has an upper boundary limit of $z = 0.5$.

Using the above equations, it can directly be seen that the Zeldovich factor is approximated by

$$z = \beta / \pi^{\frac{1}{2}} \tag{16.16}$$

The value β describes the ease with which supernuclei survive rather than decay. This is reflected in the curvature of the work to form a cluster around the nucleus size n^*. A high curvature coincides with a large β and the decay probability $Q(n)$ for supernuclei will be small. An expression for β is given by (Kashchiev 2000)

$$z = \frac{\beta}{\pi^{1/2}} = \left[\frac{-1}{2\pi kT}\left(\frac{d^2W}{dn^2}\right)_{n=n^*}\right]^{1/2} = z_0 \ln^2 S \tag{16.17}$$

The Zeldovich factor z is therefore a function of the supersaturation ratio. The factor z_0 is independent of supersaturation but mainly depends on the interfacial energies. The Zeldovich factor will usually have values around $0.001 < z < 0.5$.

16.4 Crystal Nucleation Rate Measurements

Unfortunately, due to the small size of the nuclei (roughly 1–1000 molecules large), a direct measurement of the nucleation rate is extremely hard. There are various techniques that can detect crystals in a suspension if the crystals above their detectable size are present in sufficient numbers. One could use, for instance, in situ focus beam reflectance measurements (FBRM) (Kim et al. 2013), turbidity measurements (Jiang and ter Horst 2010), or the human eye (Yang and Rasmuson 2013). These techniques all signal detection above a certain crystal size or suspension volume fraction of crystals rather than the actual number of nuclei. In order to be observed, the nuclei thus have to grow out to a detectable size. If the nucleation and growth stage are not separated in the nucleation rate measurement method, the growth rate might have a significant effect on the measured rates.

While on the one side growth is intimately connected to crystal nucleation making it hard to determine nucleation rate, on the other side the stationary cluster concentration X_n needs time to develop in the supersaturated solution. During this development period in which the cluster concentration $X_n = X_n(t)$ depends on time, the nucleation rate will be nonstationary (Kashchiev 2000). According to CNT the time during which the nucleation rate is nonstationary is strongly influenced by the attachment frequency f^*: a small attachment frequency results in a long time period of nonstationary nucleation. Usually the time period of nonstationary nucleation in solution nucleation is neglected due to the assumed high attachment frequency.

16.4.1 Double Pulse

One of the ways to circumvent the effect of the growth rate is to use the double pulse method in which a solution first undergoes a high supersaturation pulse of a specific time to nucleate crystals after which these crystals are grown out to detectable size using a lower supersaturation pulse at which no nucleation takes place. The number of crystals at the end of the experiment and the time of the high supersaturation pulse lead to the nucleation rate. This was, for instance, applied for the polypeptide hormone insulin in a relatively large volume (Nanev et al. 2010). They found that heterogeneous nucleation in the bulk solution onto foreign particles of unknown biological origin was dominant.

Such double pulse methods are easily miniaturized and done in high-throughput mode using microwells. The microwell approach was applied earlier on protein nucleation in which a large number of small volumes at equal conditions were used to determine nucleation rates of the protein lysozyme (Galkin and Vekilov 1999). The increase in the average number of crystals at a certain supersaturation ratio as a function of pulse time was determined to be the homogeneous nucleation rate. They additionally found a larger than zero intercept at zero pulse time which they attributed to heterogeneous nucleation. Interestingly, they found discretely changing nucleus sizes with supersaturation. Further double pulse experiments subsequently led to the experimental identification of the two-step nucleation mechanism for protein systems (Vekilov 2004). Microfluidic devices make it even easier to generate a large number of equally supersaturated solution droplets with the same small volume (Ildefonso et al. 2011).

Care should be taken to have a constant supersaturation during the nucleation pulse. If excessive nucleation and growth occur in the nucleation pulse, the supersaturation decreases so that at the apparent supersaturation, a smaller number of crystals are formed and a lower nucleation rate is measured than actually is prevailing at that supersaturation. In addition, the supersaturation during the growth pulse needs careful tuning: too low supersaturations lead to the dissolution of formed supernuclei; too high supersaturations lead to additional nucleation of supernuclei.

16.4.2 Induction Time

The induction time is the time period from the establishment of a constant supersaturated state to the detection of crystals in the solution. Already a long time ago, it was identified that induction time is related to the crystal nucleation rate. Induction times are usually measured in stirred solutions, similar to metastable zone width measurements. The presence of a certain volume fraction of crystals can be detected then by, e.g., turbidity techniques. However, this makes the induction time a complex function of nucleation and growth, while possibly also secondary nucleation, due to attrition of large crystals with the stirrer, is involved.

There are various methods to analyze the measured induction times and relate them to nucleation behavior. Lately the method using induction time distributions (Jiang and ter Horst 2010) has found increased interest of researchers. Figure 16.6 shows induction time distributions for butyl paraben in 90 % ethanol/water mixtures at three different supersaturation ratios (Yang and Rasmuson 2013; Yang et al. 2014). Such series of equally conditioned induction time measurements in volumes of around 1 ml and smaller shows a very broad variation that cannot be explained by experimental variations (Brandel and ter Horst 2015).

This broad variation means the observed variation must originate from the nucleation process caused by the relatively low nucleation rate. It was realized that only a single crystal is nucleated which grows out to larger sizes and then undergoes attrition, which makes the suspension that is then detected (Kadam et al. 2012). A model for this single nucleus mechanism (Fig. 16.5) is captured in the following equation (Jiang and ter Horst 2010):

$$P(t) = 1 - \exp\left(-JV\left(t - t_g\right)\right) \tag{16.18}$$

where $P(t)$, the probability that a certain induction time or lower is measured, is obtained from the induction time measurements. The growth time t_g is the time it takes for the nucleus to be detected, which is assumed constant for a certain supersaturation. The nucleus is detected only after it has grown out to macroscopic size so that collisions with the stirrer create attrition fragments. The growth of these attrition fragments fills the suspension and leads to a detection signal. If this mechanism is followed, nucleation and growth are effectively separated.

Figure 16.6 also gives a typical induction time distribution from a series of 120 induction time measurements at the same conditions for para-hydroxy acetophenone in ethanol at a supersaturation $S = 1.3$ (Brandel and ter Horst 2015). The volume used was 1 ml and the solution was stirred with a magnetic stirrer bar. In this

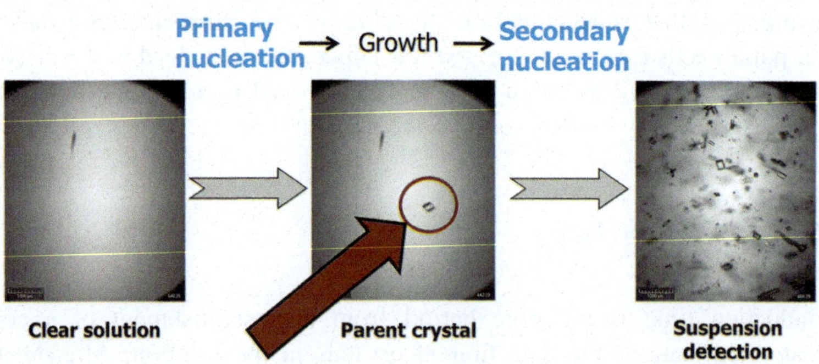

Fig. 16.5 The single nucleus mechanism in an induction time measurement

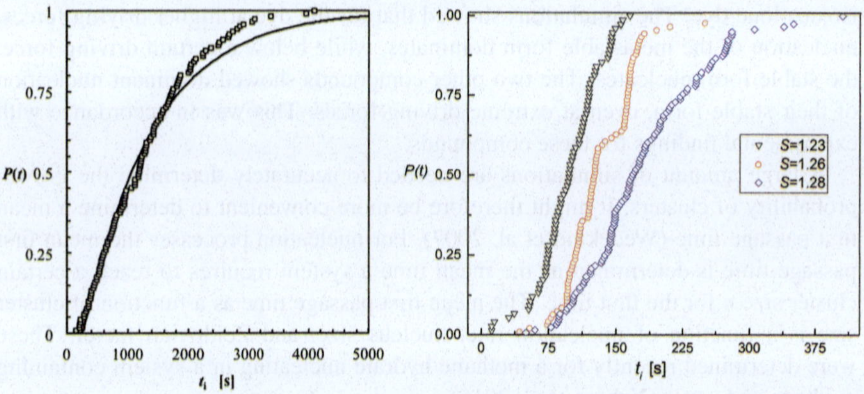

Fig. 16.6 Induction time distributions for 4-hydroxyacetophenone in ethanol (*left*) and butyl paraben in 90 % ethanol (*right*)

particular distribution, no induction times below $t = 156$ s are measured. The highest induction time was measured to be $t = 4100$ s. The induction times thus can vary more than an order of magnitude. In between the measured induction time, probability can be nicely fitted to Eq. 16.20 which results in a nucleation rate $J = 865 \pm 20\ \mathrm{m}^{-3}\mathrm{s}^{-1}$ and growth time $t_g = 263 \pm 20$ s.

It is interesting to note that such induction time measurements also enable the study of the effect of additional parameters on the nucleation rate, such as fluid shear (Forsyth et al. 2014).

16.4.3 Molecular Simulations

The probability that nucleation occurs in a supersaturated solution volume is proportional to the volume and time. Since volumes and times in computer simulations are usually small, a direct determination of nucleation rates from simulations is difficult and would require a large amount of computer power or unrealistically high supersaturation values. Therefore, large amounts of work were put in the determination of the nucleation work or nucleus size. Due to increasing computer power, also kinetic features of the process can be investigated leading to an improved mechanistic nucleation understanding on a molecular scale (Anwar and Zahn 2011).

Growth probabilities $P(n)$ or decay probabilities $Q(n) = 1 - P(n)$ can be determined using molecular simulations (ter Horst and Kashchiev 2003; Deij et al. 2007). Such growth probabilities as a function of cluster size then lead to the determination of the nucleus size at a growth probability of $P(n^*) = 1 - Q(n^*) = 0.5$. Monte Carlo simulations determining growth probabilities $P(n)$ showed the difference in nucleus size for different polymorphs nucleating under the same supersaturation conditions for paracetamol, L-glutamic acid, and a yellow isoxazolone dye. The simulations showed that for the dye at higher driving forces, nucleation of the metastable form dominates, while below a certain driving force, the stable form nucleates. The two other compounds showed dominant nucleation of their stable form, even at extreme driving forces. This was in accordance with experimental findings for these compounds.

A large amount of simulations are needed to accurately determine the growth probability of clusters. It might therefore be more convenient to determine a mean first passage time (Wedekind et al. 2007). For nucleation processes the mean first passage time is determined as the mean time a system requires to reach a certain cluster size n for the first time. The mean first passage time as a function of cluster size is a function of nucleation rate, nucleus size, and Zeldovich factor. These were determined recently for a methane hydrate nucleating in a system containing methane and water (Yuhara et al. 2015).

16.5 Molecular Interpretation of Crystal Nucleation Behavior

When using the previously derived expressions for z, f^*, and C^* in Eq. 16.4, we arrive at an expression for the stationary nucleation rate from classical nucleation theory:

$$J = zf^*C^* = AS\exp\left(-\frac{B}{\ln^2 S}\right) \tag{16.19}$$

with $A = z_0 f_0 C_0$. Note that the kinetic pre-exponential factor A and the thermodynamic exponential factor B are both a function of temperature. The factors A and B can be determined from the supersaturation-dependent nucleation rate at a constant temperature by plotting $\ln(J/S)$ against $\ln^{-2}S$. The straight line fitted through the data leads to the values of A and B.

Table 16.1 and Fig. 16.7 show data for the organic compounds ethylparaben and propylparaben in ethanol (Yang et al. 2014). At a supersaturation ratio between 1.1 and 1.5, nucleation rates were determined from induction time distributions. The nucleation rates are in the range from tens to thousands of particles per second per cubic meter. These are typical data generally obtained from induction time measurements but also double pulse measurements.

Now using the nucleation rate equation, as Eq. 16.19, from CNT enables the determination of the thermodynamic parameter B and kinetic parameter A. From the slope and intercept in Fig. 16.7, the parameters A and B were determined. Parameters A and B for ethylparaben in ethanol are an order of magnitude larger than for propylparaben in ethanol. It is interesting to see that a slight difference in the molecular structure results in a substantially different nucleation behavior. Now

Table 16.1 The nucleation rate of ethylparaben and propylparaben in ethanol (Yang et al. 2014). The values for the kinetic factor A and thermodynamic factor B were determined from the fit of Eq. 16.19 to the data. The value for B is then used to determine nucleus size n^* and nucleation work W^*

Ethylparaben in ethanol				Propylparaben in ethanol			
S	J	n^*	W^*/kT	S	J	n^*	W^*/kT
[−]	$[\mathrm{m^{-3}s^{-1}}]$	[−]	[−]	[−]	$[\mathrm{m^{-3}s^{-1}}]$	[−]	[−]
1.22	33 ± 1	71	6.9	1.12	28 ± 1	95	5.3
1.36	990 ± 20	20	2.9	1.15	61 ± 2	53	3.5
1.48	$9.65 \pm 0.61 \times 10^3$	10	1.8	1.29	$2.08 \pm 0.04 \times 10^3$	8.6	1.1
$A\ [\mathrm{m^{-3}s^{-1}}]$		24×10^3		$A\ [\mathrm{m^{-3}s^{-1}}]$		3.5×10^3	
B [−]		0.27		B [−]		0.068	

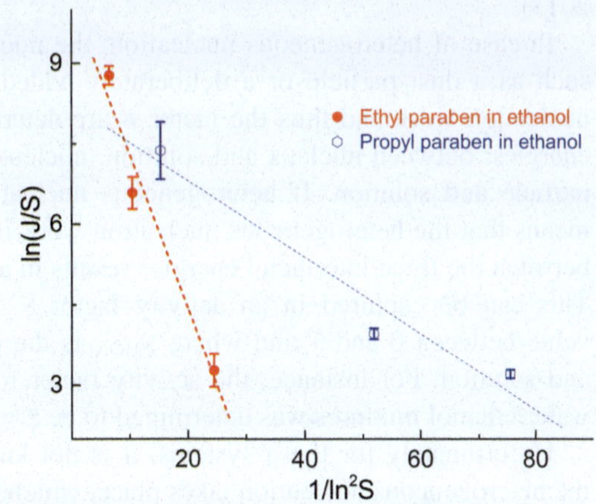

Fig. 16.7 The nucleation rate of ethylparaben and propylparaben in ethanol plotted using Eq. 16.19

that we experimentally obtained A and B, we can try to interpret these in order to gain an increased understanding of heterogeneous crystal nucleation from solution (Davey et al. 2013), also in the light of nonconventional nucleation mechanisms.

16.5.1 Exponential Thermodynamic Factor B

When parameter B is known, CNT enables us to determine nucleation work and nucleus size. As shown in Table 16.1, at the same supersaturation ratio S, the nucleation work $W^*/kT = B/\ln^2 S$ is much larger for ethylparaben than for propylparaben. The nucleation work W^* approaches $1kT$ at higher supersaturations suggesting that the barrier for nucleation vanishes for higher supersaturations and spinodal decomposition (barrierless demixing) will occur. The nucleus $n^* = 2W^*/kT \ln S$ contains several molecules at the high supersaturations going toward 100 molecules at the lower supersaturations.

CNT states that parameter B is related to the interfacial energy between crystal nucleus and solution. Since B is known, an averaged interfacial energy γ can be determined by assuming a shape factor c:

$$\gamma = \left[\frac{27k^3T^3B}{4c^3v^2}\right]^{1/3} \tag{16.20}$$

The interfacial energies determined in this way are substantially lower than the separately determined interfacial energy between a large crystal surface and the solution, for instance, by contact angle measurements. This is usually attributed to the occurrence of heterogeneous nucleation onto foreign particles rather than homogeneous crystal nucleation. It was, for instance, shown that filtration of the solution decreased the nucleation rate of isonicotinamide in ethanol (Kulkarni et al. 2013).

In case of heterogeneous nucleation, the nucleus forms onto a foreign particle such as a dust particle or a deliberately added functionalized surface. Then the nucleation work and thus the factor B are determined by at least three interfacial energies: between nucleus and solution, nucleus and foreign particle, and foreign particle and solution. If heterogeneous nucleation is occurring dominantly, this means that the heterogeneous nucleation work is lower and thus that the interplay between the three interfacial energies results in a lower effective interfacial energy. This can be captured in an activity factor $\xi = \gamma/\gamma_{HON}$ which would have a value between 0 and 1 and where γ_{HON} is the interfacial energy between crystal and solution. For instance, the activity factor for m-aminobenzoic acid in 50w% water/ethanol mixtures was determined to be $\xi = 0.27$ (Jiang and ter Horst 2010).

Unfortunately for these systems, it is not known onto which foreign particles the heterogeneous nucleation takes place, which is typically the case for nucleation in solution without the deliberate addition of functionalized surfaces. Therefore, a further interpretation of the data is not possible.

One main assumption of CNT is that the interfacial energy between bulk crystal and solution also is valid for small clusters down to sizes of several molecules. This would actually mean that the cluster structure is equal to that of macroscopically large crystals. Simulations first indicated (ten Wolde and Frenkel 1997) that the formation of a crystalline cluster is preceded by a dense, liquid-like cluster. This has now also been established in practice for protein crystal nucleation (Vekilov 2005) and small organic molecules such as glycine (Chattopadhyay et al. 2005). This was coined the two-step nucleation mechanism. Attempts have been made to establish a method enabling the interpretation of crystal nucleation rate data using the two-step nucleation mechanism (Vekilov 2010).

16.5.2 Pre-exponential Kinetic Factor

The parameter $A = z_0 f_0 C_0$ is related to the Zeldovich factor, attachment frequency of building units to the nucleus, and the concentration of heterogeneous particles.

As with the interpretation of the parameter B, due to our lack of knowledge on the foreign particles, a thorough molecular interpretation of the pre-exponential factor A is not possible: we cannot independently determine the attachment frequency f_0 and C_0 from experimental nucleation rates; we can only determine the product f_0C_0. In order to increase understanding, templates will have to be added. By assuming the same concentration of heterogeneous particles, relative attachment frequencies can be compared. This assumption for nucleation rates of para-aminobenzoic acid from various solvents (Sullivan et al. 2014) led to the conclusion that solute dimerization and desolvation are rate-determining processes in the overall nucleation pathway.

However, we can independently investigate solutions on the building units present. This information then can be linked to nucleation rate data or crystallization outcomes. Isonicotinamide self-associates differently depending on the solvent used: in nitromethane chains of isonicotinamide are dominantly present and in ethanol both chains and dimers, while there are indications that chloroform contains single molecules of isonicotinamide (Kulkarni et al. 2012). This gives rise to completely different nucleation and growth behavior from these solvents as is shown by the polymorphic outcome of isonicotinamide crystallization from the various solvents: crystallization of isonicotinamide in nitrobenzene, ethanol, and chloroform leads to, respectively, polymorph form I, form II, and possibly a new form. This tells that the building unit (dimer, chain, single molecule) dominantly present in the solution and responsible for the attachment to the nucleus during nucleation and to the crystal surfaces during growth has a large effect on these processes. The interesting review (Davey et al. 2015) that further investigates the transition state from solution to crystal discusses the molecular relation between association processes in solution and crystal nucleation.

Techniques such as infrared and Raman spectroscopy (Kulkarni et al. 2012; Davey et al. 2006; Kitamura et al. 2012) (Bernardes et al. 2014a), NMR (Hunter et al. 2012; Bernardes et al. 2014b), and sophisticated x-ray techniques (Thomason et al. 2015) were used to analyze the building units in solution. However, it should be kept in mind that polymorph crystallization behavior is determined by more than the building unit alone. Also factors such as heterogeneous particles, relative stability, and relative growth rates play an important role in determining the crystallization outcome (Davey et al. 2013) (Kitamura and Horimoto 2013).

16.5.3 Template Nucleation

Nucleation from solution occurs onto foreign particles like dust particles of which concentration C_0 and functionality (related to interfacial energy γ) are not known. There might well be a reason for the phenomenon of disappearing polymorphs (Bučar et al. 2015) sometimes encountered in industry: a change in, for instance, the solvent supply might change the heterogeneous particles involved in the nucleation process and herewith the polymorphic outcome of the industrial crystallization.

In order to circumvent the lack of control over the foreign particles, functionalized surfaces or particles can be deliberately provided. During full solvent evaporation in small solution droplets on well-defined patterned surfaces, the crystallization of mefenamic acid and sulfathiazole resulted in the nucleation of various crystal forms observed in the numerous droplets (Lee et al. 2008). In this way probability distributions of polymorph occurrence could be established. The probability distribution was observed to be a function of among others droplet size and evaporation rate. In another investigation the well-defined surfaces of self-assembled monolayers were shown to steer and promote specific building unit assembly toward targeted polymorphs of isonicotinamide (Kulkarni et al. 2014).

16.5.4 Temperature Effects

The pre-exponential factor A contains an energy barrier E_a for the incorporation of a building unit into the nucleus. The energy barrier for the incorporation has been determined to be 66 kJ/mol for rather viscous sucrose solutions in water (Dunning and Shipman 1954). Surprisingly, not many reports of this barrier exist for organic compounds nucleating in solution, probably because of the large amount of experimental work needed. In order to obtain the barrier E_a, one has to determine series of nucleation rates at a series of supersaturation ratios and temperatures.

16.5.5 Nonclassical Mechanisms

A main assumption of CNT is that clusters grow and decay by single building units since the concentration of clusters larger than one building unit is very small compared to those of one building unit. However, another nucleation pathway was identified experimentally in which nanosized pre-nucleation clusters form, which subsequently coagulate into a large cluster or single crystal (Gebauer and Cölfen 2011). This was found, for instance, for the crystal nucleation of the inorganic compound calcium carbonate.

Solute clustering has been reported for a series of other small organic molecules, including urea, citrate, aminosulfonic acids, and others (Gebauer et al. 2014). There is a lack of knowledge on the cluster formation and behavior. The study of clusters in solution surely will lead to a sound theoretical description of nonclassical nucleation mechanisms in the crystal nucleation of organic compounds (Davey et al. 2013; Gebauer et al. 2014; Jawor-Baczynska et al. 2013).

16.6 Outlook

There are indications that the crystal nucleation of small organic molecules does not always follow conventional nucleation pathways. However, CNT is still the only theory that enables the prediction of nucleation rates, needed to predict polymorphic outcome and particle size distribution of a crystallization process. Unfortunately, these nucleation rates are unreliable. This is only partly due to the underlying theory.

Another factor making nucleation rate predictions unreliable is the ill-defined nature of the system in which the crystal nucleation takes place. While usually heterogeneous nucleation occurs in industrial crystallization processes, nothing is known about the heterogeneous particles onto which nucleation occurs. Future research should be targeting to measure accurate heterogeneous crystal nucleation rates onto template particles having well-defined size distributions and surface functionality. Combined with the known solution association processes that define the crystal building blocks, this would enable a molecular level interpretation of the nucleation process of small organic molecules.

Additionally, new analytical techniques and simulation methods are being developed that enable the study of the pathway from supersaturated solutions to macroscopic crystals. These new techniques will lead to new insights into the nature of the nucleus, its size, and structure. They enable probing the stochastic process that leads from a supersaturated solution to a crystal (Nielsen and De Yoreo 2017, Chap. 18).

Progress has been made on crystal nucleation understanding in the areas of inorganic compounds, biomineralization, proteins, metals, metal organic frameworks, and small organic molecules. Although the materials are different, the process of crystal nucleation for all these materials will follow the same scientific principles. A scientific connection for crystal nucleation of all these compounds will be highly advantageous for the general understanding of crystal nucleation (De Yoreo et al. 2017, Chap. 1; Lutsko 2017, Chap. 2; Andreassen en Lewis 2017, Chap. 7; Van Driessche et al. 2017, Chap. 12).

References

Aaltonen J, Allesø M, Mirza S, Koradia V, Gordon KC, Rantanen J (2009) Solid form screening–a review. Eur J Pharm Biopharm 71:23–37

Andreassen J-P, Lewis AE (2017) Classical and non-classical theories of crystal growth. In: Benning Liane G, Gebauer D, Kellermeier M, Van Driessche Alexander ES (eds) New perspectives on mineral nucleation and growth. Springer, Cham, pp 137–154

Anwar J, Zahn D (2011) Uncovering molecular processes in crystal nucleation and growth by using molecular simulation. Angew Chem Int Ed 50:1996–2013

Bernardes CE, Ilharco LM, Da Piedade MEM (2014a) Polymorphism in 4′-hydroxyacetophenone: a vibrational analysis. J Mol Struct 1078:181–187

Bernardes CE, Lopes MLM, Ascenso JR, Da Piedade MEM (2014b) From molecules to crystals: the solvent plays an active role throughout the nucleation pathway of molecular organic crystals. Cryst Growth Des 14:5436–5441

Brandel C, ter Horst JH (2015) Measuring induction times and crystal nucleation rates. Faraday Discuss 179:199–214
Bučar D-K, Lancaster RW, Bernstein J (2015) Disappearing polymorphs revisited. Angew Chem Int Ed 54:6972–6993
Chattopadhyay S, Erdemir D, Evans JM, Ilavsky J, Amenitsch H, Segre CU, Myerson AS (2005) SAXS study of the nucleation of glycine crystals from a supersaturated solution. Cryst Growth Des 5:523–527
Chen J, Sarma B, Evans JMB, Myerson AS (2011) Pharmaceutical crystallization. Cryst Growth Des 11:887–895
Davey R, Dent G, Mughal R, Parveen S (2006) Concerning the relationship between structural and growth synthons in crystal nucleation: solution and crystal chemistry of carboxylic acids as revealed through IR spectroscopy. Cryst Growth Des 6:1788–1796
Davey RJ, Schroeder SL, Ter Horst JH (2013) Nucleation of organic crystals—a molecular perspective. Angew Chem Int Ed 52:2166–2179
Davey RJ, Back K, Sullivan R (2015) FD crystal nucleation from solutions–transition states, rate determining steps and complexity. Faraday Discuss 179:9–26
De Yoreo JJ, Sommerdijk N, Patricia D (2017) Nucleation pathways in electrolyte solutions. In: Benning Liane G, Gebauer D, Kellermeier M, Van Driessche Alexander ES (eds) New perspectives on mineral nucleation and growth. Springer, Cham, pp 1–24
Deij MA, Ter Horst JH, Meekes H, Jansens P, Vlieg E (2007) Polymorph formation studied by 3D nucleation simulations: application to a yellow isoxazolone dye, paracetamol, and L-glutamic acid. J Phys Chem B 111:1523–1530
Dunning W, Shipman A (1954) Nucleation in sucrose solutions. ro Congr Intern Ind Agr Madrid 2:1448
Forsyth C, Mulheran PA, Forsyth C, Haw MD, Burns IS, Sefcik J (2014) Influence of controlled fluid shear on nucleation rates in glycine aqueous solutions. Cryst Growth Des 15:94–102
Galkin O, Vekilov PG (1999) Direct determination of the nucleation rates of protein crystals. J Phys Chem B 103:10965–10971
Gebauer D, Cölfen H (2011) Prenucleation clusters and non-classical nucleation. Nano Today 6:564–584
Gebauer D, Kellermeier M, Gale JD, Bergström L, Cölfen H (2014) Pre-nucleation clusters as solute precursors in crystallisation. Chem Soc Rev 43:2348–2371
Hunter CA, Mccabe JF, Spitaleri A (2012) Solvent effects of the structures of prenucleation aggregates of carbamazepine. CrystEngComm 14:7115–7117
Ildefonso M, Candoni N, Veesler S (2011) Using microfluidics for fast, accurate measurement of lysozyme nucleation kinetics. Cryst Growth Des 11:1527–1530
Ismail SZ, Anderton CL, Copley RC, Price LS, Price SL (2013) Evaluating a crystal energy landscape in the context of industrial polymorph screening. Cryst Growth Des 13:2396–2406
Jawor-Baczynska A, Sefcik J, Moore BD (2013) 250 nm glycine-rich nanodroplets are formed on dissolution of glycine crystals but are too small to provide productive nucleation sites. Cryst Growth Des 13:470–478
Jiang S, Ter Horst JH (2010) Crystal nucleation rates from probability distributions of induction times. Cryst Growth Des 11:256–261
Kadam SS, Kulkarni SA, Coloma Ribera R, Stankiewicz AI, Ter Horst JH, Kramer HJ (2012) A new view on the metastable zone width during cooling crystallization. Chem Eng Sci 72:10–19
Kashchiev D (2000) Nucleation: basic theory with applications. Butterworth-Heinemann, Boston
Kim JW, Kim J, Lee KD, Koo KK (2013) Evaluation of nucleation rate by in-situ focused beam reflectance measurement in an unseeded batch cooling crystallization. Cryst Res Technol 48:1097–1105
Kitamura M, Horimoto K (2013) Role of kinetic process in the solvent effect on crystallization of BPT propyl ester polymorph. J Cryst Growth 373:151–155
Kitamura M, Umeda E, Miki K (2012) Mechanism of solvent effect in polymorphic crystallization of BPT. Ind Eng Chem Res 51:12814–12820

Kulkarni SA, Mcgarrity E, Meekes H, Ter Horst JH (2012) Isonicotinamide self-association: the link between solvent and polymorph nucleation. Chem Commun 48:4983–4985
Kulkarni SA, Kadam SS, Meekes H, Stankiewicz AI, Ter Horst JH (2013) Crystal nucleation kinetics from induction times and metastable zone widths. Cryst Growth Des 13:2435–2440
Kulkarni SA, Weber CC, Myerson AS, Ter Horst JH (2014) Self-association during heterogeneous nucleation onto well-defined templates. Langmuir 30:12368–12375
Lawton S, Steele G, Shering P, Zhao L, Laird I, Ni X-W (2009) Continuous crystallization of pharmaceuticals using a continuous oscillatory baffled crystallizer. Org Process Res Dev 13:1357–1363
Lee IS, Lee AY, Myerson AS (2008) Concomitant polymorphism in confined environment. Pharm Res 25:960–968
Lutsko JF (2017) Novel paradigms in non-classical nucleation theory. In: Benning Liane G, Gebauer D, Kellermeier M, Van Driessche Alexander ES (eds) New perspectives on mineral nucleation and growth. Springer, Cham, pp 25–42
Mullin JW (2001) Crystallization. Butterworth-Heinemann, Boston
Nanev CN, Hodzhaoglu FV, Dimitrov IL (2010) Kinetics of insulin crystal nucleation, energy barrier, and nucleus size. Cryst Growth Des 11:196–202
Nielsen MH, De Yoreo JJ (2017) Liquid phase TEM investigations of crystal nucleation, growth and transformation. In: Benning Liane G, Gebauer D, Kellermeier M, Van Driessche Alexander ES (eds) New perspectives on mineral nucleation and growth. Springer, Cham, pp 353–374
Patton JS, Byron PR (2007) Inhaling medicines: delivering drugs to the body through the lungs. Nat Rev Drug Discov 6:67–74
Sullivan R, Davey R, Sadiq G, Dent G, Back K, Ter Horst J, Toroz D, Hammond R (2014) Revealing the roles of desolvation and molecular self-assembly in crystal nucleation from solution: benzoic and p-aminobenzoic acids. Cryst Growth Des 14:2689–2696
Ten Wolde PR, Frenkel D (1997) Enhancement of protein crystal nucleation by critical density fluctuations. Science 277:1975–1978
Ter Horst JH, Kashchiev D (2003) Determination of the nucleus size from the growth probability of clusters. J Chem Phys 119:2241–2246
Ter Horst J, Kramer H, Jansens P (2002) A new molecular modeling approach to predict concomitant nucleation of polymorphs. Cryst Growth Des 2:351–356
Ter Horst JH, Schmidt C, Ulrich J (2015) Fundamentals of industrial crystallization. In: Rudolph P (ed) Handbook of crystal growth, 2nd edn. Elsevier, Boston
Thomason MJ, Seabourne C, Sattelle BM, Hembury GA, Stevens J, Scott A, Aziz EF, Schroeder SL (2015) Self-association of organic solutes in solution: a NEXAFS study of aqueous imidazole. Faraday Discuss 179:269–289
Van Driessche Alexander ES, Stawski Tomasz M, Benning Liane G, Kellermeier M (2017) Calcium sulfate precipitation throughout its phase diagram. In: Benning Liane G, Gebauer D, Kellermeier M, Van Driessche Alexander ES (eds) New perspectives on mineral nucleation and growth. Springer, Cham, pp 227–256
Vekilov PG (2004) Dense liquid precursor for the nucleation of ordered solid phases from solution. Cryst Growth Des 4:671–685
Vekilov PG (2005) Two-step mechanism for the nucleation of crystals from solution. J Cryst Growth 275:65–76
Vekilov PG (2010) The two-step mechanism of nucleation of crystals in solution. Nanoscale 2:2346–2357
Wedekind J, Strey R, Reguera D (2007) New method to analyze simulations of activated processes. J Chem Phys 126:134103
Yang HY, Rasmuson AC (2013) Nucleation of butyl paraben in different solvents. Cryst Growth Des 13:4226–4238
Yang HY, Svard M, Zeglinski J, Rasmuson AC (2014) Influence of solvent and solid-state structure on nucleation of parabens. Cryst Growth Des 14:3890–3902
Yuhara D, Barnes BC, Suh D, Knott BC, Beckham GT, Yasuoka K, Wu DT, Sum AK (2015) Nucleation rate analysis of methane hydrate from molecular dynamics simulations. Faraday Discuss 179:463–474

Chapter 17
Homogeneous Nucleation of Smoke Particles and Its Relationship with Cosmic Dust Particles

Yuki Kimura and Katsuo Tsukamoto

17.1 Nucleation of Cosmic Dust Particles

In the universe, most atoms exist in the gas phase and only about 1 % of atoms are incorporated into solid materials known as "cosmic dust particles" or simply "dust" or "grain." Despite its comparatively low abundance, cosmic dust is very important, because it forms the building material for planetary systems, acts as a substrate for the formation of molecules, and controls the energy balance in astronomical environments. Our main motivation is to understand the life cycle of materials in relation to the life cycles of stars. Knowledge of the initial state of the materials is essential for modeling of their subsequent evolution. The first solid materials were formed in gaseous ejecta from dying stars, such as asymptotic giant branch (AGB) stars or supernovae (Fig. 17.1). At this stage, no substrates were available for heterogeneous nucleation and, consequently, solid materials must have nucleated homogeneously. Homogeneous nucleation can only occur under conditions of significantly high supersaturation (Kimura et al. 2011, 2012; Kimura and Tsukamoto 2011). Under these conditions, the size of critical nuclei will be of the order of several nanometers or less. Once initial solid materials form, the

Y. Kimura (✉)
Institute of Low Temperature Science, Hokkaido University, Kita-19, Nishi-8, Kita-ku, 060-0819 Sapporo, Japan
e-mail: ykimura@lowtem.hokudai.ac.jp

K. Tsukamoto
Department of Earth and Planetary Materials Science, Graduate School of Science, Tohoku University, 980-8578 Sendai, Japan

Graduate School of Engineering, Osaka University, 565-0871 Suita, Osaka, Japan

A.E.S. Van Driessche et al. (eds.), *New Perspectives on Mineral Nucleation and Growth*,
DOI 10.1007/978-3-319-45669-0_17

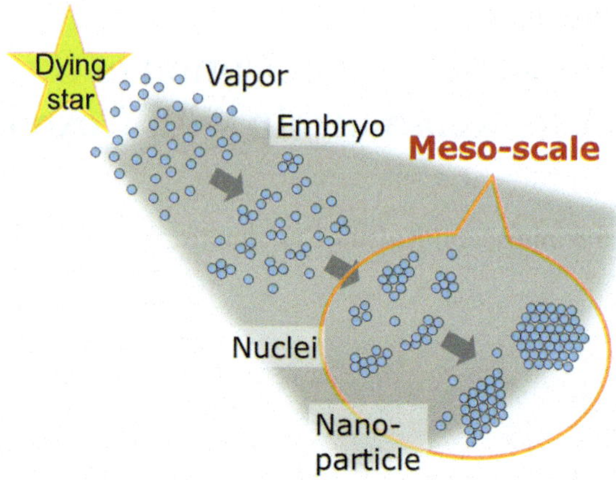

Fig. 17.1 Schematic showing the processes involved in the formation of cosmic particles in the gaseous outflow from a dying star. The size of cosmic dust particles must pass through the mesoscale to form dust with a size of the order of nanometers

remaining gaseous atoms have a much higher chance of undergoing heterogeneous nucleation on the nascent surface. Consequently, the initial nuclei govern the early stages of the growth process in the evolution of materials.

Small amounts of dust with an anomalous isotopic composition compared with solar minerals have been found in meteorites (Bernatowicz et al. 1991, 1996). Their composition suggests that they were formed before the birth of our solar system, and they are therefore referred to as "presolar grains." Spherules with a metal carbide core and a graphitic mantle are examples of presolar grains formed by heterogeneous nucleation on a primary grain. The graphitic spherules, which have a diameter of 0.43–4.2 μm and contain metal carbide crystals with a diameter of 7–90 nm, are assumed to have formed within the circumstellar envelopes of carbon-rich AGB stars (Croat et al. 2005). The metal carbide crystals, which consisted of Ti and/or Zr/Mo carbide, are generally located at the centers of individual spherules and are surrounded by well-graphitized carbon. It was therefore assumed that the metal carbide condensed before the carbon. On the basis of the condensation sequence and the sizes of the core and mantle, the conditions and environment for the formation of the grains, such as the total gas pressure, the carbon–oxygen ratio, the gas outflow velocity, and the stellar mass-loss rate, have been calculated by means of classical nucleation theory (Lodders and Fegley 1995; Sharp and Wasserburg 1995; Yamamoto and Hasegawa 1977; Draine and Salpeter 1977; Kozasa and Hasegawa 1987). However, it is difficult to explain the thickness of the graphitic mantle layer on the basis of the typical physical conditions expected in the circumstellar envelopes of carbon-rich AGB stars (Chigai et al. 2003).

Similarly, the processes involved in the formation of many other types of crystalline grains found in meteorites are unknown for reasons that might be associated with phenomena characteristic of nanoparticles and with the uncertainties associated with the physical properties of nanometer-sized particles (Kimura 2012). In the nucleation process, the size of particles passes through the mesoscale. Consequently, the physical properties and singular phenomena associated with nanoparticles must be taken into account if we are to understand the nucleation process and the processes involved in the formation of cosmic dust particles.

17.2 Characteristic Phenomena in Nucleation

In nucleation, a free-energy barrier must be overcome for atoms or molecules to form bulk materials (e.g., Lutsko 2017, Chap. 2). The size of particles, once nucleated, must pass through the mesoscale. We believe that this produces one of the major difficulties in understanding the process of nucleation. For example, let us consider the growth of a nanoparticle consisting of 44 atoms. How many steps are required for this to grow twice its initial size? Usually, at each step, an atom from a mother phase will attach to the surface of the particle and will subsequently be incorporated stochastically into the particle. The next atom then comes from elsewhere and will similarly attach to the surface of the particle. By this route, 44 steps will be needed to double the size of the initial particle. However, nanoparticles can grow by another, shorter route that involves only two steps: attachment and diffusion (Fig. 17.2).

The latter growth mode, called fusion growth, can be seen in the micrographs recorded by in situ bright-field transmission electron microscopy (TEM) shown in Fig. 17.3. These micrographs show gold nanoparticles with a size of about 5 nm lying on the silica particle shown in the bottom right-hand corner of the

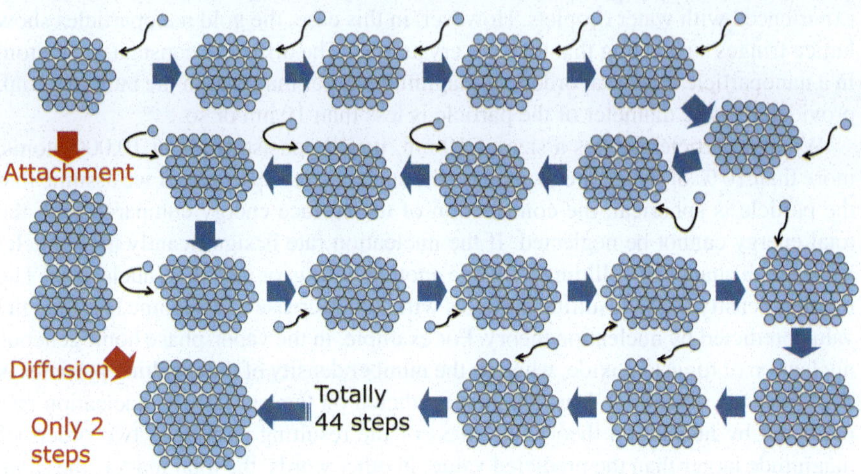

Fig. 17.2 Schematic showing two possible processes for the formation of nanoparticles. In the step-by-step process (classical model), shown by *gray arrows*, a nanoparticle grows by incorporation of individual growth units. In fusion growth, shown by *red arrows*, two nanoparticles coming into contact fuse to form a larger particle, thereby decreasing their total surface energy. Fusion growth is an example of a nonclassical model

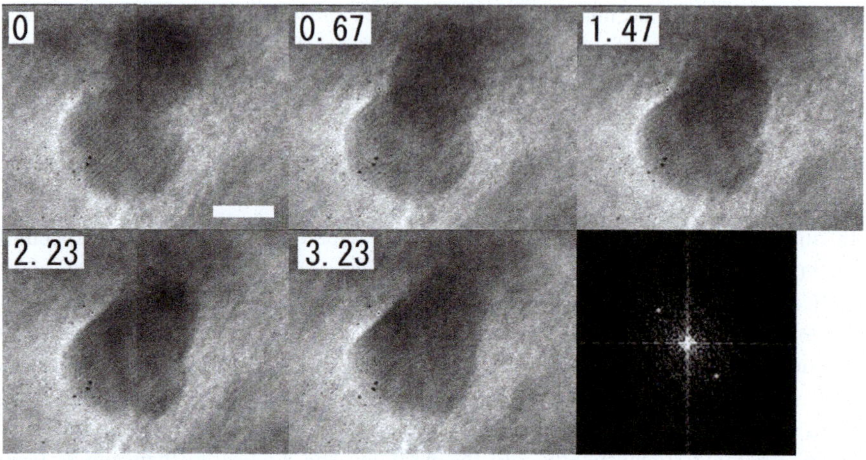

Fig. 17.3 Still micrographs of in situ bright-field TEM observations of an example of the fusion growth. The pictures are labeled with the time in seconds. Two gold particles migrate on the surface of the silica particle seen in the *bottom right-hand corner* in the images. When the two particles contact, they fuse together to decrease their total surface energy. The lattice fringes of the two gold nanoparticles indicate that the particles are crystalline. The final image shows a fast Fourier-transform (FT) diffraction pattern taken from the first image, confirming that the gold particles are crystalline. The *scale bar* corresponds to 3 nm

micrographs. Once the gold nanoparticles touch one another, they diffuse to form a larger particle, to decrease their total surface energy. This phenomenon is commonly experienced with water droplets. However, in this case, the gold nanoparticles show lattice fringes suggesting that they are crystalline. The diffusion constant of an atom in a nanoparticle is several orders of magnitude larger than that in the bulk material, provided that the diameter of the particle is less than 10 nm or so.

When a particle reaches a size of 10 nm, it will consist of about 10,000 atoms, more than 10 % of which are located on the surface of the particle. If we assume that the particle is spherical, the contribution of the surface energy compared with the total energy cannot be neglected. If the nucleation rate is significantly high, nuclei have a high chance of colliding with one another during, or just after, nucleation. The number density of the resulting particles will then decrease and become less than the value predicted by nucleation theory. For example, in the vapor-phase homogeneous nucleation of tungsten oxide, whereas the number density of the resulting particles is two orders of magnitude less than that predicted on the basis of the nucleation rate predicted by nucleation theory, the mass of the resulting particle is two orders of magnitude larger than the predicted value; in other words, the total mass is the same in the theoretical and experimental results (Kimura et al. 2011). This experiment suggests that hundreds of nuclei combined to form a particle during the nucleation process. The difference of several orders of magnitude between the experimental and the nucleation rate and that predicted by molecular-dynamics simulations and nucleation theory is generally considered to be a limitation of nucleation theory (Schmitt et al. 1982; Adams et al. 1984; Feder et al. 1966; Yasuoka and Matsumoto 1998; Toxvaerd 2001, 2003). However, some of the difference between the actual formation rate and the calculated nucleation rate might be the result of fusion growth caused by the so-called size effects of nanoparticles. Therefore, characteristic phenomena of nanoparticles need to be taken into account to understand the process of formation of cosmic dust particles.

17.3 Smoke Experiments as Model Processes for Dust Formation

When a material evaporates in an inert gas, the evaporated hot vapor subsequently cools and nucleates to form nanoparticles that flow along with the convection current generated by the hot evaporation source. The flow resembles the smoke from a cigarette. Inside the chamber, because there is no substrate near the evaporation source, the particles nucleate homogeneously in a manner similar to cosmic dust particles. Although the gas around a dying star is much less dense than the experimental atmosphere, the presence of an inert gas decreases the mean free path of the evaporant. If materials are evaporated at similar pressures (below 10^{-4} Pa) to those surrounding a dying star, the evaporated vapor will collide directly with the surface of the chamber wall, because of the long mean free path (in excess of 10^3 m), and will form a thin film on it.

The presence of an inert gas also affects the nucleation conditions, particularly the relationship between the cooling rate and the collision frequency of the evaporated atoms or molecules. Yamamoto and Hasegawa (1977) proposed a nondimensional parameter Λ, defined by the timescale for supersaturation increase τ_{sat} and the collision frequency ν. Cosmic dust particles are formed in gaseous ejecta where the supersaturation gradually increases as the temperature decreases with distance from the central star. The parameter Λ is useful in discussing the characteristics of particles formed in a gaseous atmosphere consisting of various objects or even in the laboratory. In other words, particles formed under conditions with the same Λ value can be discussed similarly and should have similar characteristics. Typical Λ values for various objects and for various experiments are summarized in Table 17.1. This shows that dust in novae can be reproduced by laboratory experiments, and dust in supernovae and evolved stars can be reproduced by experiments performed in conditions of microgravity.

Table 17.1 Comparison of Λ values for natural and laboratory conditions

Object	Nova	Supernova	Evolved star	Solar system	Laboratory experiments	Microgravity experiments
Λ	10^0–10^2	10^4	10^3–10^6	10^9	10^0–10^2	10^2–10^4

Revision of Yamamoto and Hasegawa (1977)

17.4 In Situ Observation of Nucleation by Optical Methods

Nucleation is very difficult to visualize experimentally because it is a rapid process that occurs on a nanometer scale (e.g., Nielsen and De Yoreo 2017, Chap. 18). In contrast, direct observation of the nucleation environment is relatively easy and has been achieved by in situ measurements of temperatures, concentrations, and infrared (IR) spectra during nucleation from a vapor phase (Fig. 17.4); the temperatures and concentrations were measured by using a double-wavelength Mach–Zehnder-type interferometer, and the IR spectra were recorded by using an FT-IR spectrometer.

To produce nanoparticles by homogeneous nucleation from an evaporated vapor, the mean free path of the atmosphere must be reduced by the introduction of a buffer gas; this also permits the detection of the temperature and concentration of the evaporated vapor from the changes in its refractive index, as detected by an interferometer. Our experimental setup was capable of detecting a difference in refractive index of less than 1×10^{-6}, which, for example, corresponds to the difference in temperature from 298 K to 302 K for argon gas at 1×10^{4} Pa.

When a material is evaporated by electrical heating, changes in the refractive index are induced not only by variations in the temperature but also by variations in the concentration of the evaporated vapor. Because we used a double-wavelength Mach–Zehnder-type interferometer, operating at 532–635 nm, the temperature and the concentration could be determined by a simple calculation based on the displacement of the interference fringes at the two wavelengths and the total pressure, monitored by using a pressure gauge. The FT-IR spectrometer can distinguish the crystalline structure of free-flying nanoparticles during nucleation by means of spectra obtained in the region stretching from near the evaporation source to several centimeters above it.

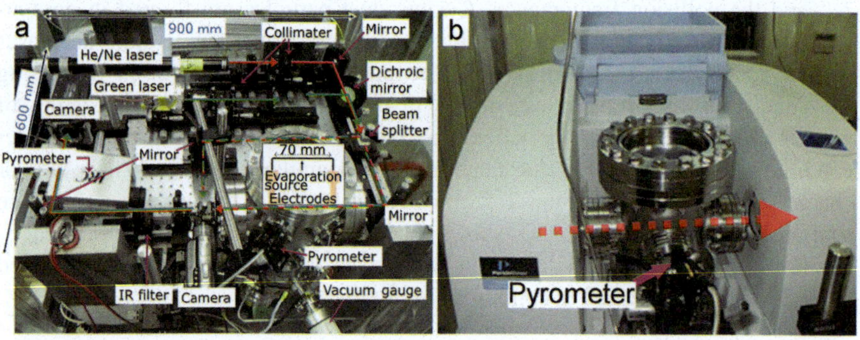

Fig. 17.4 (**a**) Photographs of (**a**) the experimental apparatus designed to produce nanoparticles that mimic cosmic dust together with the Mach–Zehnder-type interferometer and (**b**) the infrared spectrometer for in situ observation of the nucleation environment. The *red* and *green arrows* in (**a**) show the optical path of the interferometer. The *dotted arrow* in (**b**) shows the optical path of the infrared light

17.5 Interferometer Shows the Difficulty of Homogeneous Nucleation

Figure 17.5a is an example of an image of interference fringes before heating of the evaporation source. When the evaporation source was electrically heated, the interference fringes moved as a result of a decrease in the refractive index of the hotter surrounding argon gas (Fig. 17.5b). The refractive index of Ar, $n_{\mathrm{Ar(T,P)}}$, can be expressed as a function of temperature T (K) and pressure P_{Ar} (Pa) as follows:

$$n_{Ar(T,P)} - 1 = \frac{\left[n_{Ar(273.15,P_0)} - 1\right]}{1 + a\Delta T}\frac{P_{Ar}}{P_0} \tag{17.1}$$

where a is the coefficient of volume expansion (0.003663 K^{-1} in this experiment) and P_o is the pressure (101325 Pa). The shift in the fringes depends on the change in the optical path length L, which is defined as nl, where l is the physical length (70 mm in this experiment). The numbers of fringes of green (Δd_G) and red (Δd_R) that showed displacements from the positions of the starting interference fringes in Fig. 17.5a during heating are therefore given by the following equations:

$$\Delta d_G = \left[n_{GAr(Ti,Pi)} - n_{GAr(T,P_{Ar})} - n_{G_{E\left(T,P_E\right)}} + 1 \right] \frac{l}{\lambda_G} \tag{17.2}$$

and

$$\Delta d_R = \left[n_{RAr(Ti,Pi)} - n_{RAr(T,P_{Ar})} - n_{R_{E\left(T,P_E\right)}} + 1 \right] \frac{l}{\lambda_R} \tag{17.3}$$

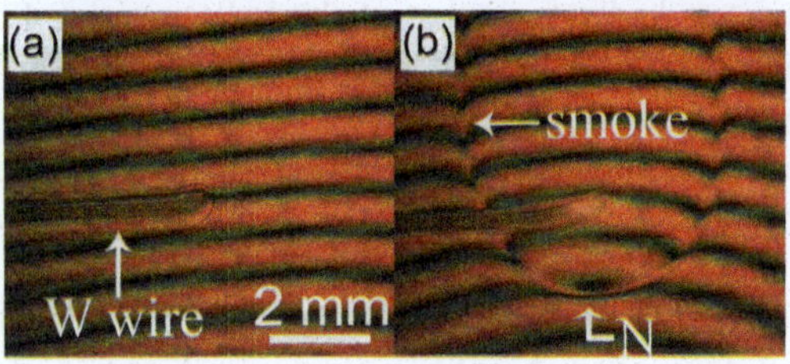

Fig. 17.5 Typical interference fringes taken along the longer axis of the tungsten wire with source temperatures of (**a**) room temperature and (**b**) ~1600 K in a mixture gas of argon (9.0×10^3 Pa) and oxygen (1.0×10^3 Pa). The tungsten wire was 0.3 mm in diameter and 70 mm long and it was oriented parallel to the optical path. The interference fringes in (**b**) suggest that homogeneous nucleation occurred around the evaporation source. Nucleated smoke particles flowed upward as a result of the convection current generated in the ambient gas

where T_i and P_i are the initial temperature and pressure before the temperature of the source was increased. The subscripts G and R show the values for the green and red lasers, respectively, and $P_{Ar} = P_{total} - P_E$, where P_E is the partial pressure of the evaporant. By using both equations, the temperature and concentration profiles can be determined simultaneously from the images in Fig. 17.5. The existence of nucleated particles can be confirmed in Fig. 17.5b, because the higher refractive index of solid particles produces the marked change in the interference fringes shown by the arrow labeled "smoke." Because of the rising convection current generated by the hot evaporation source, the evaporated vapor did not diffuse uniformly from the tungsten wire; instead, the rising tungsten oxide vapor was accelerated and downward flow of the vapor was restrained. As the result, a relatively high-concentration field was generated below the evaporation source compared in the same centric distance, i.e., the degree of supersaturation was highest to the direction below the evaporation source. Consequently, nucleation occurred below the evaporation source at the position shown by arrow N in Fig. 17.5b. The supersaturation ratio reached values as high as 10^7 for tungsten oxide and 10^4 for manganese. This high supersaturation was the result of homogeneous nucleation, because supersaturation does not increase to such high levels if heterogeneous nucleation would be occurred.

17.6 Determination of Physical Constants

The surface free energy and the sticking probability markedly affect the prediction by nucleation theory of the characteristics of a dust, such as its nucleation temperature, mineral species, size, and number density. The actual size of cosmic dust particles ranges from a few nanometers to around a hundred nanometers. Despite this, the surface free energy of the bulk material has always been used in calculations. Recent studies have, however, shown that the surface free energy of nanoparticle differs markedly from that of the bulk material. For example, in the case of rutile, which is a polymorph of titanium dioxide, the surface free energy of 14-nm-diameter particles is 30 % larger than that of the bulk material (Zhang et al. 2009). Because absolute values of the sticking probability are generally unknown, its value has often been assumed to be unity for the purposes of calculation. Actually, a higher sticking probability provides a better explanation of the amount of dust in the circumstellar environment. When a bulk experiment was performed to obtain the sticking probability for iron, it showed a value very close to unity (Tachibana et al. 2011). In the experiment, iron was evaporated onto a temperature-controlled iron substrate, and the mass of the evaporant and the adherent iron film were compared. In contrast, Michael et al. (2003) found a very low sticking probability of 10^{-5} in a nucleation experiment using zinc under microgravity conditions produced in a parabolic flight.

We also performed numerical simulations of the nonequilibrium condensation of several materials to determine their surface free energies and sticking coeffi-

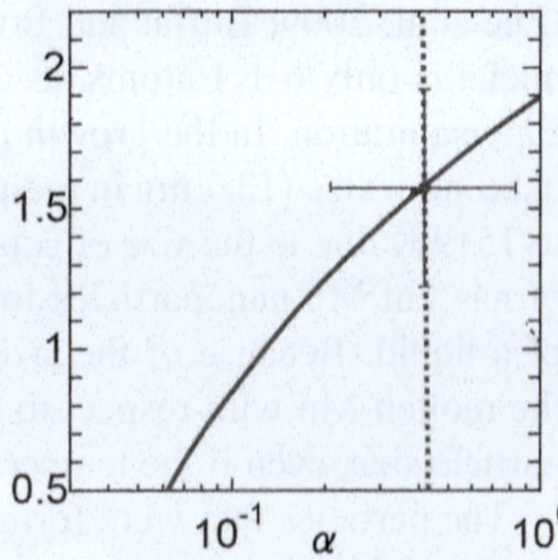

Fig. 17.6 A plot of the surface free energy σ against the sticking coefficient α, obtained by simulations using the SP nucleation model to explain the nucleation temperature (*solid line*) and the size of the resulting particles (*dashed line*) observed in experiments

cients. By in situ observations of a nucleation environment in conjunction with the application of nucleation theories, we succeeded in determining the surface free energies and sticking probabilities of nanoparticles undergoing homogeneous nucleation (Kimura et al. 2012). We considered a gaseous system that cools down with a specific timescale. By using this timescale, we calculated the equations governing the temporal evolution of the gas number density and the nucleation rate (Yamamoto and Hasegawa 1977; Tanaka et al. 2002). From these simulations, we obtain the condensation temperature and the size of the particles.

By comparing the results of the simulations with the experimental results, we were able to determine the surface fee energy and the sticking coefficient. Figure 17.6 shows possible values of the sticking coefficient α and the corresponding surface free energies σ that explain the nucleation temperature (solid line) and the size of the resulting particles (dashed line) from the experiments in terms of a semi-phenomenological (SP) nucleation model based on the results of experiments using Mn as an example. The crossing points are the surface free energy and sticking coefficient of Mn at 1106 ± 50 K, which are 1.57 ± 0.35 J m^{-2} and 0.4 (+0.42, −0.21), respectively. The error bar arises from variations in the partial pressure and timescale for cooling of Mn. Reported surface free energies of molten Mn include values of 1.1 J m^{-2} at 1573–1773 K (Turkdogan 1980) and 0.92–1.18 J m^{-2} at 1573 K (Shinozaki et al. 1998). The surface free energy of Mn nanoparticles is therefore about 30 % larger than that of molten Mn. We also performed a nucleation experiment with iron in a microgravity environment generated by using a sounding rocket. In this experiment, the sticking probability of iron was found to be as low as 10^{-4}, a value significantly lower than that of $\sim 10^{0}$ obtained by a ground-based bulk experiment (Tachibana et al. 2011).

17.7 Multistep Nucleation Processes

Homogeneous nucleation can only occur under conditions of very high supersaturation where the critical nuclei are only a few atoms in size. The melting point of a nanoparticle is generally lower than that of the corresponding bulk material

(Lee et al. 2009; Buffat and Borel 1976). In the case of Mn, the size of the critical nuclei is only 6 ± 1 atoms, as calculated from the SP model, due to the very large supersaturation. In the growth process of Mn nanoparticles from six atoms to their maximum size (13.7 nm in radius), the melting point will increase from about 800 K to 1519 K due to the size effects of the nanoparticles. Therefore, the initial structure of nascent Mn nanoparticles formed at 1106 ± 50 K can be regarded as being that of a liquid. Because of the size dependence of the melting point, supercooling of the molten Mn with respect to the melting point increases directly with increasing particle size, even if the temperature remains steady.

The particles that were formed consisted of mixtures of α-Mn and β-Mn, which are the stable phases at temperatures below 1015 K and between 1015 and 1368 K, respectively. The unit cells of α-Mn and β-Mn contain 58 and 20 atoms, respectively (JCPDS cards 32-637 and 33-887). Because the size of critical nuclei is much smaller than these values, the structure of the Mn nanoparticles is not determined at the point of nucleation. The polymorphism of the resulting Mn nanoparticles is therefore determined in the particle-growth phase and not in the vapor-to-liquid nucleation phase. This result confirms that the initial nucleation from the vapor phase does not control the resulting crystalline phase that eventually forms from the molten droplet. If that were the case, the crystal structure of the Mn nanoparticles would be determined when the crystal first nucleated homogeneously within the initial molten particle (second liquid–solid nucleation) rather than later in the growth phase. Both α-Mn and β-Mn nanoparticles can nucleate stochastically from the molten particles independently of the phase diagram; this is similar to the nucleation behavior of sodium chlorate (Kimura et al. 2014). Rapid cooling at more than 10^3 K s^{-1} may allow the preservation of the higher-temperature phase. This kind of two-step nucleation has also been reported in a molecular-dynamics simulation of the homogeneous condensation of argon vapor (Tanaka et al. 2011).

Infrared spectra recorded during the nucleation of magnesium silicate from a vapor phase also show the initial formation of a molten phase (Fig. 17.7). The initial single feature at 10 μm originates from amorphous or molten magnesium silicate. On cooling to about 500 K, the single feature becomes a double peak as a result of the presence of a crystalline phase. This crystallization is the result of supercooling of initially nucleated molten nanoparticles (Ishizuka et al. 2015). These studies for in situ observation of a nucleation environment are inchoate and will generate deeper insight and progress the understanding of nucleation in the coming decade.

17.8 Outlook

Using nucleation theories, we can calculate expected nucleation rates; however these rates are very different from those obtained by experiments or molecular-dynamics simulations. For instance, in the case of water or methanol nucleation from its vapor, the difference in nucleation rates is several orders. Hence, one of

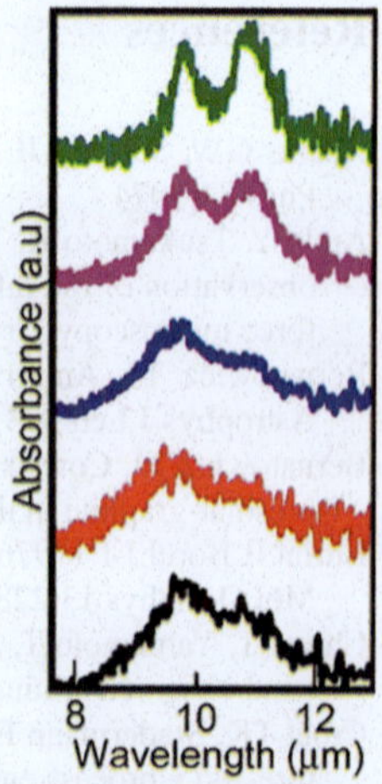

Fig. 17.7 Infrared spectra of magnesium silicate nanoparticles during nucleation recorded at distances of 2, 4, 6, 8, and 10 cm above the evaporation source (from the *bottom* to the *top*, respectively). The temperature decreases gradually above the evaporation source, and the crystallinity increases from less than 1 % at 2 cm to more than 20 % at 10 cm

the largest problems to understand nucleation is that we do not yet realize why nucleation rates are so different between theory and experiment or molecular-dynamics simulation. There are some possibilities, such as limitation of the theories, heterogeneous nucleation caused by not an ideal experiment, and occurrence of multistep nucleation. Nucleation from the vapor phase has the advantage that there is no water solution, which induces large hindrance during solution growth due to the dehydration process (e.g., DeYoreo et al. 2017, Chap. 1). In a water solution, both the crystals and growth units are usually surrounded by hydrated layers on their surfaces (Araki et al. 2014). Growth units have to eliminate/rearrange these hydrated layers before being incorporated into a crystal or nuclei. To do so, the system will try to find more thermodynamically favorable routes, which can eventually lead to "unexpected" nucleation processes.

Hence, before making a *super-unified* nucleation theory, first a more simple nucleation theory, applicable to the vapor phase, should be developed. In case of vapor-phase nucleation, binding energy of growth units seems more important than surface free energy, because a phenomenological nucleation theory, which takes into account the binding energy of a dimer, could reproduce nucleation rates in homogeneous nucleation experiments for some conditions (Kimura et al. 2012). Currently, these experiments have been performed using only one element. In the near future, we will perform these experiments for multicomponent systems, and then nucleation rates may be difficult to predict due to the occurrence of multistep nucleation events, even in a vapor phase. For instance, two different kinds of nanoparticles can merge to be a compound particle by fusion growth similarly with that shown in Fig. 17.3. Magnesium silicate (Mg_2SiO_4), which is an astronomically important mineral, may be able to form by a fusion growth of MgO and SiO_2 particles. A better understanding of the nucleation process should also provide us with new perspectives on material evolution in the universe.

References

Adams GW, Schmitt JL, Zalabsky RA (1984) The homogeneous nucleation of nonane. J Chem Phys 81:5074

Araki Y, Tsukamoto K, Takagi R, Miyashita T, Oyabu N, Kobayashi K, Yamada H (2014) Direct observation of the influence of additives on calcite hydration by frequency modulation atomic force microscopy. Cryst Growth Des 14:6254–6260

Bernatowicz TJ, Amari S, Zinner E et al (1991) Interstellar grains within interstellar grains. Astrophys J Lett 373:L73–L76

Bernatowicz TJ, Cowsik R, Gibbons PC et al (1996) Constraints on stellar grain formation from presolar graphite in the Murchison meteorite. Astrophys J 472:760–782

Buffat P, Borel J-P (1976) Size effect on the melting temperature of gold particles. Phys Rev A: At Mol Opt Phys 13:2287

Chigai T, Yamamoto T, Kaito C et al (2003) Are TiC grains a carrier of the 21 micron emission band observed around post–asymptotic giant branch objects? Astrophys J 587:771–776

Croat TK, Stadermann FJ, Bernatowicz TJ (2005) Presolar graphite from AGB stars: microstructure and *s*-process enrichment. Astrophys J 631:976–987

De Yoreo JJ, Sommerdijk NAJM, Dove PM (2017) Nucleation pathways in electrolyte solutions. In: Van Driessche AES, Kellermeier M, Benning LG, Gebauer D (eds) New perspectives on mineral nucleation and growth, Springer, Cham, pp 1–24

Draine BT, Salpeter EE (1977) Time-dependent nucleation theory. J Chem Phys 67:2230–2235

Feder J, Russell KC, Lothe I et al (1966) Homogeneous nucleation and growth of droplets in vapours. Adv Phys 15:111–178

Ishizuka S, Kimura Y, Sakon I (2015) In situ infrared measurements of free-flying silicate during condensation in the laboratory. Astrophys J 803:88

Kimura Y (2012) Phenomena of nanoparticles in relation to the solar system. In: Mann I, Meyer-Vernet N, Czechowski A (eds) Nanodust in the solar system: discoveries and interpretations. Springer, Heidelberg, pp 31–46

Kimura Y, Tsukamoto K (2011) Interferometric observation of temperature distributions in the smoke experiment. J Jpn Soc Microgravity Appl 28:S9–S12

Kimura Y, Miura H, Tsukamoto K, Li C, Maki T (2011) Interferometric in-situ observation during nucleation and growth of WO_3 nanocrystals in vapor phase. J Cryst Growth 316:196–200

Kimura Y, Tanaka KK, Miura H, Tsukamoto K (2012) Direct observation of the homogeneous nucleation of manganese in the vapor phase and determination of surface free energy and sticking coefficient. Cryst Growth Des 12:3278–3284

Kimura Y, Niinomi H, Tsukamoto K, García-Ruiz JM (2014) In situ live observation of nucleation and dissolution of sodium chlorate nanoparticles by transmission electron microscopy. J Am Chem Soc 136:1762

Kozasa T, Hasegawa H (1987) Grain formation through nucleation process in astrophysical environments II. Prog Theor Phys 77:1402–1410

Lee J, Lee J, Tanaka T et al (2009) In situ atomic-scale observation of melting point suppression in nanometer-sized gold particles. Nanotechnology 20:475706

Lodders K, Fegley B Jr (1995) The origin of circumstellar silicon carbide grains found in meteorites. Meteoritics 30:661–678

Lutsko JF (2017) Novel paradigms in nonclassical nucleation theory. In: Van Driessche AES, Kellermeier M, Benning LG, Gebauer D (eds) New perspectives on mineral nucleation and growth, Springer, Cham, pp 25–42

Michael BP, Nuth JA III, Lilleleht LU (2003) Zinc crystal growth in microgravity. Astrophys J 590:579–585

Nielsen MH, De Yoreo JJ (2017) Liquid phase TEM investigations of crystal nucleation, growth, and transformation. In: Van Driessche AES, Kellermeier M, Benning LG, Gebauer D (eds) New perspectives on mineral nucleation and growth, Springer, Cham, pp 353–371

Schmitt JL, Adams GW, Zalabsky RA (1982) Homogeneous nucleation of ethanol. J Chem Phys 77:2089–2097

Sharp C, Wasserburg G (1995) Molecular equilibria and condensation temperatures in carbon-rich gases. Geochim Cosmochim Acta 59:1633–1652

Shinozaki N, Sonoda M, Mukai K (1998) Wettability, surface tension, and reactivity of the molten manganese/zirconia–yttria ceramic system. Metall Mater Trans A 29:1121–1125

Tachibana S, Nagahara H, Ozawa K et al (2011) Kinetic condensation and evaporation of metallic iron and implications for metallic iron dust formation. Astrophys J 736(16):8

Tanaka KK, Tanaka H, Nakazawa K (2002) Non-equilibrium condensation in a primordial solar nebula: formation of refractory metal nuggets. Icarus 160:197–207

Tanaka KK, Tanaka H, Yamamoto T, Kawamura K (2011) Molecular dynamics simulations of nucleation from vapor to solid composed of Lennard–Jones molecules. J Chem Phys 134:204313

Toxvaerd S (2001) Molecular-dynamics simulation of homogeneous nucleation in the vapor phase. J Chem Phys 115:8913–8920

Toxvaerd S (2003) Molecular dynamics simulation of nucleation in the presence of a carrier gas. J Chem Phys 119:10764–10770

Turkdogan ET (ed) (1980) Physical chemistry of high temperature technology. Academic, New York

Yamamoto T, Hasegawa H (1977) Grain formation through nucleation process in astrophysical environment. Prog Theor Phys 58:816–828

Yasuoka K, Matsumoto M (1998) Molecular dynamics of homogeneous nucleation in the vapor phase. I. Lennard–Jones fluid. J Chem Phys 109:8451–8462

Zhang H, Chen B, Banfield JF (2009) The size dependence of the surface free energy of titania nanocrystals. Phys Chem Chem Phys 11:2553–2558

Chapter 18
Liquid Phase TEM Investigations of Crystal Nucleation, Growth, and Transformation

Michael H. Nielsen and James J. De Yoreo

18.1 Introduction

Many open questions in crystallization processes remain in part because traditional characterization tools do not allow for direct observations of the phenomena at relevant length and time scales. The dynamics of nucleation, including formation of the nascent particle, attachment and detachment of the fundamental units, and the transient existence and transformation of potential intermediate phases, typically occur at nanometer length scales limiting the utility of in situ optical (Hu et al. 2012), scanning electron microscopy (Verch et al. 2013), and X-ray-based imaging (Rieger et al. 2000) and spectroscopic (Lee et al. 2007) methods. In situ scanning probe microscopy has been successfully applied to nucleation and growth in many surface-mediated systems due to its capability of imaging with adequate spatial resolution (Sleutel et al. 2014). However, the early events in the formation of many materials occur rapidly relative to the time required for image acquisition. Although recent advances in high scan speed approaches have greatly improved the attainable temporal resolution (Ando et al. 2001), surface scanning techniques are limited in the samples they can characterize. Scanning probe microscopy is of little use, for example, in examining the role of organic matrices, such as globular phases of

M.H. Nielsen (✉)
Physical and Life Sciences Directorate, Lawrence Livermore National Laboratory, Livermore, CA 94550, USA

Department of Materials Science and Engineering, University of California, Berkeley, CA 94720, USA
e-mail: nielsen29@llnl.gov

J.J. De Yoreo
Physical Sciences Division, Pacific Northwest National Laboratory, Richland, WA 99352, USA

Departments of Materials Science and Engineering and of Chemistry, University of Washington, Seattle, WA 98195, USA

A.E.S. Van Driessche et al. (eds.), *New Perspectives on Mineral Nucleation and Growth*,
DOI 10.1007/978-3-319-45669-0_18

macromolecules (Smeets et al. 2015), lipid vesicles (Tester et al. 2011), amelogenin (Fang et al. 2011), and collagen (Nudelman et al. 2010), where mineralization likely occurs within the matrix. More broadly, methods that minimize surface effects are necessary for better understanding crystallization processes in solution, as surfaces often substantially change the kinetic barriers and free energies that control nucleation (Giuffre et al. 2013; Hamm et al. 2014, De Yoreo et al. 2017, Chap. 1).

Lacking techniques for direct, in situ examination of mineralization mechanisms, researchers have turned to alternative approaches to understand these phenomena. Low-resolution in situ techniques have produced indirect evidence of crystallization pathways (Bots et al. 2012; Hu et al. 2012), while ex situ imaging (Rieger et al. 2007), calorimetric (Radha et al. 2010), and spectroscopic (Lee et al. 2007) approaches have given insight into individual time slices of mineralization processes. In the absence of direct experimental data that follow the birth and development of individual particles, tenets of classical nucleation theory (CNT) originating from Gibbs (1876, 1878) have been challenged by novel theoretical frameworks. Nucleation in electrolyte solutions, as described by CNT, occurs via unstable density fluctuations of monomer attachment and detachment that, given sufficient time, stochastically build a nucleus exceeding the critical size (De Yoreo et al. 2017, Chap. 1, Andreassen and Lewis Chap. 7). Crossing this threshold makes further growth energetically favorable, and, given a sufficient reservoir of material, monomeric attachment of growth units continues unabated. An example of alternative theories to CNT is well illustrated by recent studies in $CaCO_3$ crystallization. Titration and analytical ultracentrifugation data have served as the basis for proposing that a stable, multi-ion population of "pre-nucleation clusters" exist (Gebauer et al. 2008) and that these clusters aggregate to form an amorphous precursor that subsequently transforms into one of the anhydrous crystalline phases. Supporting evidence for the existence of these clusters came from cryogenic transmission electron microscopy (cryo-TEM) (Pouget et al. 2009), although these data could not establish whether these were stable clusters or transient species arising from the density fluctuations described by CNT; nor could they verify a link between the observed clusters and later nuclei. Alternatively — or in addition — to the clusters, a metastable, bicarbonate-rich dense liquid precursor phase has been proposed based on NMR data, as forming prior to the appearance of a solid phase (Bewernitz et al. 2012, Wolf and Gower 2017, Chap. 3). In silico approaches have suggested that highly dynamic clusters form (Demichelis et al. 2011), followed by spinodal decomposition driving a liquid-liquid phase separation (Wallace et al. 2013) at moderate supersaturations. The solid phase was then reasoned to form via amorphous calcium carbonate (ACC) generation through dehydration of the denser liquid droplets. In addition, ACC has been frequently observed as the initially formed precursor (Brecevic and Nielsen 1989) that eventually ends up as one of the crystalline phases (Gal et al. 2013, Rodriguez-Blanco et al. 2017, Chap. 5). Despite many attempts, however, cryo-TEM (Pouget et al. 2009), X-ray spectroscopy (Lee et al. 2007), and in situ X-ray studies (Rieger et al. 2000; Bots et al. 2012) have failed to determine whether crystal formation from an ACC precursor occurred through dissolution-reprecipitation or direct phase transformation.

Complex mineral nucleation pathways involving cluster aggregation and transformation of precursor phases have also been observed in other systems. Calcium phosphate, for example, has been suggested, based on cryo-TEM and in situ scanning probe microscopy data, to form through the initial development of an extended network of nanometer-sized amorphous particles which then undergo sequential chemical, structural, and morphological transformations to produce the final crystalline phase (Habraken et al. 2013, Birkedal 2017, Chap. 10, Delgado-Lopez and Guagliardi 2017, Chap. 11). However, because the characterization tools lacked the capability of directly observing particle formation and evolution through the proposed pathway, the results are open to other interpretations.

Looking beyond nucleation mechanisms, processes governing crystal growth are also poorly understood. While the predictions of well-developed classical theories accurately describe growth in many systems as proceeding via monomer-by-monomer attachment, they fail to correctly portray growth in other systems. Particle-based growth mechanisms provide a clear illustration of this point. It was discovered that TiO_2 primary particles can serve as growth units and attach on specific crystallographic faces, a process that has come to be called "oriented attachment," yielding complex single crystals (Penn and Banfield 1998a, Penn et al. Chap. 14). In the years since those findings, particle-based mechanisms have been recognized as perhaps dominant in numerous biological, environmental, and synthetic systems, particularly in the cases of semiconductor nanoparticles and nanowires with complex topologies (Penn 2004). Particle attachment mechanisms need not proceed through alignment along perfect crystallographic registry. Attachment at slightly misaligned orientations (Penn and Banfield 1998b) may precede crystal relaxation that expels defects from the structure to result in a well-ordered single crystal (Li et al. 2012). Furthermore, attaching particles may not even be of a uniform phase. Instead, a metastable, or stable only at the nanoscale, phase attaches to the surface of a crystal of the stable bulk phase, followed by recrystallization of the attached particle to yield a single crystal (Baumgartner et al. 2013; Frandsen et al. 2014; Nielsen et al. 2014a, Reichel and Faivre 2017, Chap. 14). The resulting crystals from any of the above examples of particle-based mechanisms may contain scant, if any, signature of the pathway by which they were produced. Additionally, most experimental data used to infer such particle-based pathways have been produced by ex situ techniques.

With this context established, the need for in situ techniques such as liquid phase TEM, capable of resolving crystallization processes at time and length scales appropriate to catch the early stages of particle formation, is clear. The abovementioned and many similar studies, some of which may be controversial or inconclusive, suggest a multitude of mechanisms contribute to nucleation and growth depending on material system and precipitation conditions. Thus, in order to truly understand the physics behind such nonclassical mechanisms and develop a more complete picture of materials formation, identifying which processes are active is of paramount importance. Direct observation of these phenomena and the rates at which they proceed would additionally provide quantitative insight into the energetics of nucleation and growth, as formation pathways depend on energy barriers between initial and final states, changes in free energy, and gradients in interaction potentials. Therefore, the ability of liquid phase TEM to acquire both imaging and crystallographic data renders it an invaluable tool for advancing our knowledge of crystallization processes.

18.2 Liquid Phase TEM

Many of the developments in liquid phase TEM approaches can largely trace their roots back to the pioneering work at IBM (Williamson et al. 2003), which entailed the construction of a hermetically sealed liquid cell from the assembly of microfabricated silicon chips. These cells sandwiched a thin liquid layer between two electron-transparent silicon nitride windows. The cell's liquid volume was constant due to the hard epoxy seal around the cell, and electrodes fabricated onto the liquid cell provided electrochemical control over the electrolyte solution filling the cell. In the recent past, a number of alternative approaches to liquid phase TEM have appeared. Among these variants are the use of ultrathin graphene membranes rather than silicon nitride for the windows (Yuk et al. 2012), and the use of non-sealed cells in conjunction with plumbed TEM stages to allow continuous reagent flow through the liquid cell (Ring and de Jonge 2010). While initially liquid phase TEM was used by individual labs that designed and built their own systems, multiple companies now exist that offer commercial in situ TEM stages, and the technique is rapidly becoming more widespread.

Two generic liquid phase TEM platforms that represent the approaches used for the data discussed below are depicted in Fig. 18.1. One system incorporates a sealed cell (Fig. 18.1a) after the manner of the IBM cell mentioned above, while the other pumps reagent solutions through an open cell via tubes integrated into the TEM stage (Fig. 18.1c). Both cell types use as major components silicon wafers covered with silicon nitride thin films deposited on both faces. On each wafer one nitride film and the underlying silicon are etched away, to yield a suspended nitride membrane that serves as the electron-transparent window. A spacer layer of desired thickness, typically less than 500 nm, is deposited across regions of one wafer, to create a gap between the windows and define the nominal fluid layer thickness in the imaging region.

The sealed cell incorporates a few additional components, to provide solution reservoirs which facilitate sample preparation. Sealing the main body of the cell can be accomplished through a number of means, including wafer bonding during the microfabrication process (Grogan and Bau 2010), gluing with a curable epoxy to provide a hard seal (Williamson et al. 2003), and using a non-curing sealant to yield a soft seal (Zheng et al. 2009). The reactant solution is placed in one reservoir and flows through the channel between the windows to the other reservoir, which is then topped off with additional liquid. The reservoirs are sealed, yielding a liquid-filled

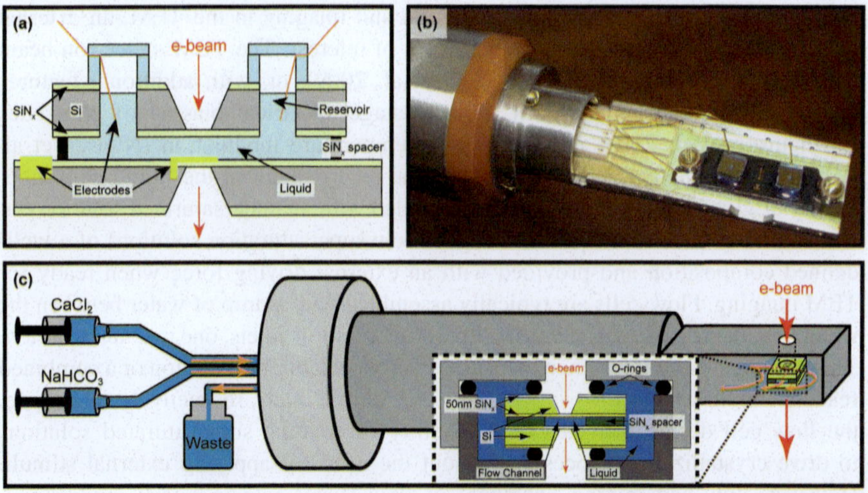

Fig. 18.1 Generic sealed cell and flow cell variants of liquid phase TEM. (**a**) Side-view schematic of hermetically sealed liquid cell. Silicon nitride membranes suspended from silicon wafers separated by a spacer layer enclose the liquid layer imaged in the TEM. Optional features such as electrodes in the imaging area may provide additional functionality. (**b**) Assembled liquid cell on a custom TEM stage. Wires leading from the cell to the internal section of the stage connect the liquid cell to external electronics. (**c**) Dual-inlet flow stage and side view of sample region (*inset*). Syringes pump reagent solutions through two separate inlets. Liquid streams combine into a single solution stream shortly before flowing through and around the liquid cell and exit the stage via an outlet tube connected to a waste container. (*Inset*) Liquid cell assembled from two silicon wafers supporting nitride membranes, separated by a spacer layer, as positioned in TEM stage with o-rings providing the barrier between liquid and vacuum environments (Figure 18.1a–b is adapted from Ref. (Nielsen et al. 2012) and reproduced with permission by The Royal Society of Chemistry. Figure 18.1c is reproduced from Ref. (Nielsen et al. 2014b) and used with permission by AAAS)

chamber that can then be placed in the vacuum of the TEM. Depending on the liquid cell's geometry the sealed cell may either fit on a standard TEM stage, or a custom stage designed to fit the cell may be necessary.

Open cells are used in conjunction with stages that incorporate tubing systems to pump one or more reagent solutions through the liquid cell. The cells have a liquid flow pathway defined by the pattern of the spacer layer, which extends across the entirety of the cell between openings that serve as inlets and outlets. These cells sit in a chamber on the TEM stage which is sealed from the microscope's vacuum by o-rings compressed between the stage and the cell. Liquid is pumped through one or multiple inlets in the stage, depending on design, and joined into a single stream near the liquid cell. This combined reagent stream flows into the cell and through the flow channel, as well as around the outside of the cell in the chamber in which the cell is placed, with an outlet tube carrying the effluent to a collection container.

The two basic approaches are suited to investigating different types of phenomena due to the characteristics of the cells and stages. Because sealed cells have no fluid exchange between sample preparation and imaging in the TEM, an external stimulus is necessary to initiate the process of interest. The TEM's electron beam may provide this driving force (Zheng et al. 2009), or, with additional features fabricated onto the cell or built into the stage, electrical biasing of electrodes (Williamson et al. 2003) (Fig. 18.1b) or temperature modulation (Nielsen et al. 2012) may alternatively serve to modify the environment in the liquid cell either locally or globally. The sealed cell is thus filled with an undersaturated solution (or, if the reaction kinetics are slow enough, a low supersaturation solution) of a well-defined composition and provided with an external driving force when ready for TEM imaging. Flow cells are typically assembled with a drop of water between the wafers, or, in the case of systems incorporating multiple inlets, one reactant solution. This volume of solution is then displaced by a supersaturated solution or a combined reagent stream which flows in from one or multiple inlets, respectively. Therefore, the flow cell design provides the capability of creating supersaturated solutions to drive crystallization processes without the need for applying external stimuli. However, flow and mixing behaviors of flow stages remain poorly understood, leading to a degree of uncertainty regarding the solution composition under electron beam illumination when mixing multiple reagent streams. Moreover, the typical resolution achieved with flow cell designs has been lower than with static cells, presumably because the pressure required to achieve flow results in a greater outward bowing of the windows and thus a thicker liquid layer. Typically, the smallest resolvable particles have sizes in the single to double digit nanometer range, depending on the fluid layer thickness.

18.3 $CaCO_3$ Formation Pathways

In this section we discuss two recent studies of $CaCO_3$ that demonstrate the utility of liquid phase TEM in understanding phenomena occurring throughout the early stages of mineralization. While many of the issues regarding the interest in this system are discussed elsewhere in this volume (Chaps. 2, 3, 4, 5, 6, 7, 8, 9, and 10), a few of the many open questions around $CaCO_3$ formation bear emphasizing here. Which of the observed phases or proposed precursors are active participants in crystallization? While calcite is the thermodynamically stable phase under typical laboratory conditions, other anhydrous and hydrated crystalline phases exist (Johnston et al. 1916; Krauss and Schriever 1930; Brooks et al. 1950), as do one (Brecevic and Nielsen 1989) or multiple (Koga and Yamane 2008; Gebauer et al. 2010; Radha et al. 2010) amorphous precursors (Fernandez-Martinez et al. 2017, Chap. 4). Additional precursors such as pre-nucleation clusters and dense liquid droplets have also been proposed (Wolf and Gower 2017, Chap. 3). With this collection of phases, does the classical picture of direct calcite nucleation from solution occurring through monomeric addition to a nascent calcite crystal accurately reflect the actual process? Does Ostwald's rule of stages (Cardew and Davey 1985; Voorhees 1985) more accurately reflect nucleation, with smaller kinetic barriers resulting in a progression of increasingly stable states from amorphous to calcite? Or are these classical ideas insufficient because they do not take into account additional precursor species or states that are crucial to the process? Furthermore, mechanistically how do organic matrices such as those found in biomineral systems affect the early stages of $CaCO_3$ formation? The two following studies sought to provide insight into these questions, utilizing the high spatial and temporal resolution provided by liquid phase TEM to directly image the mineralization process. The first study investigated $CaCO_3$ precipitation from mixtures of simple salt solutions, while the second looked at the effect of an organic additive. Viewed together, these in situ TEM studies highlight the multiple pathways available to nucleating systems and the strong control an organic matrix can exert over the nucleation processes.

18.3.1 Precipitation from Simple Salt Solutions

A dual inlet liquid phase TEM stage was used to study formation pathways of $CaCO_3$ precipitated through mixing two precursor salt solutions to create a supersaturated solution in the imaging area (Nielsen et al. 2014b). An extensive matrix of solution conditions were tested, varying initial concentrations and relative ratios of Ca^{2+}/CO_3^{2-} to account for uncertainty in reagent stream flow and mixing characteristics with the TEM stage and liquid cell. Calculated supersaturations for the various phases ranged from undersaturated to highly supersaturated, assuming perfect mixing of the reagent streams. Under at least some conditions, all of the anhydrous crystalline phases, calcite, aragonite, and vaterite, ACC were found to form. Increasing initial ion concentrations resulted in the observation of increasingly more $CaCO_3$ phases, and multiple nucleation pathways were simultaneously present. Both of these phenomena are expected from predictions of classical theories, as increasing supersaturation lowers energy barriers to make more potential phases and pathways accessible.

Two series of images following representative direct and multistep formation pathways of individual particles are shown in Fig. 18.2. In the first sequence (Fig. 18.2a–d), two calcite rhombohedra formed in the absence of any visible precursor, likely on the nitride window, and grew throughout the following minute. In parallel with the calcite, a particle with roughly circular projection formed and grew into contact with the center rhomb. Obvious diffraction contrast was not observed nor was a diffraction pattern acquired, so there is some ambiguity as to the identity of this particle as both ACC and vaterite have been observed to have similar two dimensional projections. However, prior observations indicate that ACC dissolves in the presence of calcite (Hu et al. 2012). Furthermore, as shown in the following example, ACC is rapidly consumed when in contact with crystalline phases. In this case, however, the two phases grew concurrently after coming into contact, suggesting that the second particle was vaterite rather than ACC. The second image series (Fig. 18.2e–l) follows the initial formation and extended period of growth of a single ACC particle. The ACC grew to a few microns in diameter

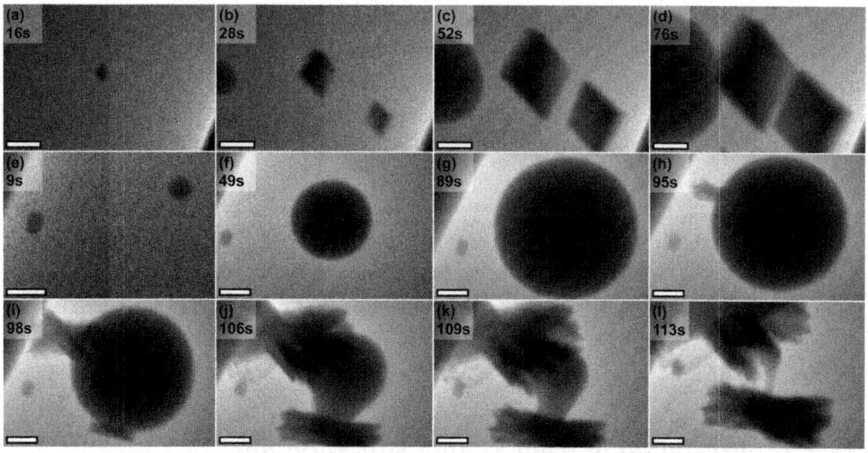

Fig. 18.2 Single- and multistep $CaCO_3$ formation pathways. Neighboring calcite rhombohedra and vaterite form in parallel from solution (**a–d**), with no observed precursor phase. Amorphous particle forms on image's *right edge* (**e**), followed by period of growth (**f**, **g**). Nucleation of aragonite on or near the ACC surface (**h**) precedes continued growth of the crystalline phase at the expense of the amorphous precursor (**i–l**). *Scale bars* are 500 nm and times listed are relative to the beginning of the respective image series (This figure is adapted from Ref. (Nielsen et al. 2014b) with permission by AAAS)

and then shrank briefly prior to the appearance of two sheaf-of-wheat bundles, a typical aragonite morphology (Andreassen et al. 2012), at or just below the ACC's surface. The aragonite then rapidly grew, consuming the amorphous precursor, with physical contact maintained between the two throughout the transformation process. In each observation of similar multistep pathways, the crystals appeared to form through heterogeneous nucleation on the precursor or just below its surface, rather than through dissolution of the ACC and crystalline re-precipitation.

A summary of the formation pathways observed by Nielsen et al. is presented in Fig. 18.3. They observed direct formation from solution of all three anhydrous crystalline phases and ACC, with no evidence of any precursor phase. Additionally, amorphous or crystalline precursors were found as initial precipitates in multistep pathways. In many experiments ACC particles led to the formation of vaterite and aragonite but, significantly, were not observed to precede the formation of calcite. Although the latter mechanism has long been assumed to take place, this data adds to the body of evidence that this process is unlikely in the absence of additives and, indeed, has never been directly observed. An apparent crystal-to-crystal transformation was additionally observed, with calcite seeming to form on and grow at the expense of an initially formed aragonite structure. However, given that TEM images are 2D projections of 3D volumes, there is some uncertainty as to whether the calcite nucleus forms on the aragonite bundle. Switching between imaging and diffraction modes throughout the nucleation and growth processes

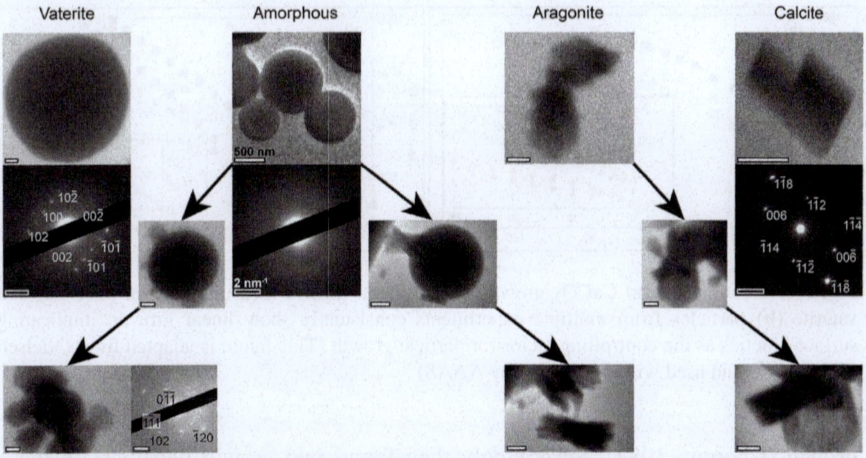

Fig. 18.3 $CaCO_3$ formation pathways. ACC and the anhydrous crystalline phases vaterite, aragonite, and calcite form directly from solution with no observed precursor in single-step processes. Multistep pathways incorporate ACC particles as a precursor to vaterite and aragonite. Crystal nucleation on existing amorphous particles leads to direct transformation from amorphous to crystal. Aragonite serves as a precursor to calcite, which forms on and grows at the expense of the aragonite, in an example of crystal-to-crystal transformation. *Scale bars* 500 nm and 2 nm^{-1} for images and diffraction data, respectively (This figure is reproduced from Ref. (Nielsen and De Yoreo 2016) and used with permission by Cambridge University Press)

allowed, at times, for unambiguous phase assignment. However, oftentimes the fluid layer thickness degraded the diffraction signal beyond use or the transformations occurred while imaging. For such situations, the distinct morphologies typically exhibited by each phase were used to infer which phases were present.

With the caveat that the precise supersaturations in the imaging area are unknown, these formation pathways were found to exist, and oftentimes to coexist, across a range of solution conditions. Increasing the solute concentrations resulted in ACC becoming progressively more predominant and more pathways observed, consistent with CNT expectations. However, while the data seem to contradict models that allow only for pathways involving aggregation of pre-nucleation clusters into amorphous precursors or dehydration of dense liquid droplets to form ACC, these results provide no evidence for or against the existence of such pathways. The resolution is inadequate to detect clusters of the size expected by the pre-nucleation cluster model, and when ACC particles are observed, the data do not definitively reveal whether they are solid or liquid.

Uncertainty about how the reagent streams mix and flow through and around the liquid cell as assembled on the TEM stage raises the question of whether the liquid stream flows through the liquid cell at all, or if diffusion alone carries solutes into the imaging region. Nielsen et al. measured lateral growth rates throughout several experiments for various precipitates, as shown in Fig. 18.4. They found that growth was linear, which signified that surface kinetics, not diffusive transport, limited

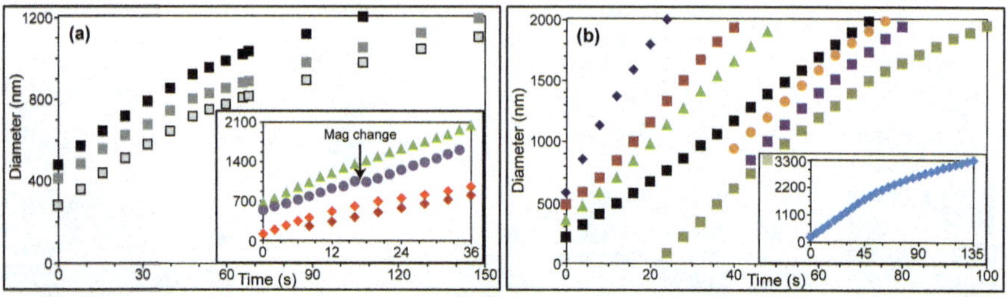

Fig. 18.4 Post-nucleation $CaCO_3$ growth rates. Lateral growth measurements of ACC (**a**) and vaterite (**b**) particles from multiple experiments consistently show linear growth, implicating surface kinetics as the controlling factor for particle growth (This figure is adapted from (Nielsen et al. 2014b) and used with permission by AAAS)

growth (Chernov 1984). Surprisingly they found two growth behaviors in these measurements: A single growth rate described the entire growth period for some experiments, while two distinct rates were found in others (evident from Fig. 18.4a and the inset to Fig. 18.4b). In this latter case, the original rate was always found to be larger than the second rate. A proposed explanation for the latter behavior is that nucleation events upstream of the imaging region partially block continued solute entrance to the liquid cell or simply decrease the solution supersaturation. This might also explain the shrinking ACC prior to the appearance of aragonite (Fig. 18.2g–h), although water expulsion from a hydrated amorphous precursor prior to formation of the crystal might also produce such a phenomenon.

18.3.2 The Effect of an Organic Additive

The formation of biominerals in living organisms is well known to be greatly influenced by organic matrices provided by these living systems (Mann 1993; Nudelman et al. 2006, Falini and Fermani 2017, Chap. 9). Many studies have used simple organic macromolecules containing chemical moieties present in biomolecules active in mineralization, to understand aspects of the complex interactions between the mineral and organic phases in biomineral development (Rao and Cölfen 2017, Chap. 8). One such examination (Smeets et al. 2015) harnessed the high-resolution imaging capability of liquid phase TEM to investigate the effect of an organic additive, polystyrene sulfonate (PSS), on $CaCO_3$ nucleation. PSS was chosen to mimic naturally occurring carbohydrates rich in sulfate groups and hypothesized to play a role in nucleating $CaCO_3$ within diverse species of marine organisms (Addadi et al. 1987; Marsh 1994). The same in situ TEM stage was used as for the data reported in Sect. 18.3.1, albeit with a different experimental approach. Rather than mixing separate reagent solutions to create supersaturated conditions, gaseous products of ammonium carbonate powder decomposition were diffused

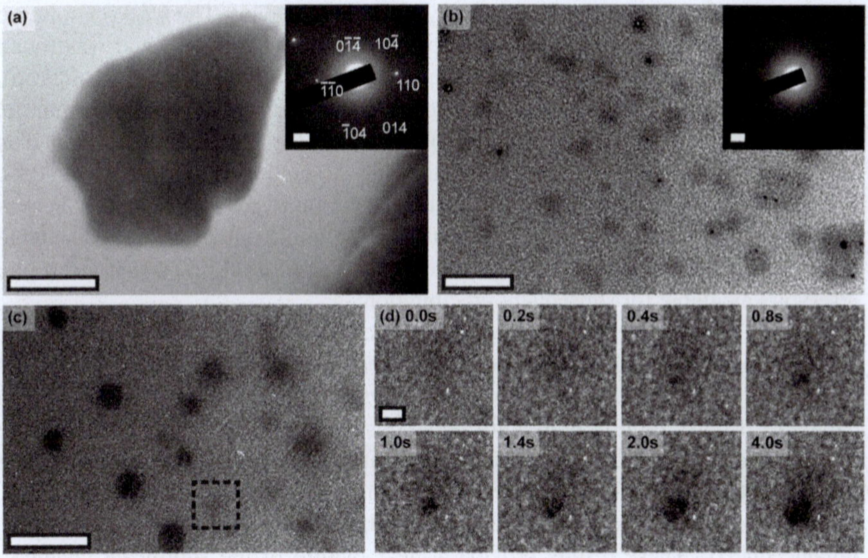

Fig. 18.5 PSS effect on $CaCO_3$ nucleation. In the absence of PSS, crystals form (**a**) with no observed precursor phase. Diffraction analysis (**a***inset*) identifies the crystals as vaterite. With the addition of PSS (**b**), the first mineral precipitate forms inside an organic matrix. Diffraction data (**b***inset*) is commensurate with amorphous nature of ACC. Addition of PSS into undersaturated calcium-bearing solutions (**c**) produces globular phase of Ca-PSS complexes. Subsequent CO_2 diffusion leads to the formation and rapid development of an ACC particle inside a Ca-PSS globule (**d**) in the highlighted region of (**c**). *Scale bars* are 1 μm (**a**), 100 nm (**b**, **c**), 10 nm (**d**), and 2 nm^{-1} (**a**, **b***insets*) (This figure has been adapted from Ref. (Smeets et al. 2015) and is used with permission by Nature Publishing Group)

through one inlet and into a liquid cell filled with calcium-bearing solution. This commonly used approach to $CaCO_3$ precipitation produces a continuously increasing supersaturation as the evolving CO_2 gas diffuses into the electrolyte solution.

When the precursor solution contained only dissolved calcium chloride (at a concentration of 1.25 mM), diffusion of CO_2 gas from the decomposing ammonium carbonate induced nucleation of vaterite (Fig. 18.5a) on the liquid cell's nitride membranes. Whereas the Nielsen et al. study discussed in Sect. 18.3.1 found all major $CaCO_3$ phases to form from the mixture of calcium and carbonate precursor solutions, Smeets et al.'s CO_2 diffusion into the electrolyte precursor produced only vaterite, though benchtop experiments using silicon nitride substrates also gave sporadic occurrences of calcite. When the same benchtop experiments were done with SiO_2 glass substrates, only calcite was observed. Consequently, the dominance of vaterite is likely due to the silicon nitride substrates, though the use of ammonium carbonate as a source of carbonate is also known to promote its appearance (Gehrke 2005). PSS inclusion (at a concentration of 7.1×10^{-3} mM) in the calcium solution, however, produced dramatically different behavior. The Ca-PSS solution was imaged prior to diffusing in the supersaturation-inducing gas and was found to contain numerous globules that formed on the nitride windows (Fig. 18.5c). Independent measurements made by means of ion-selective electrode, calorimetric, and spectroscopic methods found these objects to be Ca-PSS complexes formed as a globular phase (Trotsenko et al. 2012) through the binding of calcium to the PSS. Moreover, the amount of calcium drawn from solution to be bound in this state was determined to be approximately 56 % of the total calcium in the system. The simple addition of PSS to the electrolyte solution resulted in locally concentrated solute reservoirs available for subsequent precipitation.

Shortly after the beginning of CO_2 diffusion into the solution to increase the supersaturation in the TEM liquid cell, particles were found to nucleate within the Ca-PSS globules (Fig. 18.5d). These ACC particles, as identified from diffraction data (Fig. 18.5b), rapidly grew until they consumed the calcium contained in the globules. Following a long exposure to the continually diffusing CO_2, a much longer time than necessary to nucleate vaterite from a pure calcium chloride solution, vaterite crystals formed separate from the initially formed, and ACC-containing, organic matrix. These results provide a clear picture of the significant change a simple organic macromolecule can have on mineralization. By binding over half of the dissolved calcium in solution into Ca-PSS globules, the organic matrix produced conditions that biased the system's kinetics toward ACC formation. After the ACC consumed the bound calcium in the complexes, the PSS globules then acted as stabilizing chemical environments for the metastable ACC.

18.4 Metal Nanoparticle Nucleation and Growth

Synthetic approaches to crystal formation strive for similar levels of control over morphological development to those seen in natural biominerals such as biogenic $CaCO_3$ but fall short of the precise engineering found in nature. In a similar fashion to biomineral systems, inorganic chemists incorporate organic and inorganic additives to direct crystallization during colloidal synthesis of nanomaterials (Yin and Alivisatos 2005). As with biomineral systems, these additives can alter the energetics of crystal formation (both the surface energies and the kinetics of attachment and diffusion) to control crystal structure and promote anisotropic growth into specific architectures. Although colloidal suspensions of ligand-capped nanocrystals have been widely studied in recent years, much remains to be learned about the growth dynamics that yield nanocrystals with narrow size and shape distributions and the interactions between the organic and metal phases. By permitting high-resolution in situ imaging, liquid phase TEM provides a means of better understanding these phenomena. Recent years have seen a burgeoning body of literature using liquid phase TEM as a direct observation technique to investigate the formation and growth of individual metal nanoparticles. In this section we discuss two studies in this field. The first looks at mechanistically understanding metal nanocrystal nucleation and growth in the absence of additives, while the second investigates the effect that an organic ligand has on the development of a faceted nanocrystal.

18.4.1 Nucleation and Multiple Growth Mechanisms

Among the liquid phase TEM literature on metal nanocrystal formation, one study in particular stands out for its systematic approach to understanding nucleation and growth (Woehl et al. 2012). As with other in situ TEM investigations of metal nanoparticle formation, Woehl et al. initiated nucleation and growth with the microscope's electron beam. This study, however, sought to understand mechanistically how the electron beam's interaction with the sample induced crystallization and how changes in imaging parameters influence nanocrystal formation and growth.

A single inlet, continuous flow liquid TEM stage was used to flush an $AgNO_3$ solution into the liquid cell between experiments. The microscope was operated in scanning transmission electron microscopy (STEM) mode to facilitate a high degree of control over imaging parameters and enable precise measurement of resulting electron dose rates. The imaging magnification, pixel dwell time, and electron beam current were independently varied to investigate the effects on crystallization. Stepping through different values of these variables led to a determination of nucleation induction threshold doses, above which the concentration of aqueous electrons (elsewhere referred to as hydrated electrons (Grogan et al. 2014)), one of the many radiolytic products generated by the electron beam's interaction with the silicon nitride membranes and the liquid layer, was sufficient to induce observable crystal nucleation. They found a constant induction threshold dose irrespective of the toggled imaging parameter, suggesting that the only condition for nucleation in this system was that a certain concentration of electrons had to interact with the solution per unit time. In other words, a certain concentration of aqueous electrons created a supersaturated solution and was sufficient for initiating Ag nucleation.

Post-nucleation growth behavior was also investigated for different imaging conditions, some results of which are shown in Fig. 18.6. Imaging at high beam currents (Fig. 18.6a–c) produced a large amount of aqueous electrons which served to rapidly nucleate a large population of nearly spherical Ag nanocrystals. The depletion of the Ag ions in the imaging region led to diffusion-limited growth, as continued development of the nanocrystals was constrained by how fast the silver ions from outside of the depletion zone could diffuse to the surface of the growing crystals. The effective radius of crystals grown under these conditions (Fig. 18.6g) was found to scale as $t^{1/8}$ (Fig. 18.6i), rather than the $t^{1/3}$ expected by the Lifshitz-Slyozov-Wagner (LSW) model (Lifshitz and Slyozov 1961; Wagner 1961) for diffusion limited growth. Woehl et al. attribute the discrepancy to the high density of heterogeneously nucleated nanocrystals, as the LSW model assumes a single

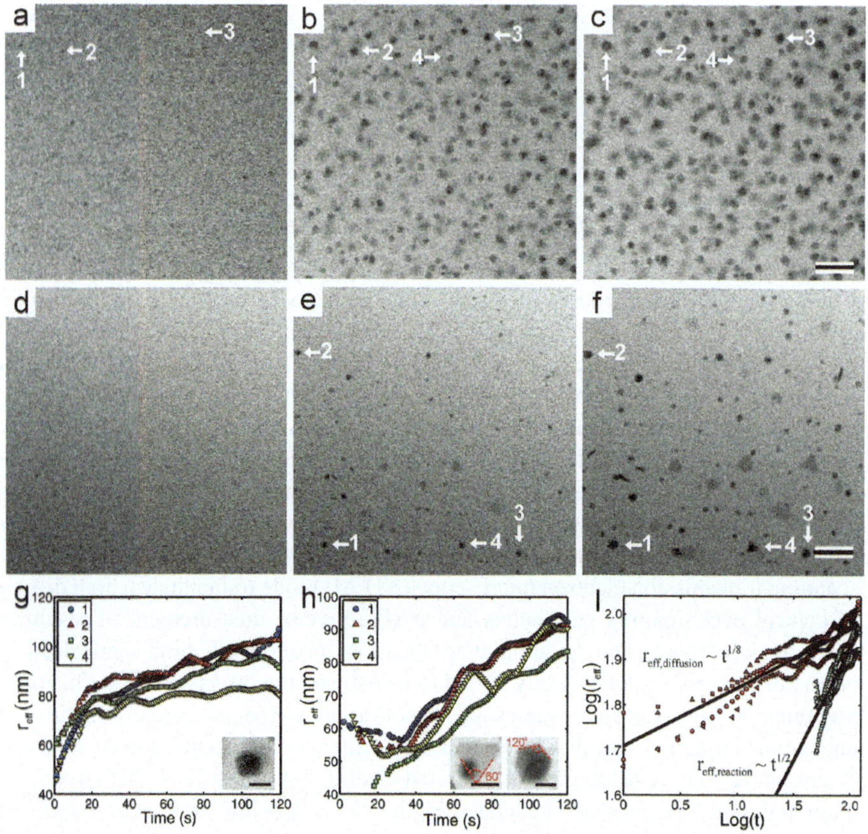

Fig. 18.6 Electron beam-dependent growth mechanisms of Ag nanocrystals. High beam currents (**a**–**c**) produce a large population of aqueous electrons which nucleate many Ag nanocrystals. Lower beam currents (**d**–**f**) yield fewer reducing agents, resulting in few nucleated Ag nanocrystals. Growth of labeled crystals in **a**–**c** (**g**) and **d**–**f** (**h**) plotted versus time. Log-log plot of effective crystal radius versus time with power law fits (**i**) suggests a high concentration of aqueous electrons produces hindered diffusion-limited growth, while a sparse concentration of aqueous electrons results in reaction-limited growth. *Scale bars* are 200 nm (**c**, **f**) and 100 nm (**g**, **h**), and sequential frames were taken at 60 s intervals (This figure is reproduced from Ref. (Woehl et al. 2012) and used with permission by the American Chemical Society)

crystal formed homogeneously in solution. Different behavior was observed when imaging at low beam currents (Fig. 18.6d–f), with the behavior being described well by reaction-limited growth that produced anisotropic, faceted crystals. Under these reaction conditions, the aqueous electron concentration does not rapidly exhaust the silver precursor concentration. Thus, growth is limited by the rate of silver ion reduction at the surface of the growing nanocrystal. Furthermore, because the growth is surface reaction limited, the precursor ions are able to explore the different faces of the crystal surface and seek out lower energy faces to produce faceted crystals. The effective radius (Fig. 18.6h) scaled as $t^{1/2}$ (Fig. 18.6i), as predicted by the LSW model.

18.4.2 Facet Development Through Ligand Attachment

A recent paper (Liao et al. 2014) has extended liquid phase TEM investigations of metal nanocrystal formation, using a high-resolution setup to resolve lattice fringes of growing Pt crystals. This resolution coupled with a fast acquisition camera allowed the authors to study how organic ligands direct growth of nascent Pt nanocrystals into well-controlled, specific morphologies. Although there is a rich literature developed from indirect observation techniques for understanding colloidal suspensions of ligand-capped nanocrystals, high-resolution imaging provided by liquid phase TEM offers the ability to track the development of individual particles so as to better understand – or confirm previously developed hypotheses about – how such colloidal systems produce nanocrystals with narrow size and shape distributions. In the abovementioned study, a sealed liquid cell was used, filled with a solution containing both the Pt precursor and ligand molecules. Crystallization was induced through the use of the TEM's electron beam to reduce the dissolved Pt precursors to metallic Pt.

Upon illumination of the liquid layer by the electron beam, Pt crystals rapidly nucleated and grew. Growth was found to occur primarily via monomer addition, with occasional coalescence events also observed, and the particles grew more or

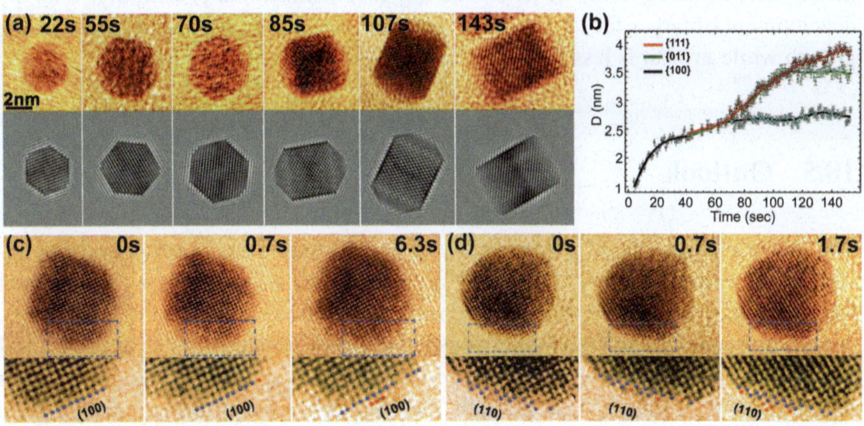

Fig. 18.7 Development of Pt nanocubes. Growth of a single Pt nanocrystal followed (**a**) as it develops into a cube. TEM image series (*top row*) matched with corresponding simulated images (*bottom row*). Growth trajectory measurements for different facets (**b**) show face-specific growth rates. Atomic layer attachment of Pt to the flat (100) (**c**) and the stepped (110) (**d**) facets followed in TEM image series. Existing surface of the growing crystal shown by *blue dots* (**c**, **d**); newly arrived Pt highlighted with *red dots* (This figure has been adapted from Ref. (Liao et al. 2014) and used with permission by AAAS)

less isotropically until their sizes reached about 5 nm. From this point onward, face-specific growth rates were observed, and the crystals developed into nanocubes. An in situ TEM image series (and the associated simulated images) follows a single particle throughout the later growth stages as the crystal develops into a nanocube (Fig. 18.7a). Measuring the growth rates (Fig. 18.7b), Liao et al. found the {110} and {111} facets grew at similar rates and outpaced {100} facet growth. The faster growing facets grew until they reached the bounding {100} facets, with the final stages of growth serving to fill in the corners of the particles yielding cubic nanocrystals. Furthermore, different growth mechanisms were observed on specific crystallographic faces. Layer-by-layer growth was found on {100} (Fig. 18.7c) and {111}, while multiple steps were visible on {110} facets (Fig. 18.7d).

Pt surface and ligand-crystal binding energies for the {100} and {111} facets were calculated so as to understand the observed differences in growth rates and development of the Pt crystals into cubes. The computations yielded a difference of 1.65 eV/nm^2 between the surface energies of the two faces passivated by the ligand. According to the Wulff construction based on thermodynamic equilibrium (Wulff 1901), these values would not result in the Pt crystals forming into cubes. Energy barriers for ligands to hop between neighboring sites on the Pt surface and to dissociate from the crystal were also calculated. Ligand movement between surface sites was found to be energetically favorable relative to dissociation from the surface. Dwell times of a ligand molecule on each of the two investigated surfaces were calculated using the hopping barriers. A ligand molecule on a {100} site was found to have a dwell time five orders of magnitude longer than a molecule on a {111} site. As such cubic nanocrystal formation was determined to be kinetically controlled. Ligand molecules on small nanocrystals are sparsely distributed across the surface, allowing Pt atoms to land on any face with similar probability. With the development of larger facets, the lower ligand mobility on {100} hinders continued growth while growth is less perturbed on the {110} and {111} surfaces.

18.5 Outlook

The above examples highlight the utility of liquid phase TEM in better understanding phenomena at the onset of crystal formation. This relatively new experimental technique has opened a window to direct observations of nucleation pathways, growth mechanisms, and the interactions between the primary crystallizing material and additives. The ability to gather crystallographic information either through lattice-resolution imaging or through electron diffraction further enhances the utility of this approach in understanding processes controlling materials formation. However, as should be expected given the relative youth of the liquid phase TEM field, there are numerous areas for future development, which will greatly enhance the technique's utility in quantitatively understanding a wider variety of experimental systems.

As mentioned in Sect. 18.2, current flow stage designs of which we are aware allow the liquid stream to flow around the outside of the cell in addition to flowing through the channel and across the imaging region within the cell. As such, flow patterns within, and mixing characteristics of, flow stages remain largely unknown. This results in some degree of uncertainty regarding the solution composition under electron beam illumination, particularly in the case of multiple inlet stages. Nielsen et al., for example, observed formation of some $CaCO_3$ phases from solutions which were expected to be undersaturated with respect to those phases, based on calculations using the starting reagent concentrations and relative flow rates of the two precursor streams (Nielsen et al. 2014b). However, these surprising results may not be caused entirely or even to a significant extent, by the uncertainty in fluid flow pathways. Another, perhaps more important, effect on the sample arises from interaction with the electron beam.

How the electron beam affects the liquid cell and its reaction solution is a significant question that has only recently begun to be addressed. Woehl et al., for example, have provided a good overview of a variety of potential artifacts that may arise during liquid phase TEM (Woehl et al. 2013). Perhaps the most important consideration is that of which radiolytic products are generated from the electron beam passing through the liquid cell, as a number of liquid phase TEM studies have utilized the electron beam to reduce solvated metal precursors to grow nanocrystals from solution. Woehl et al. discussed the various species that arise through the beam's interaction with the silicon nitride membranes and the liquid layer and identified the aqueous electron as the reducing agent for metal precursors, though other radiolytic products are also produced (Woehl et al. 2012). More recent efforts have gone toward building a predictive model for each of the species produced through electron irradiation, their spatial concentrations relative to the electron beam, lifetimes, and effect on the liquid's pH (Grogan et al. 2014; Schneider et al. 2014). Extending this model from pure water, for which it was originally developed, to complex solutions with many dissolved species will allow for more quantitative and accurate studies of crystallization processes in the future.

The above model predicts that pH changes of potentially significant amounts may be induced in the sample solution as a result of imaging in the TEM. At present there are no in situ diagnostics to measure such changes in the experimental conditions. However, in order to conduct a quantitative analysis of the energetics of crystallization, it is crucial to know solution parameters such as pH and temperature. Future developments of on-device probes that allow in situ measurement of — and control over — these and other parameters will broaden the utility of liquid phase TEM.

Although still in its early years of development, liquid phase TEM has already become an invaluable tool in understanding nucleation, growth, and control over these processes through the use of additives. Future advancements of the technique will only expand its utility in addressing such fundamental questions of materials formation. Through its unique ability to track nanoparticle motion, application of liquid phase TEM to the post nucleation phase of solutions can provide critical information on the dynamics and mechanisms of crystallization through nanoparticle assembly, such as oriented attachment (Yuk et al. 2012; Li et al. 2012) as described elsewhere in this volume (Rao and Cölfen 2017, Chap. 8, Penn et al. 2017, Chap. 13). The spatial and temporal resolutions of liquid phase TEM are also well suited to exploring the occurrence of novel crystallization pathways, such as liquid-liquid phase separation, a process that has been implicated in both protein (Galkin et al. 2002) and simple electrolyte solutions (Wallace et al. 2013). The demonstration that both macromolecular and mineral phases can be simultaneously observed during crystallization (Evans et al. 2012; Smeets et al. 2015) opens up a new approach to understanding the formation of biominerals, such as bones and teeth, in which organic scaffolds organize mineral constituents (Nudelman et al. 2010). In all of these examples, incorporation of in situ diagnostics to provide a better understanding of the solution conditions in the volume illuminated by the electron beam and the extension of current models to better represent experimental solutions would allow for quantitative studies to better understand the mechanisms underlying crystallization processes.

Acknowledgments The authors acknowledge support from the National Science Foundation under grant DMR-1312697. Additional support for this work was provided by the Laboratory Directed Research and Development Initiative on Materials Synthesis and Simulation Across Scales at the Pacific Northwest National Laboratory (PNNL). PNNL is operated by Battelle for the US Department of Energy under Contract DE-AC05-76RL01830. Work conducted at LLNL was performed under the auspices of the US DOE by LLNL under Contract DE-AC52-07NA20344.

References

Addadi L, Moradian J, Shay E, Maroudas NG, Weiner S (1987) A chemical model for the cooperation of sulfates and carboxylates in calcite crystal nucleation: relevance to biomineralization. Proc Natl Acad Sci U S A 84(9):2732–2736. doi:10.1073/pnas.84.9.2732

Ando T, Kodera N, Takai E, Maruyama D, Saito K, Toda A (2001) A high-speed atomic force microscope for studying biological macromolecules. Proc Natl Acad Sci 98(22):12468–12472. doi:10.1073/pnas.211400898

Andreassen JP, Beck R, Nergaard M (2012) Biomimetic type morphologies of calcium carbonate grown in absence of additives. Faraday Discuss 159:247–261. doi:10.1039/c2fd20056b

Baumgartner J, Dey A, Bomans PHH, Le Coadou C, Fratzl P, Sommerdijk N, Faivre D (2013) Nucleation and growth of magnetite from solution. Nat Mater 12(4):310–314. doi:10.1038/nmat3558

Bewernitz MA, Gebauer D, Long J, Cölfen H, Gower LB (2012) A metastable liquid precursor phase of calcium carbonate and its interactions with polyaspartate. Faraday Discuss 159:291–312. doi:10.1039/c2fd20080e

Bots P, Benning LG, Rodriguez-Blanco JD, Roncal-Herrero T, Shaw S (2012) Mechanistic insights into the crystallization of Amorphous Calcium Carbonate (ACC). Cryst Growth Des 12(7):3806–3814. doi:10.1021/cg300676b

Brecevic L, Nielsen AE (1989) Solubility of amorphous calcium carbonate. J Cryst Growth 98(3):504–510. doi:10.1016/0022-0248(89)90168-1

Birkedal H (2017) Phase transformations in calcium phosphate crystallization. In: Van Driessche AES, Kellermeier M, Benning LG, Gebauer D (eds) New perspectives on mineral nucleation and growth: from solution precursors to solid materials. Springer, Cham, pp 199–210

Brooks R, Clark LM, Thurston EF (1950) Calcium carbonate and its hydrates. Philos Trans R Soc London Ser A Math Phys Sci 243(861):145–167. doi:10.1098/rsta.1950.0016
Cardew PT, Davey RJ (1985) The kinetics of solvent-mediated phase transformations. Proc R Soc London Ser A-Math Phys Eng Sci 398(1815):415–428
Chernov AA (1984) Modern crystallography III, vol 36, Springer series in solid-state sciences. Springer, Berlin
De Yoreo JJ, Sommerdijk NAJM, Dove PM (2017) Nucleation pathways in electrolyte solutions. In: Van Driessche AES, Kellermeier M, Benning LG, Gebauer D (eds) New perspectives on mineral nucleation and growth, Springer, Cham, pp 1–24
Delgado-López JM, Guagliardi A (2017) Control over nanocrystalline apatite formation: what can the X-ray total scattering approach tell us. In: Van Driessche AES, Kellermeier M, Benning LG, Gebauer D (eds) New perspectives on mineral nucleation and growth, Springer, Cham, pp 211–226
Demichelis R, Raiteri P, Gale JD, Quigley D, Gebauer D (2011) Stable prenucleation mineral clusters are liquid-like ionic polymers. Nat Commun 2:590. doi:10.1038/ncomms1604
Evans JE, Jungjohann KL, Wong PCK, Chiu P-L, Dutrow GH, Arslan I, Browning ND (2012) Visualizing macromolecular complexes with In Situ liquid scanning transmission electron microscopy. Micron 43(11):1085–1090. doi:10.1016/j.micron.2012.01.018
Falini G, Fermani S (2017) Nucleation and growth from a biomineralization perspective. In: Van Driessche AES, Kellermeier M, Benning LG, Gebauer D (eds) New perspectives on mineral nucleation and growth, Springer, Cham, pp 185–198
Fang PA, Conway JF, Margolis HC, Simmer JP, Beniash E (2011) Hierarchical self-assembly of amelogenin and the regulation of biomineralization at the nanoscale. Proc Natl Acad Sci U S A 108(34):14097–14102. doi:10.1073/pnas.1106228108
Fernandez-Martinez A, Lopez-Martinez H, Wang D (2017) Structural characteristics and the occurrence of polyamorphism in amorphous calcium carbonate. In: Van Driessche AES, Kellermeier M, Benning LG, Gebauer D (eds) New perspectives on mineral nucleation and growth, Springer, Cham, pp 77–92
Frandsen C, Legg BA, Comolli LR, Zhang HZ, Gilbert B, Johnson E, Banfield JF (2014) Aggregation-induced growth and transformation of Beta-FeOOH nanorods to micron-sized Alpha-Fe2O3 spindles. CrystEngComm 16(8):1451–1458. doi:10.1039/c3ce40983j
Gal A, Habraken W, Gur D, Fratzl P, Weiner S, Addadi L (2013) Calcite crystal growth by a solid-state transformation of stabilized amorphous calcium carbonate nanospheres in a hydrogel. Angew Chem Int Ed 52(18):4867–4870. doi:10.1002/anie.201210329
Galkin O, Chen K, Nagel RL, Hirsch RE, Vekilov PG (2002) Liquid-liquid separation in solutions of normal and sickle cell hemoglobin. Proc Natl Acad Sci U S A 99(13):8479–8483. doi:10.1073/pnas.122055299
Gebauer D, Völkel A, Cölfen H (2008) Stable prenucleation calcium carbonate clusters. Science 322(5909):1819–1822. doi:10.1126/science.1164271
Gebauer D, Gunawidjaja PN, Ko JYP, Bacsik Z, Aziz B, Liu LJ, Hu YF, Bergström L, Tai CW, Sham TK, Eden M, Hedin N (2010) Proto-calcite and proto-vaterite in amorphous calcium carbonates. Angew Chem-Int Ed 49(47):8889–8891. doi:10.1002/anie.201003220
Gehrke N, Cölfen H, Pinna N, Antonietti M, Nassif N (2005) Superstructure of calcium carbonate crystals by oriented attachment. Cryst Growth Des 5(4):1317–1319. doi:10.1021/cg050051d
Gibbs JW (1876) On the equilibrium of heterogeneous substances. Trans Connecticut Acad Arts Sci 3:108–248
Gibbs JW (1878) On the equilibrium of heterogeneous substances. Trans Connecticut Acad Arts Sci 3:343–524
Giuffre AJ, Hamm LM, Han N, De Yoreo JJ, Dove PM (2013) Polysaccharide chemistry regulates kinetics of calcite nucleation through competition of interfacial energies. Proc Natl Acad Sci 110(23):9261–9266. doi:10.1073/pnas.1222162110
Grogan JM, Bau HH (2010) The nanoaquarium: a platform for In Situ transmission electron microscopy in liquid media. J Microelectromech Syst 19(4):885–894. doi:10.1109/jmems.2010.2051321

Grogan JM, Schneider NM, Ross FM, Bau HH (2014) Bubble and pattern formation in liquid induced by an electron beam. Nano Lett 14(1):359–364. doi:10.1021/nl404169a

Habraken W, Tao JH, Brylka LJ, Friedrich H, Bertinetti L, Schenk AS, Verch A, Dmitrovic V, Bomans PHH, Frederik PM, Laven J, van der Schoot P, Aichmayer B, de With G, DeYoreo JJ, Sommerdijk N (2013) Ion-association complexes unite classical and Non-classical theories for the biomimetic nucleation of calcium phosphate. Nat Commun 4:1507. doi:10.1038/ncomms2490

Hamm LM, Giuffre AJ, Han N, Tao J, Wang D, De Yoreo JJ, Dove PM (2014) Reconciling disparate views of template-directed nucleation through measurement of calcite nucleation kinetics and binding energies. Proc Natl Acad Sci 111(4):1304–1309. doi:10.1073/pnas.1312369111

Hu Q, Nielsen MH, Freeman CL, Hamm LM, Tao J, Lee JRI, Han TYJ, Becker U, Harding JH, Dove PM, De Yoreo JJ (2012) The thermodynamics of calcite nucleation at organic interfaces: classical vs. Non-classical pathways. Faraday Discuss 159:509–523. doi:10.1039/c2fd20124k

Johnston J, Merwin HE, Williamson ED (1916) The several forms of calcium carbonate. Am J Sci 41(246):473–512

Koga N, Yamane Y (2008) Thermal behaviors of amorphous calcium carbonates prepared in aqueous and ethanol media. J Therm Anal Calorim 94(2):379–387. doi:10.1007/s10973-008-9110-3

Krauss F, Schriever W (1930) The hydrates of calcium carbonate. Zeitschrift Fur Anorganische Und Allgemeine Chemie 188(1):259–260. doi:10.1002/zaac.19301880122

Lee JRI, Han TYJ, Willey TM, Wang D, Meulenberg RW, Nilsson J, Dove PM, Terminello LJ, van Buuren T, De Yoreo JJ (2007) Structural development of mercaptophenol self-assembled monolayers and the overlying mineral phase during templated $CaCO_3$ crystallization from a transient amorphous film. J Am Chem Soc 129(34):10370–10381. doi:10.1021/ja071535w

Li DS, Nielsen MH, Lee JRI, Frandsen C, Banfield JF, De Yoreo JJ (2012) Direction-specific interactions control crystal growth by oriented attachment. Science 336(6084):1014–1018. doi:10.1126/science.1219643

Liao H-G, Zherebetskyy D, Xin H, Czarnik C, Ercius P, Elmlund H, Pan M, Wang L-W, Zheng H (2014) Facet development during platinum nanocube growth. Science 345(6199):916–919. doi:10.1126/science.1253149

Lifshitz IM, Slyozov VV (1961) The kinetics of precipitation from supersaturated solid solutions. J Phys Chem Solids 19(1–2):35–50. doi:10.1016/0022-3697(61)90054-3

Mann S (1993) Molecular tectonics in biomineralization and biomimetic materials chemistry. Nature 365(6446):499–505. doi:10.1038/365499a0

Marsh ME (1994) Polyanion-mediated mineralization – assembly and reorganization of acidic polysaccharides in the Golgi system of a coccolithophorid alga during mineral deposition. Protoplasma 177(3–4):108–122. doi:10.1007/bf01378985

Nielsen MH, De Yoreo JJ (2016) Liquid cell TEM for studying environmental and biological mineral systems. In: Ross FM (ed) Liquid cell electron microscopy. Cambridge University Press, Cambridge

Nielsen MH, Lee JRI, Hu QN, Han TYJ, De Yoreo JJ (2012) Structural evolution, formation pathways and energetic controls during template-directed nucleation of $CaCO_3$. Faraday Discuss 159:105–121. doi:10.1039/c2fd20050c

Nielsen MH, Li DS, Zhang HZ, Aloni S, Han TYJ, Frandsen C, Seto J, Banfield JF, Cölfen H, De Yoreo JJ (2014a) Investigating processes of nanocrystal formation and transformation via liquid cell TEM. Microsc Microanal 20(2):425–436. doi:10.1017/s1431927614000294

Nielsen MH, Aloni S, De Yoreo JJ (2014b) In Situ TEM imaging of $CaCO_3$ nucleation reveals coexistence of direct and indirect pathways. Science 345(6201):1158–1162. doi:10.1126/science.1254051

Nudelman F, Gotliv BA, Addadi L, Weiner S (2006) Mollusk shell formation: mapping the distribution of organic matrix components underlying a single aragonitic tablet in nacre. J Struct Biol 153(2):176–187. doi:10.1016/j.jsb.2005.09.009

Nudelman F, Pieterse K, George A, Bomans PHH, Friedrich H, Brylka LJ, Hilbers PAJ, de With G, Sommerdijk N (2010) The role of collagen in bone apatite formation in the presence of hydroxyapatite nucleation inhibitors. Nat Mater 9(12):1004–1009. doi:10.1038/nmat2875

Penn RL (2004) Kinetics of oriented aggregation. J Phys Chem B 108(34):12707–12712. doi:10.1021/jp036490+

Penn RL, Banfield JF (1998a) Oriented attachment and growth, twinning, polytypism, and formation of metastable phases: insights from nanocrystalline TiO2. Am Mineral 83(9–10):1077–1082

Penn RL, Banfield JF (1998b) Imperfect oriented attachment: dislocation generation in defect-free nanocrystals. Science 281(5379):969–971. doi:10.1126/science.281.5379.969

Penn RL, Li D, Soltis JA (2017) A perspective on the particle-based crystal growth of ferric oxides, oxyhydroxides, and hydrous oxides. In: Van Driessche AES, Kellermeier M, Benning LG, Gebauer D (eds) New perspectives on mineral nucleation and growth, Springer, Cham, pp 257–274

Pouget EM, Bomans PHH, Goos JACM, Frederik PM, de With G, Sommerdijk NAJM (2009) The initial stages of template-controlled $CaCO_3$ formation revealed by Cryo-TEM. Science 323(5920):1555–1458. doi:10.1126/science.1169434

Radha AV, Forbes TZ, Killian CE, Gilbert P, Navrotsky A (2010) Transformation and crystallization energetics of synthetic and biogenic amorphous calcium carbonate. Proc Natl Acad Sci U S A 107(38):16438–16443. doi:10.1073/pnas.1009959107

Rao A, Cölfen H (2017) Mineralization schemes in the living world: mesocrystals. In: Van Driessche AES, Kellermeier M, Benning LG, Gebauer D (eds) New perspectives on mineral nucleation and growth, Springer, Cham, pp 155–184

Reichel V, Faivre D (2017) Magnetite nucleation and growth. In: Van Driessche AES, Kellermeier M, Benning LG, Gebauer D (eds) New perspectives on mineral nucleation and growth, Springer, Cham, pp 275–292

Rieger J, Thieme J, Schmidt C (2000) Study of precipitation reactions by X-ray microscopy: $CaCO_3$ precipitation and the effect of polycarboxylates. Langmuir 16(22):8300–8305. doi:10.1021/la0004193

Rieger J, Frechen T, Cox G, Heckmann W, Schmidt C, Thieme J (2007) Precursor structures in the crystallization/precipitation processes of $CaCO_3$ and control of particle formation by polyelectrolytes. Faraday Discuss 136:265–277. doi:10.1039/b701450c

Ring EA, de Jonge N (2010) Microfluidic system for transmission electron microscopy. Microsc Microanal 16(5):622–629. doi:10.1017/s1431927610093669

Rodriguez-Blanco JG, Sand KK, Benning LG (2017) ACC and vaterite as intermediates in the solution-based crystallization of $CaCO_3$. In: Van Driessche AES, Kellermeier M, Benning LG, Gebauer D (eds) New perspectives on mineral nucleation and growth, Springer, Cham, pp 93–112

Schneider NM, Norton MM, Mendel BJ, Grogan JM, Ross FM, Bau HH (2014) Electron-water interactions and implications for liquid cell electron microscopy. J Phys Chem C 118(38):22373–22382. doi:10.1021/jp507400n

Sleutel M, Lutsko J, Van Driessche AES, Duran-Olivencia MA, Maes D (2014) Observing classical nucleation theory at work by monitoring phase transitions with molecular precision. Nat Commun 5:5598. doi:10.1038/ncomms6598

Smeets PJM, Cho KR, Kempen RGE, Sommerdijk NAJM, De Yoreo JJ (2015) Calcium carbonate nucleation driven by Ion binding in a biomimetic matrix revealed by in situ electron microscopy. Nat Mater 14:394–399. doi:10.1038/nmat4193

Tester CC, Brock RE, Wu CH, Krejci MR, Weigand S, Joester D (2011) In Vitro synthesis and stabilization of Amorphous Calcium Carbonate (ACC) nanoparticles within liposomes. CrystEngComm 13(12):3975–3978. doi:10.1039/c1ce05153a

Trotsenko O, Roiter Y, Minko S (2012) Conformational transitions of flexible hydrophobic polyelectrolytes in solutions of monovalent and multivalent salts and their mixtures. Langmuir 28(14):6037–6044. doi:10.1021/la300584k

Verch A, Morrison IEG, van de Locht R, Kroger R (2013) In Situ electron microscopy studies of calcium carbonate precipitation from aqueous solution with and without organic additives. J Struct Biol 183(2):270–277. doi:10.1016/j.jsb.2013.05.017

Voorhees PW (1985) The theory of Ostwald ripening. J Stat Phys 38(1–2):231–252. doi:10.1007/bf01017860

Wagner C (1961) Theorie der Alterung von Niederschlagen durch Umlosen (Ostwald-Reifung). Z Elektrochem 65(7–8):581–591

Wallace AF, Hedges LO, Fernandez-Martinez A, Raiteri P, Gale JD, Waychunas GA, Whitelam S, Banfield JF, De Yoreo JJ (2013) Microscopic evidence for liquid-liquid separation in supersaturated $CaCO_3$ solutions. Science 341(6148):885–889. doi:10.1126/science.1230915

Williamson MJ, Tromp RM, Vereecken PM, Hull R, Ross FM (2003) Dynamic microscopy of nanoscale cluster growth at the solid-liquid interface. Nat Mater 2(8):532–536. doi:10.1038/nmat944

Woehl TJ, Evans JE, Arslan L, Ristenpart WD, Browning ND (2012) Direct in Situ determination of the mechanisms controlling nanoparticle nucleation and growth. ACS Nano 6(10):8599–8610. doi:10.1021/nn303371y

Woehl TJ, Jungjohann KL, Evans JE, Arslan I, Ristenpart WD, Browning ND (2013) Experimental procedures to mitigate electron beam induced artifacts during In Situ fluid imaging of nanomaterials. Ultramicroscopy 127:53–63. doi:10.1016/j.ultramic.2012.07.018

Wolf SE, Gower LB (2017) Challenges and perspectives of the polymer-induced liquid-precursor process: the pathway from liquid-condensed mineral precursors to mesocrystalline products. In: Van Driessche AES, Kellermeier M, Benning LG, Gebauer D (eds) New perspectives on mineral nucleation and growth, Springer, Cham, pp 43–76

Wulff G (1901) On the question of speed of growth and dissolution of crystal surfaces. Zeitschrift für Krystallographie und Mineralogie 34(5/6):449–530

Yin Y, Alivisatos AP (2005) Colloidal nanocrystal synthesis and the organic-inorganic interface. Nature 437(7059):664–670. doi:10.1038/nature04165

Yuk JM, Park J, Ercius P, Kim K, Hellebusch DJ, Crommie MF, Lee JY, Zettl A, Alivisatos AP (2012) High-resolution EM of colloidal nanocrystal growth using graphene liquid cells. Science 336(6077):61–64. doi:10.1126/science.1217654

Zheng HM, Smith RK, Jun YW, Kisielowski C, Dahmen U, Alivisatos AP (2009) Observation of single colloidal platinum nanocrystal growth trajectories. Science 324(5932):1309–1312. doi:10.1126/science.1172104

Summary and Outlook

The chapters assembled in this book illustrate the important advances that have been made in our knowledge about mineralization phenomena taking place both in nature and in the laboratory. However, there is still a long way to go until we fully and quantitatively understand how crystals nucleate and grow. Based on the insights provided in the different chapters, it seems that progress in this field is mainly hampered by two factors: (1) the lack of a comprehensive theoretical framework accounting for nucleation and growth pathways outside the scope of the standard (or "classical") models and (2) the technical limitations to probe the mechanisms governing the nucleation process at the relevant time and length scales, especially during the very early stages. The latter is a direct consequence of the intrinsically elusive nature of the nucleation event, which is a stochastic process that usually occurs at the molecular level and in fast steps. Nonetheless, a vast amount of experimental and computational data has been collected for various crystallizing systems during the past decades, giving rise to a number of alternative – often called "nonclassical" – perspectives on the mechanisms underlying the birth, growth and transformation of mineral phases. This includes a range of precursor and intermediate species that were shown to be involved at different stages of the precipitation process. All this has significantly widened our view of crystallization well beyond the horizon of established concepts that were developed during the first half of the past century, even though information on the distinct species and stages is often limited and/or only emerging.

In line with these thoughts, De Yoreo and co-authors (Chap. 1) argue that the established theoretical frameworks for nucleation and growth need to be broadened considering both thermodynamic and kinetic factors. For example, the existence of (meta)stable pre-critical clusters and metastable bulk phases should be accounted for in the free energy landscape of nucleation. However, dynamic parameters like supersaturation or temperature are equally important and may trigger complex pathways involving particles with no defined thermodynamic status or kinetically trapped non-equilibrium bulk materials; these states have to be

A.E.S. Van Driessche et al. (eds.), *New Perspectives on Mineral Nucleation and Growth*,
DOI 10.1007/978-3-319-45669-0

considered as well, along with the crucial influence of surfaces on any of the possible scenarios. By developing a mechanistic understanding of such competing and often simultaneously occurring processes, a more accurate quantitative picture of mineral formation should arise. One step in this direction was made by Lutsko (2017, Chap. 2), who proposes a new model for nucleation based on non-equilibrium dynamics, which he calls "mesoscopic nucleation theory". Although this approach provides interesting qualitative insights into the nucleation process, it has so far been focused on the "weak-noise regime", while the highly nonclassical "strong-noise" regime remains to be explored – especially now as preliminary simulation results become available. Similarly, effects caused by common modulators, such as additives, surfaces, confinement, complex fluid flow patterns (e.g. during mixing), unsteady conditions and many more, need to be described and implemented before an overarching nucleation theory can be put forward.

"Nonclassical" processes as those described above have been studied extensively in the laboratory over the past years, with one particular example being polymer-induced liquid precursors (PILP), discussed in detail by Wolf and Gower in Chap. 3. Although many aspects around these ion-enriched liquid phases can be mechanistically understood based on liquid-liquid demixing and established concepts of colloid chemistry, our insight into the different stages of the PILP process is still limited – a fact that constrains controlled applicability of the PILP concept for specific demands. Again, a proper theoretical description would certainly help to design synthesis strategies towards, e.g. nanoparticle superstructures and composite materials in a rational manner. Along the same lines, Fernandez-Martinez and co-authors (Chap. 4) emphasize the urgent need for a model of amorphous calcium carbonate that allows predicting structure-property relationships and reveal to which extent amorphous structures contain features of distinct crystalline counterparts. This is of paramount importance to understand how polymorph selection is controlled in biomineralization and how calcifying organisms tune the kinetic stability of ACC by carefully regulating conditions and/or using "impurities" like spectator ions or bio(macro)molecules. For the transformation of ACC phases into crystalline $CaCO_3$, described by Rodriguez-Blanco et al. in Chap. 5, different routes have been reported and hence there is still no "unified" picture for this step. For example, it is still unclear whether aragonite can crystallize directly from ACC with no other intermediate polymorphs being involved in the reaction. Yet, a better understanding of the transformation mechanisms of mineral phases is key to improve our ability to design new materials using biomimetic strategies. One crucial aspect is obviously to know the (crystal) structure of the relevant phases – which can be quite challenging, as shown in the case of vaterite by Demichelis et al. (2017, Chap. 6). Here, computational methods such as ab initio modelling can provide invaluable insights that are difficult, or even impossible, to gain through experimental measurements. Whether or not vaterite is a singular example of structural variability in crystalline minerals remains to be clarified, and the potential of in silico approaches for this purpose is evident. However, ab initio methods usually rely on many assumptions, and thus it is important to develop tools to improve the reliability of the results, for instance, algorithms addressing the non-generality of the approximations made or a better description of weak interactions occurring in most (bio)minerals.

In the area of crystal growth, there is convincing evidence for the existence of alternative pathways that employ nanoparticles as building units instead of ions or molecules. Even though much progress has been made over the past few years, still many central questions remain unanswered. For example, Andreassen and Lewis (Chap. 7) point out that for the assembly of nanocrystals into micron-sized superstructures, we still miss the link between supersaturation, the number of formed nanocrystals and the particular aggregation mechanisms. Furthermore, knowledge about the kinetics of nucleation at the growth front of spherulites forming at high levels of supersaturation is quite limited. Mesocrystals are well-known paradigms for systems in which the aggregation, attachment, coalescence and growth of nanoparticles lead to inherent complexity and great structural variety – although the actual oriented aggregation of preformed nanoparticles is just one of several mechanisms that can lead to mesocrystals. Rao and Cölfen (2017, Chap. 8) emphasize that all these processes are kinetically controlled and can be intimately influenced by additives. This modulation is the key to generate sophisticated mineral architectures beyond the realm of existing nucleation and growth models. Filling the gaps in our understanding of mesocrystal formation and particle-based crystallization in general should lead to new strategies for tailor-made materials and calls for detailed experimental as well as theoretical studies.

Biomineralization continues to be a rich source of inspiration to challenge, and expand, our current view of crystallization phenomena in complex systems. Today, there is compelling evidence that living organisms make use of multistep mineralization pathways, which often cannot be emulated in the laboratory. Falini and Fermani (2017, Chap. 9) argue that one of the main obstacles is our limited knowledge of the role of the various bio(macro)molecules that can be involved in the formation of biominerals (i.e. the "organic matrix"), especially at the molecular level. To solve this puzzle, we need to identify the key steps of the process: i.e. which (macro)molecules are active in which step, and how do these macromolecules influence each particular step. The availability of increasingly sophisticated characterization techniques, offering unprecedented spatial and temporal resolution, promises deeper insights in the near future, as illustrated in the contributions by Birkedal (Chap. 10) and Delgado-Lopez and Guagliardi (2017, Chap. 11). In the chapter by Birkedal, the great potential of in situ X-ray diffraction is demonstrated for the example of calcium phosphate crystallization, one of the most important processes in biomineralization. Even in the absence of any bio(macro)molecules or other additives, inherently complex behaviour is observed due to the strong pH dependence of the phosphate speciation and the fact that multiple competing mechanisms are at play. This highlights that common solution parameters (like pH, but also temperature or reactant stoichiometry) can have a profound influence on the progress and outcome of crystallization. In other words, there is usually not one universal mechanism of mineralization, but it is rather the particular conditions that decide which of the possible routes are taken. In the case of calcium phosphate, the existence and/or exact nature of possible pre-nucleation species and the steps leading to the formation of amorphous particles appear still more obscure than for $CaCO_3$, while refined structural models for the amorphous phase are just as urgently needed. Delgado-Lopez and Guagliardi (2017, Chap. 11) show that only through the use of complementary high-resolution techniques (SAXS, WAXTS and AFM in their case) – ideally with in situ capabilities to observe processes as they occur – can maximum information about the early stages of mineral formation in biological systems be obtained. Their particular studies, along with those of other groups, draw a detailed picture of how small organic molecules like citrate play a key role in the development of a complex biomaterial such as bone. Gaining insight into the *modus operandi* of larger molecules (e.g. polysaccharides or proteins) at a similarly fundamental level would certainly be a giant leap ahead in the field of biomineralization.

The high value of advanced characterization methods is a recurrent theme also in the following contributions. For calcium sulphate, a mineral of great geological and industrial importance, recent in situ scattering experiments have challenged the long-standing view of nucleation and growth for $CaSO_4$ phases from aqueous solutions, as discussed by Van Driessche et al. in Chap. 12. Even though the proposed alternative pathway seems to be able to reconcile the many observations made over the years, a truly unified model for calcium sulphate crystallization is still only emerging and awaits confirmation by further studies. Among the numerous key issues that remain unresolved, the actual structure and composition of the nanoparticulate precursor phase and the role of water in the process of phase selection deserve special attention. In a similar fashion, significant advances have been made in our understanding of the crystallization of iron (oxyhydr)oxides. However, as Penn et al. (Chap. 13) emphasize, the earliest (and smallest) precursors are still very difficult to capture analytically, limiting our knowledge on the elementary steps occurring at the nanoscale. According to Reichel and Faivre (2017, Chap. 14), this lack of understanding is the major bottleneck for improving the efficiency of biomimetic approaches to the synthesis of magnetite nanoparticles with controlled property profiles as desired for manifold applications. Despite decades of intense research, similar challenges are still encountered regarding the formation of silica and alumina colloids. In Chap. 15, Tobler et al. show that growth and aggregation processes of corresponding nanoparticles are well amenable to analyses by scattering and imaging techniques. By contrast, the onset of condensation reactions and especially the nucleation of the primary particles suffer from substantial gaps in both phenomenological and mechanistic understanding, mainly due to the lack of conclusive experimental evidence. In this respect, there is an urgent need for the development of analytical techniques that allow for an in situ and time-resolved capability to monitor silicate and/or aluminate species in solution – a notion that seems likewise attractive for characterizing the early stages of iron oxide precipitation.

It is not surprising, and perhaps even reassuring, that related communities face quite the same problems as those described above for typical mineral systems. In the crystallization of small organic molecules, nucleation rates predicted by classical nucleation theory (CNT) were also proven to be unreliable in many cases, as outlined by Yang and Ter Horst in Chap. 16. While this is certainly in part due to the assumptions made by the underlying theory, the often ill-defined description of the system in which nucleation takes place is another major limiting factor. One critical aspect in this context is the omnipresence of foreign particles, like dust, which frequently act as substrates for nucleation and yet themselves have hardly been characterized so far. To fill this gap, concerted studies should be performed to measure accurate heterogeneous nucleation rates on "template" particles with well-defined size distributions and surface functionalities. In combination with detailed analyses of solute association processes (which supposedly define the building units for crystals), this might enable a molecular interpretation of heterogeneous nucleation phenomena and their relevance in many practical situations. Similar conclusions are drawn by Kimura and Tsukamoto (Chap. 17) regarding the nucleation of smoke particles as analogues for cosmic dust, where experimentally determined nucleation rates are also very different to values predicted by CNT or molecular dynamic simulations. According to the authors, a possible solution to this problem could be to derive a new theoretical framework based on homogeneous nucleation experiments from the vapour phase, which would then be extended to more complex scenarios.

In a nutshell, the various examples of crystallizing systems described in this book underline the persisting strong interest in nucleation and growth across numerous disciplines. In the years to come, we assert that the field will strongly benefit from improved theoretical models in particular for nucleation processes (Chaps. 1, 2, 16 and 17), but also for the later stages of growth (Chaps. 7, 8 and 9) and the structure(s) of transient precursor phases (Chaps. 3, 4, 6, 9, 10, 13, 14 and 15). At the same time, it is essential to keep advancing the development of new analytical methods with highest temporal and spatial resolutions, in order to trace the key stages in the nucleation and/or transformation of mineral phases as they occur. Although X-ray scattering techniques have already proven their great potential (Chaps. 4, 5, 10, 11, 12, 13 and 15), we believe that these and similar novel methods will continue to help in the future, especially as more sensitive detectors for use in synchrotron facilities are becoming available. Another very promising approach is liquid-cell TEM, which has just recently been added to the spectrum of diagnostic tools. As shown by Nielsen and De Yoreo in Chap. 18, this technique enables direct visualization of nucleation and growth pathways at medium to high resolution in liquid media while simultaneously providing crystallographic information on the observed phases. However, at the moment there are still a number of important technical limitations and possible artefacts to consider, such as the formation of unknown radiolytic products upon exposure of the investigated solution to the electron beam. Future developments in this exciting area are directed towards the fabrication of on-device probes in order to monitor solution conditions in the irradiated volume and thus to better define the circumstances under which the observed processes actually occur. Once this is accomplished, the technique should allow for quantitative studies with great potential to foster our understanding of nucleation and crystallization mechanisms on a fundamental level.

We conclude this book by commenting on the somewhat overenthusiastic use of the term "nonclassical" in the recent literature to describe different phases and processes observed during crystallization. In our opinion, this terminology has generated a lot of confusion, and, instead of helping, it has actually slowed down our progress towards a general understanding of mineral nucleation and growth – especially when seeming differences in the end boil down to be a question of semantics. Therefore, it is important to establish common definitions and clearly explain the meaning of any "non-standard" terms used in new publications. Our major objective should be to investigate and describe the precursors and intermediates occurring during the multistep processes of the early stages of crystallization and to illustrate how they are interconnected mechanistically – rather than emphasizing what they are not from a "classical" point of view.

The manufacturer's authorised representative in the EU is Springer Nature Customer Service Centre GmbH, Europaplatz 3, 69115 Heidelberg, Germany. If you have any concerns regarding our products, please contact ProductSafety@springernature.com

Printed and bound by CPI Group (UK) Ltd, Croydon, CR0 4YY
01/10/2026
02225871-0009